Applied Muscle Biology and Meat Science

Applied Muscle Biology and Meat Science

Edited by

Min Du

Richard J. McCormick

CRC Press
Taylor & Francis Group
Boca Raton London New York

CRC Press is an imprint of the
Taylor & Francis Group, an **informa** business

CRC Press
Taylor & Francis Group
6000 Broken Sound Parkway NW, Suite 300
Boca Raton, FL 33487-2742

Printed in the United States of America on acid-free paper
10 9 8 7 6 5 4 3 2 1

International Standard Book Number-13: 978-1-4200-9272-1 (Hardcover)

Library of Congress Cataloging-in-Publication Data

Applied muscle biology and meat science / editors, Min Du and Richard J. McCormick. -- 1st ed.
p. cm.
Includes bibliographical references and index.
ISBN 978-1-4200-9272-1 (alk. paper)
1. Meat--Quality. 2. Muscles--Physiology. I. Du, Min, 1969- II. McCormick, Richard J. III. Title.

TS1955.A67 2009
664'.9--dc22 2008048196

Visit the Taylor & Francis Web site at
http://www.taylorandfrancis.com

and the CRC Press Web site at
http://www.crcpress.com

Contents

Preface

Skeletal muscle is the single largest organ mass in the vertebrate body. As a source of food, it represents most of the value in a meat-producing animal. Both the growth characteristics of muscle and the quality of the resulting meat determine the ultimate worth of a carcass. In recent years, tremendous strides have been made in the field of muscle biology, particularly in our understanding of the mechanisms controlling skeletal muscle growth and development, and the impact these factors have on meat production and quality. Many of the past difficulties that meat and animal scientists have experienced when attempting to address specific problems (e.g., stress susceptibility and poor meat quality in swine) have been due to a lack of understanding the underlying biological mechanisms driving muscle growth, metabolism, and its conversion to meat. This book is designed to provide the advanced reader with state-of-the-art knowledge about skeletal muscle and meat, and also as a platform for further investigation of specific issues. At the same time, this comprehensive review provides the newcomer with a sound background in applied muscle biology and meat science.

Chapter 1 provides a detailed analysis of muscle structure with an emphasis on muscle proteins, microstructure, membrane composition, and contraction. This discussion lays the groundwork for succeeding chapters.

An understanding of the growth and development of skeletal muscle and some of its components such as fat and connective tissue is crucial to our ability to manipulate production of the meat animal and subsequent meat quality characteristics. Chapters 2 through 7 address issues relevant to muscle growth and development. Accordingly, myogenesis and the biology and nature of satellite cells are reviewed. Adipogenesis, essential for muscle development and growth and so critical to meat quality, is discussed in Chapter 3. Chapter 4 reviews current issues related to the fetal stage of muscle development, with emphasis on the sensitivity of the fetus to maternal nutritional status and resulting fetal programming of skeletal muscle. The characteristics of muscle fiber types and mechanisms of *in vivo* protein degradation are the subjects of Chapters 5 and 6. Finally, a discussion of collagen and other matrix constituents and their contribution to muscle growth and quality concludes this part of the book.

Chapters 8 through 14 review key aspects of postmortem changes in mammalian and fish muscle that are responsible for meat quality characteristics. Thus, the first chapters in this middle section discuss, arguably, the two most important biological changes in postmortem muscle: proteolysis and glycolysis. A recurrent problem in the production of pork has been superior lean growth coupled with the development of poor quality "acid meat." Chapter 10 reviews recent research related to alterations in protein signaling pathways that are related to development of this genotype in pigs. Color has been recognized for decades as one of the premier quality traits of meat. Chapter 11 presents a comprehensive review regarding mechanisms governing meat color. Chapter 12 discusses multiple factors responsible for lipid oxidation in meat and strategies for its prevention. Recognition of the beneficial properties of specific fatty acids in meat has prompted interest in components such as ω-3 long-chain polyunsaturated fatty acids and conjugated linoleic acids. Chapter 13 reviews recent progress in nutritional management for enrichment of these meat constituents and the subsequent impact on meat quality. This section concludes with Chapter 14, a comprehensive review of the unique fish myotome and the postmortem changes it undergoes.

The penultimate chapter (Chapter 15) addresses molecular techniques that are increasingly utilized in breeding programs, including molecular mapping and marker-assisted breeding. Finally, the book concludes with an essay (Chapter 16) dedicated to animal welfare and the ethics of using animals for food, an especially timely discussion given the concerns of animal rights activists.

In summary, this book focuses on biological changes in skeletal muscle and meat. Given recent developments in energy costs and distribution, and changes in the commodities markets driven by the demand for biofuels, the challenges for animal production agriculture will only increase. Increased appreciation for the underlying biological mechanisms related to animal and meat production can only help solve the challenges. Thus, this book is suitable as a text for advanced courses in applied muscle biology and meat science; it further serves as a reference for scientists and industry personnel involved in meat science, as well as animal growth and development.

About The Editors

Dr. Min Du earned a B.S. in food science from Zhejiang University in 1990, an M.S. in meat science from China Agricultural University in 1993, and a Ph.D. in meat science from Iowa State University in 2001. He pursued post-doctoral studies in the Department of Biochemistry, University of Alberta, Canada, focusing on cell signaling events controlling skeletal muscle growth. He joined the faculty of the Animal Science Department, University of Wyoming, in 2003. His current research examines molecular mechanisms controlling skeletal muscle growth and meat quality, with a particular emphasis on fetal muscle development.

Dr. Du has published numerous peer-reviewed articles and abstracts, and has made multiple invited presentations. He is the section editor for the *Journal of Animal Science* and *Journal of Muscle Foods*, and is a reviewer for nearly 20 scientific journals and funding agencies. Dr. Du has taught in his native China and currently teaches meat science courses to undergraduate students and a unique course on cell signaling to graduate students.

Dr. Richard J. McCormick received B.S. and M.S. degrees from the Animal Science Department, University of Connecticut, and a Ph.D. from the Animal Science Department, Kansas State University. Prior to joining the faculty of the Animal Science Department of the University of Wyoming in 1985, he worked as a guest scientist at the Federal Meat Research Institute in Kulmbach, Germany. For the past 20 years, his research interests have focused on connective tissue biology in skeletal and cardiac muscle. He has published extensively on collagen biochemistry related to meat science, muscle biology, and cardiac and exercise physiology. Most recently, Dr. McCormick initiated genomic studies in models of cardiac (myocardial infarction) and lung (bovine and human high altitude disease) pathologies. He has taught numerous courses to undergraduate and graduate students in the areas of biology and food science and technology, and has lectured extensively in Europe, South America, and Asia.

Contributors

Jennifer L. Aalhus
Meat Research, Agriculture and Agri-Food Canada
Lacombe Research Centre
Lacombe, Alberta, Canada

D.U. Ahn
Animal Science Department
Iowa State University
Ames, Iowa

Marie E. Cantino
Department of Physiology and Neurobiology
University of Connecticut
Storrs, Connecticut

Min Du
Department of Animal Science and Interdepartmental Molecular and Cellular Life Sciences Program
University of Wyoming
Laramie, Wyoming

K.L. Flann
Muscle Biology Group
Department of Animal Science
University of Arizona
Tucson, Arizona

G.H. Geesink
University of New England
New South Wales, Australia

Clare A. Gill
Department of Animal Science
Texas A&M University
Texas AgriLife Research
College Station, Texas

Marion L. Greaser
Department of Animal Science
University of Wisconsin
Madison, Wisconsin

G.J. Hausman
Richard B. Russell Agricultural Research Center
USDA, ARS, APRU
Athens, Georgia

D.L. Hopkins
NSW Department of Primary Industries
Cowra, New South Wales, Australia

E.J. Lee
Animal Science Department
Iowa State University
Ames, Iowa

Richard A. Mancini
Department of Animal Science
University of Connecticut
Storrs, Connecticut

Richard J. McCormick
Department of Animal Science
University of Wyoming
Laramie, Wyoming

Warrie J. Means
Department of Animal Science
University of Wyoming
Laramie, Wyoming

K.C. Nam
Examination Division of Food & Biological Resources
Korean Intellectual Property Office
Daejeon, Korea

Niels Oksbjerg
Department of Food Science
Faculty of Agricultural Science
University of Aarhus
Aarhus, Denmark

S.P. Poulos
The Coca-Cola Company
Atlanta, Georgia

C.R. Rathbone
Muscle Biology Group
Department of Animal Science
University of Arizona
Tucson, Arizona

Joe M. Regenstein
Department of Food Science
Cornell University
Ithaca, New York

R.P. Rhoads
Muscle Biology Group
Department of Animal Science
University of Arizona
Tucson, Arizona

Penny K. Riggs
Department of Animal Science
Texas A&M University
Texas AgriLife Research
College Station, Texas

W.M. Robertson
Meat Research, Agriculture and Agri-Food Canada
Lacombe Research Centre
Lacombe, Alberta, Canada

Daniel C. Rule
Department of Animal Science
University of Wyoming
Laramie, Wyoming

Qingwu W. Shen
Department of Animal Science
University of Wyoming
Laramie, Wyoming

Darl R. Swartz
Department of Animal Science
Purdue University
West Lafayette, Indiana

Margrethe Therkildsen
Department of Food Science
Faculty of Agricultural Science
University of Aarhus
Aarhus, Denmark

Jun F. Tong
Department of Animal Science
University of Wyoming
Laramie, Wyoming

Keith R. Underwood
Department of Animal Science
University of Wyoming
Laramie, Wyoming

Jin Ye
College of Food Science
Inner Mongolia Agricultural University
Huhhot, Inner Mongolia
People's Republic of China

Peng Zhou
School of Food Science and Technology
Jiangnan University
Wuxi, Jiangsu
People's Republic of China

Meijun Zhu
Department of Animal Science, and
Interdepartmental Molecular and Cellular Life Sciences Program
University of Wyoming
Laramie, Wyoming

1 Muscle Structure and Function

Darl R. Swartz, Marion L. Greaser, and Marie E. Cantino

CONTENTS

1.1 INTRODUCTION

Muscle is the tissue that moves us about and transports the liquids within us. There are two general types of muscle — striated and smooth — with this nomenclature derived from the microscopic appearance. Striated muscle has striations while smooth does not. Striated muscle can be further subdivided into skeletal and cardiac. The common feature of the muscle types is that the cells are enriched in proteins that interact in a cyclic fashion to produce force and/or shorten the cell. These proteins are myosin in the thick filament and actin in the thin filaments. If the cell is connected to other muscle cells around a lumen, as in smooth and cardiac muscle, it will decrease the lumen volume and thus move things about within us. If it is attached to the skeleton, as for skeletal muscle, it will move the skeleton and thus allow change in our position or location. All three types of muscle are used in meat products, with smooth and cardiac mostly found in comminuted products and skeletal found in intact or near intact meat products as well as comminuted products. Skeletal muscle is by far the most important in animal agriculture and is the main focus of this chapter.

1.2 HISTOLOGY AND GROSS ANATOMY

Skeletal muscle, as the name implies, is generally attached to the skeleton. It contains the other tissue types (nerve, epithelia, connective tissue) to a much lesser extent than other tissues and organ systems. The cells of striated muscle are unique in that they are multinucleate, very large, and extremely elongate, reaching several millimeters to centimeters in length. This cylindrical and extended length of the cells results in the muscle cell being called a fiber. The parallel organization of muscle fibers within a group or fascicle gives anisotropy to the tissue or the so-called "grain" that is observed in a steak. Anisotropy imparts unique mechanical properties to near-intact muscle foods, as the resistance to breakage (such as when meat is chewed) depends on the angle of shear. The origins of this anisotropy and the striations are discussed in much greater detail in subsequent sections. Like other cells, the muscle cell is enveloped in a membrane and contains intracellular organelles such as the Golgi, mitochondria, nuclei, and an endoplasmic reticulum. The amount of these elements is much less than in other cell types as the primary function of muscle is to shorten and thus it contains very large quantities of the contractile proteins actin and myosin. Also, unlike other cells, striated muscle has an extremely low to no propensity to divide; renewal involves a myoblast-like cell called the satellite cell.

The individual muscle cells are encased in collagenous connective tissue that can be subdivided into the portion that is in intimate contact with the cell called the basement membrane and a more distal portion called the endomysium (Figure 1.1). A group or fascicle of muscle cells is surrounded by a more robust collagenous connective tissue network called the perimysium. An anatomical muscle consists of a group of fascicles and is surrounded by the epimysium. It is through the epimysium that the main product of muscle (shortening) is coupled to bone via the myotendinous junction. The collagen fibers of endo-, peri-, and epimysium intermesh such that contraction at the individual cell level is transmitted from the endomysium to the epimysium and thus to tendon and bone. The efferent and afferent nerve fibers and blood vessels penetrate the epimysium and bifurcate throughout the muscle, primarily within the perimysium. Generally, the nerve fibers and blood vessels are grouped together and subsequently bifurcate to supply each single muscle fiber with a nerve and the capillary network that feeds the muscle fibers.

The anatomical names of muscles are related to size, shape, attachments, fiber orientation, location, and action. The morphological types of muscle are related to their general form and muscle fiber orientation/organization. Muscle with fibers oriented mostly along the long axis is termed "strap" and "fusiform." Those with the fibers oriented at an oblique angle to the muscle, but mostly one angle, are termed "pennate." Those with multiple oblique angles are termed "bipennate," "multipennate," and "circumpennate." Most anatomy textbooks diagram all muscles as strap when, indeed, strap muscles are not very common. For intact to near-intact muscle foods, it is important

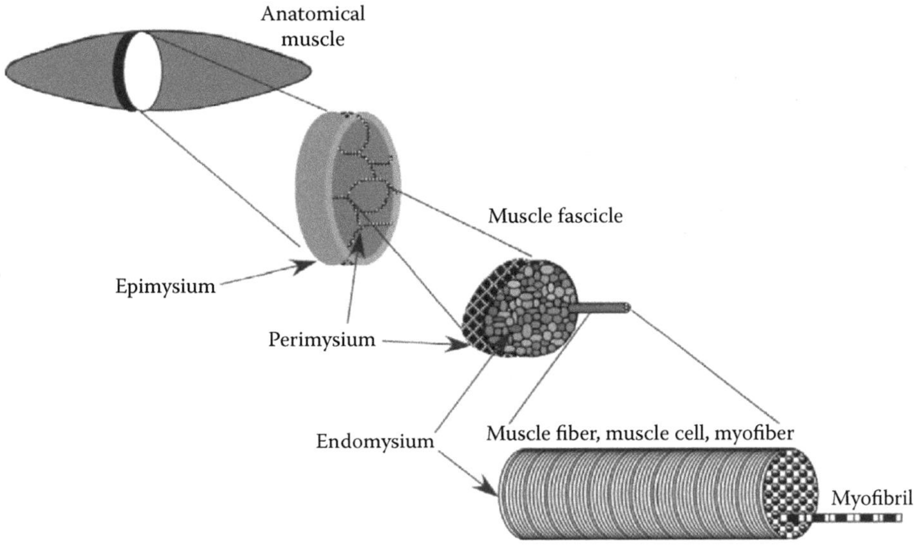

FIGURE 1.1 Organization of striated muscle and its collagenous connective tissue layers.

to understand the morphological muscle type as this may impact the tenderness. For example, the tender psoas major is a strap-type muscle. From a functional perspective, strap muscles develop the greatest extent and speed of shortening but the lowest force per cross-sectional area of the muscle (not the fiber *per se*), while circumpennate muscles develop the least extent and speed of shortening but the greatest force/cross-sectional area of the muscle. These biomechanical features arise, partially, from the fact that the speed of muscle shortening is related to the number of force-producing units in series while the force is related to the cross-sectional area of the muscle fibers for a given myosin isoform (see below). The sum of the fiber cross-sectional area for a strap muscle is equal to the muscle cross-sectional area, while for the circumpennate muscle, it is the area of a cylinder through the center of the muscle.

The microscopic appearance of a skeletal muscle cell is primarily determined by its intracellular components and their organization (Figures 1.2 and 1.3). The cell can be divided into its constituent elements, which include the outer membrane called the sarcolemma, intracellular membrane networks that include the T-tubules and sarcoplasmic reticulum, as well as mitochondria, nuclei, and myofibrils. The myofibril is a string-like structure made of repeating units called sarcomeres. The sarcomere is delimited by the α-actinin containing Z-lines to which the actin-containing thin filaments attach. Myosin-containing thick filaments are in the center of the sarcomere and form the A-band. In the center of the A-band is the M-line that contains myomesins and part of titin involved in formation of the A-band and positioning of the thick filaments in the center of the sarcomere. Thick and thin filaments interdigitate in the overlap region and the extent of overlap increases as sarcomere length decreases. The I-band includes a dense Z-line at its center and the low-density non-overlap regions of the thin filaments from the adjacent half sarcomeres on each side. Classically, the sarcomere is described as having thin-filament-containing I-bands flanking the thick-filament-containing A-band but this does not convey the functional aspect of the overlap region.

As can be seen in Figures 1.2 and 1.3, the proteins of the sarcomere are organized into a near-crystalline lattice, and the form of these proteins and their interactions with themselves and other proteins result in a highly organized structure. The fascinating feature of muscle structure and function is that one can start with most of the structures of individual proteins, organize them into their respective polymers as governed by their interprotein interactions, build the individual filaments (thick and thin) and their attachments (Z-line and M-line), build a sarcomere, and link these sarcomeres in series to form the myofibril. These myofibrils can be associated and aligned,

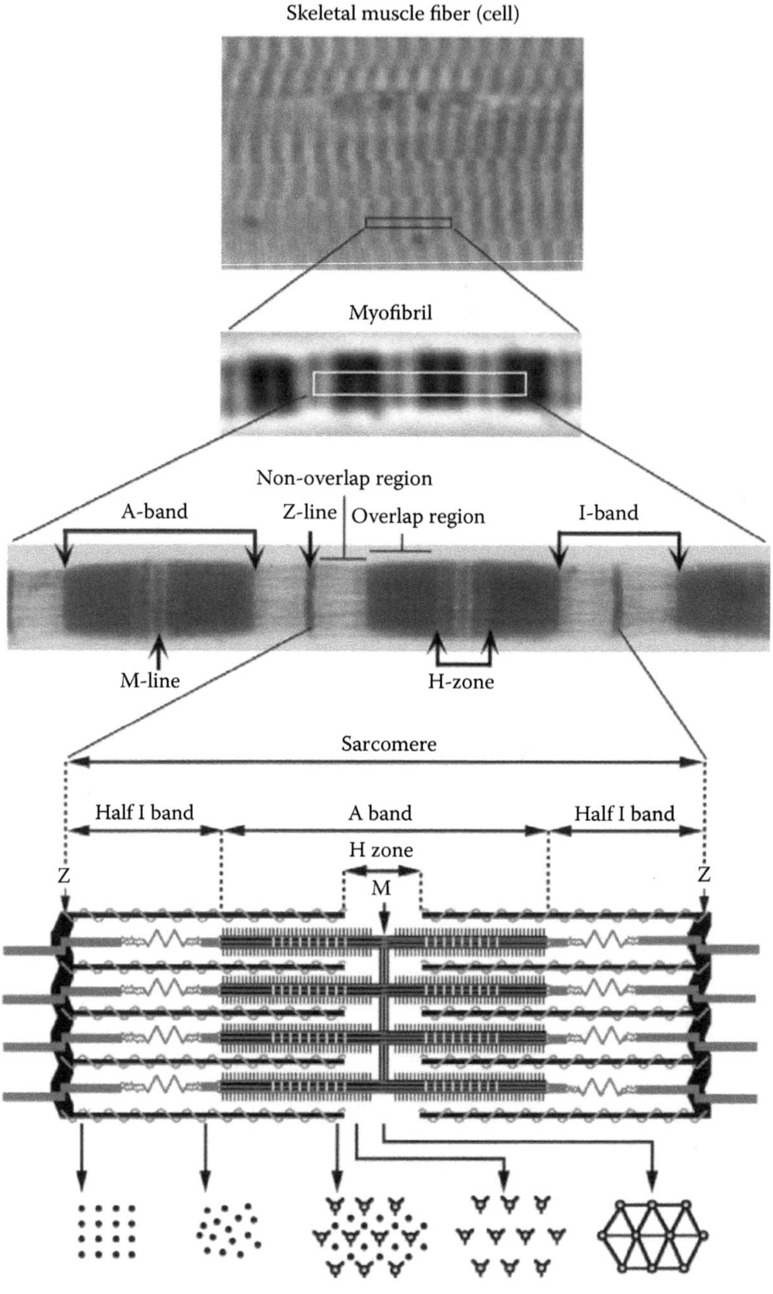

FIGURE 1.2 Vertebrate skeletal muscle fiber, myofibril, and sarcomere organization. Skeletal muscle fiber and myofibril imaged by phase-contrast microscopy (top) and single myofibril whole mount imaged with HVEM after positive staining (middle). Cartoon of sarcomeric organization in both longitudinal and cross-section (bottom).

primarily by extra-sarcomeric proteins, and the muscle cells can be arranged to generate a model of muscle organization consistent with that observed at the anatomic, light, and electron microscope (EM) levels. This molecule-to-macroscopic approach will be used for this chapter. The x-ray crystal structures available for many of the major myofibrillar proteins will be combined with electron micrographs of the ultrastructural organization to illustrate and describe the individual proteins of

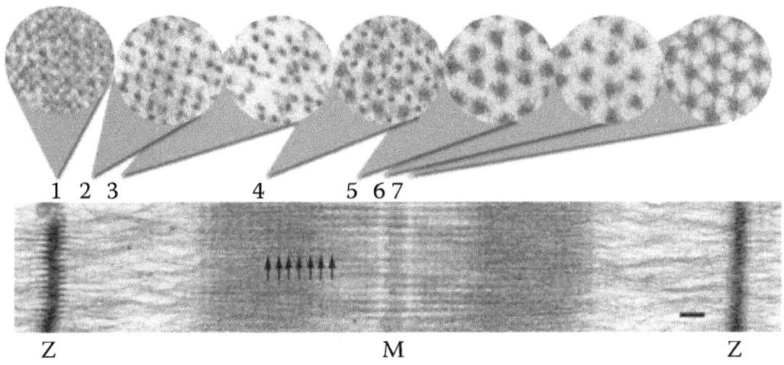

FIGURE 1.3 Filament organization within the sarcomere as seen in electron micrographs of positively stained sections of rabbit psoas muscle. Different regions of the near-crystalline lattice of the sarcomere shown in cross-section (top) and in longitudinal section (bottom) are: 1, Z-line; 2, I-band near Z-line; 3, I-band; 4, overlap region of the A-band; 5, non-overlap region of A-band; 6, bare zone of A-band adjacent to M-band; 7, M-band. Arrows note the stripes associated with MyBP-C. Bar = 100 nm.

the sarcomere, their interactions with themselves and other proteins, their function, and to build their respective subsarcomeric structures. We then build the sarcomere, the myofibril, adding some membrane elements, and describe how all these elements work together to elicit muscle contraction upon stimulation from the nerve.

1.3 MYOFIBRILLAR PROTEINS

The myofibril is an insoluble (at near-physiological salt levels) protein assembly composed of more than 30 different proteins. It is isolated by homogenizing muscle (preferably post rigor) in a buffered 0.1 M saline solution followed by cycles of low-speed centrifugation. Most of the membrane elements and cystosolic proteins remain in the supernatant while the nuclei and myofibrils are in the pellet. The proteins of the myofibril interact with each other by non-covalent bonds. Many of the individual protein-protein domain interactions are weak but the number of domains interacting is large, thus resulting in a very high overall affinity and providing stability to the protein complexes that maintain the organization of the muscle structures. The major proteins and their approximate levels in isolated myofibrils are shown in Table 1.1, and an SDS-PAGE gel of fast and slow bovine myofibrils is shown in Figure 1.4 to aid in visualizing the relative amount of the different proteins and their approximate monomeric mass. A description of the structure and function of the individual myofibrillar proteins follows, with subsequent organization of these proteins into their subsarcomeric structures. We do not focus on the numerous isoforms of the myofibrillar proteins and use isoforms differences only to highlight the functions of the proteins.

1.3.1 Myosin

The molecular motor that produces muscle force and contraction is the actin-activated ATPase, myosin. There are numerous myosin classes within this large protein family (Foth et al. 2006). The form contained in the contractile apparatus of all muscle types, and contributing nearly half of the myofibrillar protein content in vertebrate striated muscle, is myosin II. There are several muscle myosin types, differing in the speed with which they hydrolyze ATP and their regulation. We describe here some of their common structural features; additional details may be found in other reviews of muscle, myosin, and myosin filament structure (e.g., Squire 1981; Squire et al. 2005; Craig and Woodhead 2006). Myosin isoform variations in different fiber types are also described in more detail elsewhere in this volume.

TABLE 1
Major Myofibrillar Proteins

Protein	Subunits	MW (kDa)	% MF protein (w/w)
Myosin	Heterohexamer:	520	43
	Heavy chain	220	
	Essential (proximal) light chain	23	
	Regulatory (distal) light chain	20	
Actin	Homopolymer	43	22
Titin	Homohexamer?	3,200–3,700	10
Nebulin	Monomer	900	5
Troponin	Heterotrimer:	73	5
	Troponin T	31	
	Troponin I	23	
	Troponin C	18	
Tropomyosin	Homo/heterodimer:	66	5
	Alpha-tropomyosin	33	
	Beta-tropomyosin	33	
MyBP-C	Monomer	130	2
MyBP-H	Monomer	55	<1
Myomesin (1-3)	Homodimer	165–185	2
α-actinin	Homodimer	100	2
Tropomodulin	Monomer	40	<1
CapZ	Heterodimer:	68	<1
	Alpha-subunit	36	
	Beta-subunit	32	
Creatine kinase	Homodimer	43	<1
Adenylate kinase	Monomer	22	
Desmin	Homopolymer	54	<1
Synemin	Heteropolymer (w/desmin)	185	<1
Filamin	Homodimer	240	<1
Telethonin/T-cap	Monomer	19	<1
Myopalladin	Monomer	145	<1
T-cap	Monomer	19	<<1
Dystrophin	Monomer	427	<<1

Myosin II consists of two identical heavy chains and two pairs of light chains (Figure 1.5A and B), with a combined molecular weight of around 520 kDa. The C-terminal α-helical portions of each 220-kDa myosin heavy chain dimerize to form a coiled-coil rod, approximately 150 nm in length. This region is also involved in the formation of thick filaments and is discussed further below. Toward their N-terminal ends, the two heavy chains separate and each terminates in a globular region. Enzymatic digestion divides the molecule within the rod portion into light meromyosin (LMM) and heavy meromyosin (HMM). Heavy meromyosin can be further broken down to yield subfragments 1 and 2 (S1 and S2). Each S1 fragment (also referred to as the myosin head or crossbridge) includes the globular portion of the heavy chain, with sites that bind MgATP and actin, and an α-helical neck that binds two different types of light chains (Figure 1.5B), ranging from 20 to 23 kDa.

The S1 head alone is sufficient to produce movement of actin filaments *in vitro* (Toyoshima et al. 1987), confirming that muscle contraction is a direct result of conformational changes that occur within this fragment. The vertebrate myosin S1 crystal structure (Figure 1.5B), first solved in

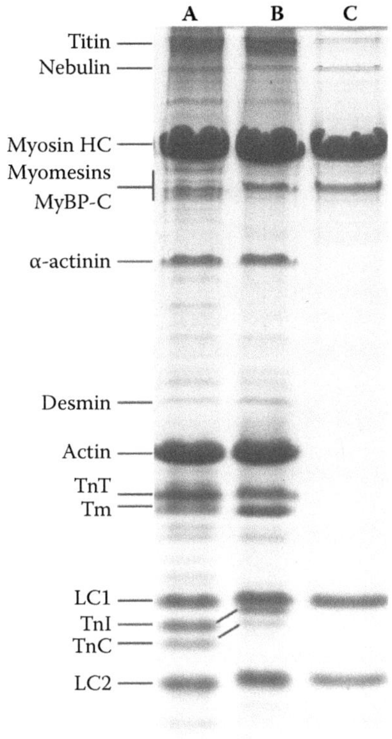

FIGURE 1.4 SDS-PAGE of myofibrillar proteins and crude myosin. Bovine cutaneous trunci myofibrils (fast muscle, lane A), bovine masseter myofibrils (slow muscle, lane B), and myosin (lane C) were subjected to SDS-PAGE.

1993 (Rayment et al. 1993), has helped identify the structural motifs important for transduction of chemical energy (Geeves et al. 2005). The globular region of S1 is known as the catalytic or motor domain because it contains the polypeptide sequences most fundamental to energy transduction in muscle. Two subdomains (upper and lower 50K) are separated by a cleft that is associated with actin binding. Adjacent to this area is a G-protein-like nucleotide binding pocket with a P-loop that coordinates binding of MgATP. Partial ATP hydrolysis and subsequent release of one or both hydrolysis products (ADP, inorganic phosphate Pi; see Figure 1.21) is closely associated with a change in the conformation of the S1 head termed the power stroke. The resulting stable complex of actin and myosin, free of nucleotide, is called the rigor complex.

The neck region of each S1 is the binding domain for two calmodulin-like light chains. The essential (or alkali) light chain binds near the N-terminal end of each neck, and is so named because it is necessary for ATPase activity and actin binding by myosin. The regulatory light chain (also referred to as lightchain 2 or the DTNB light chain) binds at the C-terminal end of the neck, nearest to S2. The functions of the light chains are not well understood, especially in vertebrate striated muscle, although recent studies are beginning to shed light on how they contribute to contractile properties (Moss and Fitzsimons 2006; Hernandez et al. 2007). Because the S1 neck acts as a lever arm during the crossbridge cycle, its structure plays a role in determining the mechanical properties of the muscle as a whole, with its length limiting the distance actin moves per ATP hydrolyzed (and therefore, muscle velocity), and its stiffness affecting the load-bearing properties of each crossbridge. The light chains are thought to not only increase the stiffness of the lever arm, but also to have important regulatory functions. Vertebrate smooth muscle and many invertebrate muscles are referred to as "myosin regulated" because their activation requires modification of the light chains by phosphorylation or calcium binding. Evidence is accumulating that a similar

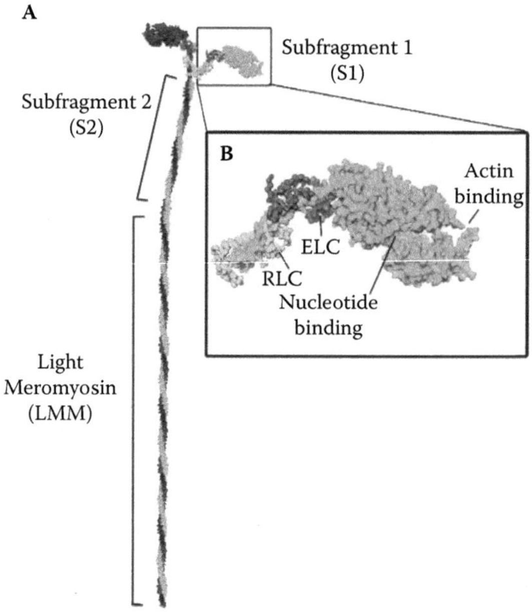

FIGURE 1.5 (See color insert following page 148.) Myosin structure in vertebrate striated muscle. The myosin molecule (A) includes two heavy chains (rendered in light and dark green) and two light chains (magenta and cyan), and can be divided into three subfragments as shown. The crystal structure of S1 (B) shows locations of light chains and binding sites for actin and nucleotides. Images were rendered with MacPyMol using pdb file 2MYS for S1 and 2FXO for S2.

mechanism, while not obligatory, plays an important role in tuning the activation of vertebrate striated muscle.

1.3.2 Myosin Binding Protein-C (MyBP-C)

Other proteins besides myosin and titin are present in the crossbridge region of thick filaments. Myosin binding protein-C (MyBP-C, also called C-protein) and a smaller related protein, myosin binding protein-H, are members of the immunoglobulin superfamily. For many years these proteins were known to bind at regular positions along the thick filaments of striated muscle (Craig and Offer 1976; Bennett et al. 1986) (Figure 1.3) but generated limited interest because they were not required for activation of contraction. The discovery that mutations in the gene for cardiac MyBP-C are associated with familial hypertrophic cardiomyopathies has stimulated renewed study of how these proteins contribute to normal muscle function, especially in the heart (De Tombe 2006; Granzier and Campbell 2006; Oakley et al. 2007).

Three isoforms of MyBP-C have been identified and are encoded by different genes. One form, MyBP-C3, is found only in cardiac muscle. MyBP-C1 and MyBP-C2 (formerly MyBP-X) are found mainly in fast and slow skeletal muscle, respectively. Each isoform consists of a single polypeptide chain (extended length approximately 50 nm), containing at least seven repeating immunoglobulin I (Ig) type domains and three fibronectin (FN) domains. These are numbered from the N-terminus as domains C0 to C10. Repeats toward the C-terminus (C7 to C10) bind to the LMM portion of myosin, and also to another myosin binding protein, titin (see below). The N-terminus of MyBP-C has affinity for the S-2 portion of myosin and for actin (Kulikovskaya et al. 2003). The cardiac isoform is larger than the skeletal, due to a proline charge-rich insert near the center of the sequence and addition of a cardiac-specific IgI domain. It also contains additional phosphorylation sites near the N-terminus.

MyBP-C binds at seven to nine locations in each half of the thick filament, separated by gaps of 43 nm, the same as the axial helical repeat of myosin heads on the filament surface (Figures 1.2, 1.3, and 1.17). Two to four molecules of MyBP-C are thought to bind at each axial position. Two models have been proposed for their arrangement. In the trimeric collar model (Winegrad 1999), three molecules of MyBP-C consecutively dimerize to form a ring around the myosin filament at each axial location. The alternative model aligns C-terminal repeats of MyBP-C parallel to the thick filament axis (Squire et al. 2003). In both models, the N-terminus can extend away from the backbone to interact with myosin S2, actin, or both. Thus far, high-resolution EM studies have been unable to rule out either model, and it is possible that the arrangement differs with muscle or fiber type (Flashman et al. 2008).

It now seems likely that MyBP-C functions in regulation, assembly, and structural stability of the contractile apparatus. Thick filament assembly and stability are promoted by interactions of the C-terminus domains of MyBP-C with titin and are impaired by its absence. Regulatory functions are associated with residues near the MyBP-C N-terminus that bind myosin S2. Studies of both intact muscle fibers and isolated myosin filaments suggest that in its unphosphorylated state, MyBP-C helps tether the S2 portion of myosin to the thick filament backbone; in cardiac muscle, phosphorylation of MyBP-C domains C1-C2 is proposed to decrease their association with myosin S2, allowing S1 to move closer to actin prior to activation (Levine et al. 2001). It is also possible that binding of its N-terminus to actin creates a strut that maintains the positions of the thin filaments midway between adjacent thick filaments in the overlap region in relaxed striated muscle (Squire et al. 2003).

1.3.3 Myomesins

Three separate genes code for related proteins found in the M line/band (Schoenauer et al. 2008). Myomesin 1 (185 kDa) is found in all striated muscles; myomesin 2 (165 kDa, also called M-protein) is found in fast twitch fibers; and myomesin 3 (162 kDa) is restricted to slow and intermediate fiber types. All three proteins have unique N-terminal regions followed by two immunoglobulin-like (Ig) domains, 5 fibronectin domain 3 like (FN3) domains, and 5 more Ig domains. BLAST sequence comparisons, however, indicate only 40% to 50% identity among the myomesins, with the longest span being less than 15 amino acids. Myomesin 1 has been proposed to be the major thick filament cross-linking protein, with a function analogous to α-actinin in the Z-line (Lange et al. 2005). The protein binds to myosin through its N-terminal domain and to titin via the most C-terminal three FN3 domains. All three myomesins form only homodimers through their carboxyl-terminal regions.

1.3.4 Creatine Kinase and Adenylate Kinase

Creatine kinase (CK) is also localized at the M-line. This enzyme catalyzes both the forward and reverse reaction of creatine phosphate + ADP to yield creatine and ATP. CK binds to myomesin through the latter's Ig domains 6 to 8 (Hornemann et al. 2003). Only the MM form of creatine kinase is found in the M-line and its binding affinity for myomesin is pH dependent, being weak at high pH (>7.5) and strong at lower pH (pH 6.0). Adenylate kinase (formerly called myokinase) interconverts 2ADP to AMP + ADP and has been found near the M-band. The interaction of adenylate kinase at the M-band has been proposed to occur through binding to Fhl2/DRAL, which in turn attaches to the carboxyl-terminal region of titin (Lange et al. 2002). The tethering of creatine kinase and adenylate kinase near the site of ATP usage appears beneficial for muscle function.

1.3.5 Actin

Actin is the most abundant protein of muscle on a molar basis, being near 600 μM within the myofilament lattice. It acts as both a structural element of the sarcomere and as a co-factor for activation of the myosin motor domain ATPase. It polymerizes to form the major component of the thin

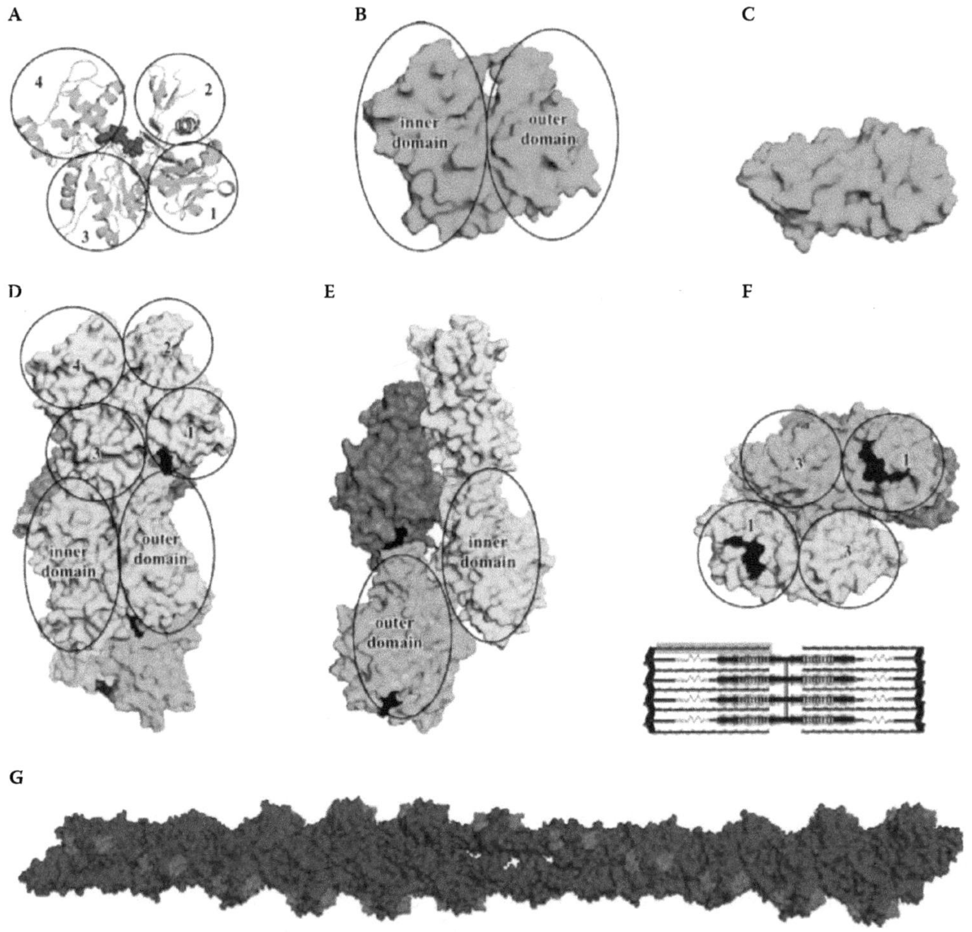

FIGURE 1.6 (See color insert following page 148.) Actin monomer and polymer structure. The crystal structure of G-actin is shown in α-carbon cartoon view (A) and surface rendered views (B and C). The subdomains are noted, as well as the inner and outer domains. Tilting of the structure 90° into the page shows the bottom (subdomains 1 and 3) and that the molecule is mostly flat (C). The crystal structure of an F-actin filament containing four monomers is shown (D through F). Each subunit is colored differently and the subdomains and domains are noted. The C-terminal three residues for all monomers is colored in black for reference. View (D) shows the filament from the front, (E) from the side, and (F) from the barbed (+) end of the filament. A smaller view of a longer filament is shown in (G) with the α-actinin binding sites colored magenta. The barbed (+) end is to the left and pointed (-) end is to the right. Note the spiral of the binding sites resulting from the helical nature of the filament. Images were adapted from Pirani et al. (2006) supplemental files.

filament in an ATP-dependent manner. The crystal structure of G-actin was solved several years ago (Kabsch et al. 1990), and different views of the structure are shown in Figure 1.6A through F. The general anatomy of the protein is more planar than spherical. It is subdivided into four subdomains numbered 1 through 4 that are named in a clockwise fashion with subdomain 2 starting in the upper right and progressing to 1, 3, and then 4 in the upper left. Subdomains 3 and 4 are called the inner domain of actin, whereas 1 and 2 are called the outer domain (Holmes et al. 1990). MgATP/ADP is bound in a cleft formed between subdomains 2 and 4. The planar nature is obvious when the structure is viewed from subdomains 1 and 3 (Figure 1.6C) or 2 and 4. Polymerization of G-actin into fibrous or F-actin is complex, being dependent on concentrations of salt, divalent cations, G-actin, and ATP (Carlier 1991). It follows a sequential process in which subdomains 2 and 4 of one monomer interact with 1 and 3 of another (yellow and cyan monomers in Figure 1.6D and E) to form a dimer. Subdomains 3 and 4 of another monomer, rotated about 170° about its vertical axis, associate with subdomain 4 of the lower G-actin and subdomain 3 of the upper G-actin of the dimer (magenta monomer in Figure 1.6D and E). If the cleft is considered the top of the molecule, dimerization involves a stacking of one monomer on the other while formation of the trimer involves addition of a monomer (after rotation of about 170°) to the side of this stack. Subsequent addition of monomers occurs by primarily adding additional monomers onto the existing trimer (green monomer Figure 1.6D and E). The process is rate limited by dimer and trimer formation. This polymerization results in the filament having polarity with the clefts all in the same direction. This end of the polymer is termed the minus (–) end of the filament because it grows more slowly than the other plus (+) end. When the filament is "decorated" with rigor myosin S1, the minus end appears pointed and is thus called the pointed end while the other end appears barbed and is called the barbed end. For most illustrations, we will orient the pointed end at the top or right of the figure.

The arrangement of actin monomers in the filament (Figure 1.6G) can be described as either a left-handed single stranded helix (sometimes referred to as "short pitch") or as a right-handed double stranded helix ("long pitch"). In electron micrographs the most prominent features are a 6 nm striation along the filaments resulting from the turns of the left-handed helix, and a 38 nm periodicity reflecting the crossover points of the two strands of the right-handed helix. The filament can be described as a twisted double strand of pearls with the twist of the two strands giving the long pitch and the twist, in going from pearl to pearl, the short-pitch. The filament, without associated proteins, is dynamic in that it can grow and shorten, depending on the conditions and G-actin level. In striated muscle, there are proteins tightly associated with either the plus end (CapZ) or minus end (tropomodulin) and along the side (nebulin and tropomyosin) that stabilize the filament.

The diameter of the filament is about 10 nm and its length in muscle is primarily determined by nebulin (see below). There are a few proteins that bind to the ends of the filament and involve both the inner and outer domains (McGough 1998). The majority of proteins bind along the filament and interact primarily with the outer domain. At the sarcomere level, these proteins include myosin, tropomyosin, troponin (TnT and TnI), α-actinin, nebulin, and possibly MyBP-C. For myosin and α-actinin, two actin monomers are bound along the long axis of the filament but to different domains of each monomer. The surface regions on F-actin involved in binding tropomyosin, myosin, and α-actinin are shown in Figure 1.7, and it is apparent that there is overlap of some of these sites. F-actin's interactions with other proteins are numerous and are discussed further in sections on the substructures of the sarcomere.

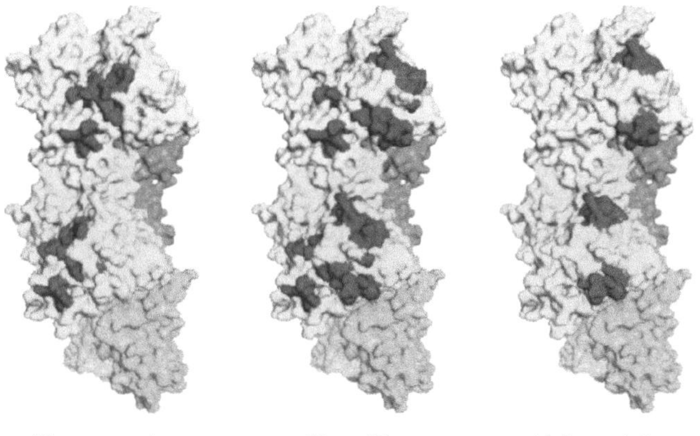

FIGURE 1.7 (See color insert following page 148.) Binding sites on the actin filament. The F-actin model from Figure 1.6 was modified by coloring the amino acid residues (dark blue) thought to be involved in the binding of tropomyosin in the blocked state, S1 in rigor, and α-actinin. Note that similar residues are involved for all three proteins. The residues involved are from McGough (1998) and images were rendered as in Figure 1.5.

1.3.6 Nebulin

Nebulin is also a large protein with a size in the 750 to 850 kDa range that is primarily associated with the thin filament (McElhinny et al. 2003). The name comes from its nebulous function at the time of its discovery. Unfortunately, the role of this protein in muscle still remains mysterious. More than 45 mutations have been detected in the human nebulin gene, and most of these lead to nemaline myopathy, a muscle disorder where microscopic rods appear in the muscle cells. Nebulin is composed of a large number of repeating domains and it is alternatively spliced. It contains an SH3 domain at the C-terminus, and this sequence is proposed to bind to a proline-rich sequence near the N-terminal end of titin in the Z-line. The proteins α-actinin, desmin, and CapZ also bind to the nebulin carboxyl end. A ~35 amino acid repeat occurs 185 times in nebulin. Each of these repeats contains a central "SDXXYK" amino acid sequence that is believed to anchor the nebulin to each actin monomer along the filament at 5.5 nm intervals. There is also a "WLKGIGW" sequence at every seventh module (22 super repeats) in the center of the 185 repeat section, and this corresponds to the 38.5 nm spacing of the troponin complexes along the thin filament. The repeating structure is thought to be α-helical. Nebulin likely follows a helical path on either side of the actin double strand (Figure 1.2, shown in green). The amino terminus binds to tropomodulin near the free end of the thin filament.

Nebulin has been proposed to function as a molecular ruler to determine the thin filament length. Support for this idea comes from the observation that the size of the nebulin correlates with the thin filament lengths when comparing different species. Thus, bovine and human thin filaments, with lengths of 1.3 microns (μm), have larger nebulins than mouse with filament lengths of 1.05 μm. The protein has also been shown to bind to calmodulin. Nebulin knockout mice have altered calcium regulation and markedly up-regulated levels of sarcolipin, a protein that alters the activity of the sarcoplasmic reticulum calcium pump. However, direct interaction of the two proteins has not been demonstrated.

1.3.7 Tropomyosin

Tropomyosin is a parallel homo- or heterodimer that is a component of the thin filament (Perry, 2001). Dimerization is via formation of an α-helical coiled coil involving hydrophobic interactions associated with its amino acid heptad repeat. The non-polar (N) and polar (P) amino acid repeat (NPPNPPP) results in a non-polar side chain at average intervals of 3.5 amino acids, very similar to the 3.6 amino acid repeat of the α helix. This results in a hydrophobic strip along one side of the α helix. Tropomyosin is a stiff, elongate molecule having a length of about 40 nm and a diameter of about 2.5 nm (Figure 1.8). Under low salt conditions, it polymerizes end-to-end involving an overlap of about 9 amino acids at both the N- and C-termini. It binds along the long axis of the actin filament, and the end-to-end interactions between the tropomyosins are such that it is akin to a linear co-polymer with F-actin. The orientation of the binding of tropomyosin to actin is such that its sequence is anti-parallel to the polarity of the actin filament (N-terminus of tropomyosin at the (–) or pointed end). While the overall binding affinity of tropomyosin to F-actin is high, its interaction with individual actin monomers is relatively weak. This allows for some minor movements of tropomyosin on the surface of F-actin that are important in regulation of contraction. The major binding sites on F-actin are on the outer domain of the filament, with other sites more toward the inner domain that are associated with un-blocking of actin for interaction with myosin (Figure 1.8). The mole ratio of actin monomers to the tropomyosin dimer is 7 to 1. Tropomyosin is involved in stabilizing the thin filament but its primary function is to regulate myosin's interaction with actin. This involves movement of tropomyosin on the surface of the actin filament that is mediated by troponin (see below). Troponin binds to tropomyosin in the C-terminal third of the molecule and slightly into the N-terminus of the overlapping tropomyosin. This binding is mostly through the TnT subunit of troponin, with TnT being anti-parallel with respect to the tropomyosin sequence. Tropomodulin binds near the N-terminal 14 amino acid residues of tropomyosin and inhibits tropomyosin end-to-end interactions.

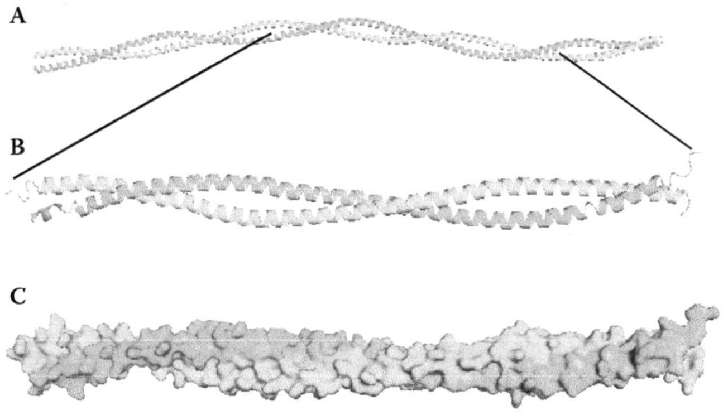

FIGURE 1.8 (See color insert following page 148.) Tropomyosin structure. A low-resolution structure of tropomyosin in α-carbon cartoon view is shown in (A), and a higher-resolution structure of part of the molecule (B). The high-resolution structure is also shown in surface view in (C). One monomer is colored cyan and the other green. Note the coiling of the monomers to form the coiled-coil dimer.

1.3.8 Troponin

Troponin (Tn) is an additional thin-filament associated protein whose primary function is to control the location of tropomyosin on the thin filament in a calcium-dependent manner and thus control contraction (Farah and Reinach 1995). It is a heterotrimeric protein, being made of subunits called TnT, TnI, and TnC. Biochemical studies over the past 40 years determined many of the interactions of the subunits with the thin filament and the subunit interactions within the heterotrimer. Troponin T primarily binds to tropomyosin (Perry 1998). Troponin T can be divided into two domains called TnT1 and TnT2. Tropomyosin binding is mainly by TnT1 whereas TnI and TnC binding is mostly by TnT2. Troponin I is inhibitory to actin activation of myosin ATPase activity and interacts primarily with actin (Perry 1999). It too can be subdivided into an N-terminal domain that is involved in structural interactions with TnC and TnT and a C-terminal domain that is involved in regulatory interactions with actin and TnC. The switching of this domain between actin and TnC is part of the molecular switch involved in regulation of contraction. Troponin C binds calcium and is considered the calcium binding switch for turning on contraction (Grabarek et al. 1992). As with TnI, TnC can also be sub-divided into a regulatory and a structural domain, but with the opposite polarity. The C-terminal globular domain is involved in structural interactions with the C-terminal region of TnT and the far N-terminal region of TnI. This domain binds two calcium ions with low micromolar affinity or two magnesium ions with low millimolar affinity. The N-terminal globular domain binds two calcium ions with micromolar affinity for the fast skeletal isoform but only one calcium for the slow/cardiac isoform. This "regulatory" calcium binding is associated with structural changes in the N-domain that result in exposure of a hydrophobic region. This calcium exposed hydrophobic region is a high affinity site for a part of the regulatory domain of TnI.

The troponin trimer forms a club-like structure with a length of about 25 nm and a diameter of about 5 nm (Flicker et al. 1982). The handle is made of the N-terminal region of TnT (TnT1), and the club part is made of the so-called core-domain of Tn, which includes the C-terminal region of TnT (TnT2) and all of TnI and TnC. Crystal structures of the core domain were recently obtained (Vinogradova et al. 2005) and are shown in Figure 1.9. There are numerous contacts between the subunits and much of TnI and TnT are not resolved in the crystal. The structural interactions that are involved in formation of the trimer are demonstrated by the coiled-coil formed between TnT and TnI, as well as the interaction of the C-terminal domain of TnC with the N-terminal domain of TnI. The C-terminal domain of TnC is held in the structure like chopsticks, with TnI being one stick and the TnIT coiled-coil being the other. In the calcium-saturated structure (Figure 1.9C–F), the N-terminal domain of TnC has structural differences from the apo state that are associated with the exposure of a

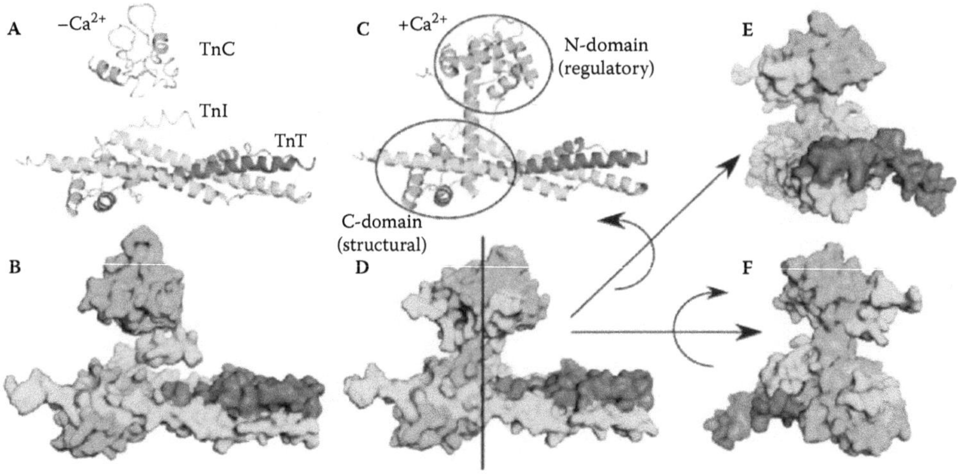

FIGURE 1.9 (See color insert following page 148.) Structure of the troponin core domain. Most of TnC (green) and part of TnI (cyan) and TnT (magenta) are shown as α-carbon cartoon structures (A and C) or surface views (B, D, E, and F) The core domain without regulatory calcium bound is shown (A and B) and with regulatory calcium bound (C through F). Views (E) and (F) are viewed rotated 90° either out of (E) or into (F) the page. Note in (A) and (C) the greater amount of structure detected in the calcium-bound state as a result of calcium-dependent regulatory interactions between TnC and TnI.

hydrophobic region. Nestled within this region is an α-helical 9 or 10 amino acid region of TnI. There is no direct competition of actin for the same amino acid residues in TnI; rather, the actin binding regions are thought to flank the TnC binding region (Luo et al. 2002). Troponin binds to the thin filament primarily via TnT, through TnT1's interaction with tropomyosin and via TnI's interaction with actin. Weaker interactions of TnT2 with actin also may be involved in binding Tn to actin (Pearlstone and Smillie 1981). There are suggestions that TnI also binds to tropomyosin at specific sites, further enhancing its interaction with the thin filament (Geeves et al. 2000; Zhou et al. 2000). The mole ratio of actin monomers to troponin heterotrimer is about 7:1, while that of the troponin heterotrimer to tropomyosin is 1:1. The so-called regulated thin filament thus has a stoichiometry of 7:1:1.

1.3.9 Tropomodulin

Tropomodulin is a low copy number protein that binds tropomyosin and actin and thus is a component of the thin filament (Fischer and Fowler 2003). It was originally discovered in red blood cells, and subsequent isoforms were found in skeletal muscle. It has an N-terminal tropomyosin binding domain and a C-terminal actin binding domain. Tropomodulin's affinity (in K_d) for tropomyosin is about 1 μM, while its affinity for actin is about 0.3 μM. As noted above, tropomodulin binds to tropomyosin's N-terminus. Its interaction with actin is such that it binds to the pointed (–) end of the actin filament, likely across the end of the filament (Fowler et al. 2003). Tropomodulin binds to the actin–tropomyosin complex with a very high affinity ($K_d < 50$ pM). Considering its interactions, one would predict a mole ratio of 2 molecules of tropomodulin per thin filament. Because of its extremely high affinity for actin–tropomyosin, it essentially prevents addition or loss of G-actin monomers from the pointed (–) end of the filament.

1.3.10 CapZ

Another low copy number protein bound to the thin filament is CapZ (Cooper and Schafer 2000). It is a heterodimer made of α- and β-subunits. It binds tightly to the barbed (+) ends of actin filaments with a K_d of about 1 nM. As the name implies, it caps actin filaments and is found at the Z-line of the sarcomere. It

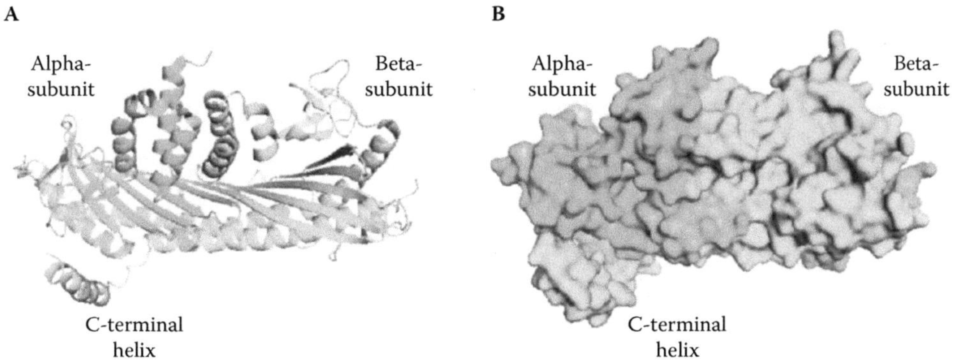

FIGURE 1.10 (See color insert following page 148.) Structure of CapZ. The α-subunit is colored green and β-subunit cyan in carbon cartoon (A) and surface (B) views. Note the similarity in general structure between the subunits and the C-terminal extension in the β-subunit that is thought to be involved in binding to the barbed (+) end of actin filaments.

also appears to bind α-actinin, and this interaction may stabilize its localization to the Z-line. CapZ was recently crystallized (Yamashita et al. 2003) and this structure is shown in Figure 1.10. The molecule is elongate, being about 9 nm long and 5 nm wide. There are extensive interactions between the subunits in the heterodimer, and the structure of the individual subunits is strikingly similar, although there is limited sequence similarity. Both subunits have a far C-terminal region that likely protrudes from the main core of the protein. The proposed interactions with the actin filaments involve binding to the barbed (+) end G-actin monomers, mediated primarily by the far C-terminal regions of CapZ (Narita et al. 2006). The likely mole ratio of CapZ to thin filaments is 1:1.

1.3.11 α-Actinin

α-actinin is an anti-parallel homodimer found in the Z-line (Otey and Carpen 2004). It has a length of about 35 nm and a diameter of about 4 nm. A model structure for α-actinin was developed from the individual structures of its domains (Tang et al. 2001) and is shown in Figure 1.11. The monomer has three major domains; near the N-terminus is a globular actin binding domain composed of a pair of calponin homology domains. This is followed by four triple α-helical, spectrin-like domain repeats. This is also called the rod domain of α-actinin. Near the C-terminal is a calmodulin-like domain, which in the muscle isoform does not bind calcium. The anti-parallel nature of the molecule results in an actin binding domain at each end, with these domains being rotated approximately 180° about the short axis of the molecule. As the name implies, the actin binding domain binds to actin, and the residues on actin involved in this binding have some overlap with residues involved in myosin and tropomyosin binding (see Figure 1.7). This likely results in competition between α-actinin and these other proteins for binding to actin, as suggested by biochemical experiments (Zeece et al. 1979). α-actinin also binds to titin and this involves the C-terminal calmodulin-like domain and spectrin repeats 2 and 3 (Young et al. 1998). Binding to nebulin occurs at the ends (Pappas et al. 2008) and to CapZ in the rod domain (Papa et al. 1999). These numerous interactions with other sarcomeric proteins result in very high affinity binding of α-actinin to the Z-line of the sarcomere. Studies with myofibrils *in vitro* demonstrate that α-actinin can readily be exchanged into the Z-line and the apparent dissociation rate is 0.01 min^{-1}, in accord with its high affinity for the various proteins at the Z-line (Swartz 1999).

1.3.12 Titin

Titin is a giant protein with a monomer size of over 3 million Daltons (for review, see Granzier and Labeit 2007). It was so named after the Greek Titans who were giants. It also is sometimes referred

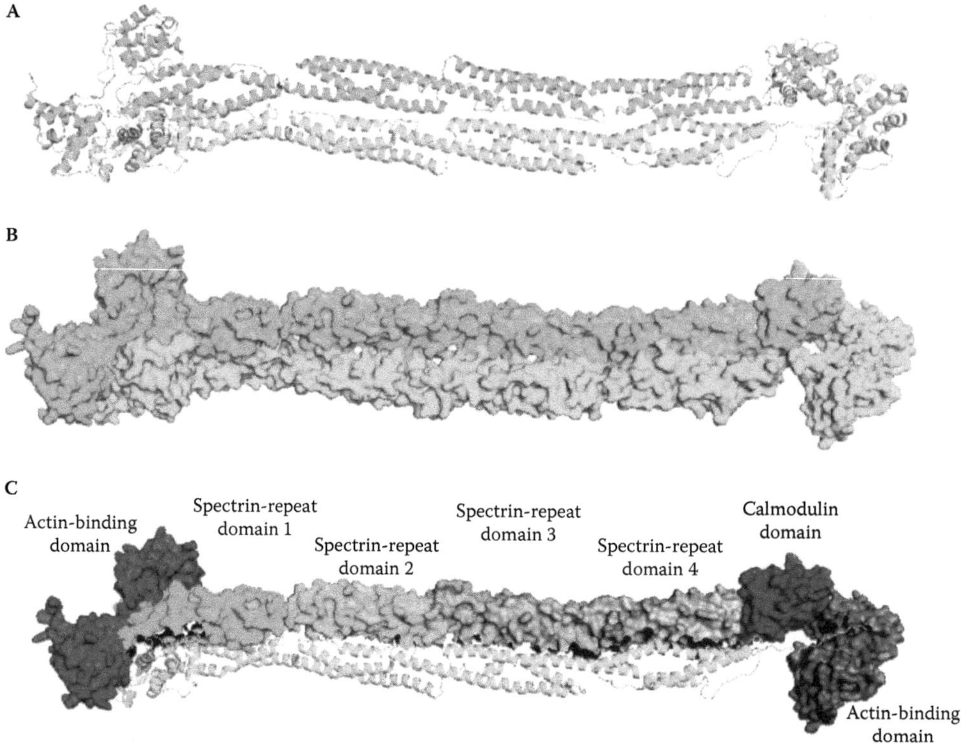

FIGURE 1.11 (See color insert following page 148.) Model structure and the domain organization of α-actinin dimer. One monomer is colored green while the other is colored cyan in the α-carbon cartoon (A) and surface (B) views. The complete domain structure of one of the monomers is shown (C) with the actin-binding domain in red, spectrin repeats in shades of green, and the calmodulin-like domain in blue. The actin-binding domain of the other monomer is shown in dark salmon to demonstrate its location opposite the calmodulin domain.

to as connectin. A single titin molecule extends from the Z-line (amino terminus) to the M-line (carboxyl terminus) in the sarcomere, a length of more than 1 μm (Figure 1.2, shown in red). Thus, the groups of titin molecules have an opposite polarity in each half sarcomere. The arrangement of titin in the myofibril has been determined based on antibody staining of permeabilized cells with monoclonal antibodies. An example is shown in Figure 1.12, where the extra mass from the monoclonal 9D10 is clearly visible in the middle I-band region and a dense line is found in the middle of each half A-band using monoclonal 856. Several studies using many different monoclonals show similar pairs of antibody stripes at equivalent positions in each half sarcomere. Titin filaments are not visible by electron microscopy in intact myofibrils except after extreme stretch where the thick and thin filaments are pulled beyond overlap and so-called "gap filaments" can be observed in the gap between the A- and I-bands. There are also structures referred to as "end filaments" that appear to be formed by aggregates of several titin molecules emerging from the ends of thick filaments after the thick filaments are mechanically removed from the sarcomere (Figure 1.13).

The structure of the protein is extremely complex, with large numbers of repeating domains. Most of the mass consists of 90 to 100 amino acid modules (as many as 280 total) that are 7 stranded β-sheets. These modules are of two types — those similar to the backbone-folding pattern of Ig and another set with similarity to the third domain of fibronectin (FN3). The portion of titin that lies in the A-band has both FN3 and Ig domains but the I-band region contains only the Ig-type domains.

Because titin is linked at both the Z-lines and the M-band, the structure must be elastic in nature to accommodate changes in sarcomere length during contraction and relaxation. In skeletal muscle, the elasticity arises from two sources. First, as a muscle is extended from its rest length,

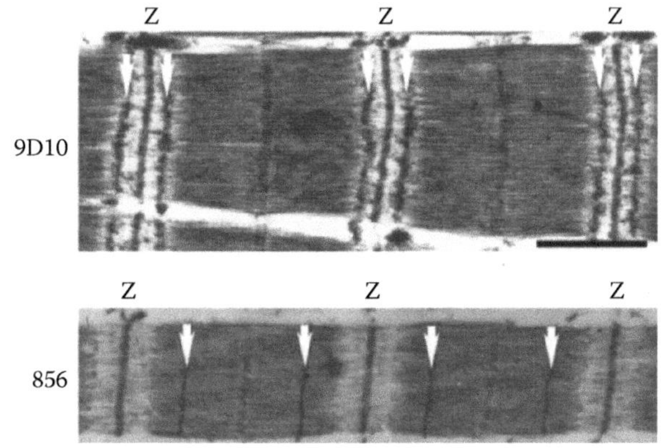

FIGURE 1.12 Titin location within the sarcomeremicron. Two sarcomeres sections of rabbit psoas muscle treated with two different monoclonal antibodies to titin prior to preparation for transmission EM. Because the antibody binds at only one position within each titin molecule, a dense band appears at only one location in each half sarcomere (white arrows). Bar = 1 μm.

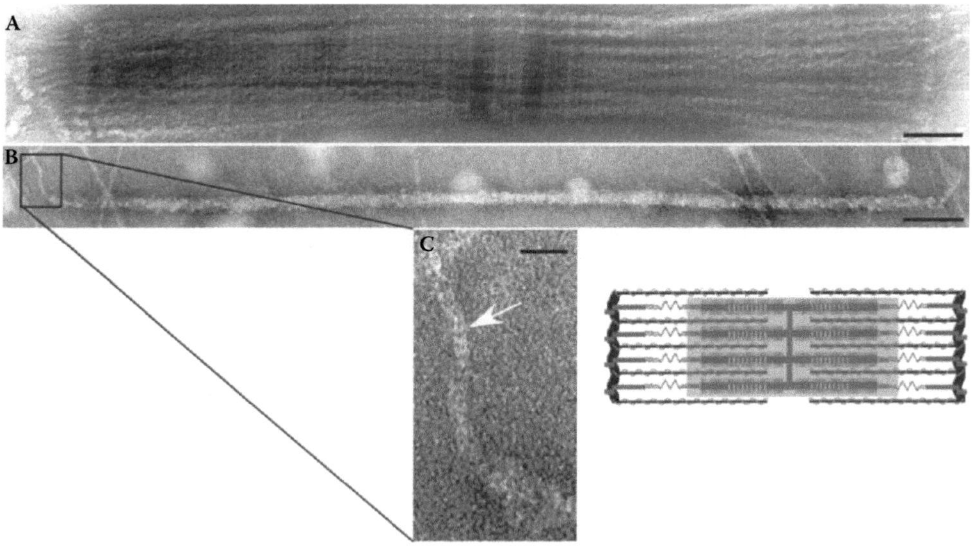

FIGURE 1.13 End filaments of the thick filament. Isolated A-band segment (A) and isolated thick filament (B) from goldfish bodywall muscle. Due to negative staining, protein structures in these electron micrographs appear as light areas on a dark background. End filaments can be seen protruding from the ends of the myosin filaments in both (A) and (B). Note the fine 4 nm striation in the end filament enlargement (C, arrow), thought to arise from the intrinsic domain repeat in this region of the titin molecule. Bars = 100 nm (A and B) or 20 nm (C).

the supercoiled titin molecules are straightened. Recent evidence indicates that this straightening occurs much like that of a carpenter's ruler that has both rigid and flexible regions. With greater extension forces, a specialized amino acid sequence called the PEVK region (positioned near the middle of the half I-band) lengthens. The PEVK is so named because more than 75% of the amino acids in this region consist of proline, glutamic acid, valine, and lysine. Most of the PEVK contains a repeating structure that includes a large number of 26 to 28 amino acid modules and several glutamic-acid-rich regions. While the Ig and FN3 modules are structurally stable in living muscle, the PEVK region is believed to be intrinsically disordered. Because this region is particularly sensitive to protease action (such as during the postmortem period), SDS gel electrophoresis often reveals two titin bands (see Figure 1.4) — the T1 (corresponding to the full-length protein) and T2 (the A-band end of the molecule).

As a muscle cell is stretched under resting conditions (i.e., with magnesium and ATP present and calcium levels less than 10^{-6} M), the resisting force gradually increases with an increase in sarcomere length. The force measured is often referred to as passive tension (see below). The increased tension with sarcomere extension also allows titin to maintain the A-band in the center of the sarcomere. Passive tension varies between different muscles, due in part to the presence of different-sized titins (between 3.2 and 3.7 MDa). There is a single titin gene in higher vertebrates, and the size differences arise through alternative splicing in the middle I-band and PEVK sequence regions.

Unlike the I-band portions discussed above, the A-band portion of titin is believed to remain at fixed length during physiological stretch. The Ig (I) and FN3 (F) domains have 6 super repeats with the pattern I-F-F-I-F-F in the so-called D (distal) region of the A-band and 11 super repeats with the pattern I-F-F-I-F-F-F-I-F-F-F in the C-region found in the middle of each half A-band. The spacing of the start position of the 11 super repeats corresponds to the spacing (43 nm) of MyBP-C in the sarcomere and is consistent with the length of 11 modules (4 nm length determined by x-ray and NMR of an isolated expressed titin module). The precise domain patterns of the titin modules have led to the view that titin serves a ruler function for the assembly of the thick filament. Recent evidence suggests that titin runs parallel to the long axis of the thick filament on the surface of the shaft. A-band titin also contains a kinase domain. It has been shown to phosphorylate telethonin (also know as T-cap) but the kinase and T-cap are half a sarcomere apart in the assembled myofibril. No other protein substrates have been identified to date.

Titin interacts with a plethora of other proteins. These include telethonin/T-cap, filamin, nebulin, α-actinin, and small ankrin 1 in the Z-line region; obscurin, actin, αß crystalline, Fhl2/DRAL, MARP, and calpain-3 in the I-band region; myosin and MyBP-C in the A-band region; and MURFs, Ca/calmodulin kinase, Nbr1, Fhl2/DRAL, lamin, calpain-3, and myomesin in the M-line region. Some of the listed proteins provide structural anchors for titin. Others have been proposed to function in signaling and/or protein turnover. However, the functional significance of most of these interactions is currently unknown.

1.3.13 Obscurin

Obscurin is a large myofibrillar protein with a confusing localization pattern (Kontrogianni-Konstantopoulos, and Bloch 2005). It appears to be bound at the Z-line in developing muscle but localizes to the M-line in adults. Similar to titin, the myomesins, and MyBP-C, obscurin contains Ig and FN domains (68 and 2, respectively). In addition, there is a calmodulin domain, a RhoGEF domain (GEF = GDP/GTP exchange factor) with an associated pleckstin homology (PH) domain, and 2 serine-threonine kinase domains. The amount of the protein is very small — on the order of one tenth as much as nebulin. Obscurin has been found to associate with the Z-line region of titin through two specific Ig domains. It also binds to small ankrin 1 and thus may help tether the sarcoplasmic reticulum to the myofibril and perhaps the myofibrils to the sarcolemma. The RhoGEF domain has been suggested to function in G protein coupled signaling.

1.4 SUBSARCOMERIC STRUCTURES

The sarcomere of the myofibril can be divided into several substructures made from the proteins described above. This includes the thin and thick filaments that make up most of the sarcomere. There are also structures involved in organization of the thin (Z-line) and thick filaments (M-band) into their respective I-bands and A-bands. The organization of these subsarcomeric structures are described below.

1.4.1 Thick Filament

Myosin II molecules associate together to form the myosin or thick filaments found in muscle cells. Whereas the globular N-terminus of myosin determines its motor activity, the C-terminal rod portion is essential for filament formation. Each LMM rod associates with rods of neighboring molecules to form the filament backbone, probably as a result of interactions between charged amino acids at ~4 nm intervals along the rod (McLachlan and Karn 1983; Straussman et al. 2005). Polymerization of myosin into thick filaments *in vitro* is salt, pH, and myosin concentration dependent (Davis 1988). Polymerization starts by formation of longitudinal dimers, association of the dimers to form the bare zone, and then addition of dimers at the ends of the newly formed bare zone. Filaments formed *in vitro* can have highly variable lengths, pointing out the importance of thick-filament associated proteins (i.e., titin) in determining *in vivo* filament lengths. In vertebrate skeletal muscle, filament lengths are quite uniform within each fiber, with around 300 myosin molecules per 1.6 μm filament. The molecules at each end of the filament face in opposite directions, with their C-terminal rods pointing toward the center (Figure 1.14). This bipolar anti-parallel arrangement results in a central bare zone devoid of projecting heads (Figure 1.14B and C). At the center of each filament are additional proteins that tether adjacent thick filaments together and form the M-band (described below). Several models of myosin rod packing have been proposed, some with and some without grouping of monomers into subfilaments (Wray 1979; Chew and Squire 1995). Structural data from vertebrate striated muscle so far do not clearly favor one particular model but EM features suggest that the rods lie nearly parallel to the filament and may surround a narrow hollow core (Squire et al. 1998; Kensler 2005). The organization of myosin in vertebrate smooth muscle filaments has not been fully characterized but probably is based on anti-parallel packing of tails in a side-polar arrangement along the filament so that heads emerge on opposite faces or sides of each filament (Xu et al. 1996). Many invertebrate thick filaments also include the protein paramyosin in their core.

Although the rod portions of the myosin molecules are packed so that some lie buried within the filament backbone, the S1 heads must all be exposed at the filament surface in order to bind to

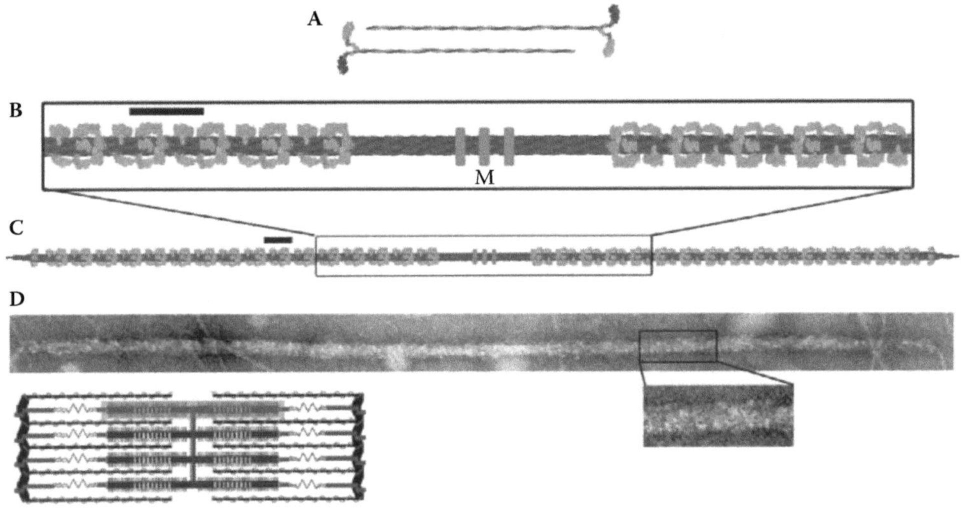

FIGURE 1.14 The filament is formed by association of myosin monomers along their LMM tails, which point in opposite directions in each half filament (A). The absence of motor domains in the center results in a bare zone (B). In vertebnrates, sequential polymerization of about 300 molecules gives rise to a bipolar filament of about 1.6 μm in length (C). Images were rendered as described in Figure 1.5 and the myosin filament arrangement was adapted from the vertebrate cardiac myosin filament structure of Al-Khayat et al. (2006). A transmission electron micrograph of a negatively stained thick filament from goldfish bodywall shows the full thick filament length (D) and at higher magnification, projections of motor domains and their apparent spiral appearance on the surface of the filament. Bar = 50 nm.

adjacent actin filaments. In relaxed muscle, the S1 heads project off the backbone in regular arrays (Figure 1.14B, C, D) that have been extensively studied by low angle x-ray diffraction and EM, and differ by muscle type (smooth vs. striated) and species (Squire et al. 2005; Craig and Woodhead 2006). For example, in insect filaments there are four myosins emerging at each level (crown), whereas in scallop there are seven. Successive crowns are evenly spaced and rotated, so that the heads form a helical pattern running along the surface of the filament. In vertebrate striated muscle, three pairs of myosin heads emerge at each crown but the spacing and rotations of successive crowns vary, with this pattern repeating every three crowns, about 43 nm.

The orientations of the individual myosin S1 heads in relaxed muscle have been the subject of many studies because they represent a starting point for the contractile event, and their highly ordered and periodic arrangement makes them accessible to analysis by x-ray diffraction (Huxley and Brown 1967; Yagi et al. 1977; Squire et al. 2006; Oshima et al. 2007) and EM image processing (Kensler and Stewart 1983; Cantino and Squire 1986; Craig et al. 1992; Al-Khayat et al. 2006; Zoghbi et al. 2008). Moreover, recent work suggests that in myosin-regulated muscles, this arrangement may help maintain the relaxed state. In smooth muscle, myosins with unphosphorylated regulatory light chains have very low ATPase activity and are therefore effectively turned off. Structural studies of smooth muscle HMM suggest that in the unphosphorylated off state, intramolecular interactions inhibit actin binding by one head and ATP hydrolysis by the other (Wendt et al. 2001). This smooth muscle interacting head motif appears consistent with the observed EM structures in relaxed thick filaments of myosin-regulated tarantula muscle (Woodhead et al. 2005). The argument is made that this arrangement is a common feature of all thick filaments that are myosin regulated.

But do such head-head interactions also occur in vertebrate striated muscle? In these systems, activation is initiated by calcium binding to troponin but thick filament protein interactions that stabilize the resting head configuration could play a secondary role in maintaining the relaxed state. Due to perturbations in the helical arrangement, the arrangement of heads in thick filaments of vertebrate striated muscle is more difficult to resolve. Structural studies of myosin filaments from vertebrate skeletal muscle have supported models with heads from the same molecules closely associated but with variations in orientation at different levels within each three-crown repeat (Kensler 2005b; Al-Khayat et al. 2006; Oshima et al. 2007). Most recently, two independent EM studies have modeled crossbridge orientations in mammalian cardiac thick filaments at 4 nm resolution or better. Both find that the smooth muscle interacting-head motif fits the observed EM images at two of three crowns in the myosin helix (Al-Khayat et al. 2008; Zoghbi et al. 2008). In cardiac muscle, it is not expected that this arrangement would prevent activation (as it may in myosin-regulated systems) but could lower its probability at low calcium, thus reducing diastolic pressure and optimizing performance of the heart. These intramolecular interactions, if shown to be widely conserved, could help explain the universal occurrence of two-headed myosins in muscle.

As noted above, the LMM portion of the α-helical rod is essential for filament formation and provides the majority of the mass found in the filament backbone. The S2 domain of the rod is more soluble, associating only weakly with the filament backbone, and single myosin molecules imaged using EM often show a bend near the LMM–HMM junction. These and other data suggest that S2 has considerable mobility *in situ*. During relaxation, it appears to lie close to the filament backbone, maintaining the ordered array of helical filaments. During activation, release of S2 from the filament surface would increase the number of target actin binding sites on adjacent filaments where S1 can bind. This mobility may also help the contractile system adapt to increases in interfilament distance that occur as the muscle shortens.

1.4.2 Thin Filament

The thin filament of skeletal muscle contains primarily F-actin. Bound along the length of the F-actin filament are nebulin, tropomyosin, and troponin (Figure 1.15). The barbed (+) end binding CapZ (Figure 1.15a) and α-actinin is located at the Z-line while the pointed (–) end binding

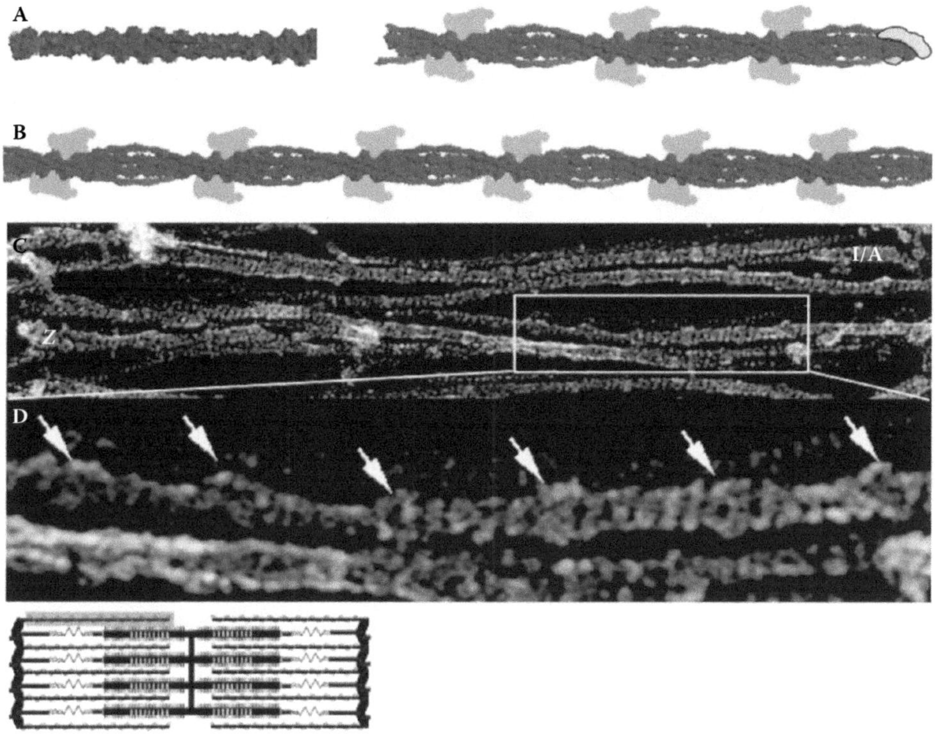

FIGURE 1.15 (See color insert following page 148.) Thin filament structure. Cartoons of the thin filament (A and B) and transmission electron micrographs (C and D) of thin filaments in the non-overlap region from plaice fin muscle after freeze fracture show the protein organization along the filament. Cartoons of the ends (Z-line and tip in the overlap region) of the thin filaments (A) show the capping of the filament. For the plus or Z-line end (leftmost), actin is colored blue and CapZ is colored red. For the minus or tropomodulin end, actin is blue, Tm is red, Tn is green, and tropomodulin is cyan. Arrows in D denote the troponin bulges. The rotation of each successive troponin pair around the filament axis has been omitted to simplify the rendering.

tropomodulin is at the opposite end. The interactions of primarily nebulin, CapZ, and tropomodulin stabilize the F-actin filament from polymerization/depolymerization (Littlefield and Fowler 1998). There may be transient interactions of MyBP-C with actin but these are not well resolved. Myosin also interacts with the thin filament in an ATP-dependent transient fashion, moving toward the barbed (+) end of the thin filament.

Thin filament length in striated mammalian muscle is not the same between muscles within a species or between species. It ranges from 1.0 to 1.3 μm in length. This relatively stable structure and its anatomy allow one to calculate the mole ratio of proteins within the thin filament. For a 1 μm thin filament, there are about 400 actin monomers, 56 troponins and tropomyosins, 2 nebulins, 2 tropomodulins, and 1 CapZ. The amount of α-actinin is dependent on Z-line width, which varies with fiber type but ranges from 2 to 8.

The azimuthal position (position with respect to the cross-section of the filament) of tropomyosin on the actin filament can change, depending on calcium level and strong (rigor) crossbridges (Vibert et al. 1997). This feature was initially inferred from x-ray diffraction studies on muscle and led to the steric blocking model of thin filament regulation (Squire and Morris 1998). More recent studies using EM reconstruction of reconstituted thin filaments (actin+tropomyosin+troponin), confirming this model and adding more detail, are discussed below. The azimuthal location of nebulin is not well resolved. Early studies suggested that the actin binding domains bind to the inner domain, primarily in subdomains 3 and 4 (Pfuhl et al. 1994). More recent studies with F-actin and a 3-module actin-binding fragment of nebulin suggest that it binds in potentially three different locations — the extreme outer domain, the outer domain, and the inner domain (Lukoyanova et al. 2002). Some of these sites are occupied by tropomyosin, so there is potential for steric clashing of these molecules on F-actin.

Some of the change in the location of tropomyosin on actin is mediated by troponin. Troponin is spaced at about 38 nm intervals along each strand of the two-stranded long-pitch helix. Troponins on opposite strands are staggered axially by about 2.5 nm (Figure 1.15A-D). The azimuthal location of troponin with respect to the actin has been difficult to determine. Recent studies using EM reconstruction and single-particle analysis have led to two different models for troponin's location with respect to actin and tropomyosin. One has troponin bound to actin+tropomyosin with the regulatory domain of TnI interacting with the actin and controlling the location of tropomyosin on the same long-pitch strand. In this model, the troponin core domain moves with the tropomyosin that it controls (Narita et al. 2001). The other model has the troponin on the opposite long-pitch strand controlling the position of tropomyosin; that is, the regulatory domain of TnI spans the short-pitch helix (across the filament) to control the location of tropomyosin (Pirani et al. 2006). This model also suggests that part of TnT spans the short pitch and further influences the location of tropomyosin. With this model, the troponin structural domain (TnIT coiled-coil and C-terminal of TnC) does not move; only the regulatory domain of TnI moves to allow for tropomyosin movement on the surface.

1.4.3 Z-Line/Z-Disc

The Z-line bisects the I-band and is a critical structural element that links the contractile activity of the individual thin filaments to contractions at the ends of the muscle cell (Vigoreaux 1994). In cross-section, it appears as a circle or disc and is sometimes called the Z-disc. At the EM level, in cross-section within the Z-disc, it appears as either a square lattice or as a basket-weave lattice (see Figures 1.2 and 1.3). For the square lattice, the corners are the actin filaments and the connections between the corners are called Z-links that are mostly made of α-actinin. For the basket weave, the "weave" are the Z-links and their points of intersection are the actin filaments. These two different appearances are thought to be the result of expanding the basket-weave lattice into the square and thus straightening out the Z-links (Yamaguchi et al. 1985). In longitudinal-section, the Z-line appears with a zigzag line connecting the actin filaments (Figure 1.16a–c). The actin filaments from the opposing half sarcomeres overlap at the Z-line (Yamaguchi et al. 1983), allowing α-actinin to link two actin filaments of opposite polarity (Figure 1.16a). The width of the Z-line is fiber type specific with fast muscles having a narrow (about 50 nm) Z-line and slow/cardiac muscles having a wide (about 140 nm) Z-line. Narrow Z-lines appear to have two Z-links while slow bovine sternomandibularis has six Z-links (Luther et al. 2002). The longitudinal spacing between the links is about 18 nm. Independent of fiber type, the width of the Z-line changes with sarcomere length, being wider at long sarcomere length than short to accommodate the constant volume nature of the filament lattice. This is readily observed in light microscopy of isolated myofibrils using phase-contrast. The Z-line is quite visible in sarcomeres at 3 μm but barely visible at 2.2 μm. However, these apparent changes in width are overestimated because of the contrast method. Additional structural proteins within the Z-line include titin and nebulin. The N-terminal of titin penetrates the Z-line and extends across the complete width. It interacts with the C-terminal, calmodulin-like domains of α-actinin through its Z-repeats. Nebulin interacts with the Z-line and penetrates to the level of the first Z-bridge (Millevoi et al. 1998).

There is emerging evidence that the Z-line is a potential nexus for several signal transduction pathways (Frank et al. 2006) and that there are up to an additional 10 proteins that localize to the Z-line or its periphery (Faulkner et al. 2001). These proteins typically are present in low copy number, many bind to α-actinin, and some are involved in various signal transduction processes and potentially myofibrillogenesis. Additional work needs to be done to document that many of these are *bona fide* structural elements of the Z-line and/or that they are transiently associated with the Z-line in a docking fashion. Some that are well characterized include γ filamin, telethonin, and myopalladin. Filamin was localized to the Z-line more than 25 years ago (Gomer and Lazarides 1981). More recent studies suggest that this localization involves both the actin binding domains of filamin and through an indirect

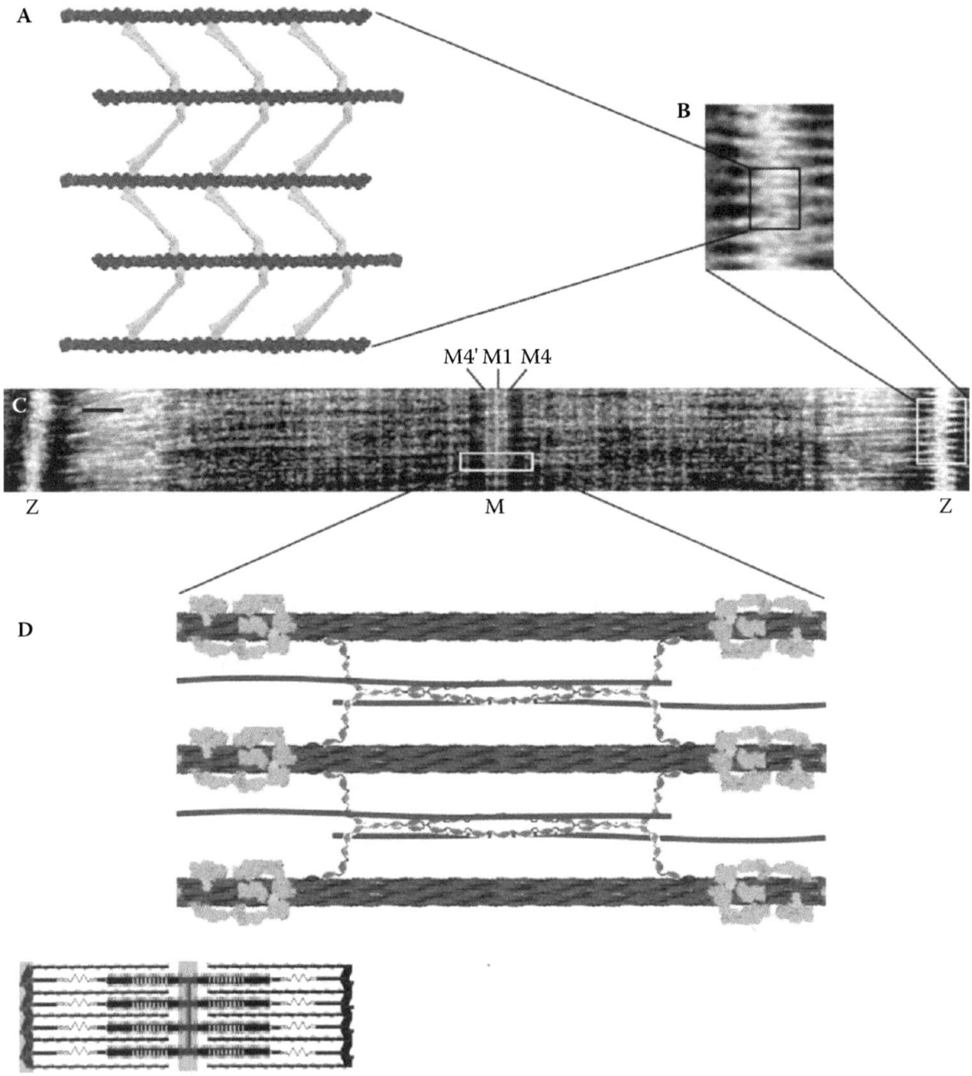

FIGURE 1.16 (See color insert following page 148.) Structure of the Z-line/Z-disc and M-band. A cartoon of the Z-line (A) was made using F-actin structures from Figure 1.8 and the α-actinin structure in Figure 1.13. F-actin is in blue, CapZ in red, and α-actinin in cyan and green. This zigzag appearance is evident in electron micrographs from negatively stained sections of frog semitendinosus muscle (B and C). The M-band as envisioned in cartoon form (D) was made using Fn and Ig domain models for myomesin and the thick filament cartoons in Figure 1.14. The general organization was interpreted from Lange et al. (2002). Myosin is colored green, titin blue, Ig domains magenta, Fn domains orange, and myosin binding domain black. Myomesin bridges are shown more widely spaced than expected for clarity. The M-lines (M4′, M1, and M4) are noted. Bar in (C) = 100 nm.

interaction with α-actinin via a protein called myotilin (Van der Ven et al. 2000). Telethonin/T-cap is localized to the Z-line through its interaction with the N-terminus of titin (see above). Myopalladin is a recently discovered Z-line 145 kDa protein that links the C-terminal Src homology domains of nebulin to the C-terminal calmodulin-like domain of α-actinin (Bang et al. 2001).

1.4.4 I-Band

The I-band was initially so-named because it is isotropic in refractive index when viewed in the polarizing microscope. In phase contrast and EM images, it is the low-density region of the sarcomere bisected by the Z-line (Figure 1.2). As such, a single I-band includes one Z-line and the region of the thin filaments on each side of the Z-line (longitudinally adjacent half sarcomeres) that are not in the overlap region. The width of the I-band decreases as the sarcomere shortens and this feature, with a lack of change in the A-band width, led to the sliding filament hypothesis for muscle contraction. The thin filaments within a half I-band have the same axial repeat or twist. In well-aligned sarcomeres in EM longitudinal sections, as shown in Figure 1.16C, this alignment of adjacent filaments produces perpendicular I-band stripes at 38 nm intervals due to the additional mass of troponin. In cross-section, the actin filaments have a square lattice near the Z-line (out to about 100 nm) and a hexagonal lattice near the A-band (see Figures 1.2 and 1.3). The center-to-center spacing between adjacent thin filaments is about 25 nm in the overlap region but varies with sarcomere length to maintain a constant volume.

1.4.5 M-Band

The electron microscopic appearance of the M-band is related to the types and distributions of myomesins (Agarkova et al. 2003). Five prominent lines spaced 22 nm apart are visible in muscle sections, and these are labeled M6, M4, M1, M4′, and M6′, with M1 at the center. The M4, M4′ lines are present in all muscles (Figure 1.16C and D) but the other lines vary with muscle type and developmental stage. The M1-band is present in fast muscles containing myomesin 2, and M6, M6′ occurs in muscles containing myomesin 3. The binding to titin and the near proximity of this interaction to the titin kinase domain suggests that the myomesins may have some involvement in titin stretch signaling. Myomesins are also thought to be the mass associated with the M-bridges observed in cross-section (Figures 1.2 and 1.3). Creatine kinase (MM isoform) also binds to the M-band to contribute additional mass.

1.4.6 A-Band

In striated muscle, thick filaments lie parallel to each other in clusters that give rise to the A-bands seen in the light microscope (Figure 1.2). The separation between thick filaments is around 48 nm in vertebrate fibers at rest, but decreases as a muscle is stretched. In most species, the filaments are connected at their centers by the M-bridges, which maintain them in a regular hexagonal array. The myomesin-dependent M-bridges and M-lines also keep the filaments in vertebrate striated muscle in close axial register, so that bare zones, myosin crowns, and myosin binding proteins are in nearly perfect alignment (Figures 1.2, 1.3, 1.16, and 1.17). In minimally disrupted A-bands, this alignment creates the appearance of stripes running across each A-band. Weak stripes can be detected

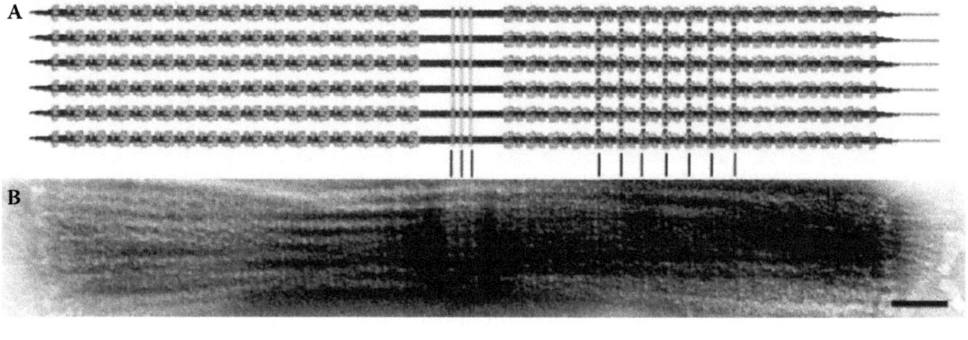

FIGURE 1.17 (See color insert following page 148.) Structure of the A-band. A cartoon of the A-band is shown (A) with thick filaments in green (from Figure 1.14C), M-line forming the M-band in cyan, MyBP-C in dark blue, and titin in red. A transmission electron micrograph of a negatively stained mechanically isolated A-band from goldfish bodywall shows the A-band structures noted in the cartoon (B). Bar = 100 nm.

every 14.3 nm, corresponding to the aligned myosin crowns, but a denser stripe occurs at every third crown. This arises, in part, from perturbation in the arrangement of myosin heads, but in the proximal two thirds of the crossbridge region, the extra density is due to the presence of myosin binding proteins. Seven to nine stripes that begin approximately 240 nm from the center of the filaments in each half A-band are the most prominent. These correspond to the locations of MyBP-C (Figures 1.3, 1.16, and 1.17) and define an area of the A-band known as the C-zone.

The appearance of the A-band as observed in EM cross-sections varies with axial position (Figure 1.3). At its center, the filaments are held in a tight hexagonal lattice by the M-bridges. These interconnections disappear in the adjacent bare zone, where thick filament profiles take on a distinctly triangular appearance in vertebrate striated muscle types. In many species (including mammals), neighboring bare zone profiles can be seen to be rotated approximately 60° relative to one another, whereas in bony fish, adjacent filaments have the same orientations (Luther et al. 1996). Distal to the bare zone is the crossbridge region. Here, the filament packing is still largely hexagonal but the backbone is studded with myosin heads projecting off the surface. The ends of the thick filaments taper and terminate, in isolated thick filaments, with so-called end-filaments (Trinick 1981), characterized by a 4 nm periodicity and presumed to be severed titin filaments (Figure 1.13). Thick filaments imaged *in situ* by EM freeze fracture deep-etch methods also appear to terminate in strands that continue into the I-band and have on their surface a 4 nm periodicity (Cantino et al. 2002). These observations support the view that titin binds to myosin in the A-band but is free for at least part of its length in the I-band.

1.5 SARCOMERE

The sarcomere consists of one A-band and two half I-bands (Figure 1.2). Its length is measured as the distance between two longitudinally adjacent Z-lines. This length can vary from about 4 μm down to less than 1 μm. Physiologically, the length range is more restricted such that it maximizes the extent of overlap between thick and thin filaments (see below). In cross-section, the sarcomere has a diameter of about 1 μm. It is typically surrounded by the intracellular membrane elements of the mitochondria, T-tubule, and sarcoplasmic reticulum (see below). At the EM level, the appearance of the cross-sectional view is determined by the location within the sarcomere (see above, and Figure 1.3). The spacing between the thin and thick filaments in the overlap region is about 25 nm and varies with sarcomere length. The sarcomere is dynamic in that it can grow and shrink in diameter by addition or subtraction of thick and thin filaments. It also has the ability to split and grow, thus adding myofibrils to the cross-section of the cell. Sarcomeres can also be added at the end of the cell to increase the overall cell length. Sarcomeres are aligned end-to-end to form the myofibril.

1.6 MYOFIBRIL-TO-MYOFIBRIL LINKAGES

A striking feature of longitudinal sections of a muscle cell at both light and electron microscope levels is the alignment of the sarcomeres across the cell (Figures 1.2 and 1.20). This alignment is the result of protein linkages between the adjacent sarcomeres at the level of the Z-line and likely the M-line. A major component of this linkage is the intermediate filament protein desmin (Paulin and Li 2004). Desmin itself does not bind to sarcomeric proteins, so its linkage to the periphery of the Z-disc involves cross-linking proteins. The two proteins likely involved in this linkage are synemin and plectin. Synemin is an intermediate filament protein that co-polymerizes with desmin via its N-terminal domain and binds to the N-terminal and rod domain of α-actinin via its C-terminal domain (Bellin et al. 2001). This affords it the ability to cross-link desmin between adjacent Z-lines. Plectin is a relatively large MW, multimodal protein that binds to actin filaments, intermediate filaments, and microtubules (Wiche, 1998). In muscle in cross-section, it is found around Z-discs in association with desmin; and in longitudinal section, the spacing between plectin-containing regions is directly proportional to sarcomere length (Hijikata et al. 1999). With its ability to bind to both intermediate filaments and actin filaments, it functions as a cross-linker between desmin and the Z-discs of adjacent myofibrils. The linking of adjacent myofibrils at the level of the Z-line is thought to enhance the mechanical stability of the muscle cell and to aid in the lateral transmission of force across the fiber to the sarcolemma.

1.7 SURFACE STRUCTURES IN MUSCLE: CORTICAL CYTOSKELETON, PLASMALEMMA, BASAL LAMINA, AND BASEMENT MEMBRANE

Surrounding the myofibril-packed muscle fiber are three structures that form the surface of muscle cells. Directly adjacent to the outermost sarcomeres is the cortical cytoskeleton (also called the sub-sarcolemmal cystoskeleton). This is surrounded by a plasma membrane, often referred to as the sarcolemma, and outside of this is the basement membrane. Forces applied to or generated within the muscle cells must be transmitted across the fragile sarcolemma by a network of proteins and protein complexes. These structures are described below and their spatial relationships are summarized. Specialized surface areas of the cell involved in neurotransmission and at the ends of the cell are discussed in a later section.

1.7.1 Cortical Cytoskeleton

Just below the sarcolemma are several intracellular structures that form the cortical cytoskeleton in striated muscle cells. Costameres are a component of the cortical cytoskeleton and are protein assemblies aligned with the Z-line of the outermost myofibrils in muscle that couple them to the sarcolemma and associated basement membrane. Other more delicate structures are seen surrounding the M-line and connected to the costameres by way of slender fibers oriented parallel to the sarcomere long axis. Of these structures, the costamere has generated the most interest because of its putative role in mechanical stability and strain-dependent cell signaling (mechanotransduction) and gene expression (Ervasti 2003). Costameres contain several cytoskeletal proteins, including γ-actin, spectrin, dystrophin, tropomodulin, and α-actinin. Most interest has focused on dystrophin, the defects in which lead to Duchenne muscular dystrophy. The rod-shaped N-terminus of skeletal muscle dystrophin forms strong lateral associations with γ-actin filaments in the costameres, while the C-terminus binds via dystrobrevin and syntrophins to a multiprotein complex of integral membrane proteins dystroglycan and sarcoglycan-sarcospan (see below). Linkage of the sarcomere to the costamere is via plectin interacting with γ-actin of the costamere and desmin of the sarcomere (Hijikata et al. 2008).

Another structure just below the sarcolemma is the focal adhesion complex. The proteins in this complex include actin, vinculin, talin, α-actinin, plectin, and the cytoplasmic domain of integrins (Critchley 2004). These proteins are also typically aligned with the Z-line, at least at low magnification. The molecular linkages from the sarcomere to the sarcolemma are thought to involve desmin at the Z-line being linked to actin filaments via plectin, much like the myofibril-to-myofibril linkages described earlier. Plectin is thought to link into the filamentous actin network in the cortical cytoskeleton. Linking these filaments to the sarcolemma involves talin, vinculin, and α-actinin that bind either directly or indirectly to the cytoplasmic domain of integrin. There is thought to be some redundancy between the dystrophin/dystroglycan complex and the focal adhesion complex in the mechanical rigidity of the sarcolemma and coupling of sarcomeric contraction to the BM (Mayer 2003).

1.7.2 Sarcolemma

The next layer is the sarcolemma. Like other plasmalemmas, the sarcolemma is a lipid bilayer in which there are embedded proteins essential for signal transduction, propagation of the action potential, ion homeostasis, force transmission, and many other functions. Regular tubular invaginations of the sarcolemma into the cell called T-tubules bring the surface membrane into close proximity to each sarcomere in the cell interior and are described in more detail below.

1.7.3 Transmembrane Proteins

Traversing the sarcolemma are transmembrane proteins that are important in structural coupling of the sarcomere to basement membrane. These proteins include the integrins and the dystroglycan complex. Integrins have a large extracellular domain that binds to laminin and/or fibronectin, and intracellular domains that bind to talin and α-actinin (Ervasti 2003). There are numerous isoforms and their expression changes with muscle development; there are also different isoforms associated with different regions of the cell (Mayer 2003). The most common form in adult muscle is $\alpha_7\beta_{1D}$ and it is aligned with Z-lines. The dystroglycan complex is a large multi-protein complex that includes the transmembrane proteins dystroglycan, sarcoglycan, and sarcospan, as well as syntrophins and α-distrobrevins on the sarcoplasmic side (Michele and Campbell 2003). Dystroglycan binds to laminin, agrin, and perlecan on its extracellular side and to dystrophin on the intracellular side. These transmembrane proteins, through their protein linkages on both the inside and outside of the cell, provide a critical mechanical link between the sarcomere and the basement membrane.

1.7.4 Basement Membrane

Exterior to the sarcolemma is a basement membrane composed of a basal lamina, connected to the sarcolemma via the dystroglycan complex and integrins, and a fibrillar reticular lamina facing the outer connective tissue layers (Figure 1.18). The basement membrane plays critical roles in both muscle development and in maintaining the integrity of muscle during activity (Sanes 2003). Although the contractile apparatus resides within the cells, the basement membrane contributes significantly to the strength and elasticity of the tissue as a whole (Tidball 1986). Several structural elements contribute to these properties. Collagen IV is the most abundant protein in the basal lamina. Its chains interact with non-collagenous proteins such as laminin. Covalently cross-linked networks of collagens and laminins are connected together by the glycoprotein, entactin/nidogen (Figure 1.18c). This network serves as a scaffolding linking structures from the reticular lamina to the sarcolemma, and providing ligands for membrane-associated receptors such as integrins and dystroglycans. Mutations in genes coding for basement membrane components tend to allow normal muscle development but result in progressive degeneration and loss of muscle integrity with use. These include defects in α2 laminin (congenital MD) and its transmembrane receptors (integrin α_7 and dystroglycan) in dystrophin (Duchenne and Becker MD), which links dystroglycan to the cortical cytoskeleton, and in dystroglycan- and dystrophin-associated sarcoglycans (limb girdle MD); the α chains of collagen VI, connecting the basal lamina to the reticular lamina (Bethlem myopathy).

1.7.5 Neuromuscular Junction

The neuromuscular junction is as the name implies: it is the junction between the axon terminal and the muscle cell (Naguib et al. 2002). The development and maintenance of the neuromuscular junction is a complex process involving both the nerve terminal and the muscle cell. Signaling events involved in this process involve acetylcholine and agrin (a secreted extracellular protein) from the nerve terminal and rapsyn and muscle-specific kinase within the muscle cell. Most adult muscle cells have just one neuromuscular junction, typically near the center of the cell. The impinging axon branches into several smaller axon terminals that reside in depressions within the muscle cell and form the motor endplate. The basement membrane of the axon terminal intermeshes with that of the sarcolemma, forming a mechanical coupling between the axon and the muscle cell. The space between the axon terminal and muscle cell is called the synaptic cleft and is about 30 nm wide. It

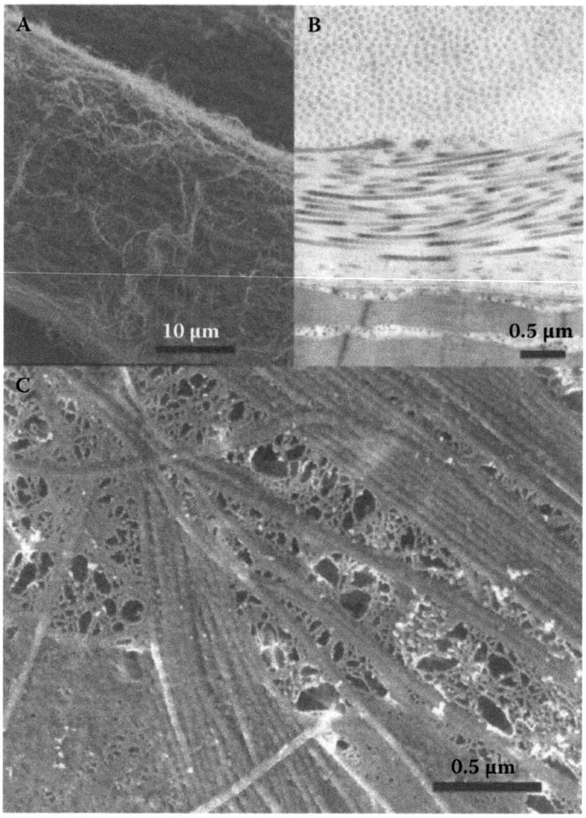

FIGURE 1.18 The surface layers of skeletal muscle. Scanning electron micrograph of a frog semitendinosus fiber (A) demonstrates the collagen fibers of the endomysium. Transmission electron micrograph of frog sartorius (B) shows the collagen fibrils in both cross- and longitudinal section in close association with the basement membrane and myofibrils. Freeze-fracture replica of the surface of chemically skinned plaice fin muscle (C) shows collagen fibrils running over remaining plasma membrane and associated cortical layer of fibers.

is enriched in acetylcholine esterase. The sarcolemma at the motor endplate is enriched in acetylcholine receptors (ligand-gated Na^+ channel). The membrane is extensively folded to increase the surface area exposed to neurotransmitter release from the axon terminal. The acetylcholine released at the axon terminal binds to the acetylcholine receptor, altering its conformation and allowing for Na^+ to pass through the membrane and depolarize the cell. The high level of acetylcholine esterase activity in the synaptic cleft quickly removes the acetylcholine from the synaptic cleft and thus from the receptor allowing the Na^+ channels to close. A single motor neuron may innervate 1 to 100 muscle fibers to control their contraction. This motor neuron, along with the group of muscle cells it controls, is called a motor unit.

1.7.6 Myotendinous Junction

The myotendinous junction is the major region of the cell where force is transmitted from the myofibrils to the connective tissue that forms the tendon (Tidball 1991). The ends of muscle cells are specialized at both the sarcomeric and sarcolemmal levels. At the sarcomere, the myofibrils terminate such that after the last Z-line, there are only thin filaments and no thick filaments or A-band (Figure 1.19). The thin filaments are of different length such that the end half sarcomere is tapered. Talin, vinculin, paxillin, and the intracellular domain of integrin are all enriched in this region to facilitate coupling of these terminal actin filaments to the extracellular matrix,

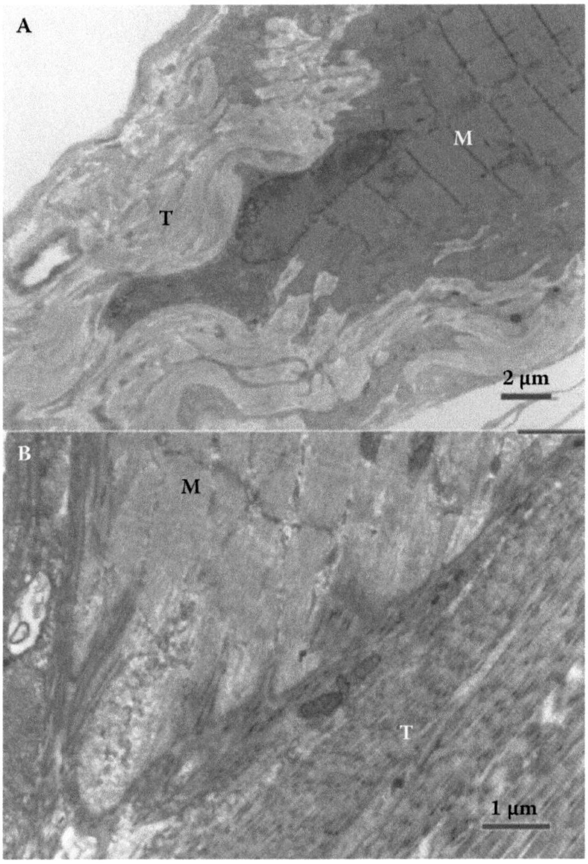

FIGURE 1.19 The myotendinous junction. A low-magnification transmission electron microscope image of the myotendinous junction in a longitudinal section of rat soleus muscle (A) demonstrates the tapering of the fiber and its sheathing by the tendon. A higher-magnification view of a longitudinal section of mouse toe muscle (B) shows that the end half sarcomere is devoid of the A-band and thick filaments. Muscle is denoted as M and tendon as T.

much like that involved in other focal adhesion regions of the cell, but much more so. The sarcolemma is folded at the end of the cell, in close association with the tapered ends of the terminal sarcomeres. The folding increases the surface area for contacts with the extracellular matrix about tenfold. As noted earlier, integrins are enriched in this region and pass through the sarcolemma. The extracellular material is enriched in fibronectin and laminin that bind to collagen fibers of the tendon. Thus, the likely molecular linkage is from the terminal actin filaments of the terminal half sarcomere to talin/vinculin to integrin to fibronectin/laminin to the collagen fibers of the tendon.

1.8 T-TUBULES AND SARCOPLASMIC RETICULUM

The T-tubules and sarcoplasmic reticulum (SR) are involved in converting sarcolemmal depolarization arising from acetylcholine release at the neuromuscular junction into an increase in cytosolic calcium. This process is called excitation-contraction coupling. The SR is also involved in uptake of calcium and thus relaxation of muscle and storage of calcium for subsequent release. Mutations in the proteins associated with these structures can result in improper calcium homeostasis and disease (Berchtold et al. 2000). The T-tubules are named as such because they are transverse with respect to the long axis of the muscle cell. They are invaginations of the sarcolemma, at regularly

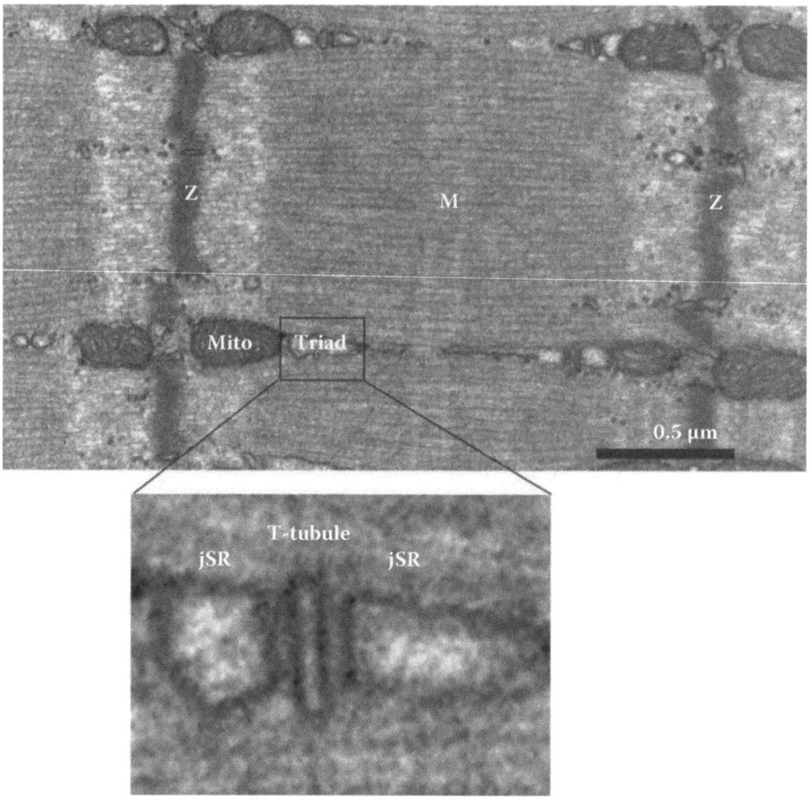

FIGURE 1.20 The location of triads and mitochondria in skeletal muscle. Transmission electron microscope image from a longitudinal section of rat soleus muscle showing the location of the mitochondria (mito) at the level of the I-band, on each side of the Z-line between the myofibrils and the triad near the I-band – A-band junction between the myofibrils. The triad (inset and at higher magnification) includes the junctional SR (jSR) and T-tubule.

spaced intervals, into and across the width of the muscle fiber. Each tube is about 20 nm in diameter and is filled with extracellular fluid. At modest magnification in cross-section, the network forms a net or mesh with each hole in the net/mesh being occupied by a myofibril. That is, the T-tubules wrap around the myofibril. In longitudinal section, the network appears as a series of holes whose location along each sarcomere depends on striated muscle type. For mammalian skeletal muscle, the T-tubules are located near the A-band – I-band junctions on either side of each half sarcomere (Figure 1.20). In avian skeletal muscle, the T-tubules are located at the Z-lines. The SR is a derivative of the smooth endoplasmic reticulum of other cell types. It is a membrane-bounded organelle found between myofibrils and has a high (near mM) calcium concentration within its lumen at rest. It does not form a continuous sheath around the myofibril; rather, it forms a network that partially surrounds the myofibril. It can be further divided into a corpular, bulbous, or junctional region in close proximity to the T-tubule that is called the terminal cisterna, and a thinner region that is more meandering and runs along the length of the sarcomere (Figure 1.20). In skeletal muscle, the SR and T-tubules are organized such that two tubules of the junctional SR lie on opposite sides of a T-tubule. In cross-section, this structure appears as a triad with the T-tubule lumen in the middle and a junctional SR lumen on each side (Figure 1.20). The close proximity of the T-tubule to the SR allows for signal transduction between proteins of these respective structures.

The major proteins involved in converting membrane depolarization into intracellular calcium release are the voltage-gated calcium channel of the T-tubule and the calcium release channel of the junctional SR. These proteins are also called the dihydropyridine receptor and the ryanodine receptor, respectively, based on the drugs that bind to them. The voltage-gated calcium channel is a heterotetrameric transmembrane protein (Dulhunty et al. 2002). Of the four subunits, the $\alpha 1$ subunit is the largest, contains the transmembrane calcium channel, and likely interacts directly with the calcium release channel in skeletal muscle. Depolarization of the T-tubule membrane is sensed by the voltage-gated channel and leads to a conformational change in the protein. The calcium release channel is a very large homotetramer that spans the membrane of the SR (Fill and Copello 2002). The monomer MW is about 520 kDa, and association into the tetramer forms the core of the calcium channel. Most of the mass of the protein is located in the sarcoplasm, facing the T-tubule. These proteins form an organized array in both the T-tubule and SR. The voltage-gated calcium channel forms a so-called tetrad, which lies opposite a calcium release channel of the SR. The ratio of calcium release channels to tetrads is 2:1, such that every other calcium release channel faces a tetrad. For those directly opposite a voltage-gated calcium channel, there are thought to be direct molecular interactions between the two complexes upon T-tubule depolarization that lead to the opening of the calcium release channel. Calcium release from the adjacent channels is thought to occur by calcium-induced calcium release or by communication between the adjacent calcium release channels. In cardiac muscle, the two channels are not directly coupled, and release is thought to be triggered entirely by calcium-induced calcium release.

Other proteins important in regulating calcium levels in both the lumen of the SR and the cytosol are the sarcoplasmic/endoplasmic reticulum calcium ATPase (SERCA) and calsequestrin. Calsequestrin is a 45 kDa, low-affinity, high-capacity calcium binding protein enriched in the corpular region of the SR (Beard et al. 2004). It has 40 to 60 calcium binding sites with a K_d of 1 to 100 μM. Upon calcium binding, it self-associates to form a homomultimer. It functions as a calcium buffer in the lumen of the SR, maintaining the free concentration of calcium in the near mM range. The SERCA or calcium ATPase is a calcium pump that transports two calcium ions from the cytosol to the lumen of the SR per ATP hydrolyzed (Periasamy and Kalyanasundaram 2007). It is a 110 kDa chemiosmotic protein pump in that it uses the energy available from ATP hydrolysis to generate a calcium gradient between the SR and the cytosol. It is a P-type ion pump in that there is transfer of the γ-phosphate of ATP to an aspartate residue near the ATP binding site. The pump can operate either way, depending on the [ATP]/[ADP]*[Pi] : [Ca^{2+}]SR lumen/[Ca^{2+}] cytosol. The kinetic pathway involves an isomerization of the enzyme from a high-affinity calcium binding protein (E1 state) with the calcium binding sites exposed to the sarcoplasm, to a low-affinity calcium binding protein (E2 state) with the calcium binding sites exposed to the SR lumen. The major conformational changes associated with the isomerization are the result of ATP binding, gamma phosphoryl transfer to the enzyme, and ADP release, allowing for isomerization from E1 to E2. The isomerization from E2 to E1 involves hydrolysis of the phosphate from the enzyme and subsequent release of the phosphate. There is a relatively high level of SERCA in the SR to rapidly reduce cytosolic calcium after calcium release from the SR, allowing the muscle to relax rapidly. The equilibrium of the reaction under physiological conditions allows the enzyme to maintain the cytosolic calcium level below 10^{-7} M.

1.9 MITOCHONDRIA

Mitochondria are membrane-bounded organelles involved in oxidative metabolism to synthesize ATP. The substrates are lipids, pyruvate, oxygen, ADP, and Pi, while the products are ATP and carbon dioxide. They are a slow but efficient machine that converts lipid and carbohydrates into a form of energy to be used by the muscle cell and primarily the acto-myosin ATPase upon activation of contraction. The volume of the cell occupied by mitochondria differs with fiber type and can be greater than 20% of the cell volume. There are differences in the mitochondrial content between different fiber types, with slow/oxidative fibers having higher mitochondrial volume than fast glycolytic fibers. Mitochondrial volume within a fiber is plastic and dependent upon the energy demands of the cell. It is often assumed that mitochondria are spheroid bodies within the cell, when in fact they form more of a reticular network. Analysis of mitochondrial shape and location within skeletal muscle shows that there are two domains where mitochondria are enriched: in the subsarcolemmal region and in an intermyofibrillar region. The intermyofibrillar region contains mostly a reticular (meandering) network of the mitochondria when observed in cross-section or a more spheroid shape in longitudinal section (Kirkwood et al. 1986). The reticulum is located around the myofibrils with a doublet of lumens centered at the Z-line (Figure 1.20). The intermyofibrillar mitochondria show higher rates of respiration and protein synthesis than the subsarcolemmal mitochondria (Cogswell et al. 1993).

1.10 NUCLEI

The nucleus of a cell is a highly organized structure that contains its genome. It is bounded by a double-membrane structure called the nuclear envelope that is contiguous with the rough endoplasmic reticulum. The nuclear envelope consists of two nuclear membranes; an underlying nuclear lamina made mostly of intermediate filament proteins called lamins, and nuclear pores. The nuclear pore is a complex protein assembly that regulates protein import and export from the nucleus. As such, it is a control point for signal transduction processes that influence transcriptional activity. Import and export of proteins is energy dependent and involves signal sequences on the protein that are recognized by the protein importin for import and exportin for export. Within the nuclear membrane is the genome with its associated proteins. This protein-DNA complex is called chromatin. The extent of packing of the chromatin is associated with its appearance at both the electron and light microscope levels. Densely packed chromatin is termed heterochromatin, while loosely packed chromatin is called euchromatin. Within a nucleus, there are regions rich in heterochromatin, typically near the nuclear envelope, with euchromatin interspersed in the other regions of the nucleus. Euchromatic chromatin is mostly the result of the dissociation of proteins from the DNA and partial unwinding of the DNA. This allows for transcription of genes. Thus, euchromatic regions of the nucleus are regions high in transcriptional activity. At the cellular level, the entire nucleus is classified as being either euchromatic (mostly containing euchromatin) or heterochromatic (mostly containing heterochromatin). There is another dense structure within the nucleus called the nucleolus. This is the site of ribosomal RNA transcription and processing as well as ribosome assembly. At the light microscope level, the shape of the nucleus is usually indicative of the shape of the cell and in most tissue types, the nucleus is in the center of the cell. The nuclei in adult skeletal muscle are usually elongate, parallel to the long axis of the cell, and located at the cell periphery. Also, skeletal muscle cells are multinucleate. More detailed analyses of the location of nuclei in adult skeletal muscle (Bruusgaard et al. 2003) show that there are regions enriched in nuclei at the motor endplate and myotendinous junction. In the other regions, the nuclei are equidistantly dispersed, and *in vivo* they are relatively immobile.

1.11 ACTOMYOSIN ATPASE AND THE CROSSBRIDGE CYCLE

Muscle contraction results from conformational changes in myosin associated with actin-activated hydrolysis of ATP. Current understanding of this biochemical process is summarized in Figure 1.21. Myosin alone is a cation-dependent ATPase. Several biochemical experiments in the 1970s led to understanding the kinetic properties of the enzyme (Geeves et al. 2005). It will hydrolyze a cation–ATP complex into ADP and Pi, with the rate of this hydrolysis being dependent upon the cation. The relative rates are $NH^{4+} > K^{+} > Ca^{2+} > Mg^{2+}$. The relative affinity of ATP for Mg^{2+} and the free concentration of Mg^{2+} make MgATP the physiological substrate for myosin. As noted above, there are conformational differences associated with the motor domain in the MgATP, Mg-ADP-Pi, MgADP, and nucleotide-free states. Myosin binds MgATP with very high affinity (Figure 1.21). This is followed by hydrolysis at the γ-phosphate, with the products still associated with the enzyme. The rate-limiting step is release of products (Pi and ADP). Product release is sequential, with Pi dissociating

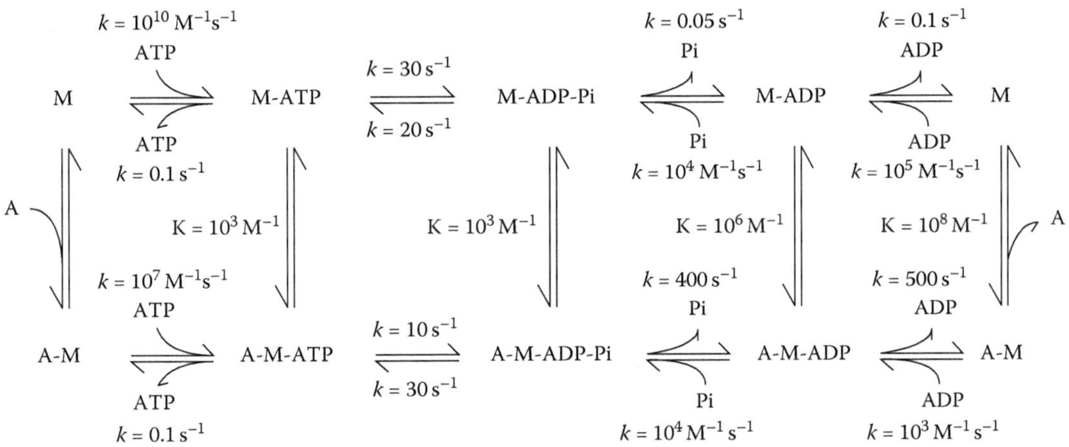

FIGURE 1.21 The myosin and acto-myosin ATPase kinetic scheme. The overall pathway was adapted from Geeves et al. (2005) and the binding and rate constants are from Gordon et al. (2000) The top pathway is for myosin (M) alone, while the bottom pathway is for acto-myosin (A-M). The reactions between the upper and lower pathways are actin (A) association (top to bottom) or dissociation (bottom to top) with myosin. The rate constants (*k*) are given for the forward (top) and reverse (bottom) of each step along the top and bottom pathways. The association constant (K) is given for actin association with myosin in the different states. Starting from rigor (A-M), the pathway in muscle is thought to be AM+ATP ↔ A_1+M-ATP ↔ A_2-M-ATP ↔ A-M-ADP-Pi ↔ A-M-ADP+Pi ↔ A-M+ADP. The subscripts on A are to note that myosin attaches to an actin further toward the Z-line after it is dissociated from the previous actin by ATP.

first and then ADP. The relative affinity of myosin for MgADP and Pi is much less than that for MgATP, favoring the hydrolysis reaction. In solution, the overall pathway is essentially irreversible (i.e., myosin cannot make ATP from an excess of Pi and ADP).

Myosin binds to actin, and the affinity is determined by the nucleotide status of myosin (Figure 1.21). The highest affinity is for myosin without bound nucleotide. This is termed the rigor state and, in the muscle, gives rise to the stiffness associated with death. The next-highest affinity is for the MgADP state. Actin binds to both MgATP and MgADP-Pi myosin rather weakly. The affinity of actin for myosin, for all myosin states, is stronger at low salt than at high salt. The main steps in the hydrolysis pathway that are influenced by interaction with actin are the product release steps, which are accelerated by several orders of magnitude compared to myosin alone. As such, actin can be considered a co-factor of the myosin ATPase that facilitates nucleotide exchange (ATP exchange for ADP). Binding of MgATP to acto-myosin (A-M) is also weaker than MgATP binding to myosin (M).

The general pathway for the crossbridge ATPase (i.e., myosin interacting or bridging with actin in muscle) is similar to the biochemical scheme but probably has additional steps (isomerizations) prior to product release (Gordon et al. 2000). The main physical difference between the crossbridge cycle and the acto-myosin cycle is that, in the muscle, myosin performs work either by moving the actin filament or by maintaining a force. In solution (biochemical system), myosin does no work. The step(s) in the pathway associated with force production are not fully resolved but most information suggests that force production occurs between ATP hydrolysis and phosphate release. Evidence for this is that isometric force rapidly declines upon a rapid increase in phosphate, thought to be associated with reversal of the power stroke (Dantzig et al. 1992). The overall rate of the crossbridge cycle is limited by the rate of MgADP dissociation and this varies with myosin isoform (Siemankowski et al. 1985). This step essentially dictates the speed of unloaded or maximal muscle shortening. The distance of the relative sliding of myosin over actin is about 10 nm/ATP. This is also called the step size of myosin.

The crossbridge can be defined as weak or strong (Brenner 1987). This is determined by the nucleotide status of myosin. With MgATP or Mg-ADP-Pi, it is a weak crossbridge, as it binds the thin filament weakly. With MgADP or no nucleotide bound, it binds strongly. This can be measured by the stiffness (related to extensibility) of muscle. Strong crossbridges, such as those found in rigor muscle, result in stiff muscle. Weak crossbridges, such as those found in relaxed muscle, result in compliant (1/stiffness) or extensible muscle. Because in muscle there is a rather exacting organization of thick and thin filaments in three dimensions, the ability of the motor domain to "find" actin is not a random event as it is in solution. This feature, to some extent, can dictate crossbridge kinetics. Fortunately, there is a slight mismatch in the helical periodicities of actin and myosin that allows myosin to find an actin to bind to in the overlap region at different sarcomere lengths. The mobility of the S2 domain and its potential modulation by MyBP-C can also influence the mobility of the motor domain, thus allowing it to find an actin in the best orientation for interaction.

1.12 REGULATION OF THE CROSSBRIDGE CYCLE

Troponin and tropomyosin are the thin-filament proteins that regulate myosin's interaction with actin in striated muscle. The steric-blocking model of thin-filament regulation was proposed in the early 1970s as the mechanism for regulation (Squire and Morris 1998). This model proposed two states, one an "off" state where tropomyosin occupies or sterically blocks the myosin binding sites on actin (Figure 1.7). With calcium, the thin filament is in the "on" state as a result of conformational changes within the thin filament. In the "on"-state, tropomyosin no longer blocks the binding sites. This allows for full activation of the myosin motor domain for interaction with actin and thus maximal ATPase and crossbridge cycling. While this model was supported by many studies, there were some observations that were not fully explained by the simple two-state model. Biochemical studies showed that the thin filament could not fully activate S1 ATPase activity when the level of S1 was low (Lehrer and Morris 1982) and that strong binding S1 (rigor or MgADP) could activate the thin filament in the absence of calcium (Bremel and Weber 1972; Murray et al. 1982). Additional biochemical studies in the late 1980s led to the proposal of the three-state, two-step model for thin-filament regulation (McKillop and Geeves 1993). Additional structural studies soon followed that supported the three-state model (Swartz et al. 1996; Vibert et al. 1997; Zhang et al. 2000; Swartz et al. 2006). The states (named B, C, and M) relate primarily to the position of tropomyosin on the actin filament, as shown in Figures 1.22 and 1.23. The reaction scheme is B<->C<->M with K_B being the equilibrium between the B- and C-states, and K_T being the equilibrium between the C- and M-states (Figure 1.23). In the B or blocked state, activation of S1 ATPase activity, S1 binding, and contraction is blocked by tropomyosin. This state is highly occupied when calcium is low. In the C or closed state, activation of S1 ATPase activity, S1 binding, and contraction are still blocked by tropomyosin. In this state, tropomyosin is shifted closer to the inner domain of actin but it still blocks some of the S1 binding regions on actin. The C-state is highly occupied when calcium is high. The M-state is associated with activation of S1 ATPase activity, S1 binding, and contraction. Tropomyosin is positioned further toward the inner domain such that it not longer blocks the sites on actin for interaction with myosin. This state is highly occupied when there is about 1 rigor S1 per 7 to 14 actins.

The position of tropomyosin is not static; rather, it fluctuates about on the surface of actin. In the absence of troponin, skeletal tropomyosin is mostly in the C-state position (Lehman et al. 2000). In the presence of troponin but absence of calcium, tropomyosin is mostly in the B-state position. Thus, troponin is thought to "latch" tropomyosin in the blocking position (Farah and Reinach 1995). The molecular mechanisms involved in latching are primarily mediated by TnI (Figure 1.23). At low calcium levels, such that the regulatory domain of TnC is devoid of calcium, the C-terminal region of TnI is mostly bound to actin. In the presence of calcium, the N-terminal regulatory domain of TnC binds calcium and undergoes a conformational change resulting in a high-affinity binding site for the C-terminal region of TnI (Figure 1.9c–f). Thus, there is competition between actin and TnC for

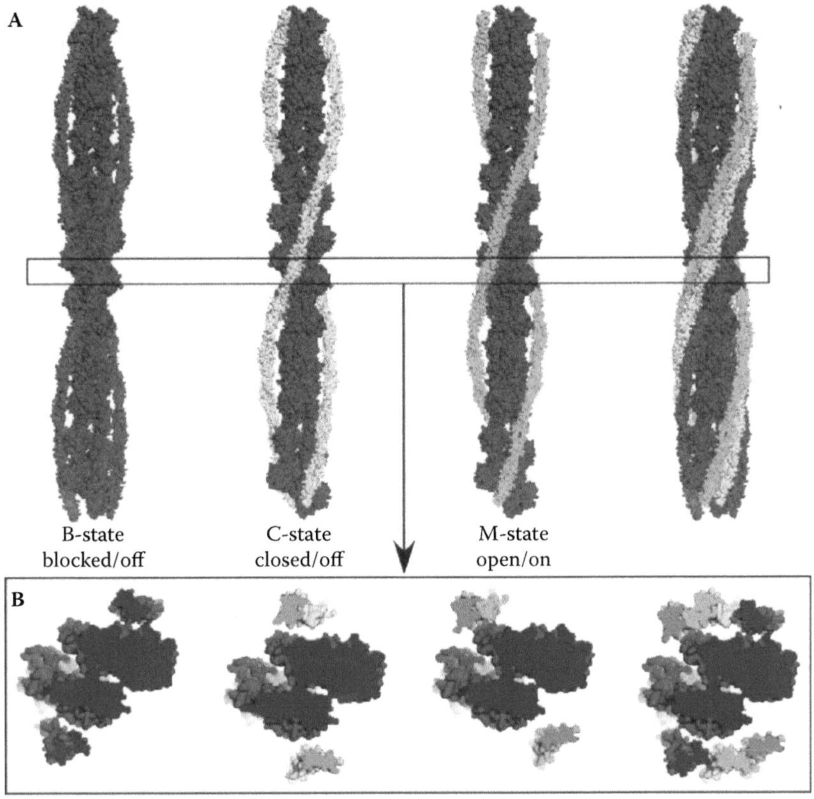

FIGURE 1.22 (See color insert following page 148.) Tropomyosin position on actin filaments associated with thin filament regulation. Longitudinal (A) and cross-sectional (B) views of F-actin with bound tropomyosin are shown with actin in blue and tropomyosin in red, yellow, or green for the B-, C-, and M-states, respectively. Note that tropomyosin is mostly on the inner domain of actin for the B-state and on the outer domain for the M-state. (The cartoons were made as in Figure 1.5 using the supplemental files in Pirani et al., 2006.)

the C-terminal region of TnI. Because TnC has a higher affinity for TnI than TnI's affinity for actin, TnI switches from being bound to actin to being bound to TnC. Thus, tropomyosin is un-latched and moves to the C-position. The equilibrium between the C- and M-states is not large, such that there is a fraction of M-state in the presence of calcium (McKillop and Geeves 1993). This small fraction allows for some myosin interaction with actin. Once there is some myosin interaction, additional actin sites near this M-state site (where the strong crossbridge is at) are also switched to the M-state, allowing additional myosin interaction (Swartz et al. 1990; Swartz et al. 1996; Zhang et al. 2000). This local effect is likely mediated by tropomyosin's relative inelasticity. The model explains well some of the earlier data on the inability of calcium to fully activate the system and the ability of strong crossbridges to activate the system in the absence of calcium. For this latter mechanism, TnI dissociation from actin is likely to occur by a conformational change in actin-tropomyosin.

Activation of contraction is a highly cooperative process. The response of the system to calcium is usually fitted to a sigmoid logistic (derivative of the Hill equation) to obtain the calcium sensitivity (calcium level for half-maximum response or pCa_{50}) and the Hill coefficient (n_H or steepness of the response). Non-cooperative systems have $n_H = 1$, while cooperative systems have a $n_H > 1$. Skeletal muscle can have an n_H of 2 to 6 in the response of isometric force to calcium concentration. The molecular mechanism responsible for this high cooperativity is not fully resolved but involves several features of the thin and thick filaments (Gordon et al. 2000). At the level of the thin filament, the features include cooperative binding of calcium by TnC, and tropomyosin spanning seven actin monomers along the long pitch in conjunction with its inelasticity. At the level of the thick filament,

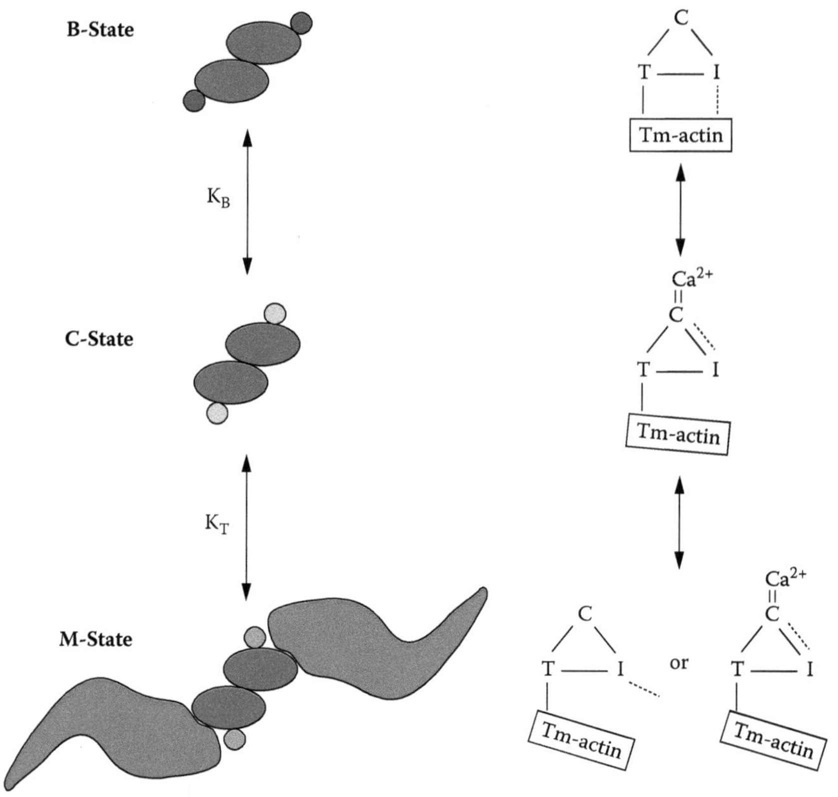

FIGURE 1.23 Thin filament regulation by tropomyosin and troponin. The position of tropomyosin associated with the different states, and the troponin subunit interactions with actin and tropomyosin are shown.

it mainly involves the nucleotide status of the motor domain and its relative mobility on the surface of the thick filament. If the crossbridge is strong (rigor or MgADP), it facilitates conversion of local actin sites into the M-state and, again, this is likely mediated by tropomyosin's inelasticity.

1.13 ACTIVATION OF CONTRACTION AND RELAXATION

Activation of contraction of a skeletal muscle cell originates in the motor cortex of the brain by signaling an action potential in a motor neuron. From here on, it involves several of the proteins discussed above, either directly or indirectly. The action potential travels down the axon to the motor endplate, resulting in the release of acetylcholine. This ligand binds to the acetylcholine receptor and opens the receptor's Na^+ channel, resulting in local membrane depolarization. The local membrane depolarization activates adjacent voltage-dependent Na^+ channels, allowing for the depolarization to propagate along the sarcolemma and down the T-tubules. Depolarization of the T-tubules results in a conformational change in the voltage-gated calcium channel of the T-tubule in the triad that is sensed by the calcium release channel of the junctional SR. The channel opens and calcium flows from the lumen of the SR into the cytosol. The calcium concentration in the cytosol increases to about 10^{-5} M. Some of this free calcium is bound by the N-terminal regulatory domain of TnC, and the conformation of this domain changes such that it has a high-affinity site for the regulatory domain of TnI. Once the C-terminal domain of TnI dissociates from actin, it is bound by TnC and remains bound by TnC because, in the calcium-saturated state, TnC has a higher affinity for TnI than actin. Upon TnI dissociation from actin, tropomyosin shifts from the B-state position, near the outer domain of the actin filament, to the C-state position, more toward the inner domain. There is also a small fraction of the thin filament that is in the M-state position that allows for full interaction of actin and myosin. Once there are a few strong crossbridges attached, more of the thin filament is switched into the M-state, allowing for completion of the crossbridge cycle and thus sarcomere shortening and transmission of this shortening to the tendon and bone causing it to move.

Relaxation involves removal of acetylcholine from the synaptic cleft and closure of the ligand-gated Na^+ channel. Membrane repolarization occurs, resulting in closure of the calcium release channel. Removal of cytosolic calcium by SERCA reduces cytosolic calcium to about 10^{-7} M. Calcium dissociates from the regulatory domain of TnC, resulting in a conformational change that weakens its interaction with TnI. If the proportion of strong crossbridges is modest, as in a physiological contraction, the fraction of M-state is small, and the remainder is in C-state. Once TnI dissociates from TnC, it associates with actin and returns tropomyosin from the C-state position to the B-state position. This, in turn, pulls the three-state equilibrium further away from M-state and eventually nearly all the thin filament is in the B-state, preventing crossbridge cycling and thus relaxing the muscle. In pathological conditions, where ATP is low, the fraction of strong crossbridges is high, and relaxation is prevented even when calcium is low. This condition can occur in the conversion of pre-rigor to post-rigor muscle.

1.14 MUSCLE MECHANICS

Muscle contraction is the result of the myosin motor domain mediated movement of the thick filaments towards the Z-line (or the pulling of the Z-line towards the center of the sarcomere). If the muscle, single fiber, myofibril, or sarcomere is held at a constant length, contraction is said to be isometric (i.e., no change in length). If the muscle, single fiber, myofibril, or sarcomere is not held, so-called unloaded, then it will shorten at its maximum velocity, primarily dictated by the Mg-ADP release step of the crossbridge cycle. If there is a constant load or weight that the muscle is lifting or working against, it will shorten at a constant rate and under these conditions the contraction is called isotonic. Measurement of the shortening velocity as a function of load results in a curve of the form of a rectangular hyperbola called the force-velocity curve. At its extremes, it defines the maximum force (the load at which no shortening occurs) and the maximum shortening velocity (where there is no load on the muscle). The general nature of this curve is the result of slower crossbridge movement when it is doing work (shortening against a load).

It is important to realize that muscle shortening is the result of the individual crossbridges acting at the level of the half sarcomere. Thus, the shortening velocity is determined by the crossbridge throw or step size times the number of cycles per second. However, the speed at which the two ends of a muscle fiber move toward each other (in units of absolute distance/time) also depends on the number of sarcomeres in series. Thus, a muscle that has 1000 sarcomeres in series will shorten 10 times faster than a muscle having 100 sarcomeres in series, as long as the motor is the same. This feature is called the sarcomere amplifier effect. This has importance with regard to overall muscle anatomy, as described previously. To express the shortening velocity as a parameter that is independent of fiber or muscle length, the velocity is often given in muscle lengths per second. Normalizing to muscle fiber lengths per second corrects for this difference as long as the sarcomere lengths are the same between the fibers of different length.

Muscle total isometric force is dependent upon the cross-sectional area of the muscle fibers within it and the sarcomere length of the myofibrils, but not the number of sarcomeres in series. The cross-sectional area component takes into account the amount of myofibrillar material contributing to the force. Because the various fiber types differ in their proportion of myofibrillar mass per unit volume, they will produce different forces per cross-sectional area. For example, slow fibers have significantly more volume of the cell occupied by mitochondria than fast fibers. The sarcomere length component of isometric force is more complicated. There is an inactive (or passive) component of this that rises in a near-exponential fashion with sarcomere length, once the length is greater than about 2.4 μm. In a single fiber that has had the sarcolemma, basement membrane, and endomysium removed, this is mainly mediated by titin (discussed above). This is considered the sarcomeric component to passive tension. At the intact single fiber level, the basement membrane and endomysium also contribute. At the muscle level, the perimysium and epimysium contribute to the passive force.

Muscle active isometric force (via cycling crossbridges) is dictated by the sarcomere length but in a much more complex way than passive force. The active isometric force–sarcomere length relationship was determined with precision in the 1960s (Gordon et al. 1966), and the general features of it are shown in Figure 1.24A for a fiber with a thin filament length of about 1.0 μm. Below about

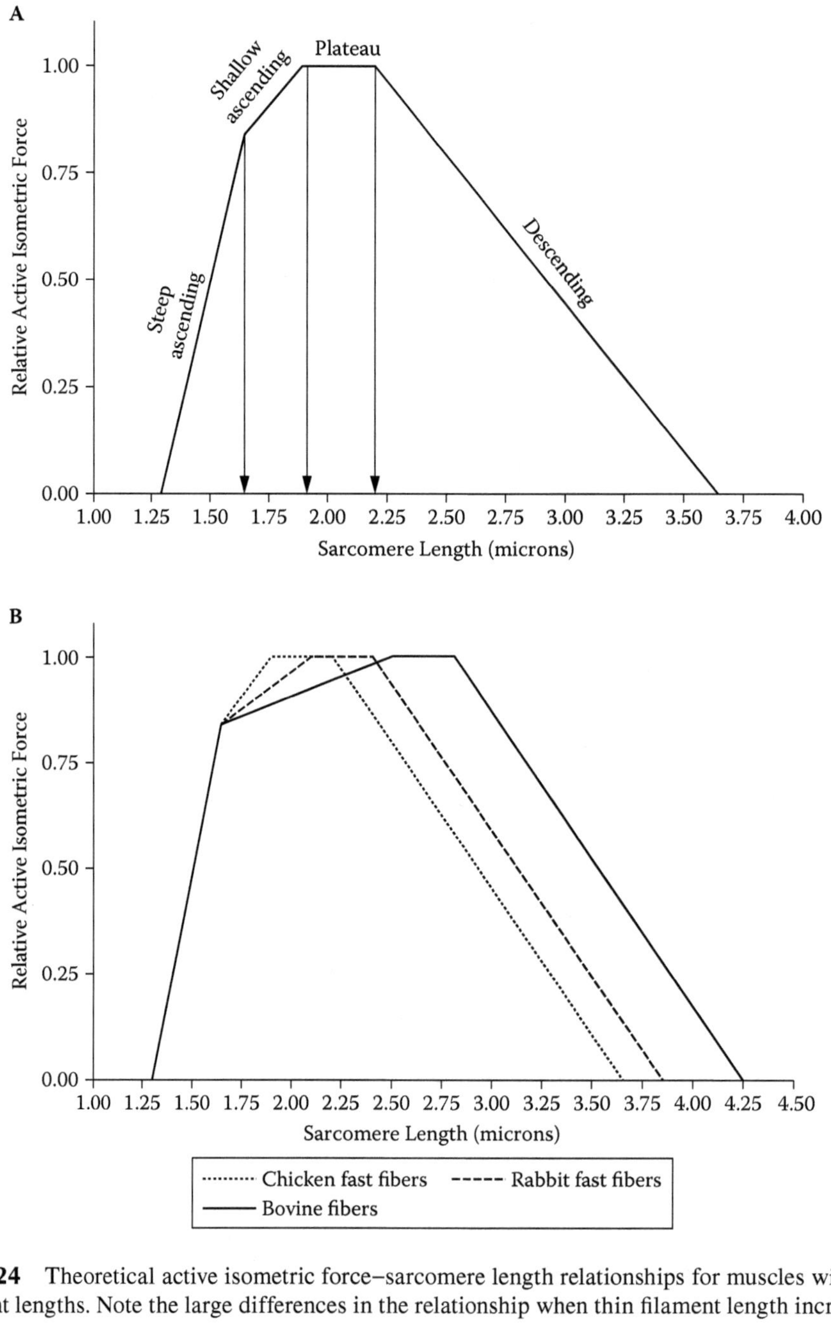

FIGURE 1.24 Theoretical active isometric force–sarcomere length relationships for muscles with different thin filament lengths. Note the large differences in the relationship when thin filament length increases.

1.3 μm, no active force is developed. At longer sarcomere length, the relation can be divided into four phases (steep ascending, shallow ascending, plateau, and descending). The ascending phase is subdivided into a steep ascending phase between 1.3 and 1.6 μm, and a shallow phase between 1.6 and 1.9 μm. This is followed by the plateau phase between 1.9 and 2.2 μm. From 2.2 to 3.65 μm is the descending phase where force declines.

The structure and organization of the thick and thin filaments within the sarcomere determine the shape of the relationship. The main structural mechanism that results in the form of the curve is that the force exerted on a thin filament is determined by the number of cycling crossbridges. This is much akin to people (crossbridges) pulling on a rope (thin filament) to move an object. Adding more people to the rope allows them to move a larger object (develop more force). The steep and shallow ascending phases are the least understood and will be discussed last. The plateau and descending phase are better understood. The sarcomere length range associated with the plateau result from the maximum extent of overlap of thick and thin filaments and thus the maximum number of crossbridges. Even though there are changes in the filament overlap between 1.8 and 2.2 μm, since there are no crossbridges in the bare zone of the A-band, there is no change in force in this region of the curve. Physiologically, skeletal muscle works in the plateau phase in most cases. At longer lengths (>2.2 μm), the tips of the thin filaments are pulled further away from the M-line, resulting in there being no actin for the motor domains closest to the M-line to interact with. Further increases in length result in fewer and fewer crossbridges on the thin filament and thus a linear decline in force with increases in sarcomere length.

The decreased force at lengths shorter than about 2.0 μm likely involves several mechanisms. For the steep ascending phase (1.3 to ~1.65 μm), there is an internal load (muscle working against itself) resulting from the thick filaments abutting the Z-lines. Either the thick filaments are forced through the Z-line or they buckle at the Z-line. Additionally, the force generated by the crossbridge is about tenfold less when it interacts with actin of the opposite polarity (Ishijima et al. 1996). In this length range, a substantial length of the thin filament from the opposing half sarcomere is present, thus, half the thin filaments in the actin-double overlap zone are of the opposite polarity. Early studies using *in vitro* motility revealed that the actin filament polarity determined the direction of movement (Toyoshima et al. 1987). Thus, for a crossbridge to form with a thin filament of the opposite polarity, it must swivel 180° about the long axis of the thick filament, likely via the S2 domain. This contortion of the crossbridge likely decreases its efficiency. In the shallow ascending phase, the main factors decreasing the force are the inefficiency of the misoriented crossbridges and a lesser contribution of an internal load, likely via forcing the thin filaments from the opposing half sarcomere through the M-line and past each other. Interestingly, the ATPase activity as a function of sarcomere length does not have an ascending phase, only a descending phase at lengths greater than about 2.2 μm (Stephenson et al. 1989). Similarly, almost all motor domains form rigor crossbridges at sarcomere lengths below the plateau range (Swartz et al. 1993).

It is important to appreciate that the exact sarcomere length–active isometric force relationship is determined by the length of the thin and thick filaments, with a minor contribution from the width of the Z-line. In vertebrate muscle, the thick filaments are 1.6 μm long, whereas the thin filaments and Z-line width differ between vertebrate species and between fiber types within some species. For fast chicken muscle, the thin filament length is 1.0 μm, for fast rabbit muscle it is about 1.1 μm, and for bovine (and human) it is about 1.3 μm (Ringkob et al. 2004). These differences in thin filament length give different predicted isometric force–sarcomere length relationships, as shown in Figure 1.24B.

The steep ascending phases are the same between the different muscles as this is mainly mediated by thick filament length. The shallow ascending phase becomes shallower with longer thin filament length, as thin filament length is involved in this phase. Here, the sarcomere length at which the shallow ascending phase intersects the plateau is equal to 2X thin filament length plus one Z-line width. The absolute width of the plateau phase is independent of thin filament length but its location in terms of sarcomere length is at longer lengths for fibers with longer thin filaments.

Finally, the slopes of the descending phase are independent of thin filament length as there is a linear dependence of force on number of crossbridges (i.e., length of the overlap region). The beginning and end of this phase is dependent upon thin filament length, with the phase being shifted to longer sarcomere lengths with longer thin filaments. This is in accord with the beginning of the descending phase starting at the sum of 2X thin filament length plus 1 Z-line width, and the end being determined by the thick filament length plus 2X thin filament length plus 1 Z-line width. These large differences in the active force–sarcomere length relationship because of differences in thin filament length are generally not appreciated.

1.15 SUMMARY

Skeletal muscle is a highly organized tissue. This organization can be traced back to the molecular level of the individual proteins that are within the sarcomere. Subsequent organization of the sarcomeres into myofibrils and myofibrils into skeletal muscle cells requires additional proteins to allow for mechanical coupling of the contractile elements to tendons and bone and thus transport us about. This transport, at the molecular level, involves a highly regulated interaction between actin and myosin. Structural features of the sarcomere dictate the mechanical properties of muscle contraction and relaxation as well as cooked meat eating quality. By understanding the properties of the major skeletal muscle proteins, it is hoped that we can gain further insight into skeletal muscle structure and function and its properties as a food.

ACKNOWLEDGMENTS

The authors would like to thank USDA (NC1131 Regional Hatch Project) and NIH (R01 HL 077196, R01 HL 073828, and R01 HL 049443) for funding. They also thank John Squire, Pradeep Luther, Michael Chew, Joseph Eichen, and Lisa Brown for their help in producing EM samples used for these figures.

REFERENCES

Agarkova, I., E. Ehler, S. Lange, R. Schoenauer, and J.C. Perriard. 2003. M-band: a safeguard for sarcomere stability? *J. Muscle Res. Cell Motil.* 24:191–203.

Al-Khayat, H.A., E.P. Morris, R.W. Kensler, and J.M. Squire. 2006. 3D structure of relaxed fish muscle myosin filaments by single particle analysis. *J. Struct. Biol.* 155:202–217.

Al-Kyahat, H.A., E.P. Morris, R.W. Kensler, and J.M. Squire. 2008. Myosin filament 3D structure in mammalian cardiac muscle. *J. Struct. Biol.* 163:117–126.

Bang, M.L., R.E. Mudry, A.S. McElhinny, K. Trombitas, A.J. Geach, R. Yamasaki, H. Sorimachi, H. Granzier, C.C. Gregorio, and S. Labeit. 2001. Myopalladin, a novel 145-kilodalton sarcomeric protein with multiple roles in Z-disc and I-band protein assemblies. *J. Cell Biol.* 153:413–427.

Beard, N.A., D.R. Laver, and A.F. Dulhunty. 2004. Calsequestrin and the calcium release channel of skeletal and cardiac muscle. *Prog. Biophys. Mol. Biol.* 85:33–69.

Bellin, R.M., T.W. Huiatt, D.R. Critchley, and R.M. Robson. 2001. Synemin may function to directly link muscle cell intermediate filaments to both myofibrillar Z-lines and costameres. *J. Biol. Chem.* 276:32330–32337.

Bennett, P., R. Craig, R. Starr, and G. Offer. 1986. The ultrastructural location of C-protein, X-protein and H-protein in rabbit muscle. *J. Muscle Res. Cell Motil.* 7:550–567.

Berchtold, M.W., H. Brinkmeier, and M. Muntener. 2000. Calcium ion in skeletal muscle: its crucial role for muscle function, plasticity, and disease. *Physiol. Rev.* 80:1215–1265.

Bremel, R.D. and A. Weber. 1972. Cooperation within actin filament in vertebrate skeletal muscle. *Nat. New Biol.* 238:97–101.

Brenner, B. 1987. Mechanical and structural approaches to correlation of cross-bridge action in muscle with actomyosin ATPase in solution. *Annu. Rev. Physiol.* 49:655–672.

Bruusgaard, J C., K. Liestol, M. Ekmark, K. Kollstad, and K. Gundersen. 2003. Number and spatial distribution of nuclei in the muscle fibres of normal mice studied *in vivo*. *J. Physiol.* 551:467–478.

Cantino, M.E., M.W. Chew, P.K. Luther, E. Morris, and J.M. Squire. 2002. Structure and nucleotide-dependent changes of thick filaments in relaxed and rigor plaice fin muscle. *J. Struct. Biol.* 137:164–175.

Cantino, M.E. and J.M. Squire.1986. Resting myosin crossbridge configuration in frog muscle thick filaments. *J. Cell Biol.* 102:610–618.

Carlier, M.F. 1991. Actin: protein structure and filament dynamics. *J. Biol. Chem.* 266:1–4.

Chew, M.W.K. and J.M. Squire. 1995. Packing of α-helical coiled-coil myosin rods in vertebrate muscle thick filaments. *J. Struct. Biol.* 115:233–249.

Cogswell, A. M., R.J. Stevens, and D.A. Hood. 1993. Properties of skeletal muscle mitochondria isolated from subsarcolemmal and intermyofibrillar regions. *Am. J. Physiol.* 264:C383–389.

Cooper, J.A. and D.A. Schafer. 2000. Control of actin assembly and disassembly at filament ends. *Curr. Opin. Cell. Biol.* 12:97–103.

Craig, R. and G. Offer. 1976. The location of C-protein in rabbit skeletal muscle. *Proc. Roy. Soc. London B* 192:451–461.

Craig, R. and J.L. Woodhead. 2006. Structure and function of myosin filaments. *Curr. Opin. Struct. Biol.* 16:204–212.

Critchley, D.R. 2004. Cytoskeletal proteins talin and vinculin in integrin-mediated adhesion. *Biochem. Soc. Trans.* 32:831–836.

Dantzig, J.A., Y.E. Goldman, N.C. Millar, J. Lacktis, and E. Homsher. 1992. Reversal of the cross-bridge force-generating transition by photogeneration of phosphate in rabbit psoas muscle fibres. *J. Physiol.* 451:247–278.

Davis, J.S. 1988. Assembly processes in vertebrate skeletal thick filament formation. *Annu. Rev. Biophys. Biophys. Chem.* 17:217–239.

de Tombe, P.P. 2006. Myosin binding protein C in the heart. *Circ. Res.* 98:1234–1236.

Dulhunty, A.F., C.S. Haarmann, D. Green, D.R. Laver, P.G. Board, and M.G. Casarotto. 2002. Interactions between dihydropyridine receptors and ryanodine receptors in striated muscle. *Prog. Biophys. Mol. Biol.* 79:45–75.

Ervasti, J.M. 2003. Costameres: the Achilles' heel of Herculean muscle. *J. Biol. Chem.* 278:13591–13594.

Farah, C.S. and F.C. Reinach. 1995. The troponin complex and regulation of muscle contraction. *FASEB J.* 9:755–767.

Faulkner, G., G. Lanfranchi, and G. Valle. 2001. Telethonin and other new proteins of the Z-disc of skeletal muscle. *IUBMB Life* 51:275–282.

Fill, M. and J.A. Copello. 2002. Ryanodine receptor calcium release channels. *Physiol. Rev.* 82:893–922.

Fischer, R.S. and V.M. Fowler. 2003. Tropomodulins: life at the slow end. *Trends Cell. Biol.* 13:593–601.

Flashman, E., L. Korkie, H. Watkins, C. Redwood, and J.C. Moolman-Smook. 2008. Support for a trimeric collar of myosin binding protein C in cardiac and fast skeletal muscle, but not in slow skeletal muscle. *FEBS Lett.* 582:434–438.

Flicker, P.F., G.N. Phillips, Jr., and C. Cohen. 1982. Troponin and its interactions with tropomyosin. An electron microscope study. *J. Mol. Biol.* 162:495–501.

Foth, B.J., M.C. Goedecke, and D. Soldati. 2006. New insights into myosin evolution and classification. *Proc. Natl. Acad. Sci.* 103:368–3686.

Fowler, V.M., N.J. Greenfield, and J. Moyer. 2003. Tropomodulin contains two actin filament pointed end-capping domains. *J. Biol. Chem.* 278:40000–40009.

Frank, D., C. Kuhn, H.A. Katus, and N. Frey. 2006. The sarcomeric Z-disc: a nodal point in signalling and disease. *J. Mol. Med.* 84:446–468.

Geeves, M.A., M. Chai, and S.S. Lehrer. 2000. Inhibition of actin-myosin subfragment 1 ATPase activity by troponin I and IC: relationship to the thin filament states of muscle. *Biochemistry.* 39:9345–9350.

Geeves, M.A. and K.C. Holmes. 2005. The molecular mechanism of muscle contraction. *Adv. Protein Chem.* 71:161–193.

Gomer, R.H. and E. Lazarides. 1981. The synthesis and deployment of filamin in chicken skeletal muscle. *Cell* 23:524–532.

Gordon, A.M., E. Homsher, and M. Regnier. 2000. Regulation of contraction in striated muscle. *Physiol. Rev.* 80:853–924.

Gordon, A.M., A.F. Huxley, and F.J. Julian. 1966. The variation in isometric tension with sarcomere length in vertebrate muscle fibres. *J. Physiol.* 184:170–192.

Grabarek, Z., T. Tao, and J. Gergely. 1992. Molecular mechanism of troponin-C function. *J. Muscle Res. Cell Motil.* 13:383–393.

Granzier, H. and S. Labeit. 2007. Structure-function relations of the giant elastic protein titin in striated and smooth muscle cells. *Muscle Nerve* 36:740–755.

Granzier, H.L. and K.B. Campbell. 2006. New insights in the role of cardiac myosin binding protein C as a regulator of cardiac contractility. *Circ. Res.* 99:795–797.

Hernandez, O.M., M. Jones, G. Guzman and D. Szczesna-Cordary. 2007. Myosin essential light chain in health and disease. *Am. J. Physiol. Heart Circ. Physiol.* 292:H1643–1654.

Hijikata, T., T. Murakami, M. Imamura, N. Fujimaki, and H. Ishikawa. 1999. Plectin is a linker of intermediate filaments to Z-discs in skeletal muscle fibers. *J. Cell Sci.* 112(Pt. 6):867–876.

Hijikata, T., A. Nakamura, K. Isokawa, M. Imamura, K. Yuasa, R. Ishikawa, K. Kohama, S. Takeda, and H. Yorifuji. 2008. Plectin 1 links intermediate filaments to costameric sarcolemma through {beta}-synemin, {alpha}-dystrobrevin and actin. *J. Cell Sci.* 121:2062–2074.

Holmes, K.C., D. Popp, W. Gebhard, and W. Kabsch. 1990. Atomic model of the actin filament. *Nature* 347:44–49.

Hornemann, T., S. Kempa, M. Himmel, K. Hayess, D. O. Furst, and T. Wallimann. 2003. Muscle-type creatine kinase interacts with central domains of the M-band proteins myomesin and M-protein. *J. Mol. Biol.* 332: 877–887.

Huxley, H.E. and W. Brown. 1967. The low-angle x-ray diagram of vertebrate striated muscle and its behavior during contraction and rigor. *J. Mol. Biol.* 30:383–434.

Ishijima, A., H. Kojima, H. Higuchi, Y. Harada, T. Funatsu, and T. Yanagida. 1996. Multiple- and single-molecule analysis of the actomyosin motor by nanometer-piconewton manipulation with a microneedle: unitary steps and forces. *Biophys. J.* 70:383–400.

Kabsch, W., H.G. Mannherz, D. Suck, E.F. Pai, and K.C. Holmes. 1990. Atomic structure of the actin:DNase I complex. *Nature* 347:37–44.

Kensler, R.W. 2005. The mammalian cardiac muscle thick filament: backbone contributions to meridional reflections. *J. Struct. Biol.* 149:313–324.

Kensler, R.W. 2005b. The mammalian cardiac muscle thick filament: Crossbridge arrangement. *J. Struct. Biol.* 149:303–312.

Kirkwood, S.P., E.A. Munn, and G.A. Brooks. 1986. Mitochondrial reticulum in limb skeletal muscle. *Am. J. Physiol.* 251:C395–C402.

Kontrogianni-Konstantopoulos, A. and R.J. Bloch. 2005. Obscurin: a multitasking muscle giant. *J. Muscle Res. Cell Motil.* 26:419–426.

Kulikovskaya, I., G. McClellan, J. Flavigny, L. Carrier, and S. Winegrad. 2003. Effect of MyBP-C binding to actin on contractility in heart muscle. *J. Gen. Physiol.* 122:761–774.

Lange, S., D. Auerbach, P. McLoughlin, E. Perriard, B.W. Schafer, J.C. Perriard, and E. Ehler. 2002. Subcellular targeting of metabolic enzymes to titin in heart muscle may be mediated by DRAL/FHL-2. *J. Cell Sci.* 115:4925–4936.

Lange, S., M. Himmel, D. Auerbach, I. Agarkova, K Hayess, D.O. Furst, J.C. Perriard, and E. Ehler. 2005. Dimerisation of myomesin: implications for the structure of the sarcomeric M-band. *J. Mol. Biol.* 345: 289–298.

Lehrer, S.S. and E.P. Morris. 1982. Dual effects of tropomyosin and troponin-tropomyosin on actomyosin subfragment 1 ATPase. *J. Biol. Chem.* 257:8073–8080.

Lehman, W., V. Hatch, V. Korman, M. Rosol, L. Thomas, R. Maytum, M.A. Geeves, J.E. Van Eyk, L.S. Tobacman, and R. Craig. 2000. Tropomyosin and actin isoforms modulate the localization of tropomyosin strands on actin filaments. *J. Mol. Biol.* 302:593–606.

Levine, R., A. Weisberg, I. Kulikovskaya, G. McClellan, and S. Winegrad. 2001. Multiple structures of thick filaments in resting cardiac muscle and their influence on cross-bridge interactions. *Biophys. J.* 81:1070–1082.

Littlefield, R. and V.M. Fowler. 1998. Defining actin filament length in striated muscle: rulers and caps or dynamic stability?. *Annu. Rev. Cell. Dev. Biol.* 14:487–525.

Lukoyanova, N., M.S. VanLoock, A. Orlova, V.E. Galkin, K. Wang, and E.H. Egelman. 2002. Each actin subunit has three nebulin binding sites: implications for steric blocking. *Curr. Biol.* 12:383–388.

Luo, Y., J. Leszyk, B. Li, Z. Li, J. Gergely, and T. Tao. 2002. Troponin-I interacts with the Met47 region of skeletal muscle actin. Implications for the mechanism of thin filament regulation by calcium. *J. Mol. Biol.* 316:429–434.

Luther, P.K., J.S. Barry, and J.M. Squire. 2002. The three-dimensional structure of a vertebrate wide (slow muscle) Z-band: lessons on Z-band assembly. *J. Mol. Biol.* 315:9–20.

Luther, P.K., J.M. Squire, and P.L. Forey. 1996. Evolution of myosin filament arrangements in vertebrate skeletal muscle. *J. Morphol.* 229:325–335.

Mayer, U. 2003. Integrins: redundant or important players in skeletal muscle? *J. Biol. Chem.* 278: 14587–14590.

McElhinny, A.S., S.T. Kazmierski, S. Labeit, and C.C. Gregorio. 2003. Nebulin: the nebulous, multifunctional giant of striated muscle. *Trends Cardiovasc. Med.* 13:195–201.
McGough, A. 1998. F-actin-binding proteins. *Curr. Opin. Struct. Biol.* 8:166–176.
McKillop, D.F. and M.A. Geeves. 1993. Regulation of the interaction between actin and myosin subfragment 1: evidence for three states of the thin filament. *Biophys. J.* 65:693–701.
McLachlan, A.D. and J. Karn. 1983. Periodic features in the amino acid sequence of nematode myosin rod. *J. Mol. Biol.* 164:605–626.
Michele, D.E. and K.P. Campbell. 2003. Dystrophin-glycoprotein complex: post-translational processing and dystroglycan function. *J. Biol. Chem.* 278:15457–15460.
Millevoi. S., K. Trombitas, B. Kolmerer, S. Kostin, J. Schaper, K. Pelin, H. Granzier, and S. Labeit. 1998. Characterization of nebulette and nebulin and emerging concepts of their roles for vertebrate Z-discs. *J. Mol. Biol.* 282:111–123.
Moss, R.L. and D.P. Fitzsimons. 2006. Myosin light chain 2 into the mainstream of cardiac development and contractility. *Circ. Res.* 99:225–227.
Murray, J.M., M.K. Knox, C.E. Trueblood, and A. Weber. 1982. Potentiated state of the tropomyosin actin filament and nucleotide-containing myosin subfragment 1. *Biochemistry.* 21:906–915.
Naguib, M., P. Flood, J.J. McArdle, and H.R. Brenner. 2002. Advances in neurobiology of the neuromuscular junction: implications for the anesthesiologist. *Anesthesiology* 96:202–231.
Narita, A., S. Takeda, A. Yamashita, and Y. Maeda. 2006. Structural basis of actin filament capping at the barbed-end: a cryo-electron microscopy study. *EMBO J.* 25:5626–5633.
Narita, A., T. Yasunaga, T. Ishikawa, K. Mayanagi, and T. Wakabayashi. 2001. Ca(2+)-induced switching of troponin and tropomyosin on actin filaments as revealed by electron cryo-microscopy. *J. Mol. Biol.* 308:241–261.
Oakley, C.E., J. Chamoun, L.J. Brown, and B.D. Hambly. 2007. Myosin binding protein-C: Enigmatic regulator of cardiac contraction. *Int. J. Biochem. & Cell Biol.* 39:2161–2166.
Oshima, K., Y. Takezawa, Y. Sugimoto, T. Kobayashi, T.C. Irving, and K. Wakabayashi. 2007. Axial dispositions and conformations of myosin crossbridges along thick filaments in relaxed and contracting states of vertebrate striated muscles by x-ray fiber diffraction. *J. Mol. Biol.* 367:275–301.
Otey, C.A. and O. Carpen. 2004. Alpha-actinin revisited: a fresh look at an old player. *Cell Motil. Cytoskeleton* 58:104–111.
Papa, I., C. Astier, O. Kwiatek, F. Raynaud, C. Bonnal, M.C. Lebart, C. Roustan, and Y. Benyamin. 1999. Alpha actinin-CapZ, an anchoring complex for thin filaments in Z-line. *J. Muscle Res. Cell Motil.* 20:187–197.
Pappas, C.T., N. Bhattacharya, J.A. Cooper, and C.C. Gregorio. 2008. Nebulin interacts with CapZ and regulates thin filament architecture within the Z-disc. *Mol. Biol. Cell* 19:1837–1847.
Paulin, D. and Z. Li. 2004. Desmin: a major intermediate filament protein essential for the structural integrity and function of muscle. *Exp. Cell Res.* 301:1–7.
Pearlstone, J.R. and L.B. Smillie. 1981. Identification of a second binding region on rabbit skeletal troponin-T for alpha-tropomyosin. *FEBS Lett.* 128:119–122.
Periasamy, M. and A. Kalyanasundaram. 2007. SERCA pump isoforms: their role in calcium transport and disease. *Muscle Nerve* 35:430–442.
Perry, S.V. 1998. Troponin T: genetics, properties and function. *J. Muscle Res. Cell Motil.* 19:575–602.
Perry, S.V. 1999. Troponin I: inhibitor or facilitator. *Mol. Cell. Biochem.* 190:9–32.
Perry, S.V. 2001 Vertebrate tropomyosin: distribution, properties and function. *J. Muscle Res. Cell Motil.* 22:5–49.
Pfuhl, M., S.J. Winder, and A. Pastore. 1994. Nebulin, a helical actin binding protein. *EMBO J.* 13:1782–1789.
Pirani, A., M.V. Vinogradova, P.M. Curmi, W.A. King, R.J. Fletterick, R. Craig, L.S. Tobacman, C. Xu, V. Hatch, and W. Lehman. 2006. An atomic model of the thin filament in the relaxed and Ca2+-activated states. *J. Mol. Biol.* 357:707–717.
Rayment, I., W.R. Rypniewski, K. Schmidt-Base, R. Smith, D.R. Tomchick, M.M. Benning, D.A. Winkelmann, G. Wesenberg, and H.M. Holden. 1993. Three-dimensional structure of myosin subfragment-1: a molecular motor. *Science* 261:50–58.
Ringkob, T.P., D.R. Swartz, and M.L. Greaser. 2004. Light microscopy and image analysis of thin filament lengths utilizing dual probes on beef, chicken, and rabbit myofibrils. *J. Anim. Sci.* 82:1445–1453.
Sanes, J.R. 2003. The basement membrane/basal lamina of skeletal muscle. *J. Biol. Chem.* 278:12601–12604.
Schoenauer, R., S. Lange, A. Hirschy, E. Ehler, J.C. Perriard, and I. Agarkova. 2008. Myomesin 3, a novel structural component of the M-band in striated muscle. *J. Mol. Biol.* 376:338–351.

Siemankowski, R.F., M.O. Wiseman, and H.D. White. 1985. ADP dissociation from actomyosin subfragment 1 is sufficiently slow to limit the unloaded shortening velocity in vertebrate muscle. *Proc. Natl. Acad. Sci. U.S.A.* 82:658–662.

Squire, J.M. 1981. *The Structural Basis of Muscular Contraction.* Plenum Press, New York.

Squire, J.M., H.A. Al-Khayat, C. Knupp, P.K. Luther. 2005. Molecular architecture in muscle contractile assemblies. *Adv. Protein Chem.* 71:18–87.

Squire, J.M., T. Bekyarova, G. Farman, D. Gore, G. Rajkumar, C. Knupp, C. Lucaveche, M.C. Reedy, M.K. Reedy, and T.C. Irving. 2006. The myosin filament superlattice in the flight muscles of flies: A-band lattice optimisation for stretch-activation? *J. Mol. Biol.* 361:823–838.

Squire, J.M., M. Cantino, M.W.K. Chew, R.C. Denny, J.J. Harford, L. Hudson, and P.K. Luther. 1998. Myosin rod packing schemes in vertebrate muscle thick filaments. *J. Struct. Biol.* 122:128–138.

Squire, J.M., P.K. Luther, and C. Knupp. 2003. Structural evidence for the interaction of C-protein (MyBP-C) with actin and sequence identification of a possible actin-binding domain. *J. Mol. Biol.* 331:713–724.

Squire, J.M. and E.P. Morris. 1998. A new look at thin filament regulation in vertebrate skeletal muscle. *FASEB J.* 12:761–771.

Stephenson, D.G., A.W. Stewart, and G.J. Wilson. 1989. Dissociation of force from myofibrillar MgATPase and stiffness at short sarcomere lengths in rat and toad skeletal muscle. *J. Physiol.* 410:351–366.

Straussman, R., J.M. Squire, A. Ben-Ya'acov, and S. Ravid. 2005. Skip Residues and Charge Interactions in Myosin II Coiled-coils: Implications for Molecular Packing. *Journal of Molecular Biology.* 353:613–628.

Swartz, D.R. 1999. Exchange of alpha-actinin in isolated rigor myofibrils. *J. Muscle Res. Cell Motil.* 20:457–467.

Swartz, D.R., M.L. Greaser, and B.B. Marsh. 1990. Regulation of binding of subfragment 1 in isolated rigor myofibrils. *J. Cell Biol.* 111:2989–3001.

Swartz, D.R., M.L. Greaser, and B.B. Marsh. 1993. Structural studies of rigor bovine myofibrils using fluorescence microscopy. II. Influence of sarcomere length on the binding of myosin subfragment 1, alpha-actinin and G-actin to rigor myofibrils. *Meat Sci.* 33:157–190.

Swartz, D.R., R.L. Moss, and M.L. Greaser, 1996. Calcium alone does not fully activate the thin filament for S1 binding to rigor myofibrils. *Biophys. J.* 71:1891–1904.

Swartz, D.R., Z. Yang, A. Sen, S.B. Tikunova, and J.P. Davis. 2006. Myofibrillar troponin exists in three states and there is signal transduction along skeletal myofibrillar thin filaments. *J. Mol. Biol.* 361:420–435.

Tang, J., D.W. Taylor, and K.A. Taylor. 2001. The three-dimensional structure of alpha-actinin obtained by cryoelectron microscopy suggests a model for Ca(2+)-dependent actin binding. *J. Mol. Biol.* 310:845–858.

Tidball, J.G. 1986. Energy stored and dissipated in skeletal muscle basement membranes during sinusoidal oscillations. *Biophys. J.* 50:1127–1138.

Tidball, J.G. 1991. Force transmission across muscle cell membranes. *J. Biomech.* 24(Suppl. 1):43–52.

Toyoshima, Y.Y., S.J. Kron., E.M. McNally, K.R. Niebling, C. Toyoshima, and J.A. Spudich. 1987. Myosin subfragment-1 is sufficient to move actin filaments *in vitro*. *Nature* 328:536–539.

Trinick, J.A. 1981. End-filaments: a new structural element of vertebrate skeletal muscle thick filaments. *J. Mol. Biol.* 151 309–314.

van der Ven, P.F., S. Wiesner, P. Salmikangas, D. Auerbach, M. Himmel, S. Kempa, K. Hayess, D. Pacholsky, A. Taivainen, R. Schroder, O. Carpen, and D.O. Furst. 2000. Indications for a novel muscular dystrophy pathway. gamma-filamin, the muscle-specific filamin isoform, interacts with myotilin. *J. Cell Biol.* 151:235–248.

Vibert, P., R. Craig, and W. Lehman. 1997. Steric-model for activation of muscle thin filaments. *J. Mol. Biol.* 266:8–14.

Vigoreaux, J.O. 1994. The muscle Z band: lessons in stress management. *J. Muscle Res. Cell Motil.* 15:237–255.

Vinogradova, M.V., D.B. Stone, G.G. Malanina, C. Karatzaferi, R. Cooke, R.A. Mendelson, and R.J. Fletterick. 2005. Ca(2+)-regulated structural changes in troponin. *Proc. Natl. Acad. Sci. U.S.A.* 102:5038–5043.

Wendt, T., D. Taylor, K.M. Trybus, and K. Taylor. 2001. Three-dimensional image reconstruction of dephosphorylated smooth muscle heavy meromyosin reveals asymmetry in the interaction between myosin heads and placement of subfragment 2. *Proc. Natl. Acad. Sci. U.S.A.* 98:4361–4366.

Wiche, G. 1998. Role of plectin in cytoskeleton organization and dynamics. *J. Cell. Sci.* 111(Pt. 17):2477–2486.

Winegrad, S. 1999. Cardiac myosin binding protein C. *Circ. Res.* 84:1117–1126.

Woodhead, J.L., F.-Q. Zhao, R. Craig, E.H. Egelman, L. Alamo, and R. Padron. 2005. Atomic model of a myosin filament in the relaxed state. *Nature* 436:1195–1199.

Wray, J.S. 1979. Structure of the backbone in myosin filaments of muscle. *Nature* 277:37–40.
Xu, J.-Q., B.A. Harder, P. Uman, and R. Craig. 1996. Myosin filament structure in vertebrate smooth muscle. *J. Cell. Biol.* 134:53–66.
Yagi, N., M.H. Ito, H. Nakajima, T. Izumi, and I. Matsubard. 1977. Return of myosin heads to thick filaments after muscle contraction. *Science.* 197:685–687.
Yamaguchi, M., M. Izumimoto, R.M. Robson, and M.H. Stromer. 1985. Fine structure of wide and narrow vertebrate muscle Z-lines. A proposed model and computer simulation of Z-line architecture. *J. Mol. Biol.* 184:621–643.
Yamaguchi, M., R.M. Robson, and M.H. Stromer. 1983. Evidence for actin involvement in cardiac Z-lines and Z-line analogues. *J. Cell. Biol.* 96:435–442.
Yamashita, A., K. Maeda, and Y. Maeda. 2003. Crystal structure of CapZ: structural basis for actin filament barbed end capping. *EMBO J.* 22:1529–1538.
Young, P., C. Ferguson, S. Banuelos, and M. Gautel. 1998. Molecular structure of the sarcomeric Z-disk: two types of titin interactions lead to an asymmetrical sorting of alpha-actinin. *EMBO J.* 17:1614–1624.
Zeece, M.G., R.M. Robson, and P.J. Bechtel. 1979. Interaction of alpha-actinin, filamin and tropomyosin with F-actin. *Biochim. Biophys. Acta* 581:365–370.
Zhang, D., K.W. Yancey, and D.R. Swartz. 2000. Influence of ADP on cross-bridge-dependent activation of myofibrillar thin filaments. *Biophys. J.* 78:3103–3111.
Zhou, X., E.P. Morris, and S.S. Lehrer. 2000. Binding of troponin I and the troponin I-troponin C complex to actin-tropomysin. Dissociation by myosin subfragment 1. *Biochemistry.* 39:1128–1132.
Zoghbi, M.E., J.L. Woodhead, R.L. Moss, and R. Craig. 2008. Three-dimensional structure of vertebrate cardiac muscle myosin filaments. *Proc. Natl. Acad. Sci. U.S.A.* 105:2386–2390.

2 Satellite Cell Biology

R.P. Rhoads, C.R. Rathbone, and K.L. Flann

CONTENTS

2.1 INTRODUCTION

Skeletal muscle is extremely responsive to environmental and physiological cues, and is able to modify growth and functional characteristics in accordance with the demands placed on it. Upon injury or trauma, skeletal muscle also retains the capacity to regenerate despite being largely composed of post-mitotic, multinucleated fibers. The plasticity of skeletal muscle results, in large part, from a population of resident stem-like cells, often referred to as satellite cells. When needed, satellite cells proceed through a terminal differentiation program culminating in fusion competency. During muscle fiber hypertrophy or repair, satellite cells are able to fuse with the existing muscle fiber for nuclei donation. When muscle fibers are lost to damage, satellite cells fuse to each other for the formation of a nascent myotube and eventual muscle fiber replacement.

This cellular basis for postnatal muscle growth and regeneration has been realized over the past several decades (Allen et al. 1979; Campion 1984). Recently, investigations into satellite cell biology have intensified, thereby expanding our knowledge of this critical skeletal muscle cell. The purpose of this chapter is to highlight our current understanding of satellite cell biology. Specific topics include the expression of molecular markers during the satellite cell "life cycle," biological signals critical to satellite cell regulation, and satellite cell contributions to skeletal muscle mass during various physiological states.

2.2 DEFINING THE SATELLITE CELL

Nearly 50 years ago, Mauro (1961) identified a peculiar cell residing in a "wedged" position between the plasma membrane and basement membrane of a frog skeletal muscle fiber by electron

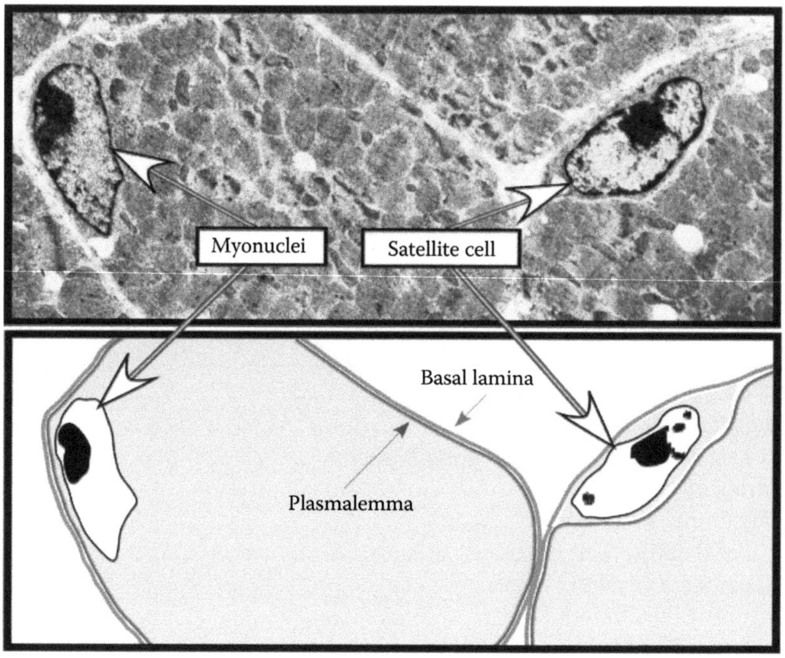

FIGURE 2.1 Identification of a satellite cell associated with a skeletal muscle fiber and comparison to myonuclei. Satellite cells reside between the plasma membrane and basement membrane of a skeletal muscle fiber. Satellite cells in adult skeletal muscle exhibit distinct morphological characteristics consisting of a substantial nuclear-to-cytoplasmic ratio, little organelle content, and a small nucleus (compared to adjacent myonuclei) with considerable amounts of heterochromatin.

microscopy. Unknown at the time, Mauro termed the cells "satellite cells" and postulated that the cells may proliferate upon skeletal muscle trauma. Satellite cells in adult skeletal muscle exhibit distinct morphological characteristics consisting of a substantial nuclear-to-cytoplasmic ratio, little organelle content, and a small nucleus (compared to adjacent myonuclei) with considerable amounts of heterochromatin, consistent with a relative lack of cellular activity (Figure 2.1; Schultz et al. 1978; Schultz and McCormick 1994; Snow 1983). As one might expect, these distinct morphological characteristics are reversed upon muscle trauma as satellite cells become progressively more active. Since the seminal observations by Mauro, elegant work by Moss and Leblond (1971) established that the sources of new myonuclei in myofibers are satellite cells. Nuclei within muscle fibers are post-mitotic, and the postnatal accumulation of DNA is due to the division and fusion of satellite cells to myofibers (Moss and Leblond 1971). Satellite cells have since been identified in the skeletal muscles of mammals (Campion et al. 1981; Dodson et al. 1987; Schmalbruch and Hellhammer 1976; Schmalbruch and Hellhammer 1977), poultry (Feldman and Stockdale 1991; McFarland et al. 1988; Yablonka-Reuveni et al. 1987), fish (Matschak and Stickland 1995) and reptiles (Kahn and Simpson, Jr. 1974).

The population of satellite cells varies among species and by muscle type (Gibson and Schultz 1982; Schmalbruch and Hellhammer 1977; Schultz and McCormick 1994). For example, the satellite cell content of slow-twitch, oxidative soleus muscle is several-fold greater than the fast-twitch, glycolytic extensor digitorum longus and tibialis anterior muscle counterparts. Shortly after birth, satellite cell populations exhibit a steady decline with advancing age from approximately 30% of total muscle nuclei in newborn animals to approximately 3% to 6% of muscle nuclei in adult skeletal muscle (Gibson and Schultz 1983; Snow 1977). This decline in satellite cell numbers coincides with skeletal muscle growth and additional accumulation of nuclei within skeletal muscle fibers (see "Satellite Cell Contribution to Muscle Mass"). In contrast to the case with young growing animals, satellite cells in adult skeletal muscle generally reside in a quiescent state, and activation of quiescent satellite cells is the pivotal event that initiates both hypertrophic growth and repair in adult skeletal muscle.

2.3 "LIFE CYCLE" OF THE SATELLITE CELL

Satellite cells are primarily inactive during most periods of an animal's adult life. This state of rest is termed "quiescence" and refers to a cell in the G_0 phase of the cell cycle. In this dormant and undifferentiated state, satellite cells exhibit limited gene expression or protein synthesis and remain in quiescence until becoming activated in response to either injury and/or mechanical stress (Allen et al. 1979; Bischoff 1986; Carlson 1973; Darr and Schultz 1989). Following activation, the satellite cells migrate out of their niche below the basal lamina, enter the G_1 phase, and are subsequently termed "muscle precursor cells" or "myoblasts." Once outside the basal lamina, satellite cells begin to proliferate, and a population of the progeny will ultimately differentiate and commit to a myogenic lineage.

Upon differentiation, these cells contribute to adult muscle mass in two unique ways. The first is through fusion with preexisting muscle fibers. The fusion of the satellite cell to a preexisting fiber results in a contribution of additional nuclei to the multinucleated muscle fiber. This addition can increase the number of nuclei on the muscle fiber, allowing for muscle growth (hypertrophy), or can replace damaged nuclei in the repair process. Furthermore, in response to injury, the satellite cell can not only replace the damaged nuclei, but also is able to replace damaged myofibers with a new regenerated muscle fiber. In this second method of maintenance, satellite cells fuse with other satellite cells, forming new myotubes and ultimately new multinucleated fibers. It is important to recognize that the formation of a new fiber does not increase the fiber number (hyperplasia) under most conditions, but rather is responsible for maintaining the original myofiber number in the muscle (Figure 2.2).

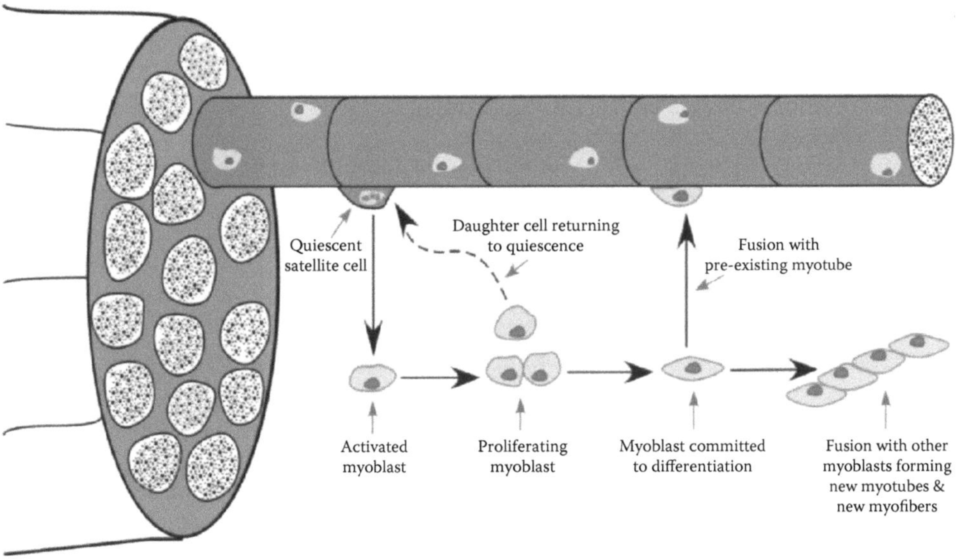

FIGURE 2.2 Life cycle of the satellite cell in adult skeletal muscle. Satellite cells remain in quiescence until becoming activated in response to either injury and/or mechanical stress. Once activated, satellite cells begin to proliferate, and a population of the progeny will ultimately differentiate and commit to a myogenic lineage. Finally, satellite cells fuse to an existing fiber or to other satellite cells for nascent myotube formation. Activated satellite cells are also able to exit the cell cycle and return to the quiescent state, a self-renewal process.

The ability of the satellite cell to maintain, repair, and regulate muscle mass is dependent upon the satellite cell's ability to respond to a variety of both intrinsic and extrinsic signals. These signals are responsible for the up-regulation and repression of various transcription factors involved in cell cycle progression and hence the regulation of myogenesis. These myogenic regulatory factors (MRFs) include muscle-specific transcription factors (MRFs) as well as a set of transcription factors called paired box proteins 3 and 7 (Pax3 and Pax7). The MRFs include myogenin (also named myogenic factor 4), myogenic determination factor 1 (MyoD), and myogenic factor 5 (Myf5), all of which have been shown to be necessary for muscle formation and presence of myogenic cells in adult skeletal muscle (Kassar-Duchossoy et al. 2004). While Pax3 and Pax7 are not restricted to myogenic lineages, they lie genetically upstream from Myf5 and MyoD, and therefore have also been shown to be critical in this well-orchestrated genetic program resulting in myogenesis (Relaix et al. 2005; Seale et al. 2000).

During quiescence, satellite cells express the transcription factors Pax7, Pax3, and possibly Myf5. Pax7 has widely been recognized as being involved in expression of survival and anti-apoptotic factors. Experiments utilizing Pax7-null mice demonstrate compromised satellite cell proliferation ultimately resulting in the loss of satellite cell numbers, indicating that Pax7 expression is vital for satellite cell survival (Kuang et al. 2006; Oustanina et al. 2004). Conversely, recent data show that over-expression of Pax7 in myogenic Pax7-null cells up-regulates MyoD while delaying myogenin expression (Olguin et al. 2007; Zammit et al. 2006). Taken together, this suggests that Pax7 has a dual role of inducing myoblast proliferation and delaying differentiation through the regulation of MyoD (Relaix et al., 2006).

Pax3, a close relative and paralog of Pax7, is essential for embryonic muscle development but its function in adult muscle is less well understood. Similar to Pax7, Pax3 has been implicated as a participant in satellite cell progression and is transiently expressed during activation (Conboy and Rando 2002). However, it is only detected in high levels in certain muscle types, such as the diaphragm (Day et al. 2007; Relaix et al. 2006), and therefore the exact role of Pax3 in adult skeletal muscle maintenance remains largely unknown. Finally, Myf5 is thought to regulate proliferation rate and homeostasis (Beauchamp et al. 2000; Ustanina et al. 2007; Zammit et al. 2006). It is expressed at relatively high levels in freshly isolated satellite cells; however, there appears to be a small population of satellite cells that remain Myf5-inactive (Day et al. 2007; Kuang et al. 2007). Therefore, it is not known if Myf5 is expressed in quiescent cells, and the exact role of Myf5 remains in question.

Upon activation of the satellite cell, expression of Pax7, Pax3, and Myf5 is retained and the cell also begins to express the transcription factor MyoD (Figure 2.3). Expression of MyoD may be required for satellite cell differentiation and is often thought of as the master myogenic transcription factor (Tapscott 2005). There are studies, however, indicating that satellite cells from MyoD-null mice are able to transition into the differentiation phase but the process is delayed both *in vitro* and *in vivo* (White et al. 2000; Yablonka-Reuveni et al. 1999).

Transcription factors needed and involved in the proliferation phase appear to be identical to those expressed in activation — Pax3, Pax7, Myf5, and MyoD (Figure 2.3). At this stage in the satellite cell program, most proliferating satellite cells will begin committing to differentiation as indicated by the up-regulation of myogenin. The requirement of myogenin expression for differentiation was confirmed using myogenin-null transgenic mice. Loss of myogenin leads to a decrease in muscle size, presumably by preventing the skeletal myoblast contribution to postnatal muscle growth (Knapp et al. 2006). During commitment of myoblasts to the differentiation phase, expression of Pax3 and Pax7 is lost. Recently it was also shown that myogenin may act to facilitate down-regulation of Pax7 during differentiation (Olguin et al. 2007).

Finally, fusion of a satellite cell to an existing fiber or to other satellite cells for nascent myotube formation leads to the cessation of all myogenic regulatory factor expression except for myogenin expression. At this point, the satellite cell is now terminally differentiated and can no longer enter back into the cell cycle (Figures 2.2 and 2.3).

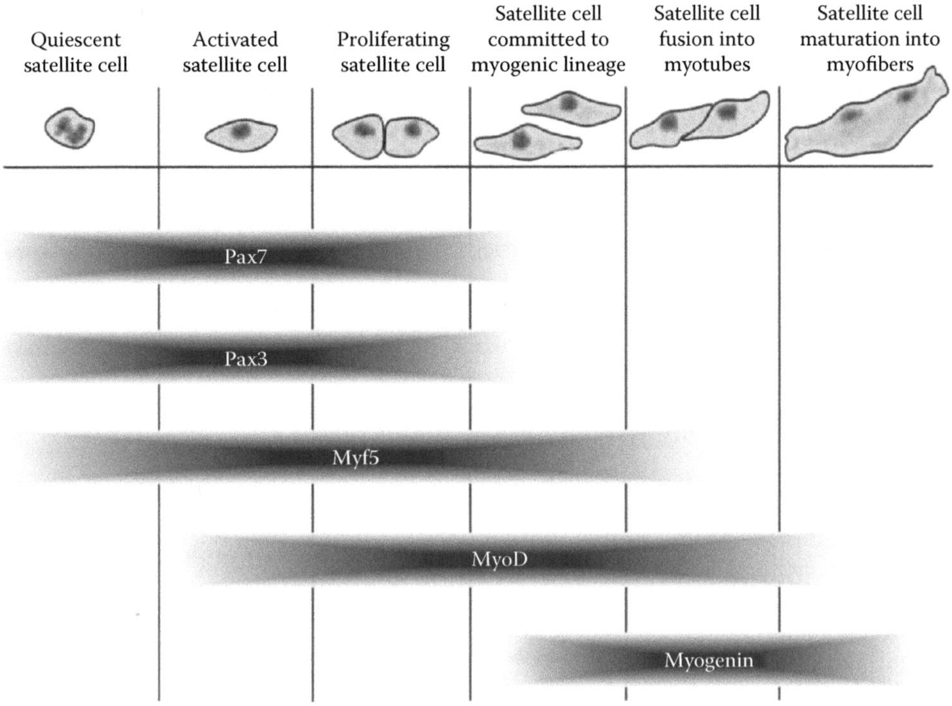

FIGURE 2.3 Myogenesis and myogenic regulatory factors (MRFs). The progression of the satellite cell and their progeny through the myogenic lineage is regulated by various transcription factors, including MRFs as well as a set of transcription factors called paired box proteins 3 and 7 (Pax3 and Pax7) shown here.

Stem cells are able to both generate differentiated progeny as well as the ability to undergo self-renewal. Importantly, satellite cells possess this quality where small populations of activated satellite cells can exit the cell cycle and return to the quiescent state. This is of obvious importance in the adult skeletal muscle for maintaining satellite cell numbers and hence the muscle's regenerative capacity. This idea was first described by Moss and Leblond (1971), who suggested either a stochastic event or asymmetrical division where one daughter cell is committed to differentiation and the second returns to quiescence. Recent studies have shown that while most of the proliferating cells will suppress Pax7 expression and up-regulate MyoD before differentiation, there is a small population of these cells that maintain Pax7, repress MyoD, and return to quiescence (Zammit et al. 2004; 2006). The molecular mechanisms regulating satellite cell self-renewal, however, remain poorly understood.

Numerous other satellite cell markers have been identified and may be useful for our future understanding of these cells. For example, a quiescent satellite cell exhibits a distinct gene expression profile including Pax7(+) and CD34(+) positive and CD45(–) and Sca1(–) negative expression (Montarras et al. 2005). Furthermore, all cells becoming myogenic must express structural proteins (α sarcomeric actin, myosin heavy chain, desmin) as well as cell adhesion proteins (neural cell adhesion molecules N-cam and M-cadherin). While Pax7 remains the most useful current marker for identifying quiescent satellite cells due to the availability of a high-quality antibody (Shefer et al. 2006), these other proteins can serve in additional capacities. For example, because Pax7 is expressed in both quiescence and proliferation phases, it is impossible to distinguish between satellite cells within these two phases through the use of Pax7 alone. However, the exact stage of satellite cell progression can be identified through the use of Pax7 and another molecular marker, such as CD34. Although not specific to satellite cells, CD34 is only expressed in quiescent satellite cells and can be used to distinguish satellite cells on isolated myofibers and aid in the identification of quiescent vs. proliferative cells (Beauchamp et al. 2000).

2.4 SATELLITE CELL CONTRIBUTION TO MUSCLE MASS

Skeletal muscle has a remarkable ability to conform to the demands placed on it. Several physiological events are capable of altering satellite cell activity, including muscle injury (due to cytotoxins, trauma, mechanical stretch, genetic defects); hypertrophic events (e.g., compensatory hypertrophy, physical training, postnatal growth); and environmental stimuli (e.g., hyperthermia, nutrition). Both increases and decreases in muscle mass are accompanied by corresponding alterations in muscle DNA content (Adams and Haddad 1996; Allen et al. 1995a; Roy et al. 1999), which suggests that the ratio of muscle DNA to its cytoplasmic volume remains constant, a concept known as the myonuclear domain (Allen et al. 1999; Cheek et al. 1971). Because myonuclei are post-mitotic, it logically follows that increases in muscle mass are dependent on satellite cells to maintain a constant myonuclear domain. Similarly, myonuclear loss, which is largely attributable to apoptosis and occurs in conjunction with an increase in protein degradation, is associated with skeletal muscle atrophy (Allen et al. 1997; Hikida et al. 1997). Satellite cells are necessary to provide the genetic machinery for increases in protein synthesis that must occur for successful regeneration and re-growth following damage and/or atrophy. Given the intimate relationship between satellite cells and the regulation of muscle mass, it is critically important to consider the influence of satellite cells under conditions where muscle mass fluctuations may occur, namely, during hypertrophy, re-growth, and repair. It is also of particular relevance to characterize satellite cells in these models when the regulation of muscle size is impaired, such as that which has been observed in old skeletal muscle.

2.4.1 HYPERTROPHIC EVENTS

The formation of skeletal muscle fibers occurs during prenatal life and absolute fiber number becomes fixed around the time of birth (Oksbjerg et al. 2004; Picard et al. 2002). For this reason, prenatal development has often been viewed as a period of hyperplasia, and postnatal muscle growth has been characterized as a period of hypertrophy due to substantial increases in protein deposition. It is interesting to note, however, that during the postnatal growth phase, muscle fibers display a substantial increase in DNA content such that skeletal muscle accumulates between 50 and 99% of the total DNA after birth, depending on the species and muscle type (Allen et al. 1979). As noted earlier, the satellite cell is responsible for skeletal muscle DNA accretion, and thus early postnatal life can be characterized as an intense period of satellite cell proliferation. On a temporal basis, early studies have demonstrated that DNA incorporation precedes the accumulation of muscle protein and that muscle fiber diameter in growing animals is directly related to the total number of myonuclei (reviewed in Allen et al. 1979; Harbison et al. 1976; Swatland 1977; Winick and Noble 1966). These observations led to the notion that DNA incorporation may be a rate-limiting step for protein accretion and postnatal muscle hypertrophy. In fact, if satellite cell activity is experimentally ablated *in vivo* by irradiation (Mozdziak et al. 1997; Rosenblatt and Parry 1993) or reduced using hindlimb unloading (Darr and Schultz 1989), the normal increases in myonuclear number and fiber volume during postnatal growth are reduced.

Several models have demonstrated the relationship between increased satellite cell activity and skeletal muscle hypertrophy. In humans, myonuclear addition accompanied myofiber hypertrophy after 16 weeks of resistance training, and an increase in the number of myonuclei per fiber, which the authors attributed to the need for myofiber hypertrophy (Petrella et al. 2008). The importance of satellite cell contribution to hypertrophy is evident when muscles are irradiated prior to the initiation of a hypertrophic event. The use of irradiation to limit the contribution of satellite cells has been employed to delineate the importance of satellite cells in the control of muscle mass apart from that of the existing myonuclei. Irradiation causes satellite cells to become apoptotic when they divide. Because myonuclei are post-mitotic, they do not divide and are presumably unaffected by the irradiation (Lewis 1954). Several studies have used this technique to demonstrate the role of satellite cells during skeletal muscle hypertrophy. As an example, ablation of the tibialis anterior caused an increase in extensor digitorum longus muscle mass and myofiber hypertrophy, with a concomitant increase in myonuclei, processes that were eliminated when irradiation was used prior to the ablation for the purpose of inhibiting satellite cell activity (Rosenblatt et al. 1994).

2.4.2 Repair and Re-growth Following Atrophy

As stated above, myonuclei eliminated during periods of atrophy and/or damage need to be replaced via the contribution of satellite cells. Removal of weight-bearing activity via tail suspension in rodents causes significant atrophy, especially in type I muscle fibers (Carlson et al. 1999; Steffen et al. 1990). Mitchell and Pavlath (2001) reported a 53% atrophy of the soleus muscle concomitant with a 35% decrease in myonuclei after a 14-day period of hindlimb suspension. When animals were allowed to ambulate normally, soleus muscle re-growth and myonuclear accretion were complete by 2 weeks, however, re-growth and myonuclear addition were inhibited when muscles were irradiated prior to the initiation of re-growth (Mitchell and Pavlath 2001), thus providing support for satellite cells during re-growth.

Limb immobilization, a model that also causes large decreases in muscle mass, may be physiologically relevant for the study of muscle re-growth because it is a common treatment for injuries such as fractures, tears, and sprains (Appell 1986; Booth and Kelso 1973; Shefer et al. 2008). A 6-week period of limb immobilization is sufficient to reduce the cross-sectional area of both type I and II fibers in the human vastus lateralis muscle by 29 and 36%, respectively (Blakemore et al. 1996). When immobilization is removed, normal healthy skeletal muscle mass successfully re-grows due, in part, to satellite cell activity. The importance of satellite cells in atrophied muscle re-growth is based on observations demonstrating enhanced re-growth following interventions that increase satellite cell activity (Chakravarthy et al. 2000; Shefer et al. 2008). Skeletal muscle repair following severe damage methods, such as crush injury, injection of myotoxins, freeze injury, and eccentric contractions, have also been shown to be associated with increased satellite cell activity to achieve successful regeneration (Granata et al. 1998; Hall-Craggs 1974; Hurme and Kalimo 1992; Pavlath et al. 1998; Rathbone et al. 2003; Schultz et al. 1985).

2.4.3 Sarcopenia

It is important to consider the involvement of satellite cells when the regulation of muscle size is impaired — more specifically, in models of re-growth, repair, and hypertrophy in skeletal muscle of aged individuals. Defects in these processes, and satellite cell function, may provide unique insight into the skeletal muscle disorder sarcopenia, a process characterized by the inevitable loss of skeletal muscle mass and strength that occurs with age. Decreases in skeletal muscle mass and function contribute to loss of independent living and an increase in disability in our growing elderly population as compared to individuals with normal muscle mass (Baumgartner et al. 1998; Greenlund and Nair 2003; Roubenoff 2000). The frailty and falls that occur subsequent to the loss of muscle mass and strength increase the chance of morbidity, thus highlighting the importance of understanding this skeletal muscle disease (Metter et al. 2002).

Cachexia — the muscle wasting that accompanies diseases including heart disease, cancer, and human immunodeficiency virus — is often associated with sarcopenia (Thomas 2007). More specifically, the interrelationship between cachexia and sarcopenia is evident when one considers the increased likelihood for such diseases with advanced age. Decreases in satellite cell function may be associated with sarcopenia; therefore, decreases in satellite cell function may also be associated with cachexia as well.

Support for the notion of satellite cell involvement in sarcopenia draws from the idea that the hypertrophy, re-growth, and repair of old skeletal muscle exhibits a similar phenotype to that seen when irradiation is used in conjunction with models of hypertrophy, re-growth, and repair in young, healthy skeletal muscle. For example, irradiation prior to synergistic ablation prevented hypertrophy (Rosenblatt et al. 1994), and aged rats failed to hypertrophy following synergistic ablation although adult rats increased muscle mass by 53% (Blough and Linderman 2000).

The inability of aged skeletal muscle to grow following immobilization may be a likely contributor of age-related atrophy. Even in young, healthy skeletal muscle, the recovery of muscle mass following immobilization often takes weeks and even months, depending on the severity of the injury. This may lead to atrophy in elderly subjects as they may not fully recover the muscle mass lost during immobilization. As an example, satellite cell function is restored in young skeletal muscle when an atrophy-inducing stimulus is removed allowing re-growth to occur (Childs et al. 2003); however, decreases in satellite cell function in old animals persist as old skeletal muscle is unable to re-grow (Chakravarthy et al. 2000). Old skeletal muscle exhibits an incomplete recovery following eccentric contraction-induced injury (Brooks et al. 2001; Rader and Faulkner 2006), similar to that observed when muscles are irradiated prior to eccentric contraction-induced injury (Rathbone et al. 2003). Together, this lends support to the idea that satellite cells in old skeletal muscle are defective. Indeed, factors that affect satellite cell activity were shown to be discordant in old as compared to young skeletal muscle, both at rest and following acute resistance exercise, providing further support for a role of satellite cell contributing to sarcopenia (Dennis et al. 2008).

Recent data demonstrate that the reestablishment of satellite cell activity restores old skeletal muscle mass, again pointing to a role for satellite cell dysfunction limiting old skeletal muscle growth. These observations indicate that satellite cells may be suitable pharmacological targets for re-growth in old skeletal muscle (Barton-Davis et al. 1998; Chakravarthy et al. 2000; Conboy et al. 2005). For example, directly applying IGF-I to the skeletal muscle of old rats rescued both the depressed satellite cell proliferative capacity and muscle re-growth from an immobilization-induced atrophy (Chakravarthy et al. 2000). Importantly, studies demonstrating defects in satellite cell activation, proliferation, and differentiation (Barani et al. 2003; Chakravarthy et al. 2000; Decary et al. 1997; Dreyer et al. 2006; Johnson and Allen 1993; Lees et al. 2006; Petrella et al. 2006; Yablonka-Reuveni et al. 1999) may help explain the similarities in phenotype between irradiation and old skeletal muscle models. It is likely that characterization of the satellite cell role in disease states and associated molecular pathways will become pharmacological targets and lead to treatments for skeletal muscle diseases including sarcopenia and cachexia.

2.4.4 Nutritional Influences

It is well recognized that plane of nutrition has profound consequences on skeletal muscle mass by influencing cellular protein accretion through the balance of protein synthesis and protein degradation within the myocyte. Equally important — although not as well understood — are satellite cell contributions to skeletal muscle growth and mass during periods of poor nutrition.

Nutritional effects on satellite cell activity can manifest during prenatal life. Greenwood et al. (1999) used a model of intrauterine growth retardation (IUGR) in sheep to demonstrate this concept. In this model, placental insufficiency causes growth restriction in the developing fetal lamb by preventing transfer of maternally derived nutrients (fetal malnutrition), resulting in a small fetal size that is approximately 50% of normal fetuses. In such cases, small fetuses display slower muscle accretion than larger fetuses despite similar numbers of myofibers during late fetal development. Associated with the lesser degree of muscle protein accretion, small fetuses have a smaller proportion of cycling cells and less overall muscle DNA as compared to large fetuses. Overall, this study demonstrates that fetal nutrition influences satellite cell activity and subsequent skeletal muscle growth.

The intense period of satellite cell activity during early postnatal life is subject to nutritional regulation. Early studies conducted in rodents and humans demonstrate that poor nutrition reduces satellite cell numbers and muscle DNA accumulation (Beermann et al. 1983; Hansen-Smith et al. 1979; Moss 1968). Subsequent studies in neonatal sheep, pigs, and poultry indicate that nutrient restriction can transiently alter *in vivo* satellite cell kinetics during this critical growth phase and potentially lead to permanent muscle growth deficits (Fiorotto et al. 1986; Greenwood et al. 2000; Halevy et al. 2003; Jeanplong et al. 2003; Moore et al. 2005). Malnutrition may affect satellite cell kinetics by depressing the anabolic growth factor, IGF-I, and stimulating an inhibitor of muscle growth, myostatin, although definitive proof is lacking (Jeanplong et al. 2003). Thus, despite our knowledge that nutritional influences can limit satellite cell dynamics and skeletal muscle growth potential, the underlying mechanisms leading to such effects are poorly characterized and require additional study, especially in species of agricultural importance.

2.5 REGULATION OF SATELLITE CELL BEHAVIOR

It is well accepted that the control of muscle mass is influenced by both anabolic and catabolic hormones and growth factors, and therefore it is no great surprise that such influences occur, at least in part, through the regulation of satellite cell activation, proliferation, and differentiation. In an effort to understand the molecular basis for satellite cell activity and contribution toward muscle mass, scientists have identified numerous factors, including hormones, growth factors and cytokines (see Table 2.1), a list that continues to grow. Although several of these anabolic and catabolic factors arrive via the circulation, it is important to understand that several such growth factors are synthesized by satellite cells and myotubes themselves, and can act in an autocrine/paracrine manner. It is beyond the scope of this chapter to examine each factor individually, and therefore only a subset of factors will be discussed.

2.5.1 Hepatocyte Growth Factor

Few growth factors have been identified that affect satellite cell activation — that is, cause satellite cells to exit quiescence (G_0) and enter the cell cycle (G_1). Bischoff (1986) made the observation that crushed muscle extract could stimulate early entry into the cell cycle, and a series of experiments determined that the growth factor crushed muscle extract was hepatocyte growth factor (HGF) (Allen et al. 1995b; Tatsumi et al. 1998). Originally identified as a growth factor synthesized in the liver and known for the ability to affect cell migration, HGF has since been shown to be synthesized by satellite cells and myotubes, and has also been shown to be located in skeletal muscle extracellular matrix, having the unique ability to stimulate satellite cell activation of quiescent satellite cells *in vitro* and *in vivo* (Allen et al. 1995b; Sheehan et al. 2000; Tatsumi et al. 1998). In addition to its role as a major player in the onset of activation, HGF increases proliferation and decreases differentiation (Leshem et al. 2000; Sheehan and Allen 1999; Zeng et al. 2002). Importantly, Tatsumi et al. (1998) demonstrated that direct intramuscular injection of HGF caused increased activation of satellite cells, thus giving physiological relevance to this growth factor as a therapeutic agent for satellite cell activation. The receptor for HGF, c-met, is present on quiescent satellite cells and can bind HGF (Allen et al. 1995b; Bottaro et al. 1991; Tatsumi et al. 1998). The release of HGF and its localization to the c-met receptor has been shown to occur within 15 minutes following injury, a response that is nitric oxide dependent (Anderson 2000; Tatsumi et al. 2002; Tatsumi and Allen 2004). HGF-induced increases in satellite cell activation occur through MAPK/ERK signaling, specifically through decreases in the cell cycle inhibitor $p27^{Kip1}$ to allow cells to withdraw from quiescence, pass the G_1:S checkpoint, and synthesize DNA (Leshem and Halevy 2002).

2.5.2 Insulin-Like Growth Factors

To date, two insulin-like growth factors have been identified: IGF-I and IGF-II. Both growth factors play an important role in satellite cell regulation through complex interactions involving multiple receptors, binding proteins, splice variants, and intracellular pathways. Unlike other growth factors, IGF-I in particular has been shown to stimulate both myoblast proliferation and differentiation, depending on the timing and intracellular conditions of the myoblast (Engert et al. 1996; Florini

TABLE 2.1
Modulation of Satellite Cell Activity by Various Factors

Factor	Proliferation	Differentiation	Migration	Ref.
Estradiol-17β	Increase			Johnson et al. 1998
FGF	Increase	Decrease	Equivocal	Allen and Boxhorn 1989
				Bischoff 1997
				Robertson et al. 1993
Ghrelin	Decrease	Increase		Filigheddu et al. 2007
				Zhang et al. 2007
HGF	Increase	Decrease	Increase	Allen et al. 1995
				Miller et al. 2000
				Bischoff 1997
IGF-I	Increase	Increase		Allen and Boxhorn 1989
				Florini et al. 1996
IGF-II	Increase	Increase		Florini et al. 1996
				Doumit et al. 1993
IL-6	Increase			Cantini et al. 1995
IL-15		Increase		Quinn et al. 1995
Leptin	Increase	Decrease		Yu et al. 2008
LIF	Increase		Increase	Barnard et al. 1994
				Austin and Burgess 1991
Myostatin	Decrease	Decrease		Joulia et al. 2003
				McCroskery et al. 2003
PDGF	Increase		Increase	Robertson et al. 1993
Testosterone	Increase			Joubert and Tobin 1989
				Johnson et al. 1998
TGF-β	Equivocal	Equivocal	Increase	Robertson et al. 1993
				Zentella and Massague 1992
				Greene and Allen 1991
TNF-	Equivocal	Decrease		Li 2003
				Miller et al. 1988
				Spurlock 1997
				Langen et al. 2004
VEGF	Increase	Increase	Increase	Germani et al. 2003
				Arsic et al. 2004
				Christov et al. 2007

et al. 1996). It is known that IGF-I production increases in response to muscle loading, (DeVol et al. 1990; DeVol et al. 1991; Yan et al. 1993; Yang et al. 1997) and can elevate within 24 hr of a single bout of resistance training (Bamman et al. 2001), and it has been shown that muscle mass can increase over twofold in transgenic mice with increased IGF-I expression (Coleman et al. 1995; Musaro et al. 2001). IGF-I exerts its effects on skeletal muscle hypertrophy through the direct regulation of satellite cells as well as through general anabolic effects, including the regulation of gene transcription and protein translation. Activated satellite cells also initiate expression of members of the IGF family and expression of a muscle-specific isoform of IGF-I, termed mechano-growth factor, has also been described as an early event in satellite cell activation (Goldspink 2005).

Binding of the IGF-I protein to its receptor (a receptor tyrosine kinase) on satellite cells results in an induced expression of myogenic regulatory factors and intracellular signaling cascades involving both mitogenic and myogenic responses (Coolican et al. 1997; Musaro and Rosenthal 1999). It has been shown that IGF-I acts through both the PI3K/AKT as well as the Ras/Raf intracellular signaling pathways (Moelling et al. 2002). Interestingly, it has been shown that the PI3K activation is necessary for myoblast differentiation and acts through p70 S6 kinase (Canicio et al. 1998; Tureckova et al. 2001). In contrast, the Ras-Raf pathway leads to signaling of extracellular response kinases (ERK), which have been shown to be involved in proliferation (Coolican et al. 1997). There is also increasing evidence that there is an interaction between IGF-I signaling and calcineurin that may co-coordinate myogenic differentiation in satellite cells (Delling et al. 2000; Friday et al. 2000; Musaro and Rosenthal 1999). While many of the details remain in question, IGF undoubtedly plays an important role in the regulation of satellite cells and skeletal muscle tissue as a whole.

IGF-II, in contrast, is thought to be involved predominately with satellite cell differentiation. It has been shown that when satellite cells are placed in low serum containing medium, IGF-II expression is activated and secretion of IGF-II increases significantly just prior to myoblast differentiation (Florini et al. 1991). In fact, introduction of either IGF-I or IGF-II antisense oligonucleotides to cultured media results in partial reduction of differentiated cells (Ewton et al. 1987; Florini et al. 1991).

2.5.3 Vascular Endothelial Growth Factor

Vascular endothelial growth factor (VEGF), a well-established regulator of the vasculature that enhances vessel permeability, endothelial cell survival, and migration (Dimmeler et al. 2000; Senger et al. 1983), has recently been implicated in the regulation of satellite cell activity. During skeletal muscle ischemia, strong VEGF gene expression and protein abundance are found in regenerating muscle fibers but not undamaged fibers, and expression appeared to be localized in putative satellite cells (Germani et al. 2003; Rissanen et al. 2002). Subsequent studies have confirmed that satellite cells express VEGF under basal conditions and in response to physiological stimuli such as contractile activity and hypoxia (Arany et al. 2008; Ciafre et al. 2007; Dehne et al. 2007; Jensen et al. 2004; Rhoads et al. 2008).

Physiological stimuli appear to induce VEGF production in satellite cells through distinct pathways. The first involves hypoxia-inducible factor (HIF), a transcription factor expressed in response to hypoxia (Semenza 1999). HIF drives the transcription of more than 70 genes involved in cell survival, metabolism, erythropoiesis, and angiogenesis, including VEGF (Semenza 2003). We and others have identified that HIF signaling pathway is operational in satellite cells leading to VEGF production during hypoxic conditions (Ciafre et al. 2007; Dehne et al. 2007; Rhoads et al. 2008). Recent work has implicated peroxisome proliferator-activated receptor-γ coactivator-1α (PGC-1α), a transcriptional coactivator involved in the regulation of cellular metabolism, as an HIF-independent pathway leading to satellite cell-driven VEGF production (Arany et al. 2008). A third report indicates that the VEGF promoter region contains sites for the myogenic transcription factor, MyoD, and are essential for MyoD-mediated regulation of VEGF gene expression (Bryan et al. 2008).

Cell responsiveness to VEGF occurs through two high-affinity receptors, known as the Fms-like tyrosine kinase (Flt-1) and the kinase insert domain-containing receptor (KDR/Flk-1)(Breen 2007). Satellite cells express both Flt-1 and KDR/Flk-1 receptors, and their levels increase upon VEGF exposure (Arsic et al. 2004; Bryan et al. 2008; Germani et al. 2003). Indeed, numerous reports have demonstrated that VEGF administration increases satellite cell migration, proliferation, and differentiation while preventing satellite cell apoptosis (Arsic et al. 2004; Bryan et al. 2008; Christov et al. 2007; Germani et al. 2003). Taken together, the data presented above indicates that satellite cell-derived VEGF production may enhance skeletal muscle myogenesis through autocrine/paracrine actions.

2.5.4 Cytokines

The plasticity of skeletal muscle may best be exemplified by the capacity to produce and respond to various cytokines. Depending on the nature of the inflammatory event and cytokine profile present, skeletal muscle will respond in a catabolic or anabolic fashion. For example, skeletal muscle breakdown during periods of infection supports processes related to survival (Frost and Lang 2008). In contrast, myotrauma and inflammation following a bout of exercise ultimately leads to muscle hypertrophy (Vierck et al. 2000). Not surprising then, studies examining the effect of various cytokines on satellite cell activity have provided mixed results and may be related to model systems (e.g., *in vivo* vs. *in vitro* experiments, primary cell culture vs. myogenic cell lines such as rat L8 and murine C2C12), dose, and time of exposure.

Equivocal data has been generated regarding the effect of tumor necrosis factor-α (TNF-α) on satellite cell proliferation. Acute TNF-α treatment (24 hr) of satellite cells induces their proliferation (Li 2003), whereas chronic exposure of C2C12 myoblasts to TNF-α (alone or in combination with IL-1) reduces proliferation and can induce apoptosis (Spurlock 1997; Stewart et al. 2004). From a differentiation standpoint, numerous reports have ascribed a negative role to TNF-α in myogenesis. TNF-α appears to inhibit myogenic differentiation through the activation of the transcription factor NF-κB and the subsequent down-regulation of MyoD and myogenin (Guttridge et al. 1999; Guttridge et al. 2000; Langen et al. 2001; Langen et al. 2004; Layne and Farmer 1999; Miller et al. 1988; Szalay et al. 1997). As noted earlier, the above observations have been made using model systems of either rodent or human origin. However, studies addressing the role of pro-inflammatory cytokines on bovine-derived myogenic cell behavior are lacking and represent a significant gap in our understanding of muscle biology in this agriculturally important species.

2.5.5 Myostatin

Myostatin (also known as growth and differentiation factor 8 (GDF8)), a member of the TGF-β superfamily, has perhaps received the most attention and is arguably the most impressive negative regulator of muscle mass (McPherron et al. 1997). This is evident in myostatin knockout mice, which have notable increases in muscle mass, and in Piedmontese and Belgian Blue cattle carrying mutations in the myostatin gene that exhibit the "double muscling" phenotype (Kambadur et al. 1997; McPherron and Lee 1997). Myostatin is associated with atrophy, aging, and muscle wasting due to disease (Carlson et al. 1999; Dasarathy et al. 2004; Gonzalez-Cadavid et al. 1998; Ma et al. 2003; Yarasheski et al. 2002). Lending further support to the notion that myostatin inhibits muscle growth and hypertrophy are the observations that the inhibition of myostatin increases muscle mass and strength in the adult animal (Wagner et al. 2005; Whittemore et al. 2003), and that myostatin over-expression causes skeletal muscle atrophy (Reisz-Porszasz et al. 2003; Zimmers et al. 2002). Given the role for satellite cells in the regulation of skeletal muscle size described above, it is no surprise that myostatin exerts its effects, at least in part, through the negative regulation of satellite cell activation, proliferation, and differentiation (Joulia et al. 2003; McCroskery et al. 2003; Thomas et al. 2000).

Similar to other TGF-β family members, myostatin first binds a type II serine/threonine kinase receptor, with a preference for ActRIIB (Rebbapragada et al. 2003). Binding of the appropriate type II receptor leads to the recruitment and phosphorylation of the type I receptor, either the ALK4 or ALK5 (TβRI) (Franzen et al. 1993; Rebbapragada et al. 2003), which leads to the phosphorylation and activation of Smad2 and Smad3, which are known as receptor-regulated Smads (R-Smads) (Nakao et al. 1997b; Rebbapragada et al. 2003; Zhu et al. 2004). Upon phosphorylation, the Smad2/3 complex heteromerizes with Smad4, a co-mediator Smad (Co-Smad), followed by the translocation of this complex to the nucleus, which causes increased transcriptional activity of TGF-β-specific genes (Nakao et al. 1997b). Also, Smad7, an inhibitory Smad (I-Smad), functions through a negative feedback mechanism as it interferes with the Smad2/3 and Smad4 complex, and through competition for activation by type I receptors (Nakao et al. 1997a). Based on the dramatic influence of myostatin on skeletal muscle mass, it is likely that further characterization of the myostatin signaling cascade will lead to effective pharmacological treatments for muscle wasting disorders including sarcopenia and cachexia (Roth and Walsh 2004; Tsuchida 2004), in addition to enhancing muscle growth in meat animals (Dayton and White 2008).

2.6 SUMMARY

This chapter described basic aspects of satellite cell biology, including the contributions of the satellite cell to skeletal muscle growth and repair. Clearly, the importance of the satellite cell in support of skeletal muscle hypertrophy is evident by either the absence of muscle accretion or atrophy when satellite cell activity becomes dysfunctional. Studies have begun to define various aspects of satellite cell behavior that are susceptible to modulation by physiological events, environmental stimuli, and biological signals. Despite these initial advances, a better understanding of satellite cell regulation is warranted to accurately understand the biological mechanism(s) underlying skeletal muscle growth and repair. Such knowledge is critical for developing novel approaches (i.e., genetic, nutritional, and therapeutic) to modulate satellite cell activity and affect skeletal muscle performance in the agricultural setting or human biomedical sector.

REFERENCES

Adams, G.R. and F. Haddad. 1996. The relationships among IGF-1, DNA content, and protein accumulation during skeletal muscle hypertrophy. *J. Appl. Physiol.* 81:2509–2516.

Allen, D.L., J.K. Linderman, R.R. Roy, A.J. Bigbee, R.E. Grindeland, V. Mukku, and V.R. Edgerton. 1997. Apoptosis: a mechanism contributing to remodeling of skeletal muscle in response to hindlimb unweighting. *Am. J. Physiol.* 273:C579–C587.

Allen, D.L., S.R. Monke, R.J. Talmadge, R.R. Roy, and V.R. Edgerton. 1995a. Plasticity of myonuclear number in hypertrophied and atrophied mammalian skeletal muscle fibers. *J. Appl. Physiol.* 78:1969–1976.

Allen, D.L., R.R. Roy, and V.R. Edgerton. 1999. Myonuclear domains in muscle adaptation and disease. *Muscle Nerve* 22:1350–1360.

Allen, R.E. and L.K. Boxhorn. 1989. Regulation of skeletal muscle satellite cell proliferation and differentiation by transforming growth factor-beta, insulin-like growth factor I, and fibroblast growth factor. *J. Cell. Physiol.* 138:311–315.

Allen, R.E., R.A. Merkel, and R.B. Young. 1979. Cellular aspects of muscle growth: myogenic cell proliferation. *J. Anim. Sci.* 49:115–127.

Allen, R.E., S.M. Sheehan, R.G. Taylor, T.L. Kendall, and G.M. Rice. 1995b. Hepatocyte growth factor activates quiescent skeletal muscle satellite cells *in vitro*. *J. Cell. Physiol.* 165:307–312.

Anderson, J.E. 2000. A role for nitric oxide in muscle repair: nitric oxide-mediated activation of muscle satellite cells. *Mol. Biol. Cell.* 11:1859–1874.

Appell, H.J. 1986. Morphology of immobilized skeletal muscle and the effects of a pre- and postimmobilization training program. *Int. J. Sports. Med.* 7:6–12.

Arany, Z., S.Y. Foo, Y. Ma, J.L. Ruas, A. Bommi-Reddy, G. Girnun, M. Cooper, D. Laznik, J. Chinsomboon, S.M. Rangwala, K.H. Baek, A. Rosenzweig, and B.M. Spiegelman. 2008. HIF-independent regulation of VEGF and angiogenesis by the transcriptional coactivator PGC-1alpha. *Nature* 451:1008–1012.

Arsic, N., S. Zacchigna, L. Zentilin, G. Ramirez-Correa, L. Pattarini, A. Salvi, G. Sinagra, and M. Giacca. 2004. Vascular endothelial growth factor stimulates skeletal muscle regeneration *in vivo*. *Mol. Ther.* 10:844–854.

Austin, L. and A.W. Burgess. 1991. Stimulation of myoblast proliferation in culture by leukaemia inhibitory factor and other cytokines. *J. Neurol. Sci.* 101:193–197.

Bamman, M.M., J.R. Shipp, J. Jiang, B.A. Gower, G.R. Hunter, A. Goodman, C.L. McLafferty, Jr., and R.J. Urban. 2001. Mechanical load increases muscle IGF-I and androgen receptor mRNA concentrations in humans. *Am. J. Physiol. Endocrinol. Metab.* 280:E383–E390.

Barani, A.E., A.C. Durieux, O. Sabido, and D. Freyssenet. 2003. Age-related changes in the mitotic and metabolic characteristics of muscle-derived cells. *J. Appl. Physiol.* 95:2089–2098.

Barnard, W., J. Bower, M.A. Brown, M. Murphy, and L. Austin. 1994. Leukemia inhibitory factor (LIF) infusion stimulates skeletal muscle regeneration after injury: injured muscle expresses lif mRNA. *J. Neurol. Sci.* 123:108–113.

Barton-Davis, E.R., D.I. Shoturma, A. Musaro, N. Rosenthal, and H.L. Sweeney. 1998. Viral mediated expression of insulin-like growth factor I blocks the aging-related loss of skeletal muscle function. *Proc. Natl. Acad. Sci. U.S.A.* 95:15603–15607.

Baumgartner, R.N., K.M. Koehler, D. Gallagher, L. Romero, S.B. Heymsfield, R.R. Ross, P.J. Garry, and R.D. Lindeman. 1998. Epidemiology of sarcopenia among the elderly in New Mexico. *Am. J. Epidemiol.* 147:755–763.

Beauchamp, J.R., L. Heslop, D.S. Yu, S. Tajbakhsh, R.G. Kelly, A. Wernig, M.E. Buckingham, T.A. Partridge, and P.S. Zammit. 2000. Expression of CD34 and Myf5 defines the majority of quiescent adult skeletal muscle satellite cells. *J. Cell. Biol.* 151:1221–1234.

Beermann, D.H., L.F. Hood, and M. Liboff. 1983. Satellite cell and myonuclei populations in rat soleus and extensor digitorum longus muscles after maternal nutritional deprivation and realimentation. *J. Anim. Sci.* 57:1618–1625.

Bischoff, R. 1997. Chemotaxis of skeletal muscle satellite cells. *Dev. Dyn.* 208(4):505–515

Bischoff, R.A. 1986. Satellite cell mitogen from crushed adult muscle. *Dev. Biol.* 115:140–147

Blakemore, S.J., P.K. Rickhuss, P.W. Watt, M.J. Rennie, and H.S. Hundal. 1996. Effects of limb immobilization on cytochrome c oxidase activity and GLUT4 and GLUT5 protein expression in human skeletal muscle. *Clin. Sci.* 91:591–599.

Blough, E.R. and J.K. Linderman. 2000. Lack of skeletal muscle hypertrophy in very aged male Fischer 344 x Brown Norway rats. *J. Appl. Physiol.* 88:1265–1270.

Booth, F.W. and J.R. Kelso. 1973. Production of rat muscle atrophy by cast fixation. *J. Appl. Physiol.* 34:404–406.

Bottaro, D.P., J.S. Rubin, D.L. Faletto, A.M. Chan, T.E. Kmiecik, G.F. Vande Woude, and S.A. Aaronson. 1991. Identification of the hepatocyte growth factor receptor as the c-met proto-oncogene product. *Science* 251:802–804.

Breen, E.C. 2007. VEGF in biological control. *J. Cell. Biochem.* 102:1358–1367.

Brooks, S.V., J.A. Opiteck, and J.A. Faulkner. 2001. Conditioning of skeletal muscles in adult and old mice for protection from contraction-induced injury. *J. Gerontol. A Biol. Sci. Med. Sci.* 56:B163–B171.

Bryan, B.A., T.E. Walshe, D.C. Mitchell, J.S. Havumaki, M. Saint-Geniez, A.S. Maharaj, A.E. Maldonado, and P.A. D'Amore. 2008. Coordinated vascular endothelial growth factor expression and signaling during skeletal myogenic differentiation. *Mol. Biol. Cell.* 19:994–1006.

Campion, D.R. 1984. The muscle satellite cell: a review. *Int. Rev. Cytol.* 87:225–251

Campion, D.R., R.L. Richardson, J.O. Reagan, and R.R. Kraeling. 1981. Changes in the satellite cell population during postnatal growth of pig skeletal muscle. *J. Anim. Sci.* 52:1014–1018.

Canicio, J., E. Gallardo, I. Illa, X. Testar, M. Palacin, A. Zorzano, and P. Kaliman. 1998. p70 S6 kinase activation is not required for insulin-like growth factor-induced differentiation of rat, mouse, or human skeletal muscle cells. *Endocrinology* 139:5042–5049.

Cantini, M., M.L. Massimino, E. Rapizzi, K. Rossini, C. Catani, L.L. Dalla, and U. Carraro. 1995. Human satellite cell proliferation *in vitro* is regulated by autocrine secretion of IL-6 stimulated by a soluble factor(s) released by activated monocytes. *Biochem. Biophys. Res. Commun.* 216:49–53.

Carlson, B.M. 1973. The regeneration of skeletal muscle. A review. *Am. J. Anat.* 137:119–149

Carlson, C.J., F.W. Booth, and S.E. Gordon. 1999. Skeletal muscle myostatin mRNA expression is fiber-type specific and increases during hindlimb unloading. *Am. J. Physiol.* 277:R601–R606.

Chakravarthy, M.V., B.S. Davis, and F.W. Booth. 2000. IGF-I restores satellite cell proliferative potential in immobilized old skeletal muscle. *J. Appl. Physiol.* 89:1365–1379.

Cheek, D.B., A.B. Holt, D.E. Hill, and J.L. Talbert. 1971. Skeletal muscle mass and cell growth: the concept of the deoxyribonucleic acid unit. *Pediat. Res.* 5:312–328.

Childs, T.E., E.E. Spangenburg, D.R. Vyas, and F.W. Booth. 2003. Temporal alterations in protein signaling cascades during recovery from muscle atrophy. *Am. J. Physiol. Cell. Physiol.* 285:C391–C398.

Christov, C., F. Chretien, R. Abou-Khalil, G. Bassez, G. Vallet, F.J. Authier, Y. Bassaglia, V. Shinin, S. Tajbakhsh, B. Chazaud, and R.K. Gherardi. 2007. Muscle satellite cells and endothelial cells: close neighbors and privileged partners. *Mol. Biol. Cell.* 18:1397–1409.

Ciafre, S.A., F. Niola, E. Giorda, M.G. Farace, and D. Caporossi. 2007. CoCl(2)-simulated hypoxia in skeletal muscle cell lines: role of free radicals in gene up-regulation and induction of apoptosis. *Free. Radic. Res.* 41:391–401.

Coleman, M.E., F. DeMayo, K.C. Yin, H.M. Lee, R. Geske, C. Montgomery, and R.J. Schwartz. 1995. Myogenic vector expression of insulin-like growth factor I stimulates muscle cell differentiation and myofiber hypertrophy in transgenic mice. *J. Biol. Chem.* 270:12109–12116.

Conboy, I.M., M.J. Conboy, A.J. Wagers, E.R. Girma, I.L. Weissman, and T.A. Rando. 2005. Rejuvenation of aged progenitor cells by exposure to a young systemic environment. *Nature* 433:760–764.

Conboy, I.M. and T.A. Rando. 2002. The regulation of Notch signaling controls satellite cell activation and cell fate determination in postnatal myogenesis. *Dev. Cell* 3:397–409.

Coolican, S.A., D.S. Samuel, D.Z. Ewton, F.J. McWade, and J.R. Florini. 1997. The mitogenic and myogenic actions of insulin-like growth factors utilize distinct signaling pathways. *J. Biol. Chem.* 272:6653–6662.

Darr, K.C. and E. Schultz. 1989. Hindlimb suspension suppresses muscle growth and satellite cell proliferation. *J. Appl. Physiol.* 67:1827–1834.

Dasarathy, S., M. Dodig, S.M. Muc, S.C. Kalhan, and A.J. McCullough. 2004. Skeletal muscle atrophy is associated with an increased expression of myostatin and impaired satellite cell function in the portacaval anastamosis rat. *Am. J. Physiol. Gastrointest. Liver Physiol.* 287:G1124–G1130.

Day, K., G. Shefer, J.B. Richardson, G. Enikolopov, and Z. Yablonka-Reuveni. 2007. Nestin-GFP reporter expression defines the quiescent state of skeletal muscle satellite cells. *Dev. Biol.* 304:246–259.

Dayton, W.R. and M.E. White. 2008. Cellular and molecular regulation of muscle growth and development in meat animals. *J. Anim. Sci.* 86:E217–E225.

Decary, S., V. Mouly, C.B. Hamida, A. Sautet, J.P. Barbet, and G.S. Butler-Browne. 1997. Replicative potential and telomere length in human skeletal muscle: implications for satellite cell-mediated gene therapy. *Hum. Gene. Ther.* 8:1429–1438.

Dehne, N., U. Kerkweg, T. Otto, and J. Fandrey. 2007. The HIF-1 response to simulated ischemia in mouse skeletal muscle cells neither enhances glycolysis nor prevents myotube cell death. *Am. J. Physiol. Regul. Integr. Comp. Physiol.* 293:R1693–R1701.

Delling, U., J. Tureckova, H.W. Lim, L.J. De Windt, P. Rotwein, and J.D. Molkentin. 2000. A calcineurin-NFATc3-dependent pathway regulates skeletal muscle differentiation and slow myosin heavy-chain expression. *Mol. Cell. Biol.* 20:6600–6611.

Dennis, R.A., B. Przybyla, C. Gurley, P.M. Kortebein, P. Simpson, D.H. Sullivan, and C.A. Peterson. 2008. Aging alters gene expression of growth and remodeling factors in human skeletal muscle both at rest and in response to acute resistance exercise. *Physiol. Genomics* 32:393–400.

DeVol, D.L., J. Novakofski, R. Fernando, and P.J. Bechtel. 1991. Varying amounts of stretch stimulus regulate stretch-induced muscle hypertrophy in the chicken. *Comp. Biochem. Physiol. A* 100:55–61.

DeVol, D.L., P. Rotwein, J.L. Sadow, J. Novakofski, and P.J. Bechtel. 1990. Activation of insulin-like growth factor gene expression during work-induced skeletal muscle growth. *Am. J. Physiol.* 259:E89–E95.

Dimmeler, S., E. Dernbach, and A.M. Zeiher. 2000. Phosphorylation of the endothelial nitric oxide synthase at ser-1177 is required for VEGF-induced endothelial cell migration. *FEBS Lett.* 477:258–262.

Dodson, M.V., E.L. Martin, M.A. Brannon, B.A. Mathison, and D.C. McFarland. 1987. Optimization of bovine satellite cell-derived myotube formation *in vitro*. *Tissue Cell* 19:159–166.

Doumit, M.E., D.R. Cook, and R.A. Merkel. 1993. Fibroblast growth factor, epidermal growth factor, insulin-like growth factors, and platelet-derived growth factor-BB stimulate proliferation of clonally derived porcine myogenic satellite cells. *J. Cell. Physiol.* 157:326–332.

Dreyer, H.C., C.E. Blanco, F.R. Sattler, E.T. Schroeder, and R.A. Wiswell. 2006. Satellite cell numbers in young and older men 24 hours after eccentric exercise. *Muscle Nerve* 33:242–253.

Engert, J.C., E.B. Berglund, and N. Rosenthal. 1996. Proliferation precedes differentiation in IGF-I-stimulated myogenesis. *J. Cell. Biol.* 135:431–440.

Ewton, D.Z., S.L. Falen, and J.R. Florini. 1987. The type II insulin-like growth factor (IGF) receptor has low affinity for IGF-I analogs: pleiotypic actions of IGFs on myoblasts are apparently mediated by the type I receptor. *Endocrinology* 120:115–123.

Feldman, J.L. and F.E. Stockdale. 1991. Skeletal muscle satellite cell diversity: satellite cells form fibers of different types in cell culture. *Dev. Biol.* 143:320–334.

Filigheddu, N., V.F. Gnocchi, M. Coscia, M. Cappelli, P.E. Porporato, R. Taulli, S. Traini, G. Baldanzi, F. Chianale, S. Cutrupi, E. Arnoletti, C. Ghe, A. Fubini, N. Surico, F. Sinigaglia, C. Ponzetto, G. Muccioli, T. Crepaldi, and A. Graziani. 2007. Ghrelin and des-acyl ghrelin promote differentiation and fusion of C2C12 skeletal muscle cells. *Mol Biol Cell* 18:986–994.

Fiorotto, M.L., R.J. Shulman, H.P. Sheng, and C. Garza. 1986. The effects of different total parenteral nutrition fuel mixes on skeletal muscle composition of infant miniature pigs. *Metabolism* 35:354–359.

Florini, J.R., D.Z. Ewton, and S.A. Coolican. 1996. Growth hormone and the insulin-like growth factor system in myogenesis. *Endocr. Rev.* 17:481–517.

Florini, J.R., D.Z. Ewton, and S.L. Roof. 1991. Insulin-like growth factor-I stimulates terminal myogenic differentiation by induction of myogenin gene expression. *Mol. Endocrinol.* 5:718–724.

Franzen, P., D.P. Ten, H. Ichijo, H. Yamashita, P. Schulz, C.H. Heldin, and K. Miyazono. 1993. Cloning of a TGF beta type I receptor that forms a heteromeric complex with the TGF beta type II receptor. *Cell* 75:681–692.

Friday, B.B., V. Horsley, and G.K. Pavlath. 2000. Calcineurin activity is required for the initiation of skeletal muscle differentiation. *J. Cell. Biol.* 149:657–666.

Frost, R.A. and C.H. Lang. 2008. Regulation of muscle growth by pathogen-associated molecules. *J. Anim. Sci.* 86:E84–E93.

Germani, A., A. Di Carlo, A. Mangoni, S. Straino, C. Giacinti, P. Turrini, P. Biglioli, and M.C. Capogrossi. 2003. Vascular endothelial growth factor modulates skeletal myoblast function. *Am. J. Pathol.* 163:1417–1428.

Gibson, M.C. and E. Schultz. 1982. The distribution of satellite cells and their relationship to specific fiber types in soleus and extensor digitorum longus muscles. *Anat. Rec.* 202:329–337.

Gibson, M.C. and E. Schultz. 1983. Age-related differences in absolute numbers of skeletal muscle satellite cells. *Muscle Nerve* 6:574–580.

Goldspink, G. 2005. Mechanical signals, IGF-I gene splicing, and muscle adaptation. *Physiology* (Bethesda) 20:232–238.

Gonzalez-Cadavid, N.F., W.E. Taylor, K. Yarasheski, I. Sinha-Hikim, K. Ma, S. Ezzat, R. Shen, R. Lalani, S. Asa, M. Mamita, G. Nair, S. Arver, and S. Bhasin. 1998. Organization of the human myostatin gene and expression in healthy men and HIV-infected men with muscle wasting. *Proc. Natl. Acad. Sci. U.S.A.* 95:14938–14943.

Granata, A.L., C. Vecchi, L. Graciotti, G. Fulgenzi, S. Maggi, and A. Corsi. 1998. Gamma irradiation can reduce muscle damage in mdx dystrophic mice. *Acta Neuropathol.* 96:564–568.

Greene, E.A. and R.E. Allen. 1991. Growth factor regulation of bovine satellite cell growth *in vitro*. *J. Anim. Sci.* 69:146–152.

Greenlund, L.J. and K.S. Nair. 2003. Sarcopenia — consequences, mechanisms, and potential therapies. *Mech. Ageing. Dev.* 124:287–299.

Greenwood, P.L., A.S. Hunt, J.W. Hermanson, and A.W. Bell. 2000. Effects of birth weight and postnatal nutrition on neonatal sheep. II. Skeletal muscle growth and development. *J. Anim. Sci.* 78:50–61.

Greenwood, P.L., R.M. Slepetis, J.W. Hermanson, and A.W. Bell. 1999. Intrauterine growth retardation is associated with reduced cell cycle activity, but not myofibre number, in ovine fetal muscle. *Reprod. Fertil. Dev.* 11:281–291.

Guttridge, D.C., C. Albanese, J.Y. Reuther, R.G. Pestell, and A.S. Baldwin, Jr. 1999. NF-kappaB controls cell growth and differentiation through transcriptional regulation of cyclin D1. *Mol. Cell. Biol.* 19:5785–5799.

Guttridge, D.C., M.W. Mayo, L.V. Madrid, C.Y. Wang, and A.S. Baldwin, Jr. 2000. NF-kappaB-induced loss of MyoD messenger RNA: possible role in muscle decay and cachexia. *Science* 289:2363–2366.

Halevy, O., Y. Nadel, M. Barak, I. Rozenboim, and D. Sklan. 2003. Early posthatch feeding stimulates satellite cell proliferation and skeletal muscle growth in turkey poults. *J. Nutr.* 133:1376–1382.

Hall-Craggs, E.C. 1974. Rapid degeneration and regeneration of a whole skeletal muscle following treatment with bupivacaine (Marcain). *Exp. Neurol.* 43:349–358.

Hansen-Smith, F.M., D. Picou, and M.H. Golden. 1979. Muscle satellite cells in malnourished and nutritionally rehabilitated children. *J. Neurol. Sci.* 41:207–221.

Harbison, S.A., D.E. Goll, F.C. Parrish, Jr., V. Wang, and E.A. Kline. 1976. Muscle growth in two genetically different lines of swine. *Growth* 40:253–283.

Hikida, R.S., S. Van Nostran, J.D. Murray, R.S. Staron, S.E. Gordon, and W.J. Kraemer. 1997. Myonuclear loss in atrophied soleus muscle fibers. *Anat. Rec.* 247:350–354.

Hurme, T. and H. Kalimo. 1992. Activation of myogenic precursor cells after muscle injury. *Med. Sci. Sports. Exerc.* 24:197–205.

Jeanplong, F., J.J. Bass, H.K. Smith, S.P. Kirk, R. Kambadur, M. Sharma, and J.M. Oldham. 2003. Prolonged underfeeding of sheep increases myostatin and myogenic regulatory factor Myf-5 in skeletal muscle while IGF-I and myogenin are repressed. *J. Endocrinol.* 176:425–437.

Jensen, L., P. Schjerling, and Y. Hellsten. 2004. Regulation of VEGF and bFGF mRNA expression and other proliferative compounds in skeletal muscle cells. *Angiogenesis* 7:255–267.

Johnson, B.J., N. Halstead, M.E. White, M.R. Hathaway, A. DiCostanzo, and W.R. Dayton. 1998. Activation state of muscle satellite cells isolated from steers implanted with a combined trenbolone acetate and estradiol implant. *J. Anim. Sci.* 76:2779–2786.

Johnson, S.E. and R.E. Allen. 1993. Proliferating cell nuclear antigen (PCNA) is expressed in activated rat skeletal muscle satellite cells. *J. Cell. Physiol.* 154:39–43.

Joubert, Y. and C. Tobin. 1989. Satellite cell proliferation and increase in the number of myonuclei induced by testosterone in the levator ani muscle of the adult female rat. *Dev. Biol.* 131:550–557.

Joulia, D., H. Bernardi, V. Garandel, F. Rabenoelina, B. Vernus, and G. Cabello. 2003. Mechanisms involved in the inhibition of myoblast proliferation and differentiation by myostatin. *Exp. Cell. Res.* 286:263–275.

Kahn, E.B. and S.B. Simpson, Jr. 1974. Satellite cells in mature, uninjured skeletal muscle of the lizard tail. *Dev. Biol.* 37:219–223.

Kambadur, R., M. Sharma, T.P. Smith, and J.J. Bass. 1997. Mutations in myostatin (GDF8) in double-muscled Belgian Blue and Piedmontese cattle. *Genome Res.* 7:910–916.

Kassar-Duchossoy, L., B. Gayraud-Morel, D. Gomes, D. Rocancourt, M. Buckingham, V. Shinin, and S. Tajbakhsh. 2004. Mrf4 determines skeletal muscle identity in Myf5:Myod double-mutant mice. *Nature* 431:466–471.

Knapp, J.R., J.K. Davie, A. Myer, E. Meadows, E.N. Olson, and W.H. Klein. 2006. Loss of myogenin in postnatal life leads to normal skeletal muscle but reduced body size. *Development* 133:601–610.

Kuang, S., S.B. Charge, P. Seale, M. Huh, and M.A. Rudnicki. 2006. Distinct roles for Pax7 and Pax3 in adult regenerative myogenesis. *J. Cell. Biol.* 172:103–113.

Kuang, S., K. Kuroda, G.F. Le, and M.A. Rudnicki. 2007. Asymmetric self-renewal and commitment of satellite stem cells in muscle. *Cell* 129:999–1010.

Langen, R.C., A.M. Schols, M.C. Kelders, E.F. Wouters, and Y.M. Janssen-Heininger. 2001. Inflammatory cytokines inhibit myogenic differentiation through activation of nuclear factor-kappaB. *FASEB J.* 15:1169–1180.

Langen, R.C., D. Van, V,A.M. Schols, M.C. Kelders, E.F. Wouters, and Y.M. Janssen-Heininger. 2004. Tumor necrosis factor-alpha inhibits myogenic differentiation through MyoD protein destabilization. *FASEB J.* 18:227–237.

Layne, M.D. and S.R. Farmer. 1999. Tumor necrosis factor-alpha and basic fibroblast growth factor differentially inhibit the insulin-like growth factor-I induced expression of myogenin in C2C12 myoblasts. *Exp. Cell. Res.* 249:177–187.

Lees, S.J., C.R. Rathbone, and F.W. Booth. 2006. Age-associated decrease in muscle precursor cell differentiation. *Am. J. Physiol. Cell. Physiol.* 290:C609–C615.

Leshem, Y. and O. Halevy. 2002. Phosphorylation of pRb is required for HGF-induced muscle cell proliferation and is p27kip1-dependent. *J. Cell. Physiol.* 191:173–182.

Leshem, Y., D.B. Spicer, R. Gal-Levi, and O. Halevy. 2000. Hepatocyte growth factor (HGF) inhibits skeletal muscle cell differentiation: a role for the bHLH protein twist and the cdk inhibitor p27. *J. Cell. Physiol.* 184:101–109.

Lewis, R.B. 1954. Changes in striated muscle following single intense doses of x-rays. *Lab. Invest.* 3:48–55.

Li, Y.P. 2003. TNF-alpha is a mitogen in skeletal muscle. *Am. J. Physiol. Cell. Physiol.* 285:C370–C376.

Ma, K., C. Mallidis, S. Bhasin, V. Mahabadi, J. Artaza, N. Gonzalez-Cadavid, J. Arias, and B. Salehian. 2003. Glucocorticoid-induced skeletal muscle atrophy is associated with upregulation of myostatin gene expression. *Am. J. Physiol. Endocrinol. Metab.* 285:E363–E371.

Matschak, T.W. and N.C. Stickland. 1995. The growth of Atlantic salmon (*Salmo salar* L.) myosatellite cells in culture at two different temperatures. *Experientia* 51:260–266.

Mauro, A. 1961. Satellite cell of skeletal muscle fibers. *J. Biophys. Biochem. Cytol.* 9:493–495.

McCroskery, S., M. Thomas, L. Maxwell, M. Sharma, and R. Kambadur. 2003. Myostatin negatively regulates satellite cell activation and self-renewal. *J. Cell. Biol.* 162:1135–1147.

McFarland, D.C., M.E. Doumit, and R.D. Minshall. 1988. The turkey myogenic satellite cell: optimization of *in vitro* proliferation and differentiation. *Tissue Cell* 20:899–908.

McPherron, A.C., A.M. Lawler, and S.J. Lee. 1997. Regulation of skeletal muscle mass in mice by a new TGF-beta superfamily member. *Nature* 387:83–90.

McPherron, A.C. and S.J. Lee. 1997. Double muscling in cattle due to mutations in the myostatin gene. *Proc. Natl. Acad. Sci. U.S.A.* 94:12457–12461.

Metter, E.J., L.A. Talbot, M. Schrager, and R. Conwit. 2002. Skeletal muscle strength as a predictor of all-cause mortality in healthy men. *J. Gerontol. A Biol. Sci. Med. Sci.* 57:B359–B365.

Miller, K.J., D. Thaloor, S. Matteson, and G.K. Pavlath. 2000. Hepatocyte growth factor affects satellite cell activation and differentiation in regenerating skeletal muscle. *Am. J. Physiol. Cell. Physiol.* 278:C174–C181.

Miller, S.C., H. Ito, H.M. Blau, and F.M. Torti. 1988. Tumor necrosis factor inhibits human myogenesis *in vitro*. *Mol. Cell. Biol.* 8:2295–2301.

Mitchell, P.O. and G.K. Pavlath. 2001. A muscle precursor cell-dependent pathway contributes to muscle growth after atrophy. *Am. J. Physiol. Cell. Physiol.* 281:C1706–C1715.

Moelling, K., K. Schad, M. Bosse, S. Zimmermann, and M. Schweneker. 2002. Regulation of Raf-Akt cross-talk. *J. Biol. Chem.* 277:31099–31106.

Montarras, D., J. Morgan, C. Collins, F. Relaix, S. Zaffran, A. Cumano, T. Partridge, and M. Buckingham. 2005. Direct isolation of satellite cells for skeletal muscle regeneration. *Science* 309:2064–2067.

Moore, D.T., P.R. Ferket, and P.E. Mozdziak. 2005. Early post-hatch fasting induces satellite cell self-renewal. *Comp. Biochem. Physiol. A Mol. Integr. Physiol.* 142:331–339.

Moss, F.P. 1968. The relationship between the dimensions of the fibres and the number of nuclei during restricted growth, degrowth and compensatory growth of skeletal muscle. *Am. J. Anat.* 122:565–571.

Moss, F.P. and C.P. Leblond. 1971. Satellite cells as the source of nuclei in muscles of growing rats. *Anat. Rec.* 170:421–435.

Mozdziak, P.E., E. Schultz, and R.G. Cassens. 1997. Myonuclear accretion is a major determinant of avian skeletal muscle growth. *Am. J. Physiol.* 272:C565–C571.

Musaro, A., K. McCullagh, A. Paul, L. Houghton, G. Dobrowolny, M. Molinaro, E.R. Barton, H.L. Sweeney, and N. Rosenthal. 2001. Localized Igf-1 transgene expression sustains hypertrophy and regeneration in senescent skeletal muscle. *Nat. Genet.* 27:195–200.

Musaro, A. and N. Rosenthal. 1999. Maturation of the myogenic program is induced by postmitotic expression of insulin-like growth factor I. *Mol. Cell. Biol.* 19:3115–3124.

Nakao, A., M. Afrakhte, A. Moren, T. Nakayama, J.L. Christian, R. Heuchel, S. Itoh, M. Kawabata, N.E. Heldin, C.H. Heldin, and P. ten Dijke. 1997a. Identification of Smad7, a TGFbeta-inducible antagonist of TGF-beta signalling. *Nature* 389:631–635.

Nakao, A., T. Imamura, S. Souchelnytskyi, M. Kawabata, A. Ishisaki, E. Oeda, K. Tamaki, J. Hanai, C.H. Heldin, K. Miyazono, and P. ten Dijke. 1997b. TGF-beta receptor-mediated signalling through Smad2, Smad3 and Smad4. *EMBO. J.* 16:5353–5362.

Oksbjerg, N., F. Gondret, and M. Vestergaard. 2004. Basic principles of muscle development and growth in meat-producing mammals as affected by the insulin-like growth factor (IGF) system. *Domest. Anim. Endocrinol.* 27:219–240.

Olguin, H.C., Z. Yang, S.J. Tapscott, and B.B. Olwin. 2007. Reciprocal inhibition between Pax7 and muscle regulatory factors modulates myogenic cell fate determination. *J. Cell. Biol.* 177:769–779.

Oustanina, S., G. Hause, and T. Braun. 2004. Pax7 directs postnatal renewal and propagation of myogenic satellite cells but not their specification. *EMBO J.* 23:3430–3439.

Pavlath, G.K., D. Thaloor, T.A. Rando, M. Cheong, A.W. English, and B. Zheng. 1998. Heterogeneity among muscle precursor cells in adult skeletal muscles with differing regenerative capacities. *Dev. Dyn.* 212:495–508.

Petrella, J.K., J.S. Kim, J.M. Cross, D.J. Kosek, and M.M. Bamman. 2006. Efficacy of myonuclear addition may explain differential myofiber growth among resistance-trained young and older men and women. *Am. J. Physiol. Endocrinol. Metab.* 291:E937–E946.

Petrella, J.K., J.S. Kim, D.L. Mayhew, J.M. Cross, and M.M. Bamman. 2008. Potent myofiber hypertrophy during resistance training in humans is associated with satellite cell-mediated myonuclear addition: a cluster analysis. *J. Appl. Physiol.*

Picard, B., L. Lefaucheur, C. Berri, and M.J. Duclos. 2002. Muscle fibre ontogenesis in farm animal species. *Reprod. Nutr. Dev.* 42:415–431.

Quinn, L.S., K.L. Haugk, and K. H. Grabstein. 1995. Interleukin-15: a novel anabolic cytokine for skeletal muscle. *Endocrinology* 136:3669–3672.

Rader, E.P. and J.A. Faulkner. 2006. Recovery from contraction-induced injury is impaired in weight-bearing muscles of old male mice. *J. Appl. Physiol.* 100:656–661.

Rathbone, C.R., J.C. Wenke, G.L. Warren, and R. B. Armstrong. 2003. Importance of satellite cells in the strength recovery after eccentric contraction-induced muscle injury. *Am. J. Physiol. Regul. Integr. Comp. Physiol.* 285:R1490–R1495.

Rebbapragada, A., H. Benchabane, J.L. Wrana, A J. Celeste, and L. Attisano. 2003. Myostatin signals through a transforming growth factor beta-like signaling pathway to block adipogenesis. *Mol. Cell. Biol.* 23: 7230–7242.

Reisz-Porszasz, S., S. Bhasin, J.N. Artaza, R. Shen, I. Sinha-Hikim, A. Hogue, T.J. Fielder, and N.F. Gonzalez-Cadavid. 2003. Lower skeletal muscle mass in male transgenic mice with muscle-specific overexpression of myostatin. *Am. J. Physiol. Endocrinol. Metab.* 285:E876–E888.

Relaix, F., D. Montarras, S. Zaffran, B. Gayraud-Morel, D. Rocancourt, S. Tajbakhsh, A. Mansouri, A. Cumano, and M. Buckingham. 2006. Pax3 and Pax7 have distinct and overlapping functions in adult muscle progenitor cells. *J. Cell. Biol.* 172:91–102.

Relaix, F., D. Rocancourt, A. Mansouri, and M. Buckingham. 2005. A Pax3/Pax7-dependent population of skeletal muscle progenitor cells. *Nature* 435:948–953.

Rhoads, R.P., R.M. Johnson, C.R. Rathbone, X. Liu, C. Temm-Grove, S.M. Sheehan, J.B. Hoying, and R.E. Allen. 2009. Satellite cell mediated angiogenesis coincides with a functional hypoxia-inducible factor (HIF) pathway. *Am. J. Physiol. Cell Physiol.* In review.

Rissanen, T.T., I. Vajanto, M.O. Hiltunen, J. Rutanen, M.I. Kettunen, M. Niemi, P. Leppanen, M.P. Turunen, J.E. Markkanen, K. Arve, E. Alhava, R.A. Kauppinen, and S. Yla-Herttuala. 2002. Expression of vascular endothelial growth factor and vascular endothelial growth factor receptor-2 (KDR/Flk-1) in ischemic skeletal muscle and its regeneration. *Am. J. Pathol.* 160:1393–1403.

Robertson, T.A., M.A. Maley, M.D. Grounds, and J.M. Papadimitriou. 1993. The role of macrophages in skeletal muscle regeneration with particular reference to chemotaxis. *Exp. Cell. Res.* 207:321–331.

Rosenblatt, J.D. and D.J. Parry. 1993. Adaptation of rat extensor digitorum longus muscle to gamma irradiation and overload. *Pflugers. Arch.* 423:255–264.

Rosenblatt, J.D., D. Yong, and D.J. Parry. 1994. Satellite cell activity is required for hypertrophy of overloaded adult rat muscle. *Muscle Nerve* 17:608–613.

Roth, S.M. and S. Walsh. 2004. Myostatin: a therapeutic target for skeletal muscle wasting. *Curr. Opin. Clin. Nutr. Metab. Care.* 7:259–263.

Roubenoff, R. 2000. Sarcopenia and its implications for the elderly. *Eur. J. Clin. Nutr.* 54:S40–S47.

Roy, R.R., S.R. Monke, D.L. Allen, and V.R. Edgerton. 1999. Modulation of myonuclear number in functionally overloaded and exercised rat plantaris fibers. *J. Appl. Physiol.* 87:634–642.

Schmalbruch, H. and U. Hellhammer. 1976. The number of satellite cells in normal human muscle. *Anat. Rec.* 185:279–287.

Schmalbruch, H. and U. Hellhammer. 1977. The number of nuclei in adult rat muscles with special reference to satellite cells. *Anat. Rec.* 189:169–175.

Schultz, E., M.C. Gibson, and T. Champion. 1978. Satellite cells are mitotically quiescent in mature mouse muscle: an EM and radioautographic study. *J. Exp. Zool.* 206:451–456.

Schultz, E., D.L. Jaryszak, and C.R. Valliere. 1985. Response of satellite cells to focal skeletal muscle injury. *Muscle Nerve* 8:217–222.

Schultz, E. and K.M. McCormick. 1994. Skeletal muscle satellite cells. *Rev. Physiol. Biochem. Pharmacol.* 123:213–257.

Seale, P., L.A. Sabourin, A. Girgis-Gabardo, A. Mansouri, P. Gruss, and M.A. Rudnicki. 2000. Pax7 is required for the specification of myogenic satellite cells. *Cell* 102:777–786.

Semenza, G.L. 1999. Regulation of mammalian O2 homeostasis by hypoxia-inducible factor 1. *Annu. Rev. Cell. Dev. Biol.* 15:551–578.

Semenza, G.L. 2003. Angiogenesis in ischemic and neoplastic disorders. *Annu. Rev. Med.* 54:17–28.

Senger, D.R., S.J. Galli, A.M. Dvorak, C.A. Perruzzi, V.S. Harvey, and H.F. Dvorak. 1983. Tumor cells secrete a vascular permeability factor that promotes accumulation of ascites fluid. *Science* 219:983–985.

Sheehan, S.M. and R.E. Allen. 1999. Skeletal muscle satellite cell proliferation in response to members of the fibroblast growth factor family and hepatocyte growth factor. *J. Cell. Physiol.* 181:499–506.

Sheehan, S.M., R. Tatsumi, C.J. Temm-Grove, and R.E. Allen. 2000. HGF is an autocrine growth factor for skeletal muscle satellite cells *in vitro*. *Muscle Nerve* 23:239–245.

Shefer, G., E. Carmeli, G. Rauner, Z. Yablonka-Reuveni, and D. Benayahu. 2008. Exercise running and tetracycline as means to enhance skeletal muscle stem cell performance after external fixation. *J. Cell. Physiol.* 215:265–275.

Shefer, G., D.P. Van de Mark, J.B. Richardson, and Z. Yablonka-Reuveni. 2006. Satellite-cell pool size does matter: defining the myogenic potency of aging skeletal muscle. *Dev. Biol.* 294:50–66.

Snow, M.H. 1977. The effects of aging on satellite cells in skeletal muscles of mice and rats. *Cell Tissue. Res.* 185:399–408.

Snow, M.H. 1983. A quantitative ultrastructural analysis of satellite cells in denervated fast and slow muscles of the mouse. *Anat. Rec.* 207:593–604.

Spurlock, M.E. 1997. Regulation of metabolism and growth during immune challenge: an overview of cytokine function. *J. Anim. Sci.* 75:1773–1783.

Steffen, J.M., R.D. Fell, T.E. Geoghegan, L.C. Ringel, and X.J. Musacchia. 1990. Age effects on rat hindlimb muscle atrophy during suspension unloading. *J. Appl. Physiol.* 68:927–931.

Stewart, C.E., P.V. Newcomb, and J.M. Holly. 2004. Multifaceted roles of TNF-alpha in myoblast destruction: a multitude of signal transduction pathways. *J. Cell. Physiol.* 198:237–247.

Swatland, H.J. 1977. Accumulation of myofiber nuclei in pigs with normal and arrested development. *J. Anim. Sci.* 44:759–764.

Szalay, K., Z. Razga, and E. Duda. 1997. TNF inhibits myogenesis and downregulates the expression of myogenic regulatory factors myoD and myogenin. *Eur. J. Cell. Biol.* 74:391–398.

Tapscott, S.J. 2005. The circuitry of a master switch: Myod and the regulation of skeletal muscle gene transcription. *Development* 132:2685–2695.

Tatsumi, R. and R.E. Allen. 2004. Active hepatocyte growth factor is present in skeletal muscle extracellular matrix. *Muscle Nerve* 30:654–658.

Tatsumi, R., J.E. Anderson, C.J. Nevoret, O. Halevy, and R.E. Allen. 1998. HGF/SF is present in normal adult skeletal muscle and is capable of activating satellite cells. *Dev. Biol.* 194:114–128.

Tatsumi, R., A. Hattori, Y. Ikeuchi, J.E. Anderson, and R.E. Allen. 2002. Release of hepatocyte growth factor from mechanically stretched skeletal muscle satellite cells and role of pH and nitric oxide. *Mol. Biol. Cell* 13:2909–2918.

Thomas, D.R. 2007. Loss of skeletal muscle mass in aging: examining the relationship of starvation, sarcopenia and cachexia. *Clin. Nutr.* 26:389–399.

Thomas, M., B. Langley, C. Berry, M. Sharma, S. Kirk, J. Bass, and R. Kambadur. 2000. Myostatin, a negative regulator of muscle growth, functions by inhibiting myoblast proliferation. *J. Biol. Chem.* 275:40235–40243.

Tsuchida, K. 2004. Activins, myostatin and related TGF-beta family members as novel therapeutic targets for endocrine, metabolic and immune disorders. *Curr. Drug Targets Immune Endocrinol. Metabol. Disord.* 4:157–166.

Tureckova, J., E.M. Wilson, J.L. Cappalonga, and P. Rotwein. 2001. Insulin-like growth factor-mediated muscle differentiation: collaboration between phosphatidylinositol 3-kinase-Akt-signaling pathways and myogenin. *J. Biol. Chem.* 276:39264–39270.

Ustanina, S., J. Carvajal, P. Rigby, and T. Braun. 2007. The myogenic factor Myf5 supports efficient skeletal muscle regeneration by enabling transient myoblast amplification. *Stem Cells* 25:2006–2016.

Vierck, J., B. O'Reilly, K. Hossner, J. Antonio, K. Byrne, L. Bucci, and M. Dodson. 2000. Satellite cell regulation following myotrauma caused by resistance exercise. *Cell. Biol. Int.* 24:263–272.

Wagner, K.R., X. Liu, X. Chang, and R.E. Allen. 2005. Muscle regeneration in the prolonged absence of myostatin. *Proc. Natl. Acad. Sci. U.S.A.* 102:2519–2524.

White, J.D., A. Scaffidi, M. Davies, J. McGeachie, M.A. Rudnicki, and M.D. Grounds. 2000. Myotube formation is delayed but not prevented in MyoD-deficient skeletal muscle: studies in regenerating whole muscle grafts of adult mice. *J. Histochem. Cytochem.* 48:1531–1544.

Whittemore, L.A., K. Song, X. Li, J. Aghajanian, M. Davies, S. Girgenrath, J.J. Hill, M. Jalenak, P. Kelley, A. Knight, R. Maylor, D. O'Hara, A. Pearson, A. Quazi, S. Ryerson, X.Y. Tan, K.N. Tomkinson, G.M. Veldman, A. Widom, J.F. Wright, S. Wudyka, L. Zhao, and N.M. Wolfman. 2003. Inhibition of myostatin in adult mice increases skeletal muscle mass and strength. *Biochem. Biophys. Res. Commun.* 300:965–971.

Winick, M. and A. Noble. 1966. Cellular response in rats during malnutrition at various ages. *J. Nutr.* 89:300–306.

Yablonka-Reuveni, Z., L.S. Quinn, and M. Nameroff. 1987. Isolation and clonal analysis of satellite cells from chicken pectoralis muscle. *Dev. Biol.* 119:252–259.

Yablonka-Reuveni, Z., M.A. Rudnicki, A.J. Rivera, M. Primig, J.E. Anderson, and P. Natanson. 1999. The transition from proliferation to differentiation is delayed in satellite cells from mice lacking MyoD. *Dev. Biol.* 210:440–455.

Yan, Z., R.B. Biggs, and F.W. Booth. 1993. Insulin-like growth factor immunoreactivity increases in muscle after acute eccentric contractions. *J. Appl. Physiol.* 74:410–414.

Yang, H., M. Alnaqeeb, H. Simpson, and G. Goldspink. 1997. Changes in muscle fibre type, muscle mass and IGF-I gene expression in rabbit skeletal muscle subjected to stretch. *J. Anat.* 190:613–622.

Yarasheski, K.E., S. Bhasin, I. Sinha-Hikim, J. Pak-Loduca, and N.F. Gonzalez-Cadavid. 2002. Serum myostatin-immunoreactive protein is increased in 60–92 year old women and men with muscle wasting. *J. Nutr. Health. Aging* 6:343–348.

Yu, T., G. Luo, L. Zhang, J. Wu, H. Zhang, and G. Yang. 2008. Leptin promotes proliferation and inhibits differentiation in porcine skeletal myoblasts. *Biosci. Biotechnol. Biochem.* 72:13–21.

Zammit, P.S., J.P. Golding, Y. Nagata, V. Hudon, T.A. Partridge, and J.R. Beauchamp. 2004. Muscle satellite cells adopt divergent fates: a mechanism for self-renewal? *J. Cell Biol.* 166:347–357.

Zammit, P.S., F. Relaix, Y. Nagata, A.P. Ruiz, C.A. Collins, T.A. Partridge, and J.R. Beauchamp. 2006. Pax7 and myogenic progression in skeletal muscle satellite cells. *J. Cell Sci.* 119:1824–1832.

Zeng, C., J.E. Pesall, K.K. Gilkerson, and D.C. McFarland. 2002. The effect of hepatocyte growth factor on turkey satellite cell proliferation and differentiation. *Poult. Sci.* 81:1191–1198.

Zentella, A. and J. Massague. 1992. Transforming growth factor beta induces myoblast differentiation in the presence of mitogens. *Proc. Natl. Acad. Sci. U.S.A.* 89:5176–5180.

Zhang, W., L. Zhao, and M.W. Mulholland. 2007. Ghrelin stimulates myocyte development. *Cell Physiol. Biochem.* 20:659–664.

Zhu, X., S. Topouzis, L.F. Liang, and R.L. Stotish. 2004. Myostatin signaling through Smad2, Smad3 and Smad4 is regulated by the inhibitory Smad7 by a negative feedback mechanism. *Cytokine* 26:262–272.

Zimmers, T.A., M.V. Davies, L.G. Koniaris, P. Haynes, A.F. Esquela, K.N. Tomkinson, A.C. McPherron, N.M. Wolfman, and S.J. Lee. 2002. Induction of cachexia in mice by systemically administered myostatin. *Science* 296:1486–1488.

3 Adipose Tissue Development in Extramuscular and Intramuscular Depots in Meat Animals

G.J. Hausman and S.P. Poulos

CONTENTS

3.1 INTRODUCTION

Adipose tissue, also known as fat, is a connective tissue of mesenchymal origin. Although adipose tissue contains various cell types, adipocytes predominate and are the primary storage site for triacylglycerols. Subcutaneous, internal, and intramuscular adipose tissue depots are all economically and physiologically important in meat animal production. Brown adipose tissue (BAT) is also present in the neonate of some species and is considered important for thermoregulation of the newborn. Advancements in meat production have attempted to shift energy use for maximal muscle accretion while limiting adipose accretion resulting in enhanced feed efficiency and improved growth rates. Marbling fat is one important meat quality factor that has suffered as a result of advances made to improve the efficiency of animal production. Also known as marbling fat, intramuscular lipid is lipid stored in adipocytes found within skeletal muscle that impacts meat flavor and tenderness. It is known that adequate adipose accumulation is essential. Adipose tissue is considered an endocrine organ that secretes or expresses many endocrine factors, including leptin, growth factors, and various cytokines. The discovery that adipose tissue is the major source of circulating leptin (OB), a hormone that influences growth, metabolism, and behavior, has provided a new perspective on adipose tissue function (Trayhurn et al. 2006). Furthermore, the expression of growth factor genes by adipose tissue and presence of the resulting proteins further indicates that adipose tissue has endocrine functions. For example, adipose tissue in the pig fetus produces more IGF-1 mRNA than all other tissues studied (Ramsay et al. 1994). It now appears that adipose tissue expresses and secretes cytokines, growth factors, and leptin beginning early in life and continuing throughout life (Hausman et al. 2006; Hausman et al. 2007). The coincidental expanse of vascular development as adipose tissue develops (Crandall et al. 1997; Hausman and Richardson

2004) adds credence to the suggestion that adipose functions as an endocrine organ. Therefore, the structural and functional aspects of adipose tissue development are important during fetal and postnatal growth and aging. And, it is critical to elucidate the factors that regulate the development of adipose tissue as evidence is mounting that adipose tissue may be related to metabolism and function of other systems. For example, little is known about the development and regulation of intramuscular adipose tissue, which is important in meat quality but may also be important in energy homeostasis.

3.2 ADIPOSE TISSUE DEVELOPMENT

3.2.1 White Adipose Tissue Development in the Fetus

The development of presumptive adipose tissue mesenchymal tissue lobules commences between 14 and 16 weeks of fetal life, depending on location (Poissonnet et al. 1984; Hausman and Richardson 1994) and is characterized by the appearance of a presumptive mesenchymal fat tissue that develops into a number of "primitive fat cell organs or lobules," which subsequently increase in number and size throughout fetal development (Table 3.1; Poissonnet et al. 1983, 1984; Hausman and Hausman 1993; Crandall et al. 1997; Martin et al. 1998). In the fetal pig, there is a cranial to caudal and dorsal to ventral gradient regarding development of presumptive vascular structures (with few or no fat cells) in subcutaneous adipose tissue (Hausman and Richardson 1994). Furthermore, in the fetal pig, the development of several layers of subcutaneous adipose tissue (outer, middle, inner) is followed postnatally by growth or expansion of each layer in a distinct manner (Anderson et al. 1972; Leymaster et al. 1991). Despite little lipid accretion, subcutaneous layer development in the fetus is predictive of the degree of fat cell development in each layer postnatally. Within the outer subcutaneous layer in the pig, hair follicle adipose lobules have been studied and identified as further specialization or compartmentalization of the subcutaneous depot (Klein et al. 2007). Development of hair follicle adipocytes is linked functionally, to a degree, to hair and hair follicle development (Klein et al. 2007). Regardless of depot or location, a most remarkable aspect of fetal adipose tissue development is the close spatial and temporal relationship between developing adipocytes and associated capillaries (Crandall et al. 1997; Hausman and Richardson 2004). Capillary density and vascular anatomy clearly distinguishes adipose tissue depots in the fetus while fat cell size is independent of depot (Crandall et al. 1997; Hausman and Richardson 2004). In fact, morphological development of the vasculature clearly precedes overt adipocyte differentiation in internal fat depots. The expression of a number of genes encoding major regulators of angiogenesis marks

TABLE 3.1
Characteristics of Fetal Adipose Tissue Development

Trait	
Key developmental traits	Depot-dependent structural differentiation of presumptive mesenchymal fat tissue (no lipid) → development of "primitive fat organs or lobules" and onset of lipid accretion
Mode of accretion or expansion	Depot-dependent hyperplasia and hypertrophy of primitive fat lobules; adipocyte hyperplasia with either significant or limited adipocyte hypertrophy
Molecular & ultrastructural markers	Leptin; few and simple mitochondria
Late fetal tissue & adipocyte morphology	Moderately vascular, mature or immature tissue with either unilocular cells, or unilocular and multilocular cells

Source: Reviews, Crandall et al. 1997; Hausman and Hausman 1993; Martin et al. 1998; Hausman and Richardson 1994; Hausman and Hausman 2004; Poissonnet et al. 1983; 1984; Hausman and Thomas, 1984; Hausman and Kauffman 1986; Mostyn et al. 2006; Muhlhauster et al. 2007; O'Connor et al. 2007; Yuen et al. 1999; 2003.

the development of fetal and neonatal adipose tissue, supporting the hypothesis that vascular and cellular development in adipose tissue are interrelated (see Tables 3.2, 3.4, and 3.5). At the tissue level, adipocyte cluster development (size) in fetal adipose tissue is clearly dependent on proximity to entry point of large arterioles throughout development (Table 3.3). Proximity to entry of large arterioles apparently dictates adipocyte number because adipocyte cluster development is primarily attributable to increases in adipocyte number because fetal adipocyte hypertrophy is minimal (Table 3.3; Hausman and Thomas 1986). Virtually nothing is known about fetal intramuscular adipose tissue development. Morphological studies of neonatal pig muscle provide evidence of a relationship between intramuscular adipocyte cluster development and proximity to larger arterioles (Figure 3.1). Furthermore, image analysis of marbling flecks in longissimus dorsi cross-sections in cattle showed that ventral aspects, which were more proximal to intercostal arteries, had fivefold larger marbling flecks with fourfold more adipocytes and larger adipocytes than more distal or dorsal aspects (Yang 2006). Therefore, blood vessel development and/or blood flow are clearly involved in intramuscular adipocyte development.

Subcutaneous depots develop before internal depots in many species except pigs. Subcutaneous adipose tissue clearly represents the major depot in pig fetuses (Hausman and Hausman 1993) and continues to predominate throughout life in several species. As subcutaneous adipose tissue continues to develop in the fetal pig, the tissue appears to develop sensitivity to hormones and begins producing cytokines. *In vivo* and *in vitro* studies of fetal adipose tissue stromal-vascular cell cultures (fetal preadipocytes) suggest this occurs between 50 and 75 days after conception (reviewed in Hausman et al. 2001; Hausman et al. 2008). During this sensitive period, the thyroid hormones, glucocorticoids, and growth hormone each markedly influence the development of key metabolic and cellular traits of fetal subcutaneous adipose tissue (Hausman et al. 2001; Hausman et al. 2008). Furthermore, *in vitro* studies demonstrate that hormone influences on fetal preadipocytes can be long-lasting (Hausman et al. 2001). Cultures of fetal muscle stromal-vascular cells have been used to demonstrate the presence of fetal intramuscular preadipocytes that respond to adipogenic stimuli (Poulos and Hausman 2006a). Although limited, *in vitro* studies suggest adipocyte development from stromal-vascular cells in skeletal muscle may differ from the development of cells from subcutaneous adipose tissue (Poulos and Hausman 2006a, b).

3.2.2 Adipose Tissue Development in the Neonate

Multilocular adipocyte morphology decreases with the appearance of unilocular cell morphology in all depots (Table 3.4). The slower rate of intramuscular adipose tissue accretion is associated with much less hypertrophy than in other depots (Table 3.4). At birth, sheep and cattle BAT and white adipose tissue (WAT) are virtually filled with adipocytes, and BAT rapidly transforms to WAT in neonatal ruminants during the neonatal period. Several characteristics distinguish intramuscular adipose tissue from other depots during postnatal development (Table 3.4).

3.2.3 Postnatal Development

The rate of accretion and mode of cellular development of WAT are depot and species dependent (Table 3.5), with adipocytes in internal depots generally being largest compared to adipocytes in the subcutaneous or intramuscular depots (Table 3.5). Plateaus in adipocyte hyperplasia during postnatal development have been estimated based on counts of fixed adipocytes of a minimal size (Allen et al. 1976). Intramuscular adipocyte hyperplasia apparently continues on with either a delayed plateau or no plateau in hyperplasia (Allen et al. 1976). Gene expression studies indicate expression of similar regulatory and transcription factors by intramuscular and other adipose depots during postnatal development (Table 3.2). Intramuscular adipocyte size and number determine, to a great extent, the variability in lipid content of skeletal muscle in several meat animals, including cattle, rabbits, and pigs (Gardan et al. 2006). Generally, lipid accretion is associated with variable degrees

TABLE 3.2
Collective Reports of Genes and Proteins Expressed during Adipose Tissue Development

	Regulatory Secreted	Regulatory Non-Secreted	Metabolism	Gene Transcription
		Fetal		
SCAT	Leptin, IGFBP-1, -2, -3, -4, -5; IGF-I, -II; IL-1, -1A, - 4, -6, -8, -12; APO-E; APO-A1; APO-R1; TGF-β; adipsin; BDNF; ANG-2; PAI-1; TNF-α	UCP-1; OBLR		C/EBPα, β, and δ; PPARγ
		Neonatal		
SCAT	Leptin; IGF-I, -II; IL-1, -1A, -4, -6, -8, -12; APO-E, -A1, -R1; Agouti; BDNF; ANG-2; PAI-1; TNFα; IGFBP-1, -3, -5; adipsin	UCP-2, -3; GHR; β 1-3- AR's; ANG-II-receptors; GR; PREF1	GLUT 4; HSL; LPL; SCD; aP2; GLUT 1, -11 β; HSD-1	C/EBPα, β, and δ; PPARγ; ADD1
		Postnatal		
IMAT	IGF-II; leptin; adiponectin; decorin; osteonectin	AFABP; NAT1 HFABP; UCP-2, -3; FABP-4; SOCS-3; BTG family, member 2; matrix Gla protein	ACC; FAS; LPL; HSL; G6PDH; CPT-1; SCD, CS; HAD; DGAT1; pyruvate dehydrogenase beta	SREBP-1; SREBP-2; PPARγ -2; PPARδ
SCAT	Leptin; EGF; IGFBP-1, -3, -5, -7; bFGF, HGF; IGF-I, -II; adipsin; TNFα; adiponectin; TGF-β; adipsin; decorin; osteonectin	AFABP; HFABP; GHR; β 1-3- AR's; OBLR; IGF-IR; eNOS; Dlk-1-C-2; UCP-2, -3	SCD; ACC; ACO; FAS; LPL; ME; HSL; GLUT 1, 4; ATP citrate lyase; VDAC; GDH; 11 βHSD-1; STA5A; 3β-HSD; 17β-HSD; aromatase	ADD1; SREBP-1; SREBP-2; PPARγ 1, -2; PPARα; C/EBPα, β; SREB1-1c
SCAT	IL-1A, -1B, -5, -6, -8, 10, -15, -18; IL-1RN; IFNG; NGF; RANTES; resistin; MCP-1; MIF; PAI-1; APO-E; APO-A1; APO-R1; ANG-2 PREF1*; UCP-1*; heat shock 70-kDa protein; SOCS-3			

* Undetectable. Additional genes/proteins reported: postnatal pig adipose tissue: cofilin, low density lipoprotein receptor, low density lipoprotein related protein and high density lipoprotein binding protein; postnatal cattle adipose tissue: type III collagen, ferritin heavy chain; postnatal cattle intramuscular adipose tissue: sorbin and SH3 domain-containing protein 1, triadin, titin, ferritin heavy chain, ribosomal proteins; fetal pig adipose – laminin and type IV collagen.

Abbreviations: IMAT, intramuscular adipose tissue, SCAT, subcutaneous adipose tissue ,11 β HSD-1-2, 11 beta-hydroxysteroid dehydrogenase 1-2;, ACC, acetyl CoA carboxylase; ACO, acyl-CoA oxidase; ADD1, adipocyte determination and differentiation-dependent factor 1; ANG-II, angiotensin II; aP2, fatty acid binding protein; β AR's, beta adrenergic receptors; bFGF, basic fibroblast growth factor; C/EBP, CCAAT enhancing binding protein; EGF, epidermal growth factor; FAS, fatty acid synthase; GAPDH, glyceraldehyde-3-phosphate dehydrogenase; GDH, glutamate dehydrogenase; GHR, growth hormone receptor; GLUT, glucose transport protein; HGF, hepatocyte growth factor; HSL, hormone sensitive lipase; IGF, insulin-like growth factor; IGF-1R, IGF-1 receptor; IGFBP, insulin-like growth factor binding protein; IMAT, intramuscular adipose tissue; LPL, lipoprotein lipase; ME, malic enzyme; NAT1, novel APOBEC-1 target-1; OBR, long form leptin receptor; PPAR, peroxisome proliferator activated receptor; PREF1, preadipocyte factor 1; SCAT, subcutaneous adipose tissue; SCD, stearoyl coenzyme A desaturase; SREBP, sterol regulatory element binding protein; TGF, transforming growth factor; TNF, tumor necrosis factor; UCP, uncoupling protein; VDAC, voltage-dependent anion channel.

Source: Reviews, Trayhurn et al. 2006; Hausman et al. 2006; 2007; Murdoch et al. 2005; Tang et al. 2007; Smith et al. 1999; Wu et al. 2007; Barber et al. 2000; Casteilla et al. 1994; McNeel et al. 1995; Hishikawa et al. 2005; Garden et al. 2008; Zhou et al. 2004.

TABLE 3.3
Adipocyte Cluster Development in Fetal Pig Perirenal Adipose Tissue Depends on Proximity to Entry Point of Large Arterioles

Fetal Age	Adipocyte Cluster Size (μm^2)		Adipocyte
(days)	Proximal	Distal	Cluster No.
70	126 ± 5	31 ± 3	10 ± 1
90	172 ± 11	47 ± 8	61 ± 3

Note: Adipocyte cluster size is primarily attributable to adipocyte number because adipocyte size is affected little by location (Hausman and Thomas 1986).

TABLE 3.4
Characteristics of Neonatal Adipose Tissue Development

Trait	Subcutaneous and Internal	Intramuscular
Tissue morphology	Mature tissue with contiguous adipocytes	Mature tissue with contiguous adipocytes
Adipocyte morphology	Decrease in multilocular cells & increase in unilocular cells	Decrease in multilocular cells & increase in unilocular cells
Mode of accretion or expansion	Depot-dependent hypertrophy-subcutaneous location; depot-dependent hyperplasia	Little to no hypertrophy early; primarily hyperplasia-muscle dependent
Rate of accretion or expansion	Rapid; subcutaneous layer and location dependent	Extremely slow; muscle independent.
Molecular & ultrastructural markers	Gene expression of leptin, transcription factors, & lipogenic enzymes	SREBP-1 expression, enzyme activities (i.e., ME, G6PDH, LPL)

Abbreviations: SREBP-1, sterol regulatory element binding protein 1; ME, malic enzyme; G6PDH, glucose-6-phosphate dehydrogenase; and LPL, lipoprotein lipase.

Source: Reviews, Hausman and Hausman 1993; Hausman et al. 2001; Hausman and Hausman 2008; Hausman and Hausman 2004; Chen et al. 2008; Gnanalingham et al. 2005; Hauser et al. 1997; Hausman and Gampion 1986; Martin et al., 1999.

of increases in fat cell number and size during the finishing phase in meat animals (Table 3.5). Intramuscular lipid stores in skeletal muscle tissue have been correlated to overall fatness (Kouba et al. 1999) and can be influenced by many factors (Table 3.6), including age, diet, gender, fasted state (Broad and Ulyatt 1980; Kim et al. 2000; Gondret et al. 2000), genetics (Kouba et al. 1999; Mourot and Kouba 1999), and physical activity (Harper and Pethick 2004b). It is important to consider that intramuscular lipid accretion can be influenced by these factors in a muscle and species dependent manner (Table 3.6). For example, heat exposure in chickens and somatotropin treatment in pigs produced depot-dependent changes in fatty acid composition in a comparison of subcutaneous and intramuscular adipose tissue (Clark et al. 1992; Ain et al. 1996). Dietary-induced obesity increases intramuscular lipid content in several species (Poulos and Hausman 2005). Although limited, there is also evidence of dietary manipulation affecting intramuscular lipid production in animals. Reduced protein diets may increase intramuscular lipid by preferentially increasing stearoyl-CoA desaturase enzyme activity in porcine skeletal muscle (Doran et al. 2006). Force-fed Muscovy ducks had an increased lipid content, larger intramuscular adipocytes, and altered fatty

TABLE 3.5
Characteristics of Postnatal Adipose Tissue Development

Postnatal Depots	Internal	Subcutaneous	Intramuscular
Relative adipocyte size	Largest	Intermediate	Smallest
Mode of accretion or expansion	Early: hyperplasia & hypertrophy Late: primarily hypertrophy	Early: hyperplasia & hypertrophy Later: hypertrophy dependent on gender & species	Early & late: muscle-dependent hyperplasia with little hypertrophy
Accretion & accretion rate	Species dependent	Dependent on location, layer, gender, and species	Slow-muscle dependent
Relative lipogenesis	Intermediate	Highest	Lowest
Developmental markers: key proteins/genes	Decorin, CDC42, beta-1-AR, cofilin	SCD, leptin	G6PDH, ME, FABP-4, AFABP, IGF-II, SREBP-1, DGAT1, Col3a1

ND = not determined.

Abbreviations: AFABP, adipocyte fatty acid-binding protein; Beta-1-AR, beta 1-adrenergic receptor; CDC42, cell division cycle 42; DGAT1, diacylglycerol acyltransferase; FABP-4, fatty acid-binding protein; G6PDH, glucose-6-phosphate dehydrogenase; IGF-II, insulin-like growth factor II; ME, malic enzyme; SCD, stearoyl-CoA desaturase; SREBP-1, sterol regulatory element binding protein-1.

Source: Reviews, Hausman and Hausman 1993; Hausman et al. 2001; Hausman and Hausman 2008; Hausman and Hausman 2004; Allen et al. 1976; Anderson et al. 1972; Anderson and Kauffman 1973; Bao et al. 2007; Childs et al. 2002; Choi et al. 2003; Damon et al. 2006; Gardan et al. 2007; Gondret and Lebret 2002; Jurie et al. 2007; Mourot and Kouba 1999; Ross et al. 2005; Sasaki et al. 2006; Schoonmaker et al. 2004; Thaller et al. 2003.

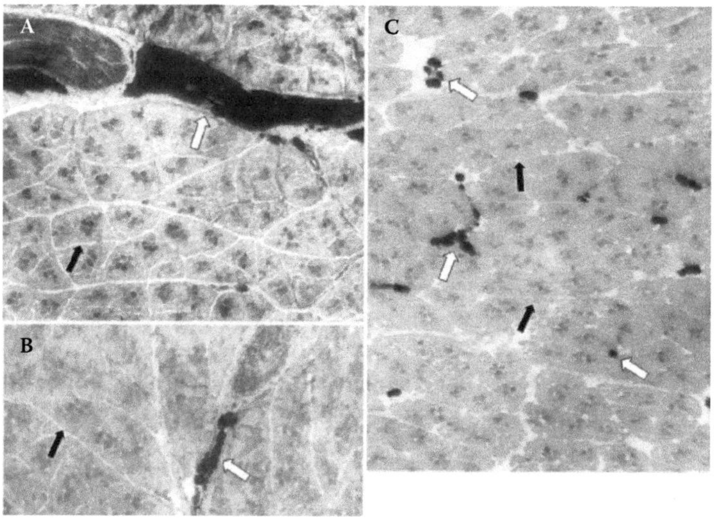

FIGURE 3.1 Lipid staining in frozen cryostat sections of the deep (A), intermediate (B), and superficial regions (C) of the semitendinosus muscle from a 14-day-old pig. Note the large cluster of adipocytes in the deep region (A, white arrow) and scattered isolated adipocytes in the superficial region (C, white arrows) of the semitendinosus muscle. Also note that there are more lipid stained muscle fibers per fasciculi (black arrows) in the deep region than in the more superficial regions of the muscle (A vs. B, C). Therefore, adipocyte development and the development of aerobic muscle fiber types are closely associated in the developing semitendinosus muscle.

TABLE 3.6
Factors that Influence Intramuscular Lipid Accretion or Marbling are Muscle and Species Dependent

Factor	Species; Muscle	Study
Age	Cattle and pigs; several muscles	Lipid accretion dependent on muscle and species (Allen et al. 1976)
	Cattle; LD	Lipid increase between 11 & 15 months; no change after 15 months (Cianzio et al. 1985)
	Cattle	Linear increase in marbling to 200–400 kg; no further increase after 420 kg (Pethick et al. 2004)
Diet	Japanese black cattle; LD	Reduced vitamin A serum levels increases marbling (Harper and Pethick 2004a)
	Holstein steer; LD	Vitamin A restriction does not influence marbling FCS and FCN (Gorocica-Buenfil et al. 2007)
	Pigs; LD and BF	Lysine deficient diet increases lipid (Essen-Gustavsson et al. 1994)
	Holstein steer; LD	Marbling FCS influenced by energy level and FCN influenced by energy source (Schoonmaker et al. 2004)
	Pigs; BF and LD	Lipid decreased with feed restriction (30–70 kg), not influenced by compensatory growth (Heyer and Lebret 2007)
	Pigs; LD, RF, deep and superficial ST	Feed restriction reduced lipid in muscle-dependent manner (ST > RF, LD) on a constant age basis (Bee et al. 2007)
	Pigs; LD	Reduced protein & energy diet reduced lipid and FCS, long-term feed restriction reduced lipid & FCN (Gondret and Lebret 2002)
	Yucatan mini-pigs; BF	Overfeeding (4–16 months) increased FCS & lipid (Guillerm-Regost et al. 2006)
Endocrine	Pigs	pGH decreases lipid while pST did not change lipid (Sorensen et al. 1996)
	Pigs; trapezius	No influence of IGF-II genotype on FCS and lipid (Gardan et al. 2006)
Birth weight	Pigs; LD and ST	Low birth weight increases ST lipid and FCS, no influence on LD (Gondret et al. 2006)
Obesity	Pigs; LD	Increased lipid. (Harbison et al. 1976; Seideman et al. 1989)
Gender	Cattle; LD	Marbling lipid highest in heifers, intermediate in steer, and lowest in bulls (Harper and Pethick 2004a)
Transgenic models	Mice	Variable influence dependent on gene (Poulos and Hausman 2005)
Pharmaceutical agents	Thiazolidinediones	Species-dependent increase in lipid (Hausman et al. 2008)

Abbreviations: BF, biceps femoris; LD, longissimus dorsi; ST, semitendinosus; RF, rectus femoris

acid composition in the pectoralis major muscle at 14 weeks of age (Zanusso et al. 2003). Angus × Hereford steers fed a high-energy concentrate beginning at 12 months of age had differences in intramuscular adiposity upon slaughter. Muscle lipid content increased by 40% when animals were fed low-protein, low-energy diets (Table 3.6). Changes in muscle lipid content could be due to age or altered muscle metabolism.

Many markers of adipocyte development / lipid accretion in several adipose depots have been identified (see Tables 3.4 and 3.5). For example, malic enzyme and glucose-6-phosphate dehydrogenase (G6PDH) have been identified as markers of lipid accretion in the intramuscular depot (see Tables 3.4 and 3.5). Cattle breeds with higher propensity for marbling have greater G6PDH and leptin expression (Bonnet et al. 2007), whereas G3PDH and lipoprotein lipase (LPL) activity did not seem to be involved in marbling accumulation. Generally, leptin gene expression is a good marker of subcutaneous adipose accretion and, as a consequence, leptin expression traits distinguish adipose depots in several species (Table 3.5). At a given age, intramuscular adipocytes have less lipogenic

TABLE 3.7
Studies of Transgenic and Gene Knockout Mouse Models Reveals a Variety of Genes Are Involved in Adipose Tissue Development

Phenotype/ Influence	Transcription Factors	Signaling Factors/ Receptors	Growth Factors/ Hormones/ Cytokines	Metabolism/ Apolipoproteins
Adipose tissue development	C/EBPα, -β (Arch 2002; Blüher 2005; Millward et al. 2007) PPARγ, -2 (Arch 2002; Blüher 2005) PPARα (Knauf et al. 2006) KLF5 (Oishi et al. 2005) SREB1-1a, -1c (Arch 2002; Blüher 2005) SRC-3 (Arch 2002; Blüher 2005)	PRLR (Flint et al. 2006) Wnt10b (Vertino et al. 2005) GHR (Flint et al. 2006)	GDF-8 (Arch 2002; Blüher 2005) IL-1Ra (Somm et al. 2005) Pref-1* (Arch 2002; Blüher 2005) TGFα* (Arch 2002; Blüher 2005)	AGT (Arch, 2002; Blüher, 2005) Plg (Arch 2002; Blüher 2005) ADAM 12-S (Arch 2002; Blüher 2005) SPARC (Arch 2002; Blüher 2005) HSL (Arch 2002; Blüher 2005)

Abbreviations: ADAM, A disintegrin and metalloproteinases; AGT, angiotensinogen; C/EBP, CCAAT/enhancer-binding protein; GDF, growth/differentiation factor; GHR, growth hormone receptor; HSL, hormone-sensitive lipase; IL, interleukin; KLF, Krüppel-like transcription factor; Plg, plasminogen; PPAR, peroxisome proliferator-activated receptor; Pref, preadipocyte factor 1; PRLR, prolactin receptor; SPARC, secreted protein acidic and rich in cysteine; SRC, steroid receptor coactivator; SREB, sterol regulatory element-binding protein; TGF, transforming growth factor.

and lipolytic activity when compared to subcutaneous and perirenal adipocytes. *In vitro* studies also showed that lipogenesis and adipocyte size were different in subcutaneous and intramuscular adipose tissue explants and stromal-vascular cells from several breeds of cattle (May et al. 1994). Gene expression for lipogenic and lipolytic enzymes follows this pattern, suggesting marked metabolic differences in adipocytes from different depots (Gardan et al. 2006). It cannot be excluded that there may be a developmental delay in metabolic efficiency, hyperplasic or hypertrophic growth that contributes to these metabolic differences; however, it remains unknown if this can be manipulated for improved performance and health outcome.

3.2.4 Adipose Tissue Studies in Transgenic and Gene Knockout Mice

Transgenic and gene knockout studies identify genes that play critical roles in adipose tissue development (Table 3.7). For example, these studies demonstrate that several groups of transcription factors have major roles in adipose tissue development (Table 3.7). These studies also show that a number of genes that encode secreted factors play roles in adipose tissue development (Table 3.7). Collectively, these studies demonstrate many potential levels of genetic influence on adipose tissue development.

3.3 INTERACTIONS BETWEEN ADIPOSE TISSUE AND SKELETAL MUSCLE

As the most energy-demanding and most energy-dense tissues in the body, skeletal muscle and adipose tissue have been the targets of many animal production efforts. Mounting evidence suggests that these tissues communicate with each other to regulate energy homeostasis, which has the potential for significant impact on the animal industry. This regulation occurs, in part, through paracrine factors released and recognized both by adipose tissue and skeletal muscles. It is now known that skeletal muscle and adipose tissue release myokines and adipokines, respectively, which are likely involved in maintaining proper energy homeostasis between the two tissues (Tomas et al. 2004; Argiles et al. 2005; Hausman et al. 2006; Hausman et al. 2007; Hausman and Hausman 2008). Interleukins that may be involved include IL-10, which appears to decreased intramuscular fatty acyl-CoA levels, and IL-6 (Kelly et al. 2004) and IL-15, which are thought to enhance skeletal muscle growth and inhibit fat deposition (Argiles et al. 2005).

Paracrine factors from adipose tissue that have been implicated in the regulation of intracellular fat content include adiponectin, plasminogen activator inhibitor-1 (PAI-1; Fibbi et al. 2002), transforming growth factor β (TGFβ; Bouche et al. 2000; Fibbi et al. 2002), leptin, and insulin-like growth factor 1 (IGF-I). Adiponectin receptors, AdipoR1 and AdipoR2, are both expressed in human skeletal muscle and appear to be regulated by glucose and lipid metabolism (Staiger et al. 2004) and inflammatory processes (Delaigle et al. 2004).

Alternatively, myokines that influence adipose tissue metabolism include those in the IGF system. The IGF system, including IGF-I (Adams 2002), IGF-II, and the IGF binding proteins (Haugk et al. 2000; Foulstone et al. 2003; Harridge 2003) are produced by skeletal muscle and may influence adipocyte growth or help signal energy requirements in skeletal muscle. Intramuscular fat content in Duroc pigs is positively correlated to serum IGF-I concentrations at 8 weeks of age (Suzuki et al. 2004). Additionally, myostatin, nerve growth factor (Toti et al. 2003), and fibroblast growth factor (Hannon et al. 1996) have been implicated in cross-talk that occurs between skeletal muscle and adipose tissue.

3.4 SUMMARY

The cellular and metabolic aspects of developing intramuscular adipose tissue and other adipose tissue depots have been studied, including examination of the expression of a number of genes. Depot-dependent or depot "marker" genes such as stearoyl-CoA desaturase and leptin for subcutaneous adipose tissue and fatty acid-binding protein 4 and glucose-6-phosphate dehydrogenase for intramuscular adipose tissue have been identified. The regulation of intramuscular adipose tissue by many factors has been examined in several species. Additional studies are necessary to determine the ontogeny and importance of adipose tissue transcription factors and other regulatory factors as influenced by depot and species. Furthermore, there is little information on the ontogeny and regulation of expression of factors expressed and secreted by intramuscular adipocytes. Additionally, a number of factors from both skeletal muscle and adipose tissue appear to be responsible for communicating the energy needs and regulating energy availability between these two tissues. However, our limited knowledge of these factors, their interactions, and the interplay between adipose tissue and skeletal muscle limits the practical application to current animal production practices.

REFERENCES

Adams, G.R. 2002. Invited Review: Autocrine/paracrine IGF-I and skeletal muscle adaptation. *J. Appl. Physiol.* 93:1159–1167.

Ain, B.H., P.A. Geraert, J.C. Padilha, and S. Guillaumin. 1996. Chronic heat exposure enhances fat deposition and modifies muscle and fat partition in broiler carcasses. *Poult. Sci.* 75:505–513.

Allen, C.E., D.C. Beitz, D.A. Cramer, and R.G. Kaufman. 1976. Biology of Fat in Meat Animals: North Central Regional Research. Publication: Number 234, University of Wisconsin, Madison.

Anderson, D.B., R.G. Kauffman, and L.L. Kastenschmidt. 1972. Lipogenic enzyme activities and cellularity of porcine adipose tissue from various anatomical locations. *J. Lipid Res.* 13:593–599.

Anderson, D.B. and R.G. Kauffman. 1973. Cellular and enzymatic changes in porcine adipose tissue during growth. *J. Lipid Res.* 14:160–168.

Arch, J.R. 2002. Lessons in obesity from transgenic animals. *J. Endocrinol. Invest.* 25:867–875.

Argiles, J.M., J. Lopez-Soriano, V. Almendro, S. Busquets, and F.J. Lopez-Soriano. 2005. Cross-talk between skeletal muscle and adipose tissue: A link with obesity? *Med. Res. Rev.* 25:49–65.

Bao, X., Y. Zeng, S. Wei, G. Wang, C. Liu, Y. Sun, Q. Chen, and H. Li. 2007. Developmental changes of Col3a1 mRNA expression in muscle and their association with intramuscular collagen in pigs. *J. Genet. Genomics* 34:223–228.

Barber, M.C., R.J. Ward, S.E. Richards, A.M. Salter, P.J. Buttery, R.G. Vernon, and M.T. Travers. 2000. Ovine adipose tissue monounsaturated fat content is correlated to depot-specific expression of the stearoyl-CoA desaturase gene. *J. Anim. Sci.* 78:62–68.

Bee, G., M. Calderini, C. Biolley, G. Guex, W. Herzog, and M.D. Lindemann. 2007. Changes in the histochemical properties and meat quality traits of porcine muscles during the growing-finishing period as affected by feed restriction, slaughter age, or slaughter weight. *J. Anim. Sci.* 85:1030–1045.

Blüher, M. 2005. Transgenic animal models for the study of adipose tissue biology. *Best. Pract. Res. Clin. Endocrinol. Metab.* 19:605–623.

Bonnet, M., Y. Faulconnier, C. Leroux, C. Jurie, I. Cassar-Malek, D. Bauchart, P. Boulesteix, D. Pethick, J.F. Hocquette, and Y. Chilliard. 2007. Glucose-6-phosphate dehydrogenase and leptin are related to marbling differences among Limousin and Angus or Japanese Black × Angus steers. *J. Anim. Sci.* 85:2882–2894.

Bouche, M., R. Canipari, R. Melchionna, D. Willems, M.I. Senni, and M. Molinaro. 2000. TGF-beta autocrine loop regulates cell growth and myogenic differentiation in human rhabdomyosarcoma cells. *FASEB J.* 14:1147–1158.

Broad, T.E. and M.J. Ulyatt. 1980. The effect of level of food intake on the incorporation of acetate into lipids and its distribution among various tissues in sheep. *Br. J. Nutr.* 44:71–79.

Casteilla, L., P. Muzzin, J.P. Revelli, D. Ricquier, and J.P. Giacobino. 1994. Expression of beta 1- and beta 3-adrenergic-receptor messages and adenylate cyclase beta-adrenergic response in bovine perirenal adipose tissue during its transformation from brown into white fat. *Biochem. J.* 297:93–97.

Chen, J., X.J. Yang, D. Xia, J. Chen, J. Wegner, Z. Jiang, and R.Q. Zhao. 2008. Sterol regulatory element binding transcription factor 1 expression and genetic polymorphism significantly affect intramuscular fat deposition in the longissimus muscle of Erhualian and Sutai pigs. *J. Anim. Sci.* 86:57–63.

Childs, K.D., D.W. Goad, M.F. Allan, D. Pomp, C. Krehbiel, R.D. Geisert, J.B. Morgan, and J.R. Malayer. 2002. Differential expression of NAT1 translational repressor during development of bovine intramuscular adipocytes. *Physiol. Genomics* 10:49–56.

Choi, K.C., S.G. Roh, D. Hishikawa, H. Miyahara, M. Kuno, H. Tsuzuki, A. Tomimatsu, Y.H. Hong, K.K. Cho, K.H. Han, and S. Sasaki. 2003. Differential expression of the nonmuscle-type cofilin gene between subcutaneous and visceral adipose tissue. *Biosci. Biotechnol. Biochem.* 67:2262–2265.

Cianzio, D.S., D.G. Topel, G.B. Whitehurst, D.C. Beitz, and H.L. Self. 1985. Adipose tissue growth and cellularity: changes in bovine adipocyte size and number. *J. Anim. Sci.* 60:970–976.

Clark, S.L., R.C. Wander, and C.Y. Hu. 1992. The effect of porcine somatotropin supplementation in pigs on the lipid profile of subcutaneous and intermuscular adipose tissue and longissimus muscle. *J. Anim. Sci.* 70:3435–3442.

Crandall, D.L., G.J. Hausman, and J.G. Kral. 1997. A review of the microcirculation of adipose tissue: anatomic, metabolic, and angiogenic perspectives. *Microcirculation* 4:211–232.

Damon, M., I. Louveau, L. Lefaucheur, B. Lebret, A. Vincent, P. Leroy, M.P. Sanchez, P. Herpin, and F. Gondret. 2006. Number of intramuscular adipocytes and fatty acid binding protein-4 content are significant indicators of intramuscular fat level in crossbred Large White × Duroc pigs. *J. Anim. Sci.* 84:1083–1092.

Delaigle, A.M., J.C. Jonas, I.B. Bauche, O. Cornu, and S.M. Brichard. 2004. Induction of adiponectin in skeletal muscle by inflammatory cytokines: *in vivo* and *in vitro* studies. *Endocrinology* 145:5589–5597.

Doran, O., S.K. Moule, G.A. Teye, F.M. Whittington, K.G. Hallett, and J.D. Wood. 2006. A reduced protein diet induces stearoyl-CoA desaturase protein expression in pig muscle but not in subcutaneous adipose tissue: relationship with intramuscular lipid formation. *Br. J. Nutr.* 95:609–617.

Essén-Gustavsson, B., A. Karlsson, K. Lundström, and A.C. Enfält. 1994. Intramuscular fat and muscle fibre lipid contents in halothane-gene-free pigs fed high or low protein diets and its relation to meat quality. *Meat Sci.* 38:269–270.

Fibbi, G., S. D'Alessio, M. Pucci, M. Cerletti, and M. Del Rosso. 2002. Growth factor-dependent proliferation and invasion of muscle satellite cells require the cell associated fibrinolytic system. *Biolog. Chem.* 383:127–136.

Flint, D.J., N. Binart, S. Boumard, J.J. Kopchick, and P. Kelly, 2006. Developmental aspects of adipose tissue in GH receptor and prolactin receptor gene disrupted mice: site-specific effects upon proliferation, differentiation and hormone sensitivity. *J. Endocrinol.* 191:101–111.

Foulstone, E.J., P.B. Savage, A.L. Crown, J.M. Holly, and C.E. Stewart, 2003. Adaptations of the IGF system during malignancy: human skeletal muscle versus the systemic environment. *Hormone Metab. Res.* 35:667–674.

Gardan, D., F. Gondret, and I. Louveau, 2006. Lipid metabolism and secretory function of porcine intramuscular adipocytes compared with subcutaneous and perirenal adipocytes. *Am. J. Physiol. Endocrinol. Metab.* 291:E372–E380.

Gardan, D., F. Gondret, K. Van den Maagdenberg, N. Buys, S. De Smet, and I. Louveau. 2008. Lipid metabolism and cellular features of skeletal muscle and subcutaneous adipose tissue in pigs differing in IGF-II genotype. *Domest. Anim. Endocrinol.* 34:45–53.

Gardan, D., I. Louveau, and F. Gondret, 2007. Adipocyte- and heart-type fatty acid binding proteins are both expressed in subcutaneous and intramuscular porcine (*Sus scrofa*) adipocytes. *Comp. Biochem. Physiol. B Biochem. Mol. Biol.* 148:14–19.

Gnanalingham, M.G., A. Mostyn, A.J. Forhead, A.L. Fowden, M.E. Symonds, and T. Stephenson. 2005. Increased uncoupling protein-2 mRNA abundance and glucocorticoid action in adipose tissue in the sheep fetus during late gestation is dependent on plasma cortisol and triiodothyronine. *J. Physiol.* 567:283–292.

Gondret, F. and B. Lebret. 2002. Feeding intensity and dietary protein level affect adipocyte cellularity and lipogenic capacity of muscle homogenates in growing pigs, without modification of the expression of sterol regulatory element binding protein. *J. Anim. Sci.* 80:3184–3193.

Gondret, F., F. Lebas, and M. Bonneau, 2000. Restricted feed intake during fattening reduces intramuscular lipid deposition without modifying muscle fiber characteristics in rabbits. *J. Nutr.* 130:228–233.

Gondret, F., L. Lefaucheur, H. Juin, I. Louveau, and B. Lebret. 2006. Low birth weight is associated with enlarged muscle fiber area and impaired meat tenderness of the longissimus muscle in pigs. *J. Anim. Sci.* 84:93–103.

Gorocica-Buenfil, M.A., F.L. Fluharty, C.K. Reynolds, and S.C. Loerch. 2007. Effect of dietary vitamin A restriction on marbling and conjugated linoleic acid content in Holstein steers. *J. Anim. Sci.* 85:2243–2255.

Guillerm-Regost, C., I. Louveau, S.P. Sébert, M. Damon, M.M. Champ, and F. Gondret. 2006. Cellular and biochemical features of skeletal muscle in obese Yucatan minipigs. *Obesity* (Silver Spring) 14:1700–1707.

Hannon, K., A.J. Kudla, M.J. McAvoy, K.L. Clase, and B.B. Olwin. 1996. Differentially expressed fibroblast growth factors regulate skeletal muscle development through autocrine and paracrine mechanisms. *J. Cell Biol.* 132:1151–1159.

Harbison, S.A., D.E. Goll, F.C. Parrish, Jr, V. Wang, and E.A. Kline. 1976. Muscle growth in two genetically different lines of swine. *Growth* 40:253–283.

Harper, G.S. and D.W. Pethick, 2004a. How might marbling begin? *Austr. J. Exper. Agric.* 44:653–662.

Harper, G.S. and D.W. Pethick. 2004b. Fatness in production animals: using genetic and environmental levers to meet consumer demand. *Asia Pac. J. Clin. Nutr.* 13(Suppl.):S37.

Harridge, S.D. 2003. Ageing and local growth factors in muscle. *Scand. J. Med. Sci. Sports* 13:34–39.

Haugk, K.L., H.M. Wilson, K. Swisshelm, and L.S. Quinn. 2000. Insulin-like growth factor (IGF)-binding protein-related protein-1: an autocrine/paracrine factor that inhibits skeletal myoblast differentiation but permits proliferation in response to IGF. *Endocrinology* 141:100–110.

Hauser, N., J. Mourot, L. De Clercq, C. Genart, and C. Remacle. 1997. The cellularity of developing adipose tissues in Pietrain and Meishan pigs. *Reprod. Nutr. Dev.* 37:617–625.

Hausman, G.J., C.R. Barb, and R.G. Dean. 2007. Patterns of gene expression in pig adipose tissue: transforming growth factors, interferons, interleukins, and apolipoproteins. *J. Anim. Sci.* 85:2445–2456.

Hausman, G.J. and D.R. Campion. 1986. Skeletal muscle cellularity and histochemistry in young lean and obese pigs. *Growth* 50:287–295.

Hausman, D.B., M. DiGirolamo, T.J. Bartness, G.J. Hausman, and R.J. Martin. 2001. The biology of white adipocyte proliferation. *Obesity Rev.* 2:239–254.

Hausman, G.J. and D.B. Hausman. 1993. Endocrine regulation of porcine adipose tissue development: cellular and metabolic aspects. In *Growth of the Pig* (Hollis, G.R., Ed.). 49.

Hausman, G.J. and D.B. Hausman. 2004. Ontogeny: adipose tissue. In *Encyclopedia of Animal Science, 1st ed.* (Pond, W.G. and Bell, A.W., Eds.). New York: Marcel Deker, Inc., 684.

Hausman, G.J. and D.B. Hausman 2008. In *Adipose Tissue in Health and Disease* (Granneman et al., Eds.).

Hausman, G.J. and R.G. Kauffman. 1986. The histology of developing porcine adipose tissue. *J. Anim. Sci.* 63:642–658.

Hausman, G.J., S.P. Poulos, T.D. Pringle, and M.J. Azain. 2008. The influence of thiazolidinediones on adipogenesis *in vitro* and *in vivo*: potential modifiers of intramuscular adipose tissue deposition in meat animals. *J. Anim. Sci.* 86:E236–E243.

Hausman, G.J., S.P. Poulos, R.L. Richardson, C.R. Barb, T. Andacht, H.C. Kirk, and R.L. Mynatt, 2006. Secreted proteins and genes in fetal and neonatal pig adipose tissue and stromal-vascular cells. *J. Anim. Sci.* 84:1666–1681.

Hausman, G.J. and R.L. Richardson. 2004. Adipose tissue angiogenesis. *J. Anim. Sci.* 82:925–934.

Hausman, G.J. and R.L. Richardson. 2004. Adipose tissue angiogenesis. *J. Anim. Sci.* 82:925–934.

Hausman, G.J. and G.B. Thomas. 1984. The development of the inner layer of backfat in fetal and young pigs. *J. Anim. Sci.* 58:1550–1560.

Hausman, G.J. and G.B. Thomas. 1986. Structural and histochemical aspects of perirenal adipose tissue in fetal pigs: relationships between stromal-vascular characteristics and fat cell concentration and enzyme activity. *J. Morphol.* 190:271–283.

Heyer, A. and B. Lebret. 2007. Compensatory growth response in pigs: effects on growth performance, composition of weight gain at carcass and muscle levels, and meat quality. *J. Anim. Sci.* 85:769–778.

Hishikawa, D., Y.H. Hong, S.G. Roh, H. Miyahara, Y. Nishimura, A. Tomimatsu, H. Tsuzuki, C. Gotoh, M. Kuno, K.C. Choi, H.G. Lee, K.K. Cho, H. Hidari, and S. Sasaki, 2005. Identification of genes expressed differentially in subcutaneous and visceral fat of cattle, pig, and mouse. *Physiol. Genomics* 21:343–350.

Jurie, C., I. Cassar-Malek, M. Bonnet, C. Leroux, D. Bauchart, P. Boulesteix, D.W. Pethick, and J.F. Hocquette. 2007. Adipocyte fatty acid-binding protein and mitochondrial enzyme activities in muscles as relevant indicators of marbling in cattle. *J. Anim. Sci.* 85:2660–2669.

Kelly, M., C. Keller, P.R. Avilucea, P. Keller, Z. Luo, X. Xiang, M. Giralt, J. Hidalgo, A.K. Saha, B.K. Pedersen, and N.B. Ruderman, 2004. AMPK activity is diminished in tissues of IL-6 knockout mice: the effect of exercise. *Biochem. Biophys. Res. Commun.* 320:449–454.

Kim, H., Y. Chi, K. Chung, K. Kim, Y. Choi, and M. Baik. 2000. Differential response of obese gene expression from fasting in bovine adipose tissues. *Biosci. Biotechnol. Biochem.* 64:2240–2242.

Klein, J., P.A. Permana, M. Owecki, G.N. Chaldakov, M. Bohm, G. Hausman, C.M. Lapiere, P. Atanassova, J. Sowinski, M. Fasshauer, D.B. Hausman, E. Maquoi, A.B. Tonchev, V.N. Peneva, K.P. Vlachanov, M. Fiore, L. Aloe, A. Slominski, C.L. Reardon, T.J. Ryan, and C.M. Pond. 2007. What are subcutaneous adipocytes really good for? *Exp. Dermatol.* 16:45–70.

Knauf, C., J. Rieusset, M. Foretz, P.D. Cani, M. Uldry, M. Hosokawa, E. Martinez, M. Bringart, A. Waget, S. Kersten, B. Desvergne, S. Gremlich, W. Wahli, J. Seydoux, N.M. Delzenne, B. Thorens, and R. Burcelin. 2006. Peroxisome proliferator-activated receptor-alpha-null mice have increased white adipose tissue glucose utilization, GLUT4, and fat mass: role in liver and brain. *Endocrinology* 147:4067–4078.

Kouba, M., M. Bonneau, and J. Noblet. 1999. Relative development of subcutaneous, intermuscular, and kidney fat in growing pigs with different body compositions. *J. Anim. Sci.* 77:622–629.

Leymaster, K.A. and H.J. Mersmann. 1991. Effect of limited feed intake on growth of subcutaneous adipose tissue layers and on carcass composition in swine. *J. Anim. Sci.* 69:2837–2843.

Martin, G.S., D.K. Lunt, K.G. Britain, and S.B. Smith, 1999. Postnatal development of stearoyl coenzyme A desaturase gene expression and adiposity in bovine subcutaneous adipose tissue. *J. Anim. Sci.* 77:630–636.

Martin, R.J., G.J. Hausman, and D.B. Hausman. 1998. Regulation of adipose cell development *in utero*. *Proc. Soc. Exp. Biol. Med.* 219:200–210.

May, S.G., J.W. Savell, D.K. Lunt, J.J. Wilson, J.C. Laurenz, and S.B. Smith. 1994. Evidence for preadipocyte proliferation during culture of subcutaneous and intramuscular adipose tissues from Angus and Wagyu crossbred steers. *J. Anim. Sci.* 72:3110–3117.

McNeel, R.L. and H.J. Mersmann. 1995. Beta-adrenergic receptor subtype transcripts in porcine adipose tissue. *J. Anim. Sci.* 73:1962–1971.

Millward, C.A., J.D. Heaney, D.S. Sinasac, E.C. Chu, I.R. Bederman, D.A. Gilge, S.F. Previs, and C.M. Croniger. 2007. Mice with a deletion in the gene for CCAAT/enhancer-binding protein beta are protected against diet-induced obesity. *Diabetes* 56:161–167.

Mostyn A, S. Sebert, J.C. Litten, K.S. Perkins, J. Laws, M.E. Symonds, and L. Clarke. 2006. Influence of porcine genotype on the abundance of thyroid hormones and leptin in sow milk and its impact on growth, metabolism and expression of key adipose tissue genes in offspring. *J. Endocrinol.* 190:631–639.

Mourot, J. and M. Kouba, 1999. Development of intra- and intermuscular adipose tissue in growing large white and Meishan pigs. *Reprod. Nutr. Dev.* 39:125–132.

Muhlhausler, B.S., J.A. Duffield, and I.C. McMillen, 2007. Increased maternal nutrition stimulates peroxisome proliferator activated receptor-gamma, adiponectin, and leptin messenger ribonucleic acid expression in adipose tissue before birth. *Endocrinology* 148:878–885.

Murdoch, G., W. Dixon, E. Okine, and R. Christopherson. 2005. Bovine tissue mRNA abundance related to acute cold exposure and acute feeding restriction. *Can. J. Anim. Sci.* 85:157–164.

O'Connor, D.M., D. Blache, N. Hoggard, E. Brookes, F.B. Wooding, A.L. Fowden, and A.J. Forhead. 2007. Developmental control of plasma leptin and adipose leptin mRNA in the ovine fetus during late gestation: role of glucocorticoids and thyroid hormones. *Endocrinology* 148:3750–3757.

Oishi, Y., I. Manabe, K. Tobe, K. Tsushima, T. Shindo, K. Fujiu, G. Nishimura, K. Maemura, T. Yamauchi, N. Kubota, R. Suzuki, T. Kitamura, S. Akira, T. Kadowaki, and R. Nagai. 2005. Krüppel-like transcription factor KLF5 is a key regulator of adipocyte differentiation. *Cell Metab.* 1:27–39.

Pethick, D.W., G.S. Harper, and V.H. Oddy. 2004. Growth, development and nutritional manipulation of marbling in cattle: a review. *Austr. J. Exper. Agric.* 44:705–715.

Poissonnet, C.M., A.R. Burdi, and F.L. Bookstein. 1983. Growth and development of human adipose tissue during early gestation. *Early Hum. Dev.* 8:1–11.

Poissonnet, C.M., A.R. Burdi, and S.M. Garn. 1984. The chronology of adipose tissue appearance and distribution in the human fetus. *Early Hum. Dev.* 10:1–11.

Poulos, S.P., and G.J. Hausman. 2005. Intramuscular adipocytes — Potential to prevent lipotoxicity in skeletal muscle. *Adipocytes* 1:79–94.

Poulos, S.P. and G.J. Hausman. 2006b. Adipogenesis in porcine stromal-vascular cell cultures derived from fetal semitendinosus muscle and subcutaneous adipose tissue. *Adipocytes* 2:29–36.

Poulos, S.P. and G.J. Hausman. 2006a. A comparison of thiazolidinedione-induced adipogenesis and myogenesis in stromal-vascular cells from subcutaneous adipose tissue or semitendinosus muscle of postnatal pigs. *J. Anim. Sci.* 84:1076–1082.

Ramsay, T.G., C.K. Wolverton, and N. C. Steele. 1994. Alteration in IGF-I mRNA content of fetal swine tissues in response to maternal diabetes. *Am. J. Physiol.* 267(5 Pt. 2):R1391–1396.

Ross, J.W., T.K. Smith, C.R. Krehbiel, J.R. Malayer, U. DeSilva, J.B. Morgan, F.J. White, M.J. Hersom, G.W. Horn, and R.D. Geisert. 2005. Effects of grazing program and subsequent finishing on gene expression in different adipose tissue depots in beef steers. *J. Anim. Sci.* 83:1914–1923.

Sasaki, Y., K. Nagai, Y. Nagata, K. Doronbekov, S. Nishimura, S. Yoshioka, T. Fujita, K. Shiga, T. Miyake, Y. Taniguchi, and T. Yamada. 2006. Exploration of genes showing intramuscular fat deposition-associated expression changes in musculus longissimus muscle. *Anim. Genet.* 37:40–46.

Schoonmaker, J.P., F.L. Fluharty, and S.C. Loerch. 2004. Effect of source and amount of energy and rate of growth in the growing phase on adipocyte cellularity and lipogenic enzyme activity in the intramuscular and subcutaneous fat depots of Holstein steers. *J. Anim. Sci.* 82:137–148.

Seideman, S.C., J.D. Crouse, and H.J. Mersmann. 1989. Carcass, muscle and meat characteristics of lean and obese pigs. *J. Anim. Sci.* 67:2950–2955.

Smith, SB., H.J. Mersmann, E.O. Smith, and K.G. Britain. 1999. Stearoyl-coenzyme A desaturase gene expression during growth in adipose tissue from obese and crossbred pigs. *J. Anim. Sci.* 77:1710–1716.

Somm, E., E. Henrichot, A. Pernin, C.E. Juge-Aubry, P. Muzzin, J.M. Dayer, M.J. Nicklin, and C.A. Meier. 2005. Decreased fat mass in interleukin-1 receptor antagonist-deficient mice: impact on adipogenesis, food intake, and energy expenditure. *Diabetes* 54:3503–3509.

Sorensen, M.T., N. Oksbjerg, N. Agergaard, and J.S. Petersen. 1996. Tissue deposition rates in relation to muscle fibre and fat cell characteristics in lean female pigs (*Sus scrofa*) following treatment with porcine growth hormone (pGH). *Compar. Biochem. Physiol. A. Physiol.* 113:91–96.

Staiger, H., S. Kaltenbach, K. Staiger, N. Stefan, A. Fritsche, A. Guirguis, C. Peterfi, M. Weisser, F. Machicao, M. Stumvoll, and H.U. Haring. 2004. Expression of adiponectin receptor mRNA in human skeletal muscle cells is related to *in vivo* parameters of glucose and lipid metabolism. *Diabetes* 53:2195–2201.

Suzuki, K., M. Nakagawa, K. Katoh, H. Kadowaki, T. Shibata, H. Uchida, Y. Obara, and A. Nishida. 2004. Genetic correlation between serum insulin-like growth factor-1 concentration and performance and meat quality traits in Duroc pigs. *J. Anim. Sci.* 82:994–999.

Tang, Z., Y. Li, P. Wan, X. Li, S. Zhao, B. Liu, B. Fan, M. Zhu, M. Yu, and K. Li. 2007. LongSAGE analysis of skeletal muscle at three prenatal stages in Tongcheng and Landrace pigs. *Genome Biol.* 8:R115.

Thaller, G., C. Kuhn, A. Winter, G. Ewald, O. Bellmann, J. Wegner, H. Zuhlke, and R. Fries. 2003. DGAT1, a new positional and functional candidate gene for intramuscular fat deposition in cattle. *Anim. Genet.* 34:354–357.

Tomas, E., M. Kelly, X. Xiang, T.S. Tsao, C. Keller, P. Keller, Z. Luo, H. Lodish, A.K. Saha, R. Unger, and N.B. Ruderman. 2004. Metabolic and hormonal interactions between muscle and adipose tissue. *Proc. Nutr. Soc.* 63:381–385.

Toti, P., M. Villanova, R. Vatti, K. Schuerfeld, M. Stumpo, L. Barbagli, A. Malandrini, and M. Costantini. 2003. Nerve growth factor expression in human dystrophic muscles. *Muscle and Nerve* 27:370–373.

Trayhurn, P., C. Bing, and I.S. Wood. 2006. Adipose tissue and adipokines — energy regulation from the human perspective. *J. Nutr.* 136:1935S–1939S.

Vertino, A.M., J.M. Taylor-Jones, K.A. Longo, E.D. Bearden, T.F. Lane, R.E. McGehee, Jr., O.A. MacDougald, and C.A. Peterson. 2005. Wnt10b deficiency promotes coexpression of myogenic and adipogenic programs in myoblasts. *Mol. Biol. Cell.* 16:2039–2048.

Wu, J., B. Wang, H. Zhang, T. Yu, and G. Yang. 2007. Different transcription profiles of SOCS-3, ob and IGF-I genes and their possible correlations in obese and lean pigs. *Acta Biochim. Biophys. Sin.* 39:305–310.

Yang, X.J., E. Albrecht, K. Ender, R.Q. Zhao, and J. Wegner. 2006. Computer image analysis of intramuscular adipocytes and marbling in the longissimus muscle of cattle. *J. Anim. Sci.* 84:3251–3258.

Yuen, B.S., I.C. McMillen, M.E. Symonds, and P.C. Owens. 1999. Abundance of leptin mRNA in fetal adipose tissue is related to fetal body weight. *J. Endocrinol.* 163:R11–R14.

Yuen, B.S., P.C. Owens, B.S. Muhlhausler, C.T. Roberts, M.E. Symonds, D.H. Keisler, J.R. McFarlane, K.G. Kauter, Y. Evens, and I.C. McMillen. 2003. Leptin alters the structural and functional characteristics of adipose tissue before birth. *FASEB J.* 17:1102–1104.

Zanusso, J., H. Remignon, G. Guy, H. Manse, and R. Babile. 2003. The effects of overfeeding on myofibre characteristics and metabolical traits of the breast muscle in Muscovy ducks (*Cairina moschata*). *Reprod. Nutr. Devel.* 43:105–115.

Zhou, J., R. Zhao, X. Wei, D. Xia, Q. Xu, and J. Chen. 2004. Developmental patterns of serum leptin levels, leptin gene expression in adipose tissue and Ob-Rb gene expression in hypothalamus of Erhualian and Large White pigs. *Sci.China C.Life Sci.* 47:190–196.

4 Fetal Programming of Skeletal Muscle Development

Min Du and Meijun Zhu

CONTENTS

4.1 FETAL STAGE IS CRUCIAL FOR SKELETAL MUSCLE DEVELOPMENT

Enhancing skeletal muscle growth is crucial for the profitability of animal agriculture and the meat industry. Effective enhancement of skeletal muscle growth depends on our understanding of the mechanisms controlling skeletal muscle growth. Skeletal muscle development can be separated into three stages: (1) embryonic, (2) fetal, and (3) postnatal. Myogenesis at the embryonic stage is characterized by the formation of primary muscle fibers, secondary muscle fibers are formed at the fetal stage, and postnatal myogenesis is mainly associated with the activation of satellite cells and their involvement in muscle growth and damage repair.

The fetal stage is particularly important for skeletal muscle development because there is no increase in the net number of muscle fibers after birth. Muscle growth is achieved later in life by increasing the muscle fiber diameter (Zhu et al. 2006). Thus, the reduction in the number of prenatal muscle fibers has irreversible effects on later postnatal life. Furthermore, the lower priority in nutrient partitioning compared to the neural system, internal organs, and bone during development makes skeletal muscle particularly vulnerable to nutrient fluctuation (Zhu et al. 2006; Zhu et al. 2008). Because skeletal muscle development and properties can be altered by maternal nutrition, skeletal muscle developmental programming plays a crucial role in predisposing offspring to accumulate fat and reduce the lean:fat ratio (Vonnahme et al. 2003; Zhu et al. 2008). Under-nutrition during gestation affects the growth performance of offspring (Dwyer et al. 1994; Osgerby et al. 2002). This chapter examines recent evidence regarding the developmental programming of skeletal muscle and discusses underlying mechanisms.

4.1.1 Embryonic Stage

4.1.1.1 Embryonic Myogenesis

Embryonic skeletal muscle cells are derived from mesenchymal stem cells. First, a portion of mesenchymal stem cells becomes myogenic progenitor cells. These cells are not yet muscle cells but are committed to become muscle cells. Embryonic myogenesis is controlled by a well-coordinated set of transcription factors, including Wnt, paired box gene Pax3 and Pax7, and myogenic regulatory factors (MRFs). Wnt signaling is crucial for mesoderm formation and mesenchymal stem cell differentiation. Within the surrounding tissues, a portion of mesenchymal stem cells to become myogenic progenitor cells express Pax3 and Pax7, which are highly related proteins and may be partially redundant. Pax3 and Pax7 induce the expression of MRFs (Hyatt et al. 2008; Maroto et al. 1997).

Myogenic precursor cells further differentiate into myoblasts and then myotubes under the control of MRFs that belong to a superfamily A of basic helix loop helix (bHLH) transcriptional regulators (Atchley and Fitch 1997). Currently, four MRFs are identified, including MyoD, Myf-5, myogenin, and MRF-4, all of which are derived from a single ancestor gene. The MRF transactivation domains are not conserved. Myogenin processes the transactivation domain in both the N- and C-termini, MyoD and MRF-4 at the N termini, and Myf5 is in the C-terminus. On the other hand, the bHLH domains of all four MRFs are highly conserved.

MRF-4 is expressed briefly during the very early stage of myogenesis, followed by the expression of MyoD and Myf-5, and myogenin is expressed later and maintained throughout fetal life. MRF-4 is also expressed later and becomes the dominant MRF postnatally. The main role of MyoD and Myf-5 is to induce the differentiation of pluripotent myogenic precursor cells into myoblasts. The functions of MyoD and Myf-5 appear compensative with each other (Maroto et al. 1997; Roth et al. 2003). MyoR and Mistl inhibit the function of MyoD and thus negatively regulate myogenesis (Lemercier et al. 1998). Myogenin is necessary for the fusion of myoblasts into myotubes. The exact mechanisms controlling myoblast fusion remain poorly defined but calcium-activated transcription factor Nuclear Factor of Activated T Cells C2 (NFATC2) has an indispensable role (Pavlath and Horsley 2003).

4.1.1.2 Primary Muscle Fibers

The amounts of primary muscle fibers developed at the embryonic stage are very limited and provide templates for the formation of secondary myofibers. Because myogenic cells are derived from myogenic progenitor cells, enhancing the proliferation of progenitor cells will generate a higher number of myogenic cells, which fuse to form more and larger muscle fibers. However, due to the fact that only a very limited number of primary myofibers are formed during the embryonic stage, their contribution to the overall muscle size and myofiber number of offspring is insignificant.

4.1.2 Fetal Stage

The majority of muscle fibers are formed during the fetal stage and, as a result, the fetal stage is crucial for skeletal muscle development. To differentiate from the primary myogenesis at embryonic stage, myogenesis at fetal stage is referred to as secondary myogenesis.

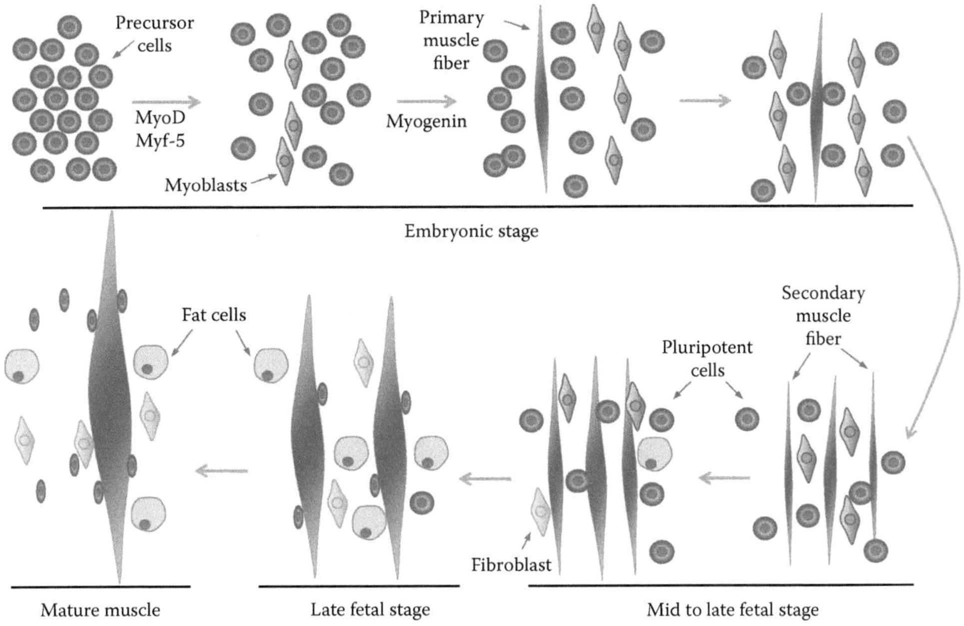

FIGURE 4.1 Fetal skeletal muscle development.

In fetal skeletal muscle, a large number of myogenic progenitor cells present. These myogenic progenitor cells are initially maintained in a proliferating, undifferentiated state and, then, differentiated into myoblasts and myofibers (Figure 4.1) (Swatland 1973). Secondary myofibers have peripherally located nuclei, which differ from primary myofibers processing centrally located nuclei (Beermann 1978; Swatland 1973). In addition, the diameter of primary muscle fibers is much larger than secondary muscle fibers (Figure 4.2).

The formation of secondary myofibers coincides with adipogenesis, which is initiated at mid-gestation in cattle and sheep but late gestation in pigs and rodents. This stage is also associated with fibrogenesis. Fibroblasts developed during this stage synthesize connective tissue, which forms perimysium and epimysium in fetal skeletal muscle at late gestation. Myocytes, adipocytes, and fibroblasts all have mesenchymal origins. Therefore, the fetal stage is crucial for skeletal muscle development in cattle and sheep. However, fetuses of rodents and pigs are born immature; in addition to fetal stage, secondary myogenesis continues into the neonatal period (Agbulut et al. 2003; Fazeli et al. 1996). To date, most studies regarding myogenesis have focused on primary myogenesis, and mechanisms regulating the formation of secondary muscle fibers, adipogenesis, and fibrogenesis, which forms the basic structure of skeletal muscle, are poorly studied. Considering that intramuscular adipogenesis and fibrogenesis forms marbling and connective tissue, which have direct impacts on meat quality, studies to understand their generation and development in fetal skeletal muscle are imperative.

4.1.3 Postnatal Stage

Postnatal skeletal muscle growth is characterized by the increase in size (hypertrophy) instead of the increase in number (hyperplasia). However, the maturity of skeletal muscle at birth is different for rodents and pigs compared to cattle and sheep. For cattle and sheep, skeletal muscle matures at late gestation (McCoard et al. 2001), but for rodents and pigs, skeletal muscle matures at the neonatal stage (Kassar-Duchossoy et al. 2005). Therefore, using birth as a marker for the stage of skeletal muscle development is not accurate and does not reflect the difference between species.

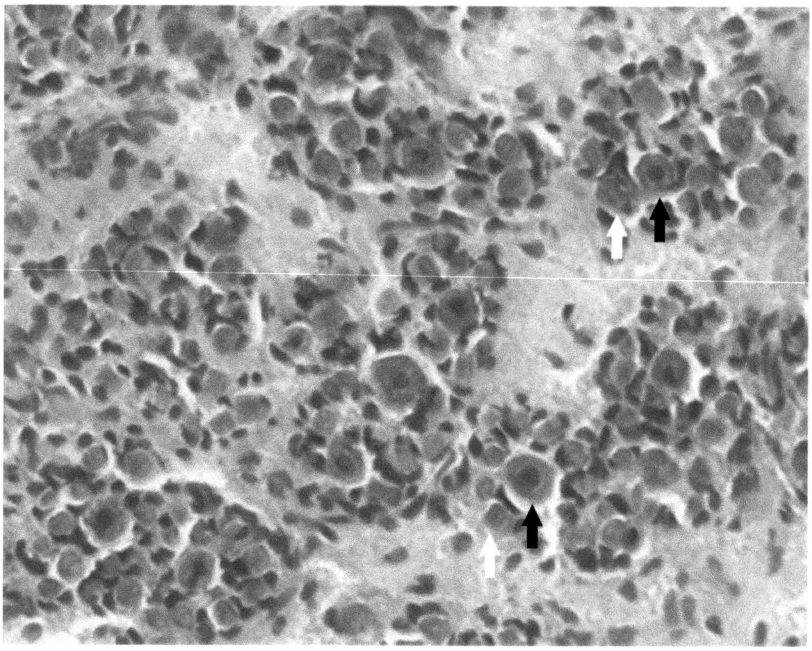

FIGURE 4.2 Primary and secondary muscle fibers in fetal skeletal muscle of sheep at day 75 of gestation. Muscle cryosection was stained with Hematoxylin and Eosin. Black arrows show the presence of primary muscle fibers. White arrows show the presence of secondary muscle fibers. Primary muscle fibers are large with centrally located nuclei, and each primary muscle fiber is surrounded by secondary muscle fibers with peripherally located nuclei.

As discussed in Chapter 2, satellite cells are inseparable from postnatal skeletal muscle growth. For fetal sheep and cattle, satellite cells are well developed and muscle structure matures at late gestation, but in rodents and pigs, satellite cells start to appear only around birth (Kassar-Duchossoy et al. 2005). In neonatal mice, satellite cells are metabolically very active but become quiescent as the animal grows older (Schultz 1976), which is quite different from satellite cells in sheep, cattle, and humans where quiescent satellite cells are formed at late gestation (McCoard et al. 2001). Therefore, when conducting studies and comparing data obtained from animals of different species, such differences in fetal skeletal muscle development need to be observed.

Chapter 2 provided detailed discussions of satellite cells and postnatal skeletal muscle growth and thus they are not repeated here.

4.2 FETAL PROGRAMMING DUE TO MATERNAL NUTRITION

4.2.1 Concept of Fetal Programming

The concept of fetal origins of adult health, known as fetal or developmental programming or the Barker hypothesis, is based on the epidemiological data linking long-term effects of low birth weight, due to malnutrition, on adult health (Barker et al. 2002). The fetal origins hypothesis states that fetal nutrient deficiency leads to disproportionate fetal growth and programs later predisposition to chronic disease states such as coronary heart diseases, stroke, diabetes, and hypertension (Barker 2001). The prevalence of type 2 diabetes increases threefold for men who weighed 5.5 pounds at birth when compared to those who had birth weights of 9.5 pounds. This relationship has been confirmed by several studies in Britain, the United States, and Sweden (Barker, 1995, 2003; Hales et al. 1991), and also numerous animal studies. Studies linking fetal programming to animal performance in livestock were initiated fairly recently. In these studies, both malnutrition and over-nutrition during gestation impact offspring growth performance (Benyshek et al. 2004; Bieswal et al. 2006; Bispham et al. 2003; Desai et al. 2005; Fernandez-Twinn et al. 2006; Fernandez-Twinn et al. 2005; King 2006; Symonds et al. 2004; Zambrano et al. 2006).

4.2.2 Maternal Nutrition during Fetal Stage Has Major Impacts on Skeletal Muscle Development

4.2.2.1 Fetal Developmental Programming of Skeletal Muscle Fiber Size, Number, and Type Composition

Maternal nutrition programs fetal development, especially fetal skeletal muscle development. The lower priority of skeletal muscle in nutrient partitioning during fetal development compared to organs, such as brain, heart, and liver, renders skeletal muscle development particularly vulnerable to nutrient alteration (Zhu et al. 2006). As discussed previously, maternal nutrition during the embryonic stage has minor effects on skeletal muscle development due to the fact that only a very limited number of myofibers are formed during this stage. Indeed, in our study in beef cattle receiving 30% nutrient restriction during 30 to 120 days of gestation, there was no effect on fetal body weight and carcass weight (Du et al. 2005). However, in sheep, 50% nutrient restriction for 24 days around periconception reduced the number of muscle fibers without affecting fetal body weight (Quigley et al. 2005). The ratio of secondary to primary fiber was reduced (Quigley et al. 2005), implying that the reduction in muscle fiber number was mainly due to the reduction in secondary myogenesis.

On the other hand, at late gestation, skeletal muscle matures (approximately day 105 of gestation for sheep and day 200 of gestation for cattle) for sheep and cattle, and nutrient restriction at this stage has no major impact on the number of muscle fibers. Maternal nutrient restriction during late gestation had no effect on muscle fiber number but reduced muscle fiber size (Greenwood et al. 1999). When comparing the fetal muscle growth of sheep with single and twin pregnancy, the competition between littermates for nutrients at late gestation impacts fetal skeletal muscle mass but such restricted nutrient supply only limits muscle fiber hypertrophy, not hyperplasia (McCoard et al. 2000).

Fetal stage, specifically mid to late gestation for sheep and cattle, and late gestation to neonatal stage for rodents and pigs, is the critical stage for programming fetal skeletal muscle development. Because the number of muscle fibers is fixed at birth, muscle fiber number at birth is of critical importance (Zhu et al. 2004). The reduction in skeletal muscle mass during fetal development therefore has long-lasting, irreversible negative physiological consequences for offspring (Stannard and Johnson 2004; Zambrano et al. 2005). In our previous studies, pregnant sheep receiving 50% nutrient restriction during days 28 to 78 of gestation reduced the total number and the ratio of secondary to primary muscle fibers (Zhu et al. 2004). Lambs obtained from these ewes were raised to 8 months and then longissimus dorsi muscle was sampled for analyses. Maternal nutrient restriction during early to mid-gestation decreased the number of muscle fibers in the offspring and increased the ratio of myosin IIx to other isoforms by 17.6 ± 4.9% ($p < 0.05$) compared with offspring of control ewes. Intramuscular triglyceride content was increased in skeletal muscle of nutrient restricted lambs. In addition, lambs from dams subjected to nutrient restriction were fatter and had a lower lean/fat ratio than their counterparts (Zhu et al. 2006). These data indicate that maternal nutrient deficiency programs skeletal muscle development and has a long-term effect on offspring performance. In another study, peri-conceptional 50% nutrient restriction of pregnant sheep reduced the total muscle fiber number of fetal muscle by about 20% (Quigley et al. 2005). Maternal nutrient restriction of ewes reduced fetal carcass weight and fat content (Luther et al. 2007). In the pig, under-nutrition *in utero* resulted in low birth weight and a decrease in muscle fiber number. The progeny of under-nourished sows had a significantly lower secondary-to-primary myofiber ratio (Dwyer et al. 1994). In guinea-pig fetuses, a similar reduction in fiber number was observed in fetuses under-nourished *in utero*

(Ward and Stickland 1991). Low-energy diets from 190 day through term decreased the calf birth weights (Houghton et al. 1990). However, another study showed no differences in calf birth weights when dams were on a low plane of nutrition during the second trimester of pregnancy, but did show decreased birth weights and body weight at day 28 when dams were placed on a low plane of nutrition during the second and third trimesters of pregnancy (Freetly et al. 2000). However, the muscle fiber number was not analyzed in this study. It is expected that second trimester nutrient restriction will reduce secondary myogenesis, and thus the number of muscle fibers. In another study, 205 day adjusted weaning weights of heifer calves whose dams were supplemented with protein during late gestation were increased (Martin et al. 2007).

In our previous study, 20 Angus × Gelbvieh rotationally crossed cows carrying female fetuses were blocked by body weight and were fed in equal numbers to either meet NRC requirements to gain weight (average = +4.25% of body weight [BW], Control, C) or fed below NRC (nutrient restricted, NR) to lose weight (average = −6.8% of BW) from day 30 to day 125 of gestation. On day 125, five C and NR cows were necropsied, and the remaining five NR cows were realimented to achieve similar BW to C cows when necropsied on day 250 of gestation. The longissimus dorsi muscle of fetuses at the 12th rib was removed, fixed, and embedded in paraffin for histochemical examination. At day 125 of gestation, maternal nutrient restriction reduced the average number of muscle fibers in muscle bundles of fetal muscle; the average number of muscle fibers per bundle from C cows was 12.2 ± 0.34 while that of NR cows was 10.2 ± 0.53 ($p < 0.05$) (Figure 4.3a). Compared to the muscle of fetuses from C cows at day 250 of gestation, maternal nutrient restriction significantly increased the volume of muscle fibers and reduced the number of muscle fibers per square area in the fetuses from NR cows; the ratio of the average cross-sectional area of fetal muscle fibers from C cows versus NR cows was 1 ± 0.07 to 1.29 ± 0.14 ($p < 0.05$, Figure 4.3b). The results showed that nutrient restriction during the early gestation (day 31 to day 125) significantly affected fetal muscle development by reducing the number of muscle fibers in each muscle bundle. This reduction in the number of muscle fibers due to maternal nutrient restriction at early gestation could not be recovered by realimentation during the late stage of gestation (day 125 to day 250), and resulted in a muscle with reduced numbers of muscle fibers of larger volume (see Figure 4.3).

In addition to muscle mass, muscle fiber type composition also affects the growth potential of offspring. Type I myofibers have high catabolism rates and are less energy efficient for growth, while Type II myofibers have low catabolism rates and higher growth efficiency. In our preliminary study in the offspring of ewes receiving 50% nutrient restriction during day 28 to day 80 of gestation, the ratio of Type IIx myofibers significantly increased in skeletal muscle of lambs that experienced maternal nutrient deficiency (Zhu et al. 2006). Therefore, maternal nutrient deficiency programs skeletal muscle to a more efficient way of growth. This result is confirmed by another study in sheep (Daniel et al. 2007).

In addition to muscle growth, maternal under-nutrition also promotes the accumulation of visceral and subcutaneous fat. Nutrient restriction during mid-gestation increases the fatness of offspring and decreases the muscle mass (Zhu et al. 2006). Runt piglets experienced maternal nutrient restriction during fetal stage, and these piglets are known to be fatter and have lower growth efficiency compared to normal littermates (Wigmore and Stickland 1983).

In addition to maternal under-nutrition, maternal over-nutrition also programs fetal muscle development. In our previous study, ewes were fed 150% of nutrient requirements to day 75 of gestation. The semitendinosus muscle weight was higher in over-nourished fetuses compared to control fetuses. Although no difference in the muscle fiber number was observed, over-nutrition reduced muscle fiber density in fetuses (Zhu et al. 2008). Over-feeding of sows from conception to day 50 of gestation increased offspring fatness at birth, and the muscle fiber type composition of progeny was altered due to maternal nutritional management (Bee 2004). Rats born to mothers fed the high fat diet either during gestation alone or during both gestation and lactation exhibited a 25% reduction in muscle cross-sectional area with approximately 20% fewer fibers compared with pups fed a balanced chow diet (Bayol et al. 2005).

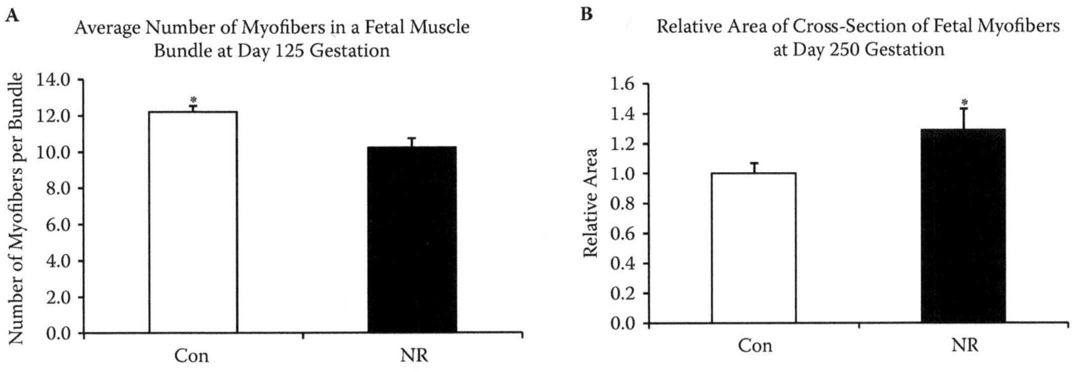

FIGURE 4.3 Fetal muscle fiber number (day 125 of gestation) and average cross-section area (day 250 of gestation) of cows receiving either 100% (Con, □) or 70% (NR, ■) nutrient requirements during day 30 to 125 of gestation. *$p < 0.05$, n = 5.

4.2.2.2 Fetal Developmental Programming of Intramuscular Adipocytes and Fibroblasts

Intramuscular adipocytes form marbling in meat while fibroblasts synthesize collagens, both of which are closely associated with meat quality. Skeletal muscle cells and adipocytes are both derived from mesoderm pluripotent stem cells. At mid-gestation, there are a large number of pluripotent cells in fetal muscle that can differentiate either into myogenic cells or adipogenic cells. The formation of adipocytes from the mesoderm begins around mid-gestation (Feve 2005; Gnanalingham et al. 2005; Muhlhausler et al. 2006). Adipose tissue growth in later life is due to both hypertrophy and hyperplasia (Feve 2005). However, new adipocytes generated later in life are mainly located in visceral and retroperitoneal fat depots and also in subcutaneous fat depots, with very few located in intramuscular fat (Faust et al. 1978; Miller et al. 1984; Valet et al. 2002). Thus, adipogenesis inside muscle during the fetal stage has a dominant effect on the number of adipocytes existing inside muscle. Enhanced adipogenesis in fetal muscle of over-nourished mothers produces a large number of adipocytes in skeletal muscle, which predisposes offspring muscle to accumulate intramuscular fat due to the hypertrophy of existing adipocytes. Mechanisms controlling adipogenesis in fetal muscle *in vivo* are poorly defined although there are numerous *in vitro* cell culture studies (Rosen and MacDougald 2006). In these studies, pre-adipocytes 3T3 and mesenchymal stem cells 10T1/2 cell lines are commonly used. These studies have demonstrated that peroxisome proliferators-activated receptor γ (PPARγ) and CCAAT-enhancer-binding proteins (C/EBPs) are crucial intracellular factors controlling adipogenesis. Their expression leads to adipogenesis from pluripotent cells (Cho and Jefcoate 2004; Giri et al. 2006).

We conducted a preliminary study to assess the impact of maternal over-nutrition on the adipogenesis and lipid content in fetal muscle. From 60 days before conception to day 75 of gestation, multiparous ewes were fed either a highly palatable diet at 100% (Con) of the National Research Council (NRC) recommendations or 150% (OB) of the NRC recommendations (NRC 1985). Fetal longissimus dorsi muscle was collected for biochemical analyses. Data showed that maternal over-nutrition altered fetal skeletal muscle development by increasing extracellular space and adipogenesis (Zhu et al. 2008), strongly suggesting that adipogenesis and fibrogenesis might be altered due to maternal over-nutrition. In fetal muscle at mid-gestation, there are no mature adipocytes (Casteilla et al. 1987; Lomax et al. 2007). However, the lack of mature adipocytes does not exclude the cells that have progressed through differentiation to the point of being equipped at the molecular level to accumulate triacylglycerols. Indeed, the initial events in adipogenesis start in mid-gestation (Feve 2005; Gnanalingham et al. 2005; Muhlhausler et al. 2006). PPARγ is a key regulator of adipogenesis. The expression of PPARγ leads to adipogenic differentiation from pluripotent cells, and PPARγ is highly expressed in adipose tissue (Spiegelman et al. 2000). Both mRNA and protein of PPARγ were higher in OB fetal skeletal muscle (Tong et al. 2008). Because PPARγ is a marker of adipocyte differentiation, these data indicated enhanced adipogenesis in OB fetal muscle. Of course, PPARγ is also expressed in muscle cells but the level of expression is very low (Verma et al. 2004; Vidal-Puig et al. 1996). To further evaluate adipogenesis in fetal muscle, the content of a pre-adipocyte marker, preadipocyte factor-1 (Pref-1), was analyzed (Kim et al. 2007; Smas and Sul 1993). Preadipocyte factor-1 is exclusively expressed in pre-adipocytes, not mature adipocytes and other cells (Fahrenkrug et al. 1999; Mei et al. 2002). Its expression was higher in over-nourished fetal skeletal muscle compared to Con fetal skeletal muscle, showing that a higher number of pluripotent cells had committed to adipogenesis in over-nourished fetuses (Tong et al. 2008). These data clearly show that adipogenesis was enhanced in over-nourished fetal skeletal muscle. In another study, cattle were fed 100% nutrient requirements (Con), 70% of nutrient requirements (NR), and 70% nutrient requirement plus supplementation of ruminal bypass protein (Prot) from day 60 to day 180 of gestation. The offspring steers of Prot had lower marbling compared to Con steers, and the subcutaneous fat was lower (unpublished data), showing maternal protein supplementation down-regulates adipogenesis in fetal skeletal muscle.

Nutritional management to affect adipogenesis in fetal muscle can also be extended to the postnatal stage. The abundance of pluripotent cells in skeletal muscle declines as animals develop. Therefore, the effectiveness of nutritional management to enhance adipogenesis in skeletal muscle (marbling) is: fetal stage > neonatal stage > early weaning stage (150 to 250 days of age) > weaning and older stages. After 250 days of age, nutritional supplementation will be largely ineffective in increasing the number of intramuscular adipocytes due to the depletion of pluripotent cells, although the size of existing intramuscular adipocytes can be increased, a reason for the enhancement in marbling during fattening. Currently, the marbling window, the window for nutritional management of cattle to enhance marbling, is between 150 and 250 days after birth (early weaning stage) (Corah and McCully 2007; Pyatt et al. 2005a, b; Wertz et al. 2002; Wertz et al. 2001). There are studies indicating that early weaning and feeding a grain, especially corn-based, diet increases marbling (Pyatt et al. 2005a, b; Wertz et al. 2002; Wertz et al. 2001). Nutritional supplementation of calves earlier than 150 day of age is more difficult due to the fact that calves are nursing and consuming significant quantities of milk prior to 150 days of age. On the other hand, fetal stages provide a unique, and perhaps the best, window for nutritional management. This is due to at least three advantages: (1) an abundant amount of pluripotent cells exist in fetal muscle and therefore it is expected that nutritional manipulations will be very effective because diverging pluripotent cells in fetal muscle to adipogenesis will dramatically increase marbling; (2) fetal nutrients are derived from maternal circulation, allowing maternal supplemented nutrients to be effectively transferred to the fetus, thus promoting adipogenesis in fetal muscle; and (3) this stage is a very convenient stage for cow/calf producers to nutritionally manage their cows.

Limited studies also indicate that maternal nutrition affects the collagen content in skeletal muscle. Runt piglets experienced maternal nutrient restriction. Compared to their counterparts, grown runt pigs have higher concentrations of collagen in skeletal muscle (Karunaratne et al. 2005). More studies on the association among maternal nutrition, fibrogenesis, and collagen accumulation in offspring muscle are apparently needed.

4.3 MECHANISMS UNDERLYING FETAL SKELETAL MUSCLE PROGRAMMING

4.3.1 Wnt Signaling and Fetal Muscle Development

The Wnt family has more than ten members that act by binding to Frizzled (Fzd) receptors on target cells (Polesskaya et al. 2003), which in turn activate β-catenin-dependent and/or -independent signaling pathways (Huelsken and Birchmeier 2001). Binding of Wnt to Fzd activates Disheveled (Dvl), leading to the phosphorylation and inactivation of glycogen synthase kinase-3β (GSK-3β). The increased phosphorylation of GSK-3β results in its target β-catenin stabilization and entering the nucleus, thereby acting in a complex with members of the T cell factor/Lymphoid enhancer factor (TCF/LEF) family of transcription factors to activate target genes (Dierick and Bejsovec 1999; Hecht and Kemler 2000), including Myf5 and MyoD (Cossu and Borello 1999).

β-Catenin plays an essential role in the canonical Wnt signaling pathway that regulates the embryonic, postnatal, and oncogenic growth of many tissues (Armstrong and Esser 2005; Mermelstein et al. 2007). It has been suggested that β-catenin regulates the expression of transcription factors Pax3 as well as Gli (Borycki et al. 2000; Capdevila et al. 1998). Pax3 is essential for skeletal myogenesis and acts upstream of MyoD during skeletal muscle development, while Gli factors play an essential role in Myf-5 expression (Gustafsson et al. 2002; Ridgeway and Skerjanc 2001). The expression of MRFs is controlled by signals such as Wnts and Sonic hedgehog (Shh) (Munsterberg et al. 1995; Petropoulos and Skerjanc 2002; Stern et al. 1995) via Pax3 and Pax7 (Kassar-Duchossoy et al. 2005). Overexpression of β-catenin in myocytes is sufficient to support cellular growth by a Wnt-independent mechanism (Haq et al. 2003). Activation of the Wnt signaling pathway enhances myogenesis and inhibits adipogenesis in cultured mesenchymal stem cells (Shang et al. 2007). Blocking the β-catenin pathway reduced the total number of myocytes (Pan et al. 2005; Yamanouchi et al. 2007). In our previous study in ewes that were fed 150% of nutrient requirements to day 75 of gestation (Tong et al. 2008), Wnt/β-catenin signaling was down-regulated in OB fetal muscle, which should be at least partially responsible for the down-regulation of myogenesis, but up-regulation of adipogenesis (Tong et al. 2008; Zhu et al. 2008).

4.3.2 Epigenetic Modifications

Because myogenesis, adipogenesis, and fibrogenesis from mesenchymal stem cells are controlled by the expression of one or more crucial genes, maternal nutrition might change fetal muscle development through epigenetic modifications. The development of the human body starts from one fertilized cell, which generates an increasing body of diversified cells sharing the same copy of genes. Silencing genes while allowing other genes to express is essential for maintaining the diversity of cells, which is accomplished through epigenetic modifications (Schuettengruber et al. 2007). By definition, epigenetic heritance occurs when the modifications, such as cytosine methylation and histone modifications, pass from parent cells to daughter cells (Bernstein et al. 2007). Depending on the nature of modifications, epigenetic modifications have different degrees of plasticity. Histone modification usually only passes for several cell generations (Rando and Verstrepen 2007), but histone modifications can guide DNA methylation, leading to stable alterations in gene expression (Vire et al. 2006). DNA methylation passes on for more than 100 cell generations; considering that only 44 cell generations are needed to form a human body, epigenetic modification through DNA methylation is very stable (Reik 2007).

Polycomb group proteins (PcG) and trithorax group proteins (trxG) regulate histone methylation, which leads to other epigenetic modifications during cell differentiation (Reik 2007). PcG and trxG regulate the methylation of histone H3 through binding to PcG and trxG response elements in the genome. PcG group proteins possess H3K27-specific trimethylase activity, which mediates gene expression repression, whereas trxG complexes have H3K4 trimethylase activity, which mediates activation of genes (Schuettengruber et al. 2007). The crucial development is the demonstration that PcG-mediated gene repression leads to DNA methylation of the targeted genes (Vire et al. 2006). The PcG protein EZH2 (Enhancer of Zeste homolog 2) interacts with DNA methyltransferases and serves as a recruitment platform for DNA methyltransferases, thus converting plastic histone modifications to stable DNA methylation (Vire et al. 2006). DNA methylation leads to the silence of genes through several mechanisms: (1) recruitment of histone deacetylases, which removes histone acetylation (because acetylation of the lysine residues at the histone neutralizes its positive charges, de-acetylation increases the affinity between histones and DNA, thereby inhibiting gene expression); (2) DNA methylation can directly interfere with the binding of transcription factors; and (3) DNA methylation leads to the formation of inactive chromatin structure.

Currently, no studies are available linking maternal nutrition to epigenetic modifications in fetal muscle. However, indirect evidences support epigenetic modification in key genes controlling fetal muscle development. Maternal under-nutrition permanently changed the insulin/insulin-like growth factor-1 signaling in fetal muscle (Ozanne et al. 1996), very likely through epigenetic modifications. Maternal diet alters the expression of PPARs in fetal muscle through DNA methylation (Rees et al. 2008). In another study, maternal cocaine administration caused epigenetic modification of a key protein kinase gene in rat heart (Zhang et al. 2007).

4.4 UNDER-NUTRITION DURING GESTATION WIDELY EXISTS IN THE BEEF INDUSTRY AND HAS IMPLICATION ON BEEF QUALITY

4.4.1 Cows in Rocky Mountain Area Frequently Experience Malnutrition during Gestation due to Drought

Animal production is the main component of agriculture in the Rocky Mountain states, with cattle as the dominant livestock. Drought is a recurring meteorological phenomenon in these states and surrounding areas, which results in arid and semi-arid conditions that cause a significant reduction in forage production (NWS 1988–1989; USGS 2004). A persistent drought has been observed since 2000 in Wyoming, Colorado, Utah, New Mexico, Nevada, Arizona, and part of California (USGS 2004). The limited availability of forage and its low protein content affect the overall production of ruminant animals (Hess et al. 1994; Smith 2003). Due to the seasonal nature of cow reproduction, cows in the Rocky Mountains area frequently experience nutrient restriction during early to mid-gestation, especially protein deficiency, which occurs due to the limited forage available on small farms and ranches, poor forage quality, and also the lack of nutritional supplements, which are rarely employed on small farms and ranches during this gestation stage (Enk et al. 2001; Jensen et al. 2002; Thomas and Kott 1995), although such supplementation to cows is often applied at late-stage gestation (last 60 to 90 days before calves) (Hall 1997). As discussed previously, such nutrient deficiency during gestation programs fetal skeletal muscle development, likely including a reduction in muscle fiber numbers and an increase in fatness.

4.4.2 Grass-Feeding and Fetal Programming of Skeletal Muscle Development

With the increase in grain prices, grass feeding is becoming more and more attractive (Roosevelt 2006). Most grass (forage)-fed beef producers are also cow/calf operators, where they grow calves to market weight and send them to slaughter directly. Cows/calves commonly graze on rangelands and are fed a forage-based diet. Grass-fed beef contains higher ratios of unsaturated fatty acids, especially n-3 fatty acids, which are good for health (Mann et al. 2003; Ponnampalam et al. 2006). In addition, grass-fed beef is produced naturally and has minimal environmental impacts. As a result, both consumer and producer's interest in grass-fed beef is increasing nationwide. Furthermore, grass-fed beef production is likely to become increasingly attractive to ranchers if grain prices are continuously high in the future. However, the poor flavor of grass-fed beef is the major problem hampering its wide acceptance (Mandell et al. 1998; Sitz et al. 2005). The flavor of beef steak mainly comes from intramuscular fat (marbling), and grass-fed beef lacks marbling, which weakens its flavor and reduces its juiciness. Therefore, if we can effectively enhance marbling in grass-fed beef, we can significantly improve its eating quality, further expanding this segment of the beef industry.

Grass-fed cattle can only be fed grass, and thus we cannot use dietary supplementation of calves and steers to enhance marbling. However, marbling is correlated with the number of adipocytes in skeletal muscle. Increased adipogenesis in fetal muscle leads to an increase in intramuscular adipocytes. These intramuscular adipocytes accumulate fat during postnatal growth, forming marbling. In grain-fed beef, cattle are fed a high-energy, grain-based diet that fattens the animal and increases the size of intramuscular adipocytes, thus enhancing marbling. Because grass-fed beef cattle are not fed a grain-based diet, it is less likely that adipocyte size will increase during the growing and "finishing" phase. Therefore, maternal nutritional management to increase adipogenesis in fetal muscle is an alternative, maybe the only choice to enhance marbling. Our preliminary studies have demonstrated that fetal stage is crucial for the formation of intramuscular adipocytes, and their formation (adipogenesis) can be effectively enhanced through maternal nutritional supplementation (Tong et al. 2008). Because grass-fed beef producers are also cow/calf operators, they are in a unique position to manage maternal nutrition in an effort to enhance intramuscular fat in calves.

Another approach to enhance adipogenesis in fetal muscle is to improve pasture conditions. In our previous study, cows were grazed on either improved pasture (IP, n = 8) or native range pasture (NP, n = 7) from 120 to 150 through 180 to 210 days of gestation. The chemical ether extract of the longissimus dorsi muscle at the 12th rib tended ($p = 0.06$) to be higher in the IP when compared with the steers born to mothers gestated on NP (unpublished data). Adipose tissue cell number per field tended ($p = 0.09$) to be greater for IP steers than for NR steers (unpublished data). Although the number of cows used in this preliminary study was very small, these data nevertheless suggest that intramuscular adipogenesis and marbling can be increased through improving the gestational plane of nutrition — in this case, improving pasture condition.

4.5 SUMMARY

Available studies clearly show that maternal nutrition affects fetal muscle development, which has long-term effects on the properties of offspring skeletal muscle and meat quality. Maternal undernutrition during fetal stage reduces skeletal muscle fiber number, alters muscle fiber composition, increases fatness, and reduces the growth performance of progeny. On the other hand, maternal over-nutrition promotes adipogenesis in fetal muscle, which might enhance marbling in offspring meat. Future studies should focus on the identification of underlying mechanisms associated with fetal developmental programming of skeletal muscle development, such as analyzing epigenetic modifications in crucial genes regulating myogenesis, adipogenesis, and fibrogenesis, and the resulting skeletal muscle growth and development and meat quality.

REFERENCES

Agbulut, O., P. Noirez, F. Beaumont, and G. Butler-Browne. 2003. Myosin heavy chain isoforms in postnatal muscle development of mice. *Biol. Cell* 95:399–406.

Armstrong, D.D. and K.A. Esser. 2005. Wnt/beta-catenin signaling activates growth-control genes during overload-induced skeletal muscle hypertrophy. *Am. J. Physiol. Cell Physiol.* 289:C853–859.

Atchley, W.R. and W.M. Fitch. 1997. A natural classification of the basic helix-loop-helix class of transcription factors. *Proc. Natl. Acad. Sci. U.S.A.* 94:5172–5176.

Barker, D.J. 1995. Fetal origins of coronary heart disease. *BMJ* 311:171–174.

Barker, D.J. 2001. A new model for the origins of chronic disease. *Med. Health Care Philos.* 4:31–35.

Barker, D.J. 2003. Coronary heart disease: a disorder of growth. *Horm. Res.* 59(Suppl. 1):35–41.

Barker, D.J., J.G. Eriksson, T. Forsen, and C.Osmond. 2002. Fetal origins of adult disease: strength of effects and biological basis. *Int. J. Epidemiol.* 31:1235–1239.

Bayol, S.A., B.H. Simbi, and N.C. Stickland. 2005. A maternal cafeteria diet during gestation and lactation promotes adiposity and impairs skeletal muscle development and metabolism in rat offspring at weaning. *J. Physiol.* 567:951–961.

Bee, G. 2004. Effect of early gestation feeding, birth weight, and gender of progeny on muscle fiber characteristics of pigs at slaughter. *J. Anim. Sci.* 82:826–836.

Beermann, D.H., Cassens R.G., Hausman, G.J. 1978. A second look at fiber type differentiation in porcine skeletal muscle. *J. Anim. Sci.* 46:125–132.

Benyshek, D.C., C.S. Johnston, and J.F. Martin. 2004. Post-natal diet determines insulin resistance in fetally malnourished, low birthweight rats (F1) but diet does not modify the insulin resistance of their offspring (F2). *Life Sci.* 74:3033–3041.

Bernstein, B.E., A. Meissner, and E.S. Lander. 2007. The mammalian epigenome. *Cell* 128:669–681.

Bieswal, F., M.T. Ahn, B. Reusens, P. Holvoet, M. Raes, W.D. Rees, and C. Remacle. 2006. The importance of catch-up growth after early malnutrition for the programming of obesity in male rat. *Obesity (Silver Spring)* 14:1330–1343.

Bispham, J., G.S. Gopalakrishnan, J. Dandrea, V. Wilson, H. Budge, D.H. Keisler, F. Broughton Pipkin, T. Stephenson, and M.E. Symonds. 2003. Maternal endocrine adaptation throughout pregnancy to nutritional manipulation: consequences for maternal plasma leptin and cortisol and the programming of fetal adipose tissue development. *Endocrinology* 144:3575–3585.

Borycki, A., A.M. Brown, and C.P. Emerson, Jr. 2000. Shh and Wnt signaling pathways converge to control Gli gene activation in avian somites. *Development* 127:2075–2087.

Capdevila, J., C. Tabin, and R.L. Johnson. 1998. Control of dorsoventral somite patterning by Wnt-1 and beta-catenin. *Dev. Biol.* 193:182–194.

Casteilla, L., C. Forest, J. Robelin, D. Ricquier, A. Lombet, and G. Ailhaud. 1987. Characterization of mitochondrial-uncoupling protein in bovine fetus and newborn calf. *Am. J. Physiol.* 252:E627–636.

Cho, Y.C. and C.R. Jefcoate. 2004. PPARgamma1 synthesis and adipogenesis in C3H10T1/2 cells depends on s-phase progression, but does not require mitotic clonal expansion. *J. Cell Biochem.* 91:336–353.

Corah, L. and M. McCully. 2007. Declining quality grades: a review of factors reducing marbling deposition in beef cattle. *Certified Angus Beef News Releases-Research.*

Cossu, G. and U. Bore'llo. 1999. Wnt signaling and the activation of myogenesis in mammals. *EMBO J.* 18:6867–6872.

Daniel, Z.C., J.M. Brameld, J. Craigon, N.D. Scollan, and P.J. Buttery. 2007. Effect of maternal dietary restriction during pregnancy on lamb carcass characteristics and muscle fiber composition. *J. Anim. Sci.* 85:1565–1576.

Desai, M., D. Gayle, J. Babu, and M.G. Ross. 2005. Programmed obesity in intrauterine growth-restricted newborns: modulation by newborn nutrition. *Am. J. Physiol. Regul. Integr. Comp. Physiol.* 288:R91–R96.

Dierick, H. and A. Bejsovec. 1999. Cellular mechanisms of wingless/Wnt signal transduction. *Curr. Top. Dev. Biol.* 43:153–190.

Du, M., M.J. Zhu, W.J. Means, B.W. Hess, and S.P. Ford. 2005. Nutrient restriction differentially modulates the mammalian target of rapamycin signaling and the ubiquitin-proteasome system in skeletal muscle of cows and their fetuses. *J. Anim. Sci.* 83:117–123.

Dwyer, C.M., N.C. Stickland, and J.M. Fletcher. 1994. The influence of maternal nutrition on muscle fiber number development in the porcine fetus and on subsequent postnatal growth. *J. Anim. Sci.* 72:911–917.

Enk, T.A., H.D. Picton, and J.S. Williams. 2001. Factors limiting a bighorn sheep population in Montana following a dieoff. *Northwest Sci.* 73:280–291.

Fahrenkrug, S.C., B.A. Freking, and T.P. Smith. 1999. Genomic organization and genetic mapping of the bovine PREF-1 gene. *Biochem. Biophys. Res. Commun.* 264:662–667.

Faust, I.M., P.R. Johnson, J.S. Stern, and J. Hirsch. 1978. Diet-induced adipocyte number increase in adult rats: a new model of obesity. *Am. J. Physiol.* 235:E279–286.

Fazeli, S., D.J. Wells, C. Hobbs, and F.S. Walsh. 1996. Altered secondary myogenesis in transgenic animals expressing the neural cell adhesion molecule under the control of a skeletal muscle alpha-actin promoter. *J. Cell Biol.* 135:241–251.

Fernandez-Twinn, D.S., S. Ekizoglou, A. Wayman, C.J. Petry, and S.E. Ozanne. 2006. Maternal low-protein diet programs cardiac beta-adrenergic response and signaling in 3-mo-old male offspring. *Am. J. Physiol. Regul. Integr. Comp. Physiol.* 291:R429–436.

Fernandez-Twinn, D.S., A. Wayman, S. Ekizoglou, M.S. Martin, C.N. Hales, and S.E. Ozanne. 2005. Maternal protein restriction leads to hyperinsulinemia and reduced insulin-signaling protein expression in 21-mo-old female rat offspring. *Am. J. Physiol. Regul. Integr. Comp. Physiol.* 288:R368–373.

Feve, B. 2005. Adipogenesis: cellular and molecular aspects. *Best Pract. Res. Clin. Endocrinol. Metab.* 19:483–499.

Freetly, H.C., C.L. Ferrell, and T.G. Jenkins. 2000. Timing of realimentation of mature cows that were feed-restricted during pregnancy influences calf birth weights and growth rates. *J. Anim. Sci.* 78:2790–2796.

Giri, S., R. Rattan, E. Haq, M. Khan, R. Yasmin, J.S. Won, L. Key, A.K. Singh, and I. Singh. 2006. AICAR inhibits adipocyte differentiation in 3T3L1 and restores metabolic alterations in diet-induced obesity mice model. *Nutr. Metab. (London)* 3:31–51.

Gnanalingham, M.G., A. Mostyn, M.E. Symonds, and T. Stephenson. 2005. Ontogeny and nutritional programming of adiposity in sheep: potential role of glucocorticoid action and uncoupling protein-2. *Am. J. Physiol. Regul. Integr. Comp. Physiol.* 289:R1407–R1415.

Greenwood, P.L., R.M. Slepetis, J.W. Hermanson, and A.W. Bell. 1999. Intrauterine growth retardation is associated with reduced cell cycle activity, but not myofibre number, in ovine fetal muscle. *Reprod. Fertil. Dev.* 11:281–291.

Gustafsson, M.K., H. Pan, D.F. Pinney, Y. Liu, A. Lewandowski, D.J. Epstein, and C.P. Emerson, Jr. 2002. *Myf5* is a direct target of long-range Shh signaling and Gli regulation for muscle specification. *Genes Dev.* 16:114–126.

Hales, C.N., D.J. Barker, P.M. Clark, L.J. Cox, C. Fall, C. Osmond, and P.D. Winter. 1991. Fetal and infant growth and impaired glucose tolerance at age 64. *BMJ* 303:1019–1022.

Hall, J.B. 1997. The Cow-Calf Manager: Winter Feeding and Supplements. Virginia Polytechnic Institute and State University, Blacksburg.

Haq, S., A. Michael, M. Andreucci, K. Bhattacharya, P. Dotto, B. Walters, J. Woodgett, H. Kilter, and T. Force. 2003. Stabilization of beta–catenin by a Wnt-independent mechanism regulates cardiomyocyte growth. *Proc. Natl. Acad. Sci. U.S.A.* 100:4610–4615.

Hecht, A. and R. Kemler. 2000. Curbing the nuclear activities of beta-catenin. Control over Wnt target gene expression. *EMBO Rep.* 1:24–28.

Hess, B.W., K.K. Park, L.J. Krysl, M.B. Judkins, B.A. McCracken, and D.R. Hanks. 1994. Supplemental protein for beef cattle grazing dormant intermediate wheatgrass pasture: effects on nutrient quality, forage intake, digesta kinetics, grazing behavior, ruminal fermentation, and digestion. *J. Anim. Sci.* 72:2113–2123.

Houghton, P.L., R.P. Lemenager, L.A. Horstman, K.S. Hendrix, and G.E. Moss. 1990. Effects of body composition, pre- and postpartum energy level and early weaning on reproductive performance of beef cows and preweaning calf gain. *J. Anim. Sci.* 68:1438–1446.

Huelsken, J. and W. Birchmeier. 2001. New aspects of Wnt signaling pathways in higher vertebrates. *Curr. Opin. Genet. Dev.* 11:547–553.

Hyatt, J.P., G.E. McCall, E.M. Kander, H. Zhong, R.R. Roy, and K.A. Huey. 2008. Pax3/7 expression coincides with MyoD during chronic skeletal muscle overload. *Muscle Nerve* 38:861–866.

Jensen, K.B., D.A. Johnson, K.H. Asay, and K.C. Olson. 2002. Seasonal-accumulated growth and forage quality of range grasses for fall and winter grazing. *Canad. J. Plant Sci.* 82:329–336.

Karunaratne, J.F., C.J. Ashton, and N.C. Stickland. 2005. Fetal programming of fat and collagen in porcine skeletal muscles. *J. Anat.* 207:763-768.

Kassar–Duchossoy, L., E. Giacone, B. Gayraud-Morel, A. Jory, D. Gomes, and S. Tajbakhsh. 2005. Pax3/Pax7 mark a novel population of primitive myogenic cells during development. *Genes Dev.* 19:1426–1431.

Kim, K.A., J.H. Kim, Y. Wang, and H.S. Sul. 2007. Pref-1 (preadipocyte factor 1) activates the MEK/extracellular signal-regulated kinase pathway to inhibit adipocyte differentiation. *Mol. Cell Biol.* 27:2294–2308.

King, J.C. 2006. Maternal obesity, metabolism, and pregnancy outcomes. *Annu. Rev. Nutr.* 26:271–291.

Lemercier, C., R.Q. To, R.A. Carrasco, and S.F. Konieczny. 1998. The basic helix-loop-helix transcription factor MIST1 functions as a transcriptional repressor of MyoD. *EMBO J.* 17:1412–1422.

Lomax, M.A., F. Sadiq, G. Karamanlidis, A. Karamitri, P. Trayhurn, and D.G. Hazlerigg. 2007. Ontogenic loss of brown adipose tissue sensitivity to beta-adrenergic stimulation in the ovine. *Endocrinology* 148:461–468.

Luther, J., R. Aitken, J. Milne, M. Matsuzaki, L. Reynolds, D. Redmer, and J. Wallace. 2007. Maternal and fetal growth, body composition, endocrinology, and metabolic status in undernourished adolescent sheep. *Biol. Reprod.* 77:343–350.

Mandell, I.B., J.G. Buchanan-Smith, and C.P. Campbell. 1998. Effects of forage vs grain feeding on carcass characteristics, fatty acid composition, and beef quality in Limousin-cross steers when time on feed is controlled. *J. Anim. Sci.* 76:2619–2630.

Mann, N.J., E.N. Ponnampalam, Y. Yep, and A.J. Sinclair. 2003. Feeding regimes affect fatty acid composition in Australian beef cattle. *Asia Pac. J. Clin. Nutr.* 12(Suppl.):S38.

Maroto, M., R. Reshef, A.E. Munsterberg, S. Koester, M. Goulding, and A.B. Lassar. 1997. Ectopic Pax-3 activates MyoD and Myf-5 expression in embryonic mesoderm and neural tissue. *Cell* 89:139–148.

Martin, J.L., K.A. Vonnahme, D.C. Adams, G.P. Lardy, and R.N. Funston. 2007. Effects of dam nutrition on growth and reproductive performance of heifer calves. *J. Anim. Sci.* 85:841–847.

McCoard, S.A., W.C. McNabb, M.J. Birtles, P.M. Harris, S.N. McCutcheon, and S.W. Peterson. 2001. Immunohistochemical detection of myogenic cells in muscles of fetal and neonatal lambs. *Cells Tissues Organs* 169:21–33.

McCoard, S.A., W.C. McNabb, S.W. Peterson, S.N. McCutcheon, and P.M. Harris. 2000. Muscle growth, cell number, type and morphometry in single and twin fetal lambs during mid to late gestation. *Reprod. Fertil. Dev.* 12:319–327.

Mei, B., L. Zhao, L. Chen, and H.S. Sul. 2002. Only the large soluble form of preadipocyte factor-1 (Pref-1), but not the small soluble and membrane forms, inhibits adipocyte differentiation: role of alternative splicing. *Biochem. J.* 364:137–144.

Mermelstein, C.S., D.M. Portilho, F.A. Mendes, M.L. Costa, and J.G. Abreu. 2007. Wnt/beta-catenin pathway activation and myogenic differentiation are induced by cholesterol depletion. *Differentiation* 75:184–192.
Miller, W.H., Jr., I.M. Faust, and J. Hirsch. 1984. Demonstration of *de novo* production of adipocytes in adult rats by biochemical and radioautographic techniques. *J. Lipid Res.* 25:336–347.
Muhlhausler, B.S., J.A. Duffield, and I.C. McMillen. 2006. Increased maternal nutrition stimulates peroxisome proliferator activated receptor-γ, adiponectin and leptin mrna expression in adipose tissue before birth. *Endocrinology* 148:2878–2885.
Munsterberg, A.E., J. Kitajewski, D.A. Bumcrot, A.P. McMahon, and A.B. Lassar. 1995. Combinatorial signaling by sonic hedgehog and Wnt family members induces myogenic bHLH gene expression in the somite. *Genes Dev.* 9:2911–2922.
NRC. 1985. *National Research Council. Nutrient Requirements of Sheep,* 6th ed. National Academy Press, Washington, D.C.
NWS. 1988–1989. National Water Summary — Floods and Droughts: Wyoming.
Osgerby, J.C., D.C. Wathes, D. Howard, and T.S. Gadd. 2002. The effect of maternal undernutrition on ovine fetal growth. *J. Endocrinol.* 173:131–141.
Ozanne, S.E., C.L. Wang, N. Coleman, and G.D. Smith. 1996. Altered muscle insulin sensitivity in the male offspring of protein-malnourished rats. *Am. J. Physiol.* 271:E1128-1134.
Pan, W., Y. Jia, J. Wang, D. Tao, X. Gan, L. Tsiokas, N. Jing, D. Wu, and L. Li. 2005. Beta-catenin regulates myogenesis by relieving I-mfa-mediated suppression of myogenic regulatory factors in P19 cells. *Proc. Natl. Acad. Sci. U.S.A.* 102:17378–17383.
Pavlath, G.K. and V. Horsley. 2003. Cell fusion in skeletal muscle — central role of NFATC2 in regulating muscle cell size. *Cell Cycle* 2:420–423.
Petropoulos, H. and I.S. Skerjanc. 2002. Beta-catenin is essential and sufficient for skeletal myogenesis in P19 cells. *J. Biol. Chem.* 277:15393–15399.
Polesskaya, A., P. Seale, and M.A. Rudnicki. 2003. Wnt signaling induces the myogenic specification of resident CD45+ adult stem cells during muscle regeneration. *Cell* 113:841–852.
Ponnampalam, E.N., N.J. Mann, and A.J. Sinclair. 2006. Effect of feeding systems on omega-3 fatty acids, conjugated linoleic acid and trans fatty acids in Australian beef cuts: potential impact on human health. *Asia Pac. J. Clin. Nutr.* 15:21–29.
Pyatt, N.A., L.L. Berger, D.B. Faulkner, P.M. Walker, and S.L. Rodriguez-Zas. 2005a. Factors affecting carcass value and profitability in early-weaned Simmental steers. I. Five-year average pricing. *J. Anim. Sci.* 83:2918–2925.
Pyatt, N.A., L.L. Berger, D.B. Faulkner, P.M. Walker, and S.L. Rodriguez-Zas. 2005b. Factors affecting carcass value and profitability in early-weaned Simmental steers. II. Days on feed endpoints and sorting strategies. *J. Anim. Sci.* 83:2926–2937.
Quigley, S.P., D.O. Kleemann, M.A. Kakar, J.A. Owens, G.S. Nattrass, S. Maddocks, and S.K. Walker. 2005. Myogenesis in sheep is altered by maternal feed intake during the peri-conception period. *Anim. Reprod. Sci.* 87:241–251.
Rando, O.J. and K.J. Verstrepen. 2007. Timescales of genetic and epigenetic inheritance. *Cell* 128:655–668.
Rees, W.D., C.J. McNeil, and C.A. Maloney. 2008. The roles of PPARs in the fetal origins of metabolic health and disease. *PPAR Res.* 2008:459030.
Reik, W. 2007. Stability and flexibility of epigenetic gene regulation in mammalian development. *Nature* 447:425–432.
Ridgeway, A.G. and I.S. Skerjanc. 2001. Pax3 is essential for skeletal myogenesis and the expression of Six1 and Eya2. *J. Biol. Chem.* 276:19033–19039.
Roosevelt, M. 2006. The grass-fed revolution. Beef raised wholly on pasture, rather than grain-fed in feedlots, may be better for your health — and for the planet. *Time* 167:76–78.
Rosen, E.D. and O.A. MacDougald. 2006. Adipocyte differentiation from the inside out. *Nat. Rev. Mol. Cell Biol.* 7:885–896.
Roth, J.F., N. Shikama, C. Henzen, I. Desbaillets, W. Lutz, S. Marino, J. Wittwer, H. Schorle, M. Gassmann, and R. Eckner. 2003. Differential role of p300 and CBP acetyltransferase during myogenesis: p300 acts upstream of MyoD and Myf5. *EMBO J.* 22:5186–5196.
Schuettengruber, B., D. Chourrout, M. Vervoort, B. Leblanc, and G. Cavalli. 2007. Genome regulation by polycomb and trithorax proteins. *Cell* 128:735–745.
Schultz, E. 1976. Fine structure of satellite cells in growing skeletal muscle. *Am. J. Anat.* 147:49–70.
Shang, Y.C., C. Zhang, S.H. Wang, F. Xiong, C.P. Zhao, F.N. Peng, S.W. Feng, M.J. Yu, M.S. Li, and Y.N. Zhang. 2007. Activated beta-catenin induces myogenesis and inhibits adipogenesis in BM-derived mesenchymal stromal cells. *Cytotherapy* 9:667–681.

Sitz, B.M., C.R. Calkins, D.M. Feuz, W.J. Umberger, and K.M. Eskridge. 2005. Consumer sensory acceptance and value of domestic, Canadian, and Australian grass-fed beef steaks. *J. Anim. Sci.* 83:2863–2868.

Smas, C.M. and H.S. Sul. 1993. Pref-1, a protein containing EGF-like repeats, inhibits adipocyte differentiation. *Cell* 73:725–734.

Smith, M.A. 2003. Sare Professional Development Program Annual Report for Wyoming. http://wsare.usu.edu/prodev/statereports/WY02.pdf.

Spiegelman, B.M., P. Puigserver, and Z. Wu. 2000. Regulation of adipogenesis and energy balance by PPAR-γ and PGC-1. *Int. J. Obes. Relat. Metab. Disord.* 24(Suppl. 4):S8–10.

Stannard, S.R. and N.A. Johnson. 2004. Insulin resistance and elevated triglyceride in muscle: more important for survival than "thrifty" genes? *J. Physiol.* 554:595–607.

Stern, H.M., A.M. Brown, and S.D. Hauschka. 1995. Myogenesis in paraxial mesoderm: preferential induction by dorsal neural tube and by cells expressing Wnt-1. *Development* 121:3675–3686.

Swatland, H.J. 1973. Muscle growth in the fetal and neonatal pig. *J. Anim. Sci.* 37:536–545.

Symonds, M.E., S. Pearce, J. Bispham, D.S. Gardner, and T. Stephenson. 2004. Timing of nutrient restriction and programming of fetal adipose tissue development. *Proc. Nutr. Soc.* 63:397–403.

Thomas, V.M. and R.W. Kott. 1995. A review of Montana winter range ewe nutrition research. *Sheep & Goat Res. J.* 11:17–24.

Tong, J., M.J. Zhu, K.R. Underwood, B.W. Hess, S.P. Ford, and M. Du. 2008. AMP-activated protein kinase and adipogenesis in sheep fetal skeletal muscle and 3T3-L1 cells. *J. Anim. Sci.* 86:1296–1305.

USGS. 2004. Climatic Fluctuations, Drought, and Flow in the Colorado River Basin.

Valet, P., G. Tavernier, I. Castan-Laurell, J.S. Saulnier-Blache, and D. Langin. 2002. Understanding adipose tissue development from transgenic animal models. *J. Lipid Res.* 43:835–860.

Verma, N.K., J. Singh, and C.S. Dey. 2004. PPAR-γ expression modulates insulin sensitivity in C2C12 skeletal muscle cells. *Br. J. Pharmacol.* 143:1006–1013.

Vidal-Puig, A., M. Jimenez-Linan, B.B. Lowell, A. Hamann, E. Hu, B. Spiegelman, J.S. Flier, and D.E. Moller. 1996. Regulation of ppar gamma gene expression by nutrition and obesity in rodents. *J. Clin. Invest.* 97:2553–2561.

Vire, E., C. Brenner, R. Deplus, L. Blanchon, M. Fraga, C. Didelot, L. Morey, A. Van Eynde, D. Bernard, J.M. Vanderwinden, M. Bollen, M. Esteller, L. Di Croce, Y. de Launoit, and F. Fuks. 2006. The polycomb group protein ezh2 directly controls DNA methylation. *Nature* 439:871–874.

Vonnahme, K.A., B.W. Hess, T.R. Hansen, R.J. McCormick, D.C. Rule, G.E. Moss, W.J. Murdoch, M.J. Nijland, D.C. Skinner, P.W. Nathanielsz, and S.P. Ford. 2003. Maternal undernutrition from early- to mid-gestation leads to growth retardation, cardiac ventricular hypertrophy, and increased liver weight in the fetal sheep. *Biol. Reprod.* 69:133–140.

Ward, S.S. and N.C. Stickland. 1991. Why are slow and fast muscles differentially affected during prenatal undernutrition? *Muscle Nerve* 14:259–267.

Wertz, E., L.L. Berge, P.M. Walker, D.B. Faulkner, F.K. McKeith, and S. Rodriguez-Zas. 2001. Early weaning and postweaning nutritional management affect feedlot performance of Angus x Simmental heifers and the relationship of 12th rib fat and marbling score to feed efficiency. *J. Anim. Sci.* 79:1660–1669.

Wertz, A.E., L.L. Berger, P.M. Walker, D.B. Faulkner, F.K. McKeith, and S.L. Rodriguez-Zas. 2002. Early-weaning and postweaning nutritional management affect feedlot performance, carcass merit, and the relationship of 12th-rib fat, marbling score, and feed efficiency among Angus and Wagyu heifers. *J. Anim. Sci.* 80:28–37.

Wigmore, P.M. and N.C. Stickland. 1983. Muscle development in large and small pig fetuses. *J. Anat.* 137(Pt. 2):235–245.

Yamanouchi, K., T. Hosoyama, Y. Murakami, and M. Nishihara. 2007. Myogenic and adipogenic properties of goat skeletal muscle stem cells. *J. Reprod. Dev.* 53:51–58.

Zambrano, E., C.J. Bautista, M. Deas, P.M. Martinez-Samayoa, M. Gonzalez-Zamorano, H. Ledesma, J. Morales, F. Larrea, and P.W. Nathanielsz. 2006. A low maternal protein diet during pregnancy and lactation has sex- and window of exposure-specific effects on offspring growth and food intake, glucose metabolism and serum leptin in the rat. *J. Physiol.* 571:221–230.

Zambrano, E., P.M. Martinez-Samayoa, C.J. Bautista, M. Deas, L. Guillen, G.L. Rodriguez-Gonzalez, C. Guzman, F. Larrea, and P.W. Nathanielsz. 2005. Sex differences in transgenerational alterations of growth and metabolism in progeny (F2) of female offspring (F1) of rats fed a low protein diet during pregnancy and lactation. *J. Physiol.* 566:225–236.

Zhang, H., A. Darwanto, T.A. Linkhart, L.C. Sowers, and L. Zhang. 2007. Maternal cocaine administration causes an epigenetic modification of protein kinase cepsilon gene expression in fetal rat heart. *Mol. Pharmacol.* 71:1319–1328.

Zhu, M.J., S.P. Ford, W.J. Means, B.W. Hess, P.W. Nathanielsz, and M. Du. 2006. Maternal nutrient restriction affects properties of skeletal muscle in offspring. *J. Physiol.* 575:241–250.

Zhu, M.J., S.P. Ford, P.W. Nathanielsz, and M. Du. 2004. Effect of maternal nutrient restriction in sheep on the development of fetal skeletal muscle. *Biol. Reprod.* 71:1968–1973.

Zhu, M.J., B. Han, J. Tong, C. Ma, J.M. Kimzey, K.R. Underwood, Y. Xiao, B.W. Hess, S.P. Ford, P.W. Nathanielsz, and M. Du. 2008. AMP-activated protein kinase signalling pathways are down regulated and skeletal muscle development impaired in fetuses of obese, over-nourished sheep. *J. Physiol.* 586:2651–2664.

5 Muscle Fiber Characteristics and Their Relation to Meat Quality

Jennifer L. Aalhus, W.M. Robertson, and Jin Ye

CONTENTS

5.1 INTRODUCTION

From the time that muscles were first recognized as being composed of individual fibers, research to understand fiber characteristics and roles in meat quality has been undertaken. Early work focused on the development of techniques to discern the histological, metabolic, and subcellular composition of the different types of fibers. This work was followed by a rapid expansion in the literature attempting to relate muscle fiber characteristics to meat quality traits. However, the existence of genetic defects such as porcine stress syndrome, the difficulty in obtaining representative sampling in large muscles combined with tedious histochemical techniques, and uncontrolled variations in rigor shortening in the predominantly used longissimus muscle have added significant complexity.

In broad terms, selection for increased lean muscle mass has resulted in an increase in the proportion of fast, glycolytic fibers, whereas improved fresh meat quality traits are associated with higher proportions of slow, oxidative fibers. In contrast, better processed meat characteristics are associated with higher proportions of fast, glycolytic fibers. Improved techniques for representative sampling combined with phenomic and genomic biomarker selection should lead to advances in selection that can meet these seemingly diverse goals. However, animals are made of muscles, not meat, and our continued understanding of the inherent characteristics of individual muscles and their direction toward appropriate end use will be necessary.

The present overview thus focuses on muscle fiber types and the difference in their morphology, subcellular composition, physiological functions, and biochemical activity. Their variation within and among muscles is discussed in relation to locomotion, in addition to species differences in fiber type composition and morphology among historically domesticated and newly domesticated livestock. Finally, these fiber type differences are discussed in the context of their effects on meat-quality attributes.

5.2 MUSCLE FIBER TYPES

Histochemical muscle fiber typing is used to identify characteristic functional and metabolic properties of individual muscle cells, and was originally based on the scientific premise that fiber types in mature animals constituted distinct species of fibers (Swatland 1994). Classification of muscle fiber types initially began with the recognition that visually red or white muscles had distinct differences in their physiological function, with dark red muscles having a postural role with slower contraction speed and a greater resistance to fatigue. While this color difference was due to higher myoglobin content in the sarcoplasm and greater aerobic potential of the fibers in red muscles, a later observation of a weak myofibrillar ATPase activity associated with slow contraction speeds (Engel and Irwin 1967) formed the basis for distinguishing fiber types histochemically. Thus, on the basis of contraction speed (Padykula and Herman 1955; Guth and Samaha 1970), fibers were categorized as slow or fast twitch fibers. The premise of the myosin ATPase stain is that the speed at which a fiber contracts is directly related to the rate at which myosin splits ATP, which is proportional to the myosin ATPase activity (Swatland 1994). This relationship between contraction speed and histochemical myofibrillar ATPase activity has since been confirmed by Zeman and Wood (1980). In slow twitch fibers, the myosin ATPase is acid stable and alkalilabile, whereas in fast twitch fibers, the myosin ATPase is alkali stable and acid labile (Guth and Samaha 1970). Thus, using acid preincubation conditions, slow twitch fibers stain darkly and fast twitch fibers do not stain. The reverse is true under alkali pre-incubation conditions.

The association of slow twitch fibers with higher myoglobin and aerobic capacity led to the early designation of these fibers as "slow red fibers." Similarly, the pale colored, fast twitch fibers had limited aerobic capacity and were designated as "fast white fibers." Histological techniques to identify the metabolic capacity of individual fibers soon followed and were based on the localization of enzymes involved in oxidative phosphorylation, which were associated with the inner membrane of the mitochondria (Lehninger 1989). Hence, stains designed to localize the activity of these enzymes, also localized the mitochondria within a cell, giving an indication of the oxidative capacity of the cell, and included NADH dehydrogenase (Bancroft 1977) and succinate dehydrogenase (SDH) (Humason 1972).

As the use of these stains on serial muscle sections proliferated, it soon became obvious that two categories of fibers (red and white) did not fully describe the range of fiber types observed, and an intermediate fiber type — having a fast contraction speed, high glycolytic capacity, and intermediate or high oxidative capacity — was described (Cassens et al. 1968). Since then, many researchers have described other types of fibers based on their histochemical staining patterns, and it is now clear that the metabolic capacity of fibers should be viewed as a continuous variable (Schmalbruch and Kamieniecka 1975; Suzuki et al. 1985). While the characterization into functional types (slow twitch and fast twitch) has been thought of as a discontinuous variable, Swatland (1994) provides a convincing argument that the physiological differentiation of muscle fibers is dynamic and that a certain "fiber type" merely reflects the functional and metabolic characteristics of a muscle fiber at any given time (Guth and Yellin 1971). The fact that fiber types can be influenced through environmental or genetic means has implications concerning the quantity of muscle and quality of meat in agriculturally important meat species.

Many taxonomic systems have been developed to describe the various fiber types, with the primary break generally done on the basis of functional characteristics of the myosin heavy chain (slow and fast isoforms) and the secondary break on metabolic characteristics. Hence, in human medicine the dominant nomenclature in use classifies slow twitch fibers as Type I and fast twitch fibers as Type II (Brooke and Kaiser 1970). Alphabetic subtypes are utilized to account for metabolic capacity, with Type IIb fibers representing fast twitch fibers with limited oxidative capacity and Type IIa and IIx fibers representing fast twitch fibers with intermediate oxidative and high glycolytic capacity (Klont et al. 1998).

A second major classification system often used in agricultural literature designates fast twitch fibers as alpha (α) fibers and slow twitch fibers as beta (β) fibers, with the aerobic metabolic activity

TABLE 5.1
Functional and Metabolic Characteristics of Muscle Fibers and Their Corresponding Systems of Nomenclature

Fiber Characteristics			
Contraction Speed	Slow	Fast	Fast
Metabolic Oxidative Capacity	High	High or intermediate	Low
Metabolic Anaerobic Capacity	Low	High	High
		Corresponding Nomenclature	
	Red	Intermediate	White
	Type I	Type IIa	Type IIb
	Type I red	Type II red	Type II white
	β R	α R	α W
	SO	FOG	FG

designated as strong (R) or weak (W) (Ashmore and Doerr 1971). Under this system, fast white fibers are designated αW, slow red fibers as βR, and intermediate fibers as αR. For the purposes of the present chapter, this classification system will be primarily used.

In addition, in the field of physical education and athletics, descriptive terminology is often used, with slow red fibers designated as slow, oxidative (SO); fast white fibers designated as fast, glycolytic (FG); and intermediate fibers designated as fast, oxidative, glycolytic (FOG) (Peter et al. 1972). Early nomenclature in animal science fields relied on the descriptors red, white, and intermediate, based on their relation to myoglobin content and meat color. Although the different methods of histochemical classification do not always totally agree with each another in practice (Klont et al. 1998), a simplistic approximation of their functional and metabolic similarities is included in Table 5.1.

5.3 SUBCELLULAR AND BIOCHEMICAL CHARACTERISTICS OF MUSCLE FIBER TYPES

Although there is a continuum of fiber types within muscle, an understanding of the different subcellular functional characteristics associated with the major fiber types is of value in understanding their role in meat science. βR fibers typically have small diameters, a high capillary density, a high myoglobin content, a high mitochondrial density, larger mitochondria, a small volume of sarcoplasmic reticulum, a high triglyceride store, and a high activity of oxidative enzymes (Ogata and Yamasaki 1985; Essén-Gustavsson et al. 1992); all are adaptations to enhance oxidative metabolism in these cells. βR fibers also have limited capacity for anaerobic metabolism, having low phosphocreatinine stores, low glycolytic enzyme activity (Fox and Mathews 1981), and lower buffering capacity (Rao and Gault 1989). Interestingly, relatively high levels of glycogen are found in both βR and αW fibers, although the size, distribution, and utilization of the glycogen granules may vary among fiber types (Marchand et al. 2002). Indeed, studies of recruitment patterns of the different fiber types shows glycogen content decreasing sooner and to a greater extent in βR fibers during endurance activities. A similar recruitment pattern of αW fibers is seen during short, high-intensity activities (Gollnick et al. 1973a; Gollnick et al. 1973b). Finally, βR fibers have thicker Z-lines (Marchand et al. 2002) and are associated with a lower density of connective tissue (Swatland 1994).

In comparison, αW fibers are adapted for anaerobic metabolism, having high levels of phosphocreatine, glycogen, and glycolytic enzyme activity (Fox and Mathews 1981). αW fibers also have extensive sarcoplasmic reticulum, membranous vesicles that provide binding sites for the glycolytic enzymes and efficiently sequester calcium ions (Wanson and Drochmans 1972; Goldstein et al.

1985; Xu and Becker 1998). When released from the sarcoplasmic reticulum, calcium ions activate anaerobic energy production through phosphorylase b kinase and act as the trigger for muscle contraction (Swatland 1994). Because a source of oxygen is not crucial to their energy production, αW fibers have low levels of myoglobin, fewer capillaries, and larger fiber diameters. In addition, αW fibers have limited amounts of triglyceride energy stores, and the mitochondria and associated oxidative enzymes required for aerobic metabolism (Fox and Mathews 1981; Ogata and Yamasaki 1985; Essén-Gustavsson et al. 1992). αW fibers have less dense Z-lines and are associated with a greater density of connective tissue (Swatland 1994; Marchand et al. 2002).

Intermediate αR fibers resemble αW fibers as they have the fast twitch myosin isoform and similar adaptations for glycolytic metabolism. However, they also are capable of aerobic metabolism and, as a result, generally have an intermediate fiber diameter and the myoglobin, mitochondria, and oxidative enzyme activities to support aerobic metabolism. Under certain conditions, such as during growth or a change in physical activity, the physiological functions of fibers can change, suggesting a degree of developmental or genetic plasticity (Swatland 1994).

5.4 FIBER TYPE DIFFERENCES AMONG MUSCLES

Individual muscles within the body have unique functions in posture and locomotion, and these functions dictate the muscle architecture, attachment, size, contractile speed, and fatigability (Davies 1981). Hence, there are fiber type adaptations for locomotion that may have subsequent influences on meat quality.

Postural muscles, such as those that act over the joints in limbs to resist flexion, tend to have higher proportions of slow twitch, oxidative fibers (Ogata and Yamasaki 1985). In contrast, muscles used for rapid acceleration or propulsion have a higher proportion of fast twitch, glycolytic fibers. However, in meat-producing animals, with the exception of the masseter muscle, which has been shown to have exclusively slow twitch, oxidative fibers (Johnson et al. 1986), muscles typically have both a propulsive and postural role. In these cases, the arrangement of muscle fibers within the muscle varies, such that the density of slow twitch, oxidative fibers lies deeper in the muscle nearest the joint over which the muscle acts. A greater proportion of fast twitch, glycolytic fibers can be found near the periphery and are recruited during more explosive activity. Interestingly, this arrangement also locates slow twitch, oxidative fibers nearer to the blood supply and fast twitch, glycolytic fibers in a position to generate more torque. A similar pattern has been observed in whole muscles, with the deepest muscles of the limbs having a higher proportion of slow twitch, oxidative fibers and the superficial muscles having a higher proportion of fast twitch, glycolytic fibers (Armstrong et al. 1987). Typically, as animal size increases, the density of slow twitch, oxidative fibers in the deep part of the muscle increases, as well as the extent of these fibers throughout the muscle, resulting in a higher proportion of slow twitch fibers in large animals than in smaller species (Davies 1981).

Of relevance to meat production is the fact that the architecture of the connective tissue and fiber orientation is also adapted for locomotory efficiency (Davies 1981). The proximal attachment of the forelimb of animals, in contrast to the hindlimb, has no bony attachment to the vertebral column, and posture is maintained through muscular attachment, particularly the serratus ventralis. Not surprisingly then, meat scientists report that the collagen content of muscles of the beef forequarter is significantly greater than that of the hindquarter (Casey et al. 1985). Adaptations to muscle shape allow sarcomeres to be arranged in series or in parallel, dictating the range of motion and the total force produced, respectively. Fusiform muscles, similar to the semitendinosus, have sarcomeres arranged in series, allowing a maximal range of motion and minimal strength. Pennate muscles, such as the longissimus thoracis et lumborum, have fibers arranged at an angle, limiting the range of contraction while increasing the force of contraction. Subsequently, pennate muscles require an additional collagenous framework to link the sarcomeres in parallel during force generation. In the case of the longissimus thoracis et lumborum, the tendinous sheet on the surface of the muscle (i.e., the fascia thoracolumbalis) fulfills this role. In other pennate muscles (e.g., the gastrocnemius muscle), the tendons are central to the muscle. Adapting this knowledge to meat, central deposits of connective tissue can be avoided during fabrication to improve value, such as in the fabrication of flat iron steaks from the infraspinatus (Von Seggern et al. 2005). A unique feature of the semitendinosus is an unusually high elastin content (Bendall 1967), contributing to its locomotory role in hip extension by providing elasticity to rapidly return to its starting configuration and thought to contribute to increased toughness (Janz et al. 2006).

Although the majority of meat research has been conducted on the longissimus thoracis et lumborum muscle, several researchers have profiled the histochemical differences among different muscles in domestic meat animals, including cattle (Hunt and Hedrick 1977; Johnston et al. 1981; Manabe et al. 1988; Rao and Gault 1989; Kirchofer et al. 2002), pigs (Beecher et al. 1965; Ruusunen and Puolanne 1997; Karlsson et al. 1999; Gil et al. 2001), and sheep (Quali et al. 1988; Suzuki 1995; Sazili et al. 2005). Further studies have characterized differences among muscles in other species that have a less traditional role as a meat animal, including horses (Raub et al. 1985; Van den Hoven, et al. 1985), buffalo (Veletto et al. 2002), and ostrich (Veletto and Crasto 2004). Still others have used biochemical attributes as indicators of fiber type differences within muscles of cattle (Talmant et al. 1986), pigs (Laborde et al. 1985; Monin et al. 1987), and sheep (Briand et al. 1981a; Briand et al. 1981b). Regardless of the method used, there exists an abundance of information regarding fiber type differences among and within muscles.

In cattle, the most extensive work comparing fiber types among muscles was conducted by Kirchofer and co-workers (2002), classifying 12 muscles of the beef round and 26 muscles of the beef chuck. They used a simultaneous staining method (Solomon and Dunn 1988) and classified muscles into red (more than 40% βR fibers), white (more than 40% αW fibers), and muscles not meeting these criteria as intermediate. Using these criteria, nine of twelve muscles from the round were classified as white, including the adductor, biceps femoris, gluteus medius, gracilis, pectineus, rectus femoris, semimembranosus, semitendinosis, and vastus intermedius muscles. The remaining three muscles (the vastus medialis, vastus lateralis, and sartorius) were classified as intermediate. The predominance of white fibers in the large propulsive muscles of the hindlimb is not an unexpected adaptation for locomotion, although domestication and genetic selection for muscle mass is likely to also play a role. The muscles of the forelimb, playing a larger role in posture than propulsion, had significantly fewer αW fibers overall. In the chuck, only seven muscles were categorized as white, nine as intermediate, and ten as red, including many of the muscles having a large postural role in stabilization of the forelimb (e.g., trapezius, biceps brachii, brachialis). The portion of the longissmus thoracis et lumborum assessed, in the thoracic region, was classified as a white muscle, agreeing with the findings of other researchers (Hunt and Hedrick 1977; Johnston et al. 1981) although the results ranged from 43% to 54% αW. At least some of this range may be attributable to sampling from different locations, as not only does a gradient of fiber type occur from deep to superficial regions of the muscle, but can also vary along the length of the muscle. The reported range in the proportion of αW fibers at different locations along the loin was from 46% to 59% in Japanese steers (Morita et al. 2000), with the more anterior location having a lower proportion (i.e., 46%), similar to the proportion reported for the anterior location of the longissimus dorsi (i.e., 43%; Kirchofer et al. 2002). When assessing fiber type composition of the bovine semimembranosus and semitendinosus muscles, Hunt and Hedrick (1977) sampled from both inner and outer portions of the muscle. Despite distinct visual differences in color, the variation in fiber type between the inner and outer regions of the semimembranosus muscle was much less than the difference between the inner and outer regions of the semitendinosus muscle where no visual difference in color was perceptible in beef. In the semitendinosus muscle, the inner portion of the muscle had 36% βR and 39% αW fibers, compared to 12% βR and 67% αW fibers in the outer portion. Thus, the importance of standardizing sampling locations or determining the number of muscle samples required to obtain reliable estimations (Cerisuelo et al. 2007) when histochemically evaluating muscle tissue becomes abundantly obvious.

Unlike the detailed large studies on histological classification of different muscles from cattle, studies in pigs have been more limited with regard to the number of muscles assessed. The early study by Beecher et al. (1965) determined the fiber type of seven muscles; however, this early work classified the muscle fibers on the basis of a Sudan Black stain and did not histologically classify both on the basis of myosin isoforms and oxidative capacity. They reported that both forelimb muscles (i.e., the serratus ventralis and trapezius) were red muscles, having greater than 40% red fibers. From the hindlimb, the rectus femoris was classified as red, and the gluteus medius was classified as white, containing less than 30% red fibers. Using this classification, the longissimus dorsi was also classified as a white muscle. In both the semitendinosus and the biceps femoris, the visually distinct red and white areas of the muscle were stained separately. Hence, the light portion of the semitendinosus and the biceps femoris (outside portion) were classified as white muscle, and the dark portion of the semitendinosus and the biceps femoris (inside portion) were classified as red muscle. In a more recent summary of the literature (Karlsson et al. 1999), a high proportion of αW have been reported in the longissimus dorsi, gluteus medius, rectus femoris, biceps femoris, vastus lateralis, and semimembranosus, all hindlimb muscles with the exception of the longissimus dorsi. Muscles containing a high proportion of βR and αR fibers (slow twitch, oxidative, and intermediate) included the trapezius, vastus intermedius, triceps brachii, infrasinatus, and supraspinatus, all muscles of the forelimb with the exception of the vastus intermedius. Of notable difference was the classification of the vastus intermedius as a red muscle in pork and as a white muscle in cattle, which may result from a true species difference or from differences in sampling location or histological methods.

In sheep, Sazil et al. (2005) studied the histologically determined myosin isoforms in the longissimus dorsi, tensor faciae latae, semitendinosus, supraspinatus, and trapezius muscles. However, they did not include histological measures of metabolic type, which would have allowed further classification into αR and αW fibers. The highest proportion of βR (slow myosin isoform) fibers were found in the trapezius with 45.3%, classifying this muscle as a "red" muscle, similar to findings in cattle and pigs. The supraspinatus had 35.8% βR fibers, similar to the 35.7% reported in cattle (Kirchofer et al. 2002). In contrast, the longissimus dorsi, tensor fasciae latae, and the semitendinosus all had very high proportions of αR and αW fibers (fast myosin isoform), with low proportions of βR (<15%). A separate metabolic characterization of ovine muscles (Briand et al. 1981a) suggests the tensor fascia latae and semitendinosus muscles would be classified as white muscles having a high glycolytic enzyme activity and a high myosin ATPase activity. The longissimus dorsi would be classified as an intermediate muscle having a high myosin ATPase activity and both high oxidative and high glycolytic activity. Both the supraspinatus and infraspinatus in sheep were extensively investigated (Suzuki 1995) and would be broadly classified as red muscles due to their high proportion of slow, oxidative fibers; however, there were marked variations in fiber type at different locations within the muscles, with more slow oxidative fibers in the deep, caudal region of the supraspinatus and in the medial location of the infraspinatus. In one further study (Ouali et al. 1988), the diaphragm, supraspinatus, triceps brachii caput longum, longissimus dorsi, and tensor fascia latae were characterized into four fiber types on the basis of myosin ATPase activity, SDH and α-glycerophosphate dehydrogenase activity in lambs of unknown maturity, and weight endpoint. Their results agree with the characterization of the tensor fascia latae as a white muscle and the longissimus dorsi as an intermediate muscle. However, both the supraspinatus and triceps brachii caput longum would be classified as intermediate muscles in their study, which likely reflects a location effect. The diaphragm, expectedly, was classified as a red muscle.

Fiber type variation among muscles of the horse have been reported for the longissimus dorsi, gluteus medius, triceps brachii, biceps femoris, semitendinosus, and soleus muscles on a single Dutch Saddle Horse mare (Van den Hoven et al. 1985) and a single female Welsh pony (Raub et al. 1985). Both studies reported significant variation in fiber type within a muscle, with a tendency for a higher proportion of Type-I aerobic fibers toward the deeper and cranial parts of the muscles, agreeing with the postural and propulsive functions within a muscle. Because single-muscle biopsies from the musculature of horses was being used to evaluate performance and training programs, these results led to the conclusion that the characterization of an entire muscle cannot be done on the basis of a single needle biopsy. Unfortunately, this observation has equal application in meat science.

The longissimus dorsi, semimembranosus, and triceps brachii caput longum from nine Italian buffalo cows were studied and related to motor function in the animal (Veletto et al. 2002). The highest proportion of βR fibers were found in the longissimus (44%) and the lowest proportion in the triceps brachii (35%). Conversely, the highest proportion of αW fibers were found in the triceps brachii (31%) and the lowest proportion in the longissimus (27%). Of the four muscles examined from the leg of an ostrich (Veletto and Crasto 2004), only the gastrocnemius had all three muscle fiber types, while in the other three muscles (tibialis cranialis caput tibiale, tibialis cranialis caput femorale, and fibularis longus tendo caudalis), αW fibers were absent. The complete absence of a class of fiber type is unusual in comparison to the mixed fiber type generally seen in muscles from quadripedal meat species. However, even in the most detailed studies (Kirchofer et al. 2002), muscles of the lower limb of quadripeds are considered low value trim meat and are generally not assessed in detail. The gastrocnemius in the ostrich had a higher proportion of αW fibers than βR and αR fibers.

5.5 FIBER TYPE DIFFERENCES AMONG SPECIES

As noted previously, muscle fiber type results can be influenced by the muscle examined, the location within that muscle and the histological methods applied, as well as the stage of maturity, individual or breed genetics, the level and type of hormones, and the level of activity (Swatland 1994; Karlsson et al. 1999). Nevertheless, there are interesting species differences to note, which will be constrained in this section to the longissimus thoracis et lumborum (longissimus dorsi). The majority of literature in meat science focuses on this muscle because it is a high value cut, is easily accessible in the carcass and is used in the assessment of carcass grades for many species in many countries, and is a large muscle allowing the opportunity to perform many analyses on the same muscle.

Although a great deal of data exist in the literature, we have been fortunate to have had the opportunity to assess the fiber type of a number of species processed through the same research abattoir, utilizing similar sampling and histological techniques. Samples for fiber typing were collected from the superficial portion of the longissimus lumborum, approximately 4 inches posterior to the last rib. Muscle sections (11-μm thickness) were stained using a combined ATPase/SDH stain according to Solomon and Dunn (1988). Pre-incubation pH, dependent on species, was adjusted to optimize the staining results with cattle (*Bos taurus*), bison (*Bison bison bison*), and musk oxen (*Ovibos moschatus*) pre-incubated at pH 4.15, wapiti (*Cervus elaphus*) at pH 4.25, and swine (*Sus scrofa domesticus*) at pH 4.30. Images for fiber count and area determinations were captured and measurement analyses performed on an axioscope (Zeiss, West Germany) equipped with a Sony DXC 930 color video camera (Sony Corporation, Japan) and Image Pro-Plus software V4.0 (Media Cybernetics, Silver Spring, MD). The results of these analyses are presented in Table 5.2 with the corresponding photomicrographs in Figure 5.1. Literature values for other species not fiber typed in our laboratory include: sheep (*Ovis aries*; βR 12–14%, αR 37–67%, αW 21–49%) (Ouali et al. 1988; Solomon and Lynch 1988; Carpenter and Cockett 2000); horse (*Equus caballus*; βR 5–15%, combined αR and αW 85–95%) (Raub et al. 1985; Van den Hoven et al. 1985); buffalo (*Bubalus bubalis* L.; βR 44%, αR 29%, αW 27%) (Veletto et al. 2002); and reindeer (*Rangifer tarandus*; βR 15%, αR 39%, αW 46%) (Essén-Gustavsson and Rehbinder 1985).

Most meat animals show a mixed pattern of fiber type distribution within the longissimus lumborum; however, porcine muscle is unique in that the concentric arrangement of aerobic and anaerobic fibers arising from prenatal development of primary and secondary fetal fibers is preserved after birth (Swatland 1994). As a result of domestication and selection (Ashmore and Doerr 1971; Ashmore et al. 1972), cattle and pigs have a much higher proportion of αW fibers than the less domesticated bison. Both the wapiti and musk oxen populations studied have similar, low proportions of αW fibers and both were from nondomesticated herds. In pigs, continued selection for lean growth potential has resulted in significant myofiber hypertrophy to the point that abnormally large,

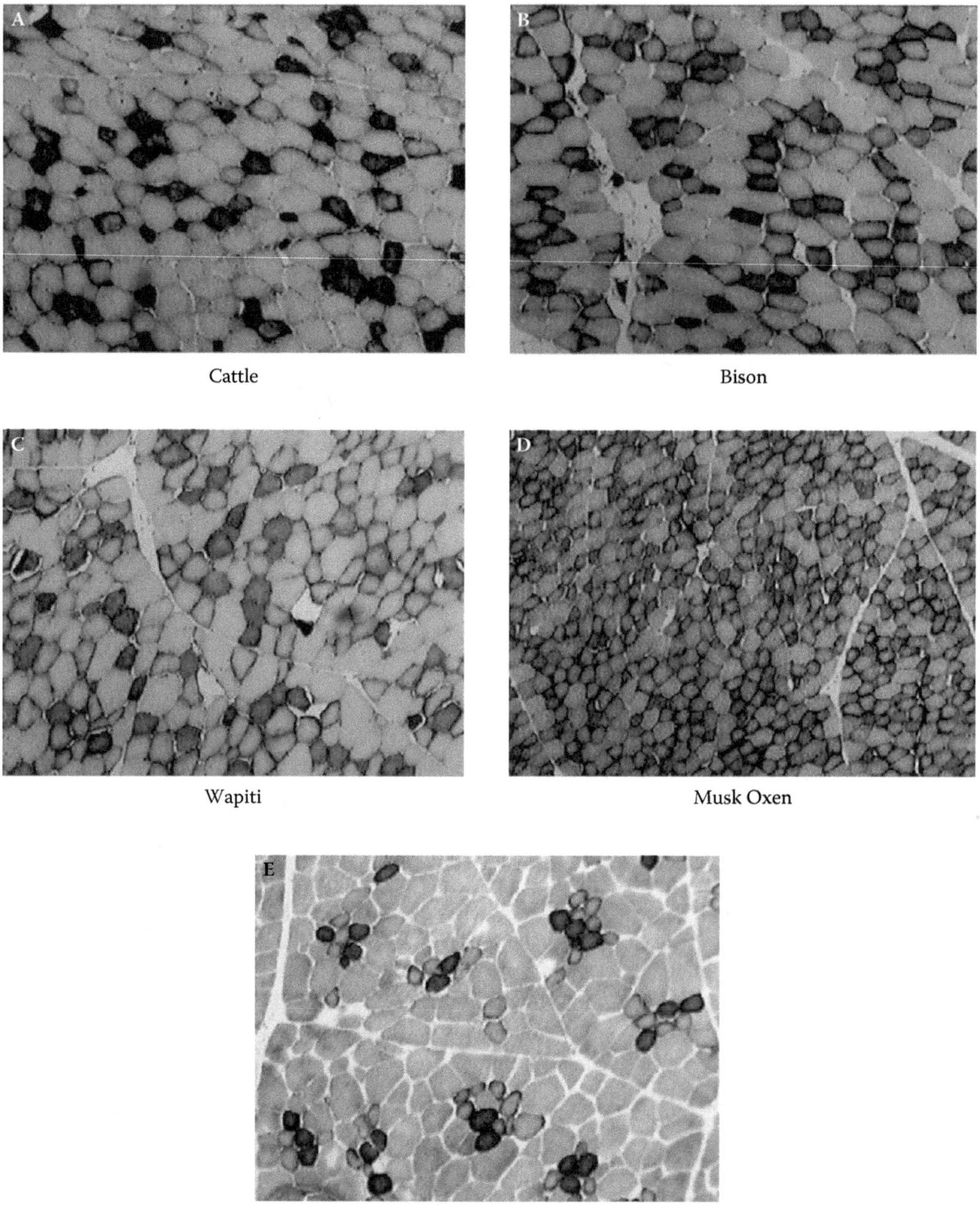

FIGURE 5.1 (See color insert following page 148.) Combined ATPase and SDH stain used to differentiate β R (brown), α R (blue), and α W (light) fibers in the *longissimus lumborum* muscle (80X magnification) of five different species.

swollen, degenerative fibers, so-called "giant" fibers (Bader 1987; Aalhus et al. 1997) are sometimes observed in postmortem muscle of domestic pigs, but not in wild-type pigs (Rehfeldt et al. 2008). Ashmore et al. (1972) suggested that the capacity for diffusion and cellular exchange of metabolites and waste products relative to energy expenditure must limit the ultimate size that a fiber can normally attain, and giant fibers probably represent this extreme in domestic pigs.

TABLE 5.2
The Proportion and Areas of Different Fiber Types in the in the *Longissimus lumborum* Muscle from Five Different Species

	Species				
	Cattle	Bison	Wapiti	Musk Oxen	Swine
			Proportion of different fibers (%)		
β R	28.0	28.0	17.6	12.4	12.6
α R	25.3	36.4	51.6	57.4	17.0
α W	46.8	35.6	30.8	30.2	70.5
			Mean fiber area (μm^2)		
β R	2843	3255	3919	2044	3834
α R	3593	3311	4492	2468	4234
α W	5939	5561	7476	3812	7358

Of note was the comparable fiber size of wapiti to swine, which was not expected because myofiber size is generally positively correlated with liveweight when domesticated species are not included (hence bison > wapiti > musk oxen) (Stickland 1979). However, the wapiti were male animals, harvested in the fall, and the unexpectedly large fiber size may result from hypertrophy as a result of hormonal influences during the breeding season. As a proportion of the αW fiber size, the cross-sectional area of the βR fibers ranged from 48% in cattle to 58% in bison, similar to the range reported in the longissimus muscle of diverse East African game animals, including giraffe, hartebeest, wildebeest, oryx, gerenuk, and dik-dik (i.e., 54% to 62%) (Cerisuelo et al. 2007). Although the size of these fibers remains relatively proportional throughout a wide range of species, it is unknown if this holds true for muscles other than the longissimus thoracis et lumborum.

Not only did the nondomesticated species studied have low populations of αW fibers, but they also had low populations of βR fibers, with a high proportion of intermediate, αR fibers. In a comparison between the horse and cattle, Karlström et al. (1994) suggested that the larger proportion of αR fibers in the horse were an adaptation to a higher level of activity compared to cattle. Thus, the higher proportion of αR fibers is likely a result of the higher activity in the nondomesticated wapiti and musk oxen. In addition, the higher proportion of intermediate αR fibers is likely advantageous to a wild species because these fibers can be transformed to meet changing functional demands (Swatland 1994) that might occur with environmental or habitat changes. The intermediate proportion of αR fibers in the bison (higher than in cattle and swine but lower than in wapiti and musk oxen) likely reflects their more recent domestication and the prevalence of an extensive management system for this species.

5.6 RELATIONSHIP OF FIBER TYPES TO MEAT QUALITY

One of the first goals of domestication is to use selection as a means to increase the muscle mass in order to increase lean meat yield. In theory, this increase in muscle mass would best be accomplished through proportional increases in both muscle cell hyperplasia pre- or perinatally and hypertrophy (Mascarello et al. 1992; Lefaucheur et al. 1995; Rehfeldt et al. 2000). However, growth in muscle mass has been shown to be mainly through an increase in fiber size, not number (Rehfeldt et al. 2008) and through fiber-type shifts from αR to αW fibers (Essén-Gustavsson and Lindholm 1984; Weiler et al. 1995). Understandably then, the effects of selection for lean growth are muscle specific, with muscles having a higher proportion of βR fibers being less affected than muscles with a higher proportion of αR and αW fibers, hence more effects in the propulsive than postural muscles. This can lead to consequences in accurately predicting lean yields based on measurements of a single muscle, without continuously adjusting lean yield prediction equations to reflect ongoing selection and management practices (Aalhus et al. 1990). Early work suggested that the effects of selection and domestication on lean meat yield may simultaneously promote qualitative alterations to meat quality (Ashmore et al. 1972) based on the inherent characteristics associated with different fiber types. Since then, an extensive body of literature has been generated examining the effects of fiber type on meat quality characteristics.

Coming back to the original historical reason for developing an interest in the characteristics of muscle fibers, meat color is the first quality characteristic assessed by consumers and can influence purchase decisions, and is significantly related to fiber type composition (Henckel et al. 1997). Species differences in fiber composition result in differences in myoglobin content, with the content in the αW fiber dominant longissimus of pigs being lowest, followed by sheep and cattle (Lawrie 1998). In addition, the rates of myoglobin oxygenation are fastest in pork, intermediate in lamb, and slowest in beef (Haas and Bratzler 1965). Not only are these species differences in fiber type responsible for the species difference in muscle color, particularly in swine the variation in fiber type within a muscle can result in distinct visual demarcations referred to as "two-toning." Lawrie (1998) reported that the concentration of myoglobin may be several hundred-fold different over distances of 1 cm within the same muscle, reflecting an abrupt change in fiber type. This is particularly obvious in the muscles of the hindlimb in pigs, such as the semitendinosus and biceps femoris, and can present problems in visual acceptability in both fresh and cured products. Furthermore, fiber type differences among different muscles can affect the tendency of their pigments to oxidize to metmyoglobin under postmortem conditions. Muscles with higher amounts of βR and αR fibers have greater intrinsic cytochrome oxidase and succinate dehydrogenase enzyme activities at their surface, which results in a greater oxygen uptake and a faster rate of metmyoglobin formation. Both bovine and porcine psoas major muscles tend to form metmyoglobin faster than the longissimus dorsi when exposed to air (Lawrie 1998). Bison longissimus muscle in oxygen-permeable retail packaging has been observed to undergo a very rapid browning, unrelated to microbial spoilage (Janz and Aalhus 2006; Pietrasik et al. 2006), which may result from the much higher proportion of intermediate fibers in this species.

A more serious condition related to fiber type composition that has effects on multiple quality characteristics but that also affects color is the deleterious pale, soft, and exudative (PSE) condition of meat that has garnished extensive research efforts. PSE develops when muscle pH falls rapidly postmortem, and the combination of acidic conditions while muscle temperature remains high results in protein denaturation and precipitation of sarcoplasmic proteins onto myofibrillar proteins, thereby altering the biophysical properties of the meat (Wismer-Pedersen 1959; Bendall and Wismer-Pedersen 1962; Offer 1991; Klont et al. 1998; Karlsson et al. 1999). In addition to pale color, this meat has a poor water holding capacity and soft, doughy texture. While this condition most often occurs in pork, it does occur occasionally in beef in the deep muscles of the hip, but in beef it is usually due to insufficient cooling rates rather than overly fast glycolysis (Aalhus et al. 1998). Research has led to the discovery of genetic conditions, the porcine stress syndrome ryanodine receptor defect, and the Rendement Napole effect, which had inadvertently been selected for in the pursuit of lean muscle mass and which were found to play a significant role in the development of PSE. Despite the ability now to control these conditions through genetic testing and selection (Fujii et al. 1991; Otsu et al. 1991; Milan, et al. 1995; Milan et al. 1996), PSE is still a prevalent meat quality defect.

Understandably, now that genetic defects can be isolated (Klont et al. 1998), the role of muscle fiber types in postmortem metabolism and in the development of PSE is of active research interest (Henckel et al. 1997; Chang et al. 2003; Fiedler et al. 2003; Ryu and Kim 2005; Choi et al. 2006; Choi et al. 2007; Franck et al. 2007; Scheffler and Gerrard 2007; Choe et al. 2008; Rehfeldt et al. 2008). Clearly, based on our knowledge of the metabolic characteristics of fiber types, αW fibers would be expected to be more actively recruited in the anaerobic postmortem environment, resulting in a more rapid utilization of glycogen, accumulation of lactate, acceleration of pH decline, and a higher extent of protein denaturation (lower protein solubility) with subsequent effects on meat color and drip loss, and water holding capacity. These effects have been demonstrated in practice (Ryu and Kim 2005; Choi et al. 2006; Choi et al. 2007; Choe et al. 2008). Conversely, muscle with a high proportion of βR fibers and low proportions of αW fibers have been shown to be associated with low levels of lactate accumulation at 45 min postmortem, a higher protein solubility and corresponding darker muscle with lower drip losses (Choe et al. 2008). Further, Franck et al. (2007) suggest that not only is the high glycolytic metabolism of the αW fibers to blame for the onset of the PSE condition, but the low capillary density associated with these fibers may lead to lactic acid trapping during stressful pre-slaughter events. Consequently, they have proposed a PSE index that combines both the numerical density of αW fibers and the numerical density of capillaries as a means of determining the susceptibility of various muscles and different genotypes to developing the PSE condition. Although a dichotomy seems to exist between selection for lean meat yield and meat quality, Karlsson et al. (1999) have suggested it should be possible to include muscle fiber characteristics for improved meat quality in selection schemes while preserving optimal production traits.

Muscle fiber type relationships to the development to the dark, firm, and dry (DFD) meat quality defect have also been reported. DFD meat occurs when low glycogen stores at slaughter result in reduced postmortem glycolysis, a lack of lactate accumulation, and a high ultimate pH (generally over 5.9 compared to the normal range of 5.4 to 5.6). The resulting meat is dark in color due to both retention of water associated with proteins above their isoelectric point and the resulting tight packing density of the fibers, which presents a barrier to light diffusion (Lawrie 1998). In addition to being visually unappealing, high-pH dark cutting meat is susceptible to a more rapid conversion to metmyoglobin due to the surviving activity of the cytochrome enzymes and to early bacterial spoilage due to the utilization of amino acids as a substrate in the absence of glucose in glycogen depleted tissues (Lawrie 1998). While DFD meat results from pre-slaughter stress, the observation that it is most severe in the longissimus, semitendinosus, semimembranosus, adductor, and gluteus medius while sparing other muscles (Tarrant and Sherington 1980) has led to the exploration of the role muscle fiber types play. In both pigs (Maltin et al. 1997) and cattle (Zerouala and Strickland 1991), elevated muscle pH has been associated with the total oxidative capacity and the area of slow oxidative βR fibers. Swatland (1994) noted that glycogen depletion from slow-twitch fibers is less labile and takes longer to recover than glycogen stores in fast-twitch fibers and this may be related to differences in pro- and macroglycogen levels (Alonso et al. 1995; Rosenvold et al. 2003) in the different fibers. Macroglycogen is primarily metabolized during aerobic exercise and proglycogen mainly under anaerobic conditions, and the difference between the two glycogen pools has been reported to be related to the rate of anaerobic metabolism (Rosenvold et al. 2003). In the longissimus muscle of pigs, where there is a considerably higher proportion of αW fibers than in cattle, a tendency toward DFD meat was seen when 30% or more of the αW fibers were glycogen depleted (Karlsson et al. 1994). Because 83% of the fibers were αW fibers in the longissimus of the pigs from this study, it is reasonable that these fibers would play a contributing role in meat quality in these animals. While fasting-induced glycogen depletion has been shown to be both muscle-type and fiber-type dependent (Wittmann et al. 1994), aggressive behavior has been shown to have the greatest effects in fast-twitch muscles (Fernandez et al. 1994; Swatland 1994). Clearly, fiber types play a role in the development of DFD but their role must be interpreted within the context of the type of pre-slaughter stress, the species of animal, the muscles recruited to respond to the stress, and the individual fiber type composition of the muscles.

A prevalent problem in beef, but less so in pork, has been a lack of tenderness; this has garnished significant research efforts (see Ouali 1990 and Koohmaraie 1996 for reviews). In pork, Maltin et al. (1997) found that the fiber diameter of intermediate αR fibers significantly contributed to the variation in objective measures of meat tenderness in the longissimus muscle. Similarly, Karlsson et al. (1994) reported a negative relationship between the proportion of fast-twitch fibers and shear force value in the longissimus. In beef cattle, the relationship between fiber type and tenderness was historically limited (Melton et al. 1974; Melton et al. 1975; Calkins et al. 1981) to the point that even extensive studies to examine the causative factors influencing palatability ignore fiber type completely (Rhee et al. 2004). However, the fact that instruments are being developed to predict longissimus tenderness on the basis of color (Wulf et al. 1997), and color is related to the proportion of fiber types, suggests there is a relationship between muscle fiber composition and tenderness.

However, just as early work to relate fiber type differences to meat quality in pork was confounded by the presence of unknown genetic factors at the time, the fact that there has been less success in relating fiber type to tenderness in beef cattle is likely due to the propensity to use the longissimus muscle, which is unrestrained during the postmortem period and can undergo variable shortening, which can have significant effects on tenderness (Thompson et al. 2006). In swine, the high proportion of αW fibers in the longissimus which are highly adapted to anaerobic metabolism, undergoes a more rapid onset of rigor avoiding the temperature-induced loss of membrane integrity that can cause cold shortening (toughening) in other species. The higher proportion of βR fibers in longissimus muscles of cattle are more susceptible to shortening when unrestrained in the postmortem period. Because the extent of shortening can vary from carcass to carcass (Koohmaraie et al. 2002) based on rates of temperature decline, glycolytic rate, carcass weight, and conformation in the hanging carcass, attempts to make associations between fiber type and tenderness can become quite complicated in beef carcasses. Added to this is the fact that the longissimus is a very large muscle and, as previously indicated, fiber type can change along the length and from superficial to deep locations; hence, small samples removed for fiber typing may be poorly related to quality traits that are evaluated over a greater area of the muscle. This, of course, is a problem of representative sampling in the longissimus, which applies equally to the variation in quality measurements throughout a muscle (Reuter et al. 2002; Belew et al. 2003; Janz et al. 2006) as it does to readily available histological and biochemical techniques. This difficulty was also identified by Karlsson et al. (1994), who recommend the need to define predictor muscles, optimal sampling locations, and more rapid and less expensive methods of muscle fiber type determination. Some interesting techniques have been developed (Choi et al. 2006; Choi et al. 2007) by collecting multiple cryostat sections and using these as the sample to determine myosin isoforms using electrophoretic separation. This combination of techniques allows a much broader sampling of the fiber type populations across a muscle without having to go through tedious and time-consuming histological examinations. Clearly, these issues have broad relevance and need to be considered in the current flurry of research to develop genomic and proteomic predictors of meat quality.

Although the direct relationships of muscle fiber type with quality characteristics have been inconsistent, muscle fiber type has been shown to be related to intramuscular fat and thereby other palatability characteristics. While it is expected that a greater amount of intramuscular fat would be found in red fibers than in white due to their role in aerobic metabolism, Essén-Gustavsson et al. (1992) confirmed in longissimus samples from pigs that all Type I (βR) fibers contained triglycerides, whereas only 26% of Type IIa (αR) and 1% of Type IIb (αW) fibers contained triglycerides. Further, they found a negative correlation between intramuscular lipid and the area of αW fibers. In beef carcasses of different maturity, increases in oxidative fibers (βR and αR) were correlated with increases in marbling, which in turn was associated with improved tenderness in aged beef loins (Calkins et al. 1981). Significant negative correlations ($r > -.30$) were also reported by Ozawa et al. (2000) between fat content and both the proportion and relative area of αR fibers in Japanese black cattle. These authors also confirmed that when intermediate αR fibers had positive correlations, αW fibers showed the converse, and vice versa. Similarly, in pigs, positive correlations were found between the marbling score and the frequency of oxidative fibers (Fiedler et al. 2003), and further detailed fatty acid analyses indicated low omega-6 to omega-3 fatty acid ratios were also related to a higher percentage of oxidative fibers. Summarizing the effects of numerous studies, Taylor (2004) indicated that the percentage of βR fibers was very important in the perception of flavor and juiciness due to the higher lipid content in these fibers.

Clear association of the effects of fiber size on meat tenderness, while seeming to be a logical relationship, has also been elusive. In practice, systems to evaluate fresh meat texture are actually evaluating the size and density of muscle fiber bundles, because individual fibers are not discernable by the human eye. Early attempts to relate the texture of muscle bundles to palatability traits were variable, with some reporting large bundles of small fibers were more tender (Sartorious and Child 1938), others reporting that small bundles were associated with tender meat (Brady 1937; Hammond 1940), and yet others reporting no difference (Melton et al. 1974). Similarly, correlations of individual muscle fiber diameter or cross-sectional areas with quality traits were not recently reported for either cattle (Ozawa et al. 2000) or pigs (Ryu and Kim 2005). However, certain models have been presented (Koohmaraie et al. 2002; Taylor 2004) to demonstrate the relationship of fiber size to tenderness/toughness, including double muscled cattle, callipyge sheep, and dietary administration of various β-adrenergic agonists. In addition, Taylor (2004) includes the examples of fiber size differences within and between muscles correlating with texture and the lower tenderness in muscles with a higher proportion of αW fibers. The complicating contribution of sarcomere length, postmortem proteolysis, and collagen concentration to these examples make a direct relationship between tenderness and fiber size difficult to ascertain. Taylor (2004) concluded that the relationship between fiber diameter and tenderness is not strong, explaining only 12% of the variation in a study between muscle types in cattle and 7% of the texture differences in porcine longissimus muscle.

While much of the research in the literature has concentrated on the effects of fiber type on fresh meat quality characteristics, an extensive review by Xiong (1994) focused entirely on the functional properties of muscle fiber types in meat processing. Protein functionality includes the hydration, surface properties, binding, and rheological behavior, including more specific functional properties such as myofibril swelling, protein solubility, and gelation. Overall, myofibrillar proteins from white or fast-twitch fibers usually exhibit superior functionalities in comparison with those from red or slow-twitch fibers under identical conditions of pH, temperature, heating, and ionic strength. In contrast, the opposite finding seems true for fresh meat quality characteristics, where red or slow-twitch fibers are generally associated with superior characteristics. In quadruped meat animals, where there is generally a wide range of muscles with different inherent characteristics, the move toward more extensive knowledge of individual muscle characteristics and a more targeted endpoint use seems a reasonable approach (Molina et al. 2005; Von Seggern et al. 2005).

5.7 SUMMARY

The fiber type composition of meat animals is generally mixed and varies within a muscle, among individual muscles, among breeds, and among species. Composition can also be affected by stage of maturity, level of activity, and hormone status. The overwhelming consequence of domestication and selection for lean muscle mass has been muscle cell hypertrophy rather than hyperplasia, and a metabolic shift toward fast-twitch, glycolytic (αW) fibers. Research efforts to determine the effect of fiber type on meat quality characteristics have been confounded by overriding effects of genetic defects, representative sampling, and uncontrolled variations in rigor shortening in the predominantly used longissimus muscle. Nevertheless, broadly concluded, a higher proportion of slow, oxidative (βR) fibers is associated with an improvement in fresh meat quality characteristics. In contrast, improved functional properties for processing applications are associated with a higher proportion of fast-twitch fibers.

Given the solid and growing meat science database on the importance of muscle fiber types in meat quality, the rapid advancements in biomarker techniques (Rothschild 2004; Bendixen 2005) will undoubtedly contribute to genetic selection that includes meat quality traits in addition to the more traditional selection for lean yield and other production traits. However, it is important to keep in mind the fact that animals are composed of numerous muscles, all having specific functions in the living animal. While we can seek to improve the characteristics overall, it will be necessary to target the fiber type diversity inherent in different muscles to the most appropriate end use.

ABBREVIATIONS

ATP: Adenosine triphosphatase
NADH: Nicotinamide adenine dinucleotide
SDH: Succinate dehydrogenase
PSE: Pale, soft, and exudative
DFD: Dark, firm, and dry

REFERENCES

Aalhus, J.L., D.R. Best, F. Costello, and A.L. Schaefer. 1997. The effects of porcine somatotropin on muscle fiber morphology and meat quality of pigs of known stress susceptibility. *Meat Sci.* 45(3):283–295.

Aalhus, J.L., D.R. Best, A.C. Murray, and S.D.M. Jones. 1998. A comparison of the quality characteristics of pale, soft and exudative beef and pork. *J. Muscle Foods* 9(3):267–280.

Aalhus, J.L., S.D. M. Jones, A.L. Schaefer, A.K. W. Tong, W.M. Robertson, J.K. Merrill, and A.C. Murray. 1990. The effect of ractopamine on performance, carcass composition and meat quality of finishing pigs. *Can. J. Anim. Sci.* 70(3):943–952.

Alonso, M.D., J. Lomako, W.M. Lomako, and W.J. Whelan. 1995. A new look at the biogenesis of glycogen. *FASEB J.* 9(12):1126–1137.

Armstrong, R.B., M.D. Delp, E.F. Goljan, and M.H. Laughlin. 1987. Distribution of blood flow in muscles of miniature swine during exercise. *J. Appl. Physiol.y* 62(3):1285–1298.

Ashmore, C.R. and L. Doerr. 1971. Comparative aspects of muscle fiber types in different species. *Exp. Neuro.* 31(3):408–418.

Ashmore, C.R., G. Tompkins, and L. Doerr. 1972. Postnatal development of muscle fiber types in domestic animals. *J. Anim. Sci.* 34(1):37–41.

Bader, R.J. 1987. Histological findings in skeletal muscles of pigs with different stress susceptibility. *J. Vet. Med., Ser. A,* 34(6):452–464.

Bancroft, J.D. 1977. Enzyme histochemistry. In *Theory and Practice of Histological Techniques*. New York, NY: Churchill Livingston.

Beecher, G.R., R.G. Cassens, W.G. Hoekstra, and E.J. Briskey. 1965. Red and white fiber content and associated post-mortem properties of seven porcine muscles. *J. Food Sci.* 30:969–975.

Belew, J.B., J.C. Brooks, D.R. McKenna, and J.W. Savell. 2003. Warner-Bratzler shear evaluations of 40 bovine muscles. *Meat Sci.* 64(4):507–512.

Bendall, J.R. 1967. The elastin content of various muscles of beef animals. *J. Sci. Food Agric.* 18(12):553–558.

Bendall, J.R. and J. Wismer-Pedersen. 1962. Some properties of the fibrillar proteins of normal and watery pork muscle. *J. Food Sci.* 27:144–159.

Bendixen, E. 2005. The use of proteomics in meat science. *Meat Sci.* 71(1):138–149.

Brady, D.E. 1937. A study of the factors influencing tenderness and texture of beef. *Proc. Am. Soc. Anim. Product.* 30:246–250.

Briand, M., A. Talmant, Y. Briand, G. Monin, and R. Durand. 1981a. Metabolic types of muscle in sheep. 1. Myosin ATPase glycolytic and mitochondrial enzyme activities. *Eur. J. Appl. Physiol. Occupat. Physiol.* 46(4):347–358.

Briand, M., A. Talmant, Y. Briand, G. Monin, and R. Durand. 1981b. Metabolic types of muscle in sheep 2. Lactate dehydrogenase ec-1.1.2.3 activity and lactate dehydrogenase iso enzyme distribution. *Eur. J. Applied Physiol. Occupat. Physiol.* 46(4):359–366.

Brooke, M.H. and K.K. Kaiser. 1970. Muscle fiber types: how many and what kind? *Archives Neurol.* 23(4):369–379.

Calkins, C.R., T.R. Dutson, G.C. Smith, Z.L. Carpenter, and G.W. Davis. 1981. Relationship of fiber type composition to marbling and tenderness of bovine muscle. *J. Food Sci.* 46(3):708–710.

Carpenter, C.E. and N.E. Cockett. 2000. Histology of longissimus muscle from 2-week-old and 8-week-old normal and callipyge lambs. *Can. J. Anim. Sci.* 80(3):511–514.

Casey, J.C., A.R. Crosland, and R.L.S. Patterson. 1985. Collagen content of meat carcasses of known history. *Meat Sci.* 12(4):189–203.

Cassens, R.G., C.C. Cooper, W.G. Moody, and E.J. Briskey. 1968. Histochemical differentiation of fiber types in developing porcine muscle. *J. Anim. Morphol.* 15:135.

Cerisuelo, A., R. Sala, G. Nürnberg, M. Baucells, and C. Rehfeldt. 2007. How many muscle samples are required to obtain reliable estimations of muscle fiber characteristics from pig longissimus muscle? *Meat Sci.* 76(3):583–587.

Chang, K.C., N. Da Costa, R. Blackley, O. Southwood, G. Evans, G. Plastow, J.D. Wood, and R.I. Richardson. 2003. Relationships of myosin heavy chain fiber types to meat quality traits in traditional and modern pigs. *Meat Sci.* 64(1):93–103.

Choe, J.H., Y.M. Choi, S.H. Lee, H.G. Shin, Y.C. Ryu, K.C. Hong, and B.C. Kim. 2008. The relation between glycogen, lactate content and muscle fiber type composition, and their influence on postmortem glycolytic rate and pork quality. *Meat Sci.* 80(2):355–362.

Choi, Y.M., Y.C. Ryu, and B.C. Kim. 2006. Effect of myosin heavy chain isoforms on muscle fiber characteristics and meat quality in porcine longissimus muscle. *J. Muscle Foods* 17(4):413–427.

Choi, Y M., Y.C. Ryu, and B.C. Kim. 2007. Influence of myosin heavy- and light chain isoforms on early postmortem glycolytic rate and pork quality. *Meat Sci.* 76(2):281–288.

Davies, A.S. 1981. *Quadrupedal Mechanics: Anatomical Principles of the Musculoskeletal System.* Palmerston North, NZ: Massey University Press.

Engel, W.K. and R.L. Irwin. 1967. A histochemical-physiological correlation of frog skeletal muscle fibers. *Am. Journal Physiol.* 213(2):511–518.

Essén-Gustavsson, B. and A. Lindholm. 1984. Fiber types and metabolic characteristics in muscles of wild boars, normal and halothane sensitive Swedish Landrace pigs. *Compar. Biochem. Physiol. – A Physiol.* 78(1):67–71.

Essén-Gustavsson, B. and C. Rehbinder. 1985. Skeletal muscle characteristics of reindeer (Rangifer tarandus L.). *Compar. Biochem. Physiol., A Compar. Physiol.* 82(3):675–679.

Essén-Gustavsson, B., K. Karlström, and K. Lundström. 1992. Muscle fiber characteristics and metabolic response at slaughter in pigs of different halothane genotypes and their relation to meat quality. *Meat Sci.* 31(1):1–11.

Fernandez, X., M. Meunier-Salaun, and P. Ecolan. 1994. Glycogen depletion according to muscle and fiber types in response to dyadic encounters in pigs (*Sus scrofa domesticus*) — Relationships with plasma epinephrine and aggressive behaviour. *Compar. Biochem. Physiol. – A Physiol.* 109(4):869–879.

Fiedler, I., K. Nürnberg, T. Hardge, G. Nürnberg, and K. Ender. 2003. Phenotypic variations of muscle fiber and intramuscular fat traits in Longissimus muscle of F_2 population Duroc × Berlin Miniature Pig and relationships to meat quality. *Meat Sci.* 63(1):131–139.

Fox, E.L. and D.K. Mathews. 1981. *The Physiological Basis of Physical Education in Athletics. 3rd ed.* Toronto, ON, Canada: Saunders College Publishing.

Franck, M., P. Figwer, C. Godfraind, M.T. Poirel, A. Khazzaha, and M.M. Ruchoux. 2007. Could the pale, soft, and exudative condition be explained by distinctive histological characteristics? *J. Anim. Sci.* 85(3):746–753.

Fujii, J., K. Otsu, F. Zorzato, S. De Leon, V.K. Khanna, J.E. Weiler, P.J. O'Brien, and D.H. MacLennan. 1991. Identification of a mutation in porcine ryanodine receptor associated with malignant hyperthermia. *Science* 253(5018):448–451.

Gil, F., O. Lopez Albors, J.M. Vazquez, R. Latorre, G. Ramirez Zarzosa, and F. Moreno. 2001. The histochemical profiles of fiber types in porcine skeletal muscle. *Histol. Histopathol.* 16(2):439–442.

Goldstein, M.A., D.L. Murphy, W.B. van Winkle, and M.L. Entman. 1985. Cytochemical studies of a glycogen-sarcoplasmic reticulum complex. *J. Muscle Res. Cell Motility* 6(2):177–187.

Gollnick, P.D., R.B. Armstrong, C.W. Saubert IV, W.L. Sembrowich, R.E. Shepherd, and B. Saltin. 1973. Glycogen depletion patterns in human skeletal muscle fibers during prolonged work. *Pflügers Archiv.* 344(1):1–12.

Gollnick, P.D., R.B. Armstrong, W.L. Sembrowich, R.E. Shepherd, and B. Saltin. 1973. Glycogen depletion pattern in human skeletal muscle fibers after heavy exercise. *J. Appl. Physiol.* 34(5):615–618.

Guth, L. and F.J. Samaha. 1970. Procedure for the histochemical demonstration of actomyosin ATPase. *Exp. Neurol.* 28(2):365–367.

Guth, L. and H. Yellin. 1971. The dynamic nature of the so-called "fiber types" of mammalian skeletal muscle. *Exp. Neurol.* 31(2):277–300.

Haas, M.C. and L.J. Bratzler. 1965. Determination of myoglobin oxygenation rates in pork, beef, and lamb by Munsell and reflectance colorimetry. *J. Food Sci.* 30:64–68.

Hammond, J.J. 1940. Some factors affecting the quality and composition of meat. *J. Soc. Chem. Ind.* 59:521.

Henckel, P., N. Oksbjerg, E. Erlandsen, P. Barton-Gade, and C. Bejerholm. 1997. Histo- and biochemical characteristics of the longissimus dorsi muscle in pigs and their relationships to performance and meat quality. *Meat Sci.* 47(3-4):311–321.

Humason, G.L. 1972. *Animal Tissue Techniques. 3rd ed.* San Francisco, CA: W.H. Freeman and Company.

Hunt, M.C. and H.B. Hedrick. 1977. Profile of fiber types and related properties of five bovine muscles. *J. Food Sci.* 42:513–517.

Janz, J.A.M. and J.L. Aalhus. 2006. Meat quality, bacteriology and retail case life of bison Longissimus Lumborum following spray chilling. *J. Muscle Foods* 17(3):330–342.

Janz, J.A.M., J.L. Aalhus, M.E.R. Dugan, and M.A. Price. 2006. A mapping method for the description of Warner-Bratzler shear force gradients in beef Longissimus thoracis et lumborum and Semitendinosus. *Meat Sci.* 72(1):79–90.

Janz, J.A.M., J.L. Aalhus, and M.A. Price. 2006. The effect of epimysial connective tissue on factors related to tenderness of beef semitendinosus. *J. Muscle Foods* 17(1):43–55.

Johnson, S.K., W.J.P. White, and R.A. Lawrie. 1986. Observations on the 3-methylhistidine content of bovine, ovine and porcine muscles. *Meat Sci.* 18(3):235–239.

Johnston, D.M., W.G. Moody, J.A. Boling, and N.W. Bradley. 1981. Influence of breed type, sex, feeding systems, and muscle bundle size on bovine fiber type charactersitics. *J. Food Sci.* 46(6):1760–1765.

Karlsson, A.H., B. Essén-Gustavsson, and K. Lundström. 1994. Muscle glycogen depletion pattern in halothane-gene-free pigs at slaughter and its relation to meat quality. *Meat Sci.* 38(1):91–101.

Karlsson, A.H., R.E. Klont, and X. Fernandez. 1999. Skeletal muscle fibers as factors for pork quality. *Livestock Product. Sci.* 60(2–3):255–269.

Karlström, K., B. Essén-Gustavsson, and A. Lindholm. 1994. Fiber type distribution, capillarization and enzymatic profile of locomotor and nonlocomotor muscles of horses and steers. *Acta Anatomica* 151(2):97–106.

Kirchofer, K.S., C.R. Calkins, and B.L. Gwartney. 2002. Fiber-type composition of muscles of the beef chuck and round. *J. Anim. Sci.* 80(11):2872–2878.

Klont, R.E., L. Brocks, and G. Eikelenboom. 1998. Muscle fiber type and meat quality. *Meat Sci.* 49(Suppl.1):S219–229.

Koohmaraie, M. 1996. Biochemical factors regulating the toughening and tenderization processes of meat. *Meat Sci.* 43(Suppl. 1):193–201.

Koohmaraie, M., M.P. Kent, S.D. Shackelford, E. Veiseth, and T.L. Wheeler. 2002. Meat tenderness and muscle growth: is there any relationship? *Meat Sci.* 62(3):345–352.

Laborde, D., A. Talmant, and G. Monin. 1985. Enzyme, metabolic and contractile activities of 30 muscles in swine; relationship with final postmortem pH. *Reprod. Nutrit. Dev.* 25(4A):619–628.

Lawrie, R.A. 1998. *Lawrie's Meat Science. 6th ed.* Cambridge, England: Woodhead Publishing.

Lefaucheur, L., F. Edom, P. Ecolan, and G.S. Butler-Browne. 1995. Pattern of muscle fiber type formation in the pig. *Dev. Dynamics* 203(1):27-41.

Lehninger, A.L. 1989. *Biochemistry: The Molecular Basis of Cell Structure and Function, 4th ed.* London, U.K.: Worth Publishers.

Maltin, C.A., C.C. Warkup, K.R. Matthews, C.M. Grant, A.D. Porter, and M.I. Delday. 1997. Pig muscle fiber characteristics as a source of variation in eating quality. *Meat Sci.* 47(3-4):237–248.

Manabe, N., T. Ishii, and T. Ishibashi. 1988. Histochemical fiber type composition and fiber size in skeletal muscles of the growing cattle, sheep and swine. *Memoirs of the College of Agriculture, Kyoto University* 131:27–36.

Marchand, I., K. Chorneyko, M. Tarnopolsky, S. Hamilton, J. Shearer, J. Potvin, and T.E. Graham. 2002. Quantification of subcellular glycogen in resting human muscle: granule size, number, and location. *J. Appl. Physiol.* 93(5):1598–1607.

Mariani, P., K. Lundstrom, U. Gustafsson, A.C. Enfalt, R.K. Juneja, and L. Andersson. 1996. A major locus (RN) affecting muscle glycogen content is located on pig chromosome 15. *Mammalian Genome* 7 (1):52–54.

Mascarello, F., M.L. Stecchini, A. Rowlerson, and E. Ballocchi. 1992. Tertiary myotubes in postnatal growing pig muscle detected by their myosin isoform composition. *J. Anim. Sci.* 70(6):1806–1813.

Melton, C.C., M.E. Dikeman, H.J. Tuma, and D.H. Kropf. 1975. Histochemical relationships of muscle biopsies with bovine muscle quality and composition. *J. Anim. Sci.* 40(3):451–456.

Melton, C., M.E. Dikeman, H.J. Tuma, and R.R. Schalles. 1974. Histological relationships of muscle biopsies to bovine meat quality and carcass composition. *J. Anim. Sci.* 38(1):24–31.

Milan, D., P. Le Roy, N. Woloszyn, J.C. Caritez, J.M. Elsen, P. Sellier, and J. Gellin. 1995. The RN locus for meat quality maps to pig chromosome 15. *Genet. Select. Evolut.* 27(2):195–199.

Milan, D., N. Woloszyn, M. Yerle, P. Le Roy, M. Bonnet, J. Riquet, Y. Lahbib-Mansais, J.C. Caritez, A. Robic, P. Sellier, J.M. Elsen, and J. Gellin. 1996. Accurate mapping of the "acid meat" RN gene on genetic and physical maps of pig chromosome 15. *Mammalian Genome* 7(1):47–51.

Molina, M.E., D.D. Johnson, R.L. West, and B.L. Gwartney. 2005. Enhancing palatability traits in beef chuck muscles. *Meat Sci.* 71(1):52–61.
Monin, G., A. Mejenes-Quijano, A. Talmant, and P. Sellier. 1987. Influence of breed and muscle metabolic type on muscle glycolytic potential and meat pH in pigs. *Meat Sci.* 20(2):149–158.
Morita, S., H. Iwamoto, Y. Fukumitsu, T. Gotoh, S. Nishimura, and Y. Ono. 2000. Heterogeneous composition of histochemical fiber types in the different parts of M longissimus thoracis from Mishima (Japanese native) steers. *Meat Sci.* 54(1):59–63.
Offer, G. 1991. Modelling of the formation of pale, soft and exudative meat: effects of chilling regime and rate and extent of glycolysis. *Meat Sci.* 30(2):157–184.
Ogata, T. and Y. Yamasaki. 1985. Scanning electron-microscopic studies on the three-dimensional structure of sarcoplasmic reticulum in the mammalian red, white and intermediate muscle fibers. *Cell Tissue Res.* 242(3):461–467.
Otsu, K., V.K. Khanna, A.L. Archibald, and D.H. MacLennan. 1991. Cosegregation of porcine malignant hyperthermia and a probable causal mutation in the skeletal muscle ryanodine receptor gene in backcross families. *Genomics* 11(3):744–750.
Ouali, A. 1990. Meat tenderization: possible causes and mechanisms. A review. *J. Muscle Foods* 1(2):129–165.
Ouali, A., M. Zabari, A. Lacourt, A. Talmant, G. Monin, J. Kopp, and C. Valin. 1988. Usefulness of various biochemical and histochemical characteristics as indices of muscle type in lamb carcasses. *Meat Sci.* 24(4):235–247.
Ozawa, S., T. Mitsuhashi, M. Mitsumoto, S. Matsumoto, N. Itoh, K. Itagaki, Y. Kohno, and T. Dohgo. 2000. The characteristics of muscle fiber types of longissimus thoracis muscle and their influences on the quantity and quality of meat from Japanese Black steers. *Meat Sci.* 54(1):65–70.
Padykula, H.A. and E. Herman. 1955. Factors affecting the activity of adenosine triphosphatase and other phosphatases as measured by histochemical techniques. *J. Histochem. Cytochem.* 3:161–169.
Peter, J.B., R.J. Barnard, V.R. Edgerton, C.A. Gillespie, and K.E. Stempel. 1972. Metabolic profiles of three fiber types of skeletal muscle in guinea pigs and rabbits. *Biochemistry* 11(14):2627–2633.
Pietrasik, Z., J.S. Dhanda, P.J. Shand, and R.B. Pegg. 2006. Influence of injection, packaging, and storage conditions on the quality of beef and bison steaks. *J. Food Sci.* 71(2):S110–S118.
Rao, M.V. and N.F.S. Gault. 1989. The influence of fiber-type composition and associated biochemical characteristics on the acid buffering capacities of several beef muscles. *Meat Sci.* 26(1):5–18.
Raub, R.R., P.J. Bechtel, and L.M. Lawrence. 1985. Variation in the distribution of muscle fiber types in equine skeletal muscles. *J. Equine Vet. Sci.* 5(1):34–37.
Rehfeldt, C., I. Fiedler, G. Dietl, and K. Ender. 2000. Myogenesis and postnatal skeletal muscle cell growth as influenced by selection. *Livestock Product. Sci.* 66(2):177–188.
Rehfeldt, C., M. Henning, and I. Fiedler. 2008. Consequences of pig domestication for skeletal muscle growth and cellularity. *Livestock Sci.* 116(1–3):30–41.
Reuter, B.J., D.M. Wulf, and R.J. Maddock. 2002. Mapping intramuscular tenderness variation in four major muscles of the beef round. *J. Anim. Sci.* 80(10):2594–2599.
Rhee, M.S., T.L. Wheeler, S D. Shackelford, and M. Koohmaraie. 2004. Variation in palatability and biochemical traits within and among eleven beef muscles. *J. Anim. Sci.* 82(2):534–550.
Rosenvold, K., B. Essén-Gustavsson, and H.J. Andersen. 2003. Dietary manipulation of pro- and macroglycogen in porcine skeletal muscle. *J. Anim. Sci.* 81(1):130–134.
Rothschild, M.F. 2004. Porcine genomics delivers new tools and results: This little piggy did more than just go to market. *Genet. Res.* 83(1):1–6.
Ruusunen, M. and E. Puolanne. 1997. Comparison of histochemical properties of different pig breeds. *Meat Sci.* 45(1):119–125.
Ryu, Y.C. and B.C. Kim. 2005. The relationship between muscle fiber characteristics, postmortem metabolic rate, and meat quality of pig longissimus dorsi muscle. *Meat Sci.* 71(2):351–357.
Sartorious, M.J. and A.M. Child. 1938. Effect of coagulation of press fluid, shear force, muscle-cell diameter and composition of beef muscle. *J. Food Res.* 3:619–626.
Sazili, A.Q., T. Parr, P.L. Sensky, S.W. Jones, R.G. Bardsley, and P.J. Buttery. 2005. The relationship between slow and fast myosin heavy chain content, calpastatin and meat tenderness in different ovine skeletal muscles. *Meat Sci.* 69(1):17–25.
Scheffler, T.L. and D.E. Gerrard. 2007. Mechanisms controlling pork quality development: the biochemistry controlling postmortem energy metabolism. *Meat Sci.* 77(1 Spec. Iss.):7–16.
Schmalbruch, H. and Z. Kamieniecka. 1975. Histochemical fiber typing and staining intensity in cat and rat muscles. *J. Histochem. Cytochem.* 23(6):395–401.

Solomon, M.B. and M.C. Dunn. 1988. Simultaneous histochemical determination of three fiber types in single sections of ovine, bovine and porcine skeletal muscle. *J. Anim. Sci.* 66(1):255–264.

Solomon, M.B. and G.P. Lynch. 1988. Biochemical, histochemical and palatability characteristics of young ram lambs as affected by diet and electrical stimulation. *J. Anim. Sci.* 66(8):1955–1962.

Stickland, N.C. 1979. Comparative aspects of muscle fiber size and succinic dehydrogenase distribution in the longissimus dorsi muscle of several species of East African mammals. *Acta Anatomica* 105(4):381–385.

Suzuki, A. 1995. Differences in distribution of myofiber types between the supraspinatus and infraspinatus muscles of sheep. *Anatomical Record* 242(4):483–490.

Suzuki, A., T. Tsuchiya, S. Ohwada, and H. Tamate. 1985. Distribution of myofiber types in thigh muscles of chickens. *J. Morphol.* 185(2):145–154.

Swatland, H.J. 1994. *Structure and Development of Meat Animals and Poultry*. Lancaster, PA: Technomic Publishing Company, Inc.

Talmant, A., G. Monin, M. Briand, M. Dadet, and Y. Briand. 1986. Activities of metabolic and contractile enzymes in 18 bovine muscles. *Meat Sci.* 18(1):23–40.

Tarrant, P.V. and J. Sherington. 1980. An investigation of ultimate pH in the muscles of commercial beef carcasses. *Meat Sci.* 4(4):287–297.

Taylor, R.G. 2004. Muscle fiber types and meat quality. In *Encyclopedia of Meat Sciences*, W.K. Jensen, Ed. Oxford, U.K.: Elsevier.

Thompson, J.M., D. Perry, B. Daly, G.E. Gardner, D.J. Johnston, and D.W. Pethick. 2006. Genetic and environmental effects on the muscle structure response post-mortem. *Meat Sci.* 74(1):59–65.

Van den Hoven, R., T. Wensing, H.J. Breukink, A.E. Meijer, and T.A. Kruip. 1985. Variation of fiber types in the triceps brachii, longissimus dorsi, gluteus medius, and biceps femoris of horses. *Am. J. Vet. Res.* 46(4):939–941.

Velotto, S. and A. Crasto. 2004. Histochemical and morphometrical characterization and distribution of fiber types in four muscles of ostrich (*Struthio camelus*). *Anatomia, Histologia, Embryologia* 33(5):251–256.

Velotto, S., G. Guida, M. Marino, G. Mase, and A. Crast. 2002. Histomorphometrical and comparative analysis of three muscles of Buffalo (*Bubalus bubalis* L.). *Ital. J. Anat. Embryol.* 107(4):233–242.

Von Seggern, D.D., C.R. Calkins, D.D. Johnson, J.E. Brickler, and B.L. Gwartney. 2005. Muscle profiling: characterizing the muscles of the beef chuck and round. *Meat Sci.* 71(1):39–51.

Wanson, J.C. and P. Drochmans. 1972. Role of the sarcoplasmic reticulum in glycogen metabolism: binding of phosphorylase phosphorylase kinase and primer complexes to the sarco vesicles of rabbit skeletal muscle. *J. Cell Biol.* 54(2):206–224.

Weiler, U., H.J. Appell, M. Kremser, S. Hofacker, and R. Claus. 1995. Consequences of selection on muscle composition. A comparative study on gracilis muscle in wild and domestic pigs. *Anatomia, Histologia, Embryologia* 24(2):77–80.

Wismer-Pedersen, J. 1959. Quality of pork in relation to rate of pH change post mortem. *J. Food Sci.* 24(6):711–727.

Wittmann, W., P. Ecolan, P. Levasseur, and X. Fernandez. 1994. Fasting-induced glycogen depletion in different fibre types of red and white pig muscles — relationship with ultimate pH. *J. Sci. Food Agric.* 66(2):257–266.

Wulf, D.M., S.F. O'Connor, J.D. Tatum, and G.C. Smith. 1997. Using objective measures of muscle color to predict beef longissimus tenderness. *J. Anim. Sci.* 75(3):684–692.

Xiong, Y.L. 1994. Myofibrillar protein from different muscle fiber types: implications of biochemical and functional properties in meat processing. *Crit. Rev. Food Sci. Nutrit.* 34(3):293–320.

Xu, Kai Y. 1998. Ultrastructural localization of glycolytic enzymes on sarcoplasmic reticulum vesicles. *J. Histochem. Cytochem.* 46(4):419–427.

Zeman, R.J. and D.S. Wood. 1980. Correlative histochemical and physiological measurements in single skinned muscle fibers: heterogeneity of Ca sensitivity in type I fibers. *J. Histochem. Cytochem.* 28(7):714–715.

Zerouala, A.C. and N.C. Stickland. 1991. Cattle at risk for dark-cutting beef have a higher proportion of oxidative muscle fibers. *Meat Sci.* 29(3):263–270.

6 Muscle Protein Turnover

Margrethe Therkildsen and Niels Oksbjerg

CONTENTS

6.1 INTRODUCTION

Muscle protein accretion is the difference between the rates of muscle protein synthesis and degradation. The growth rate of meat-producing animals is related to the muscle protein accretion — that is, the higher the protein accretion, the higher the growth rate of the animal — one of the most economically important traits in meat production. In addition, optimization of muscle protein accretion on minimal feed intake will minimize the environmental load, and thus be in favor of a sustainable production. Finally, there seems to be a link between muscle protein accretion *in vivo* and tenderness development postmortem, as the enzymes involved in the degradation of muscle protein *in vivo* also affect the proteolysis postmortem, which is essential for the tenderization (Goll et al. 1998; Koohmaraie et al. 2002). Consequently, focus on muscle protein turnover and how it can be manipulated is very important to optimize meat production both from a producer and consumer point of view.

The muscle protein pool is not static, but dynamic with continuous turnover of proteins, which proceed throughout life even when growth has ceased. The rate of muscle protein turnover is regulated through protein synthesis and protein degradation, and the rate of these processes in general exceeds the actually protein accretion rate. Thus, for example, in a 273-kg bull calf, muscle protein synthesis was measured to be 361 g/d (grams per day) and the muscle protein degradation to be 269 g/d, resulting in a net muscle protein accretion of 72 g/d (Morgan et al. 1993a). The reason for this energy-consuming process is to allow the muscle fibers to adapt acutely and rapidly to environmental changes such as exercise and nutrition, and to prevent the accumulation of proteins containing errors and proteins not in use (Goldberg and Dice 1974; Asgar and Bhatti 1987; Reeds 1989). Because muscle protein turnover is relatively large compared with the actual muscle protein accretion in growing animals, minor adaptations in either protein synthesis or degradation can have a profound influence on the net muscle protein accretion (Goll et al. 1989). Individual proteins have

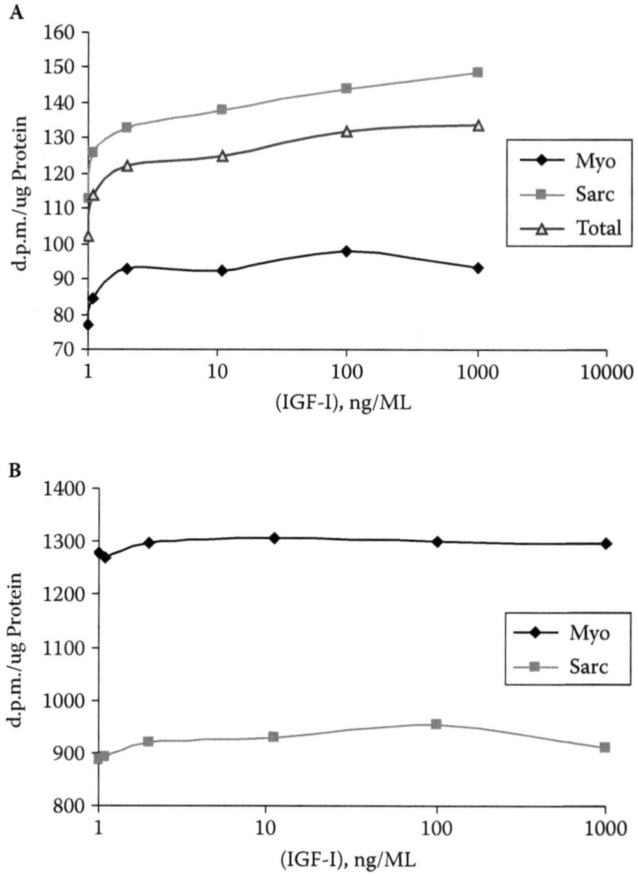

FIGURE 6.1 The influence of IGF-I addition on protein synthesis and degradation of various proteins in porcine satellite cell derived myotube cultures. (A) Protein synthesis. Myogenic satellite cells were seeded in 24-well plates and grown to 80% confluence in 20% serum (10% Fetal Calf Serum [FCS] and 10% horse serum). Thereafter, the cells differentiated in 10% FCS for 24 hours and in 4% FCS for another 48 hours. Then, increasing amounts of IGF-I were added to the medium, together with 1 μCi of [3,5 ^{3}H]tyrosine for 5 hours. Subsequently, cells were harvested by trypsination and the pellet was separated into sarcoplasmic and myofibrillar proteins. The concentration of proteins were measured by the bicinchoninic acid (BCA) procedure and counted in a β-counter. The rate of protein synthesis was calculated as d.p.m.(disintegrations per min)/μg protein. IGF-I increased synthesis of both myofibrillar and sarcoplasmic protein in a dose-dependent manner up to 5 ng/ml. Furthermore, the rate of synthesis was approximately 60% higher in sarcoplasmic proteins compared to myofibrillar proteins. (B) Degradation. The cells were grown and treated in the same way as in (A) except that they were loaded previously with 2 μCi/well for 48 hours prior to the addition of increasing levels of IGF for another 24-hour period. Then, cultures were harvested by trypsination, separated in myofibrillar and sarcoplasmic proteins, and counted in a β-counter and expressed as d.p.m./μg protein. The more activity remaining in the protein pools, the lower the protein degradation. IGF-I had no significant influence on protein degradation. However, protein degradation of sarcoplasmic proteins was higher than in myofibrillar proteins.

different turnover rates, so estimated values of protein synthesis and degradation are generally average turnover rates of the combination of individual proteins in the pool studied (e.g., skeletal muscles) (Sugden and Fuller, 1991). In agreement we found in porcine satellite myotube cultures that the protein turnover was two-fold higher in sarcoplasmic proteins compared to myofibrillar proteins (see Figure 6.1).

Lobley et al. (1980), McCarthy et al. (1983), Gopinath and Kitts (1984), Reecy et al. (1996), and Lobley et al. (2000) have measured muscle protein fractional synthesis (FSR, g protein synthesized/100 g protein/day × 100) and breakdown (FBR, g protein degraded/100 g protein/day × 100) rates in growing cattle in the range from 0.6% to 4%. From these results Gopinath and Kitts (1984) calculated a half-life of 29 days for the myofibril proteins actin and myosin. The protein turnover rates are generally higher in other tissues; thus, Lobley et al. (1980) measured turnover rates in growing heifers of 53%, 32%, and 6% in the gastrointestinal tract, in the liver, and in the hide, respectively.

6.2 MUSCLE PROTEIN TURNOVER

6.2.1 Muscle Protein Synthesis

Muscle protein synthesis is the result of the transcription of DNA to mRNA and the translation of mRNA to protein. In these processes, a range of enzymes are involved, and thus the activity of these enzymes together with the amount of DNA and the availability of amino acids determine the synthesis capacity.

Regulation of protein synthesis in the muscle fiber can be at the transcriptional stage (long-term) and/or at the translational stage (short-term), in the synthesis and/or breakdown of RNA and in the activity of existing ribosomes (Millward et al. 1973; Garlick et al. 1983; Reeds 1989). Millward and Waterlow (1978) suggested that the regulation of protein synthesis at the transcriptional stage occurs through the rate of RNA synthesis and not through increased degradation of RNA. Protein synthesis and degradation at the muscle fiber level depend on the availability of substrates (amino acids) and on the availability of energy (ATP), as protein synthesis is an energy-consuming process. Furthermore, protein turnover is under endocrine control mainly by insulin, growth hormone/insulin-like growth factor I (IGF-I)/IGF-binding proteins (GH/IGF-I/IGFBPs), the glucocorticoids, thyroid hormones, and the sex hormones. The different hormones seem to act on different sites of the transcription–translation pathway (Reeds 1989; Reeds and Davis 1992).

Hormones may act either directly on the synthesis and degradation, or influence protein turnover indirectly by affecting satellite cell proliferation. Thus, during growth, the amount of muscle DNA increases, although myonuclei are mitotically inactive. However, the satellite cell, situated between the basement membrane and the sarcolemma can divide and one or both daughter cells may fuse with existing fibers and thereby add DNA to the fiber. This is important as the DNA provides the machinery for protein synthesis not only of structural proteins, but also of regulatory proteins like proteolytic enzymes. Thus, changes in satellite cell behavior may affect muscle protein turnover. At slaughter, between 60 and 80% of muscle DNA originates from satellite cell division, thus emphasizing the importance of the satellite cell in postnatal growth.

6.2.2 Muscle Protein Degradation

Proteolytic enzymes are responsible for protein degradation in the muscle fibers (Goldberg and Dice 1974; Pontremoli and Melloni 1986), and the rate of degradation varies largely in response to the physiological demands of the fibers (Goldberg 1969a, b). Three proteolytic enzyme systems have been suggested to play a major role in the proteolysis of muscle protein: the calpains, the proteasome, and the cathepsins (Goll et al. 1989). There is now significant evidence to suggest that muscle protein degradation occurs in a sequential way (Calkins and Seideman 1988; Taylor et al. 1995; Price and Stevens 1999; Robert et al. 1999; Houbak et al. 2008), the calpain system acting on disassembling the myofibrillar matrix followed by the action of the proteasome and the cathepsins on the constituent myofibrillar proteins and the sarcoplasmic proteins.

6.3 REGULATION OF MUSCLE PROTEIN TURNOVER

The hormonal regulation of muscle protein turnover has been reviewed by Reeds (1989), Sugden and Fuller (1991), Reeds and Davis (1992), and Lobley, (1998); therefore, only a short presentation of the role of the hormones is given below.

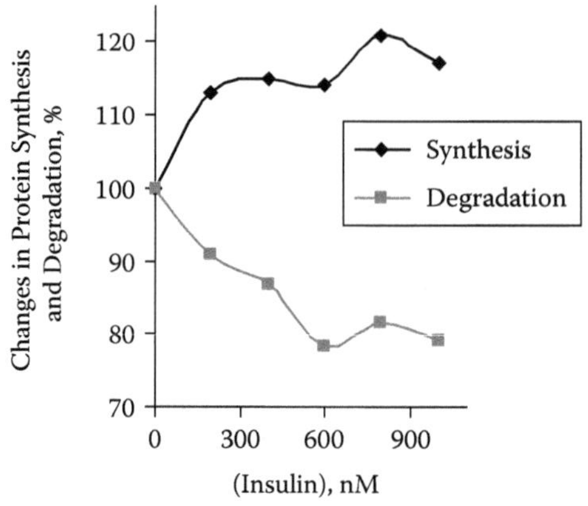

FIGURE 6.2 The influence of insulin on protein synthesis and degradation in porcine satellite cell derived myotube cultures. Porcine satellite cells were grown and differentiated as described in Figure 6.1. The rate of synthesis was calculated based on total proteins and expressed as d.p.m./µg protein. Results are presented relative to control. The rate of protein degradation was calculated as [TCA soluble d.p.m. released to the medium (F)/F+TCA insoluble d.p.m. to the medium + d.p.m. TCA precipitated in the monolayer] × 100 (Pedersen et al. 2001), and expressed relative to control. Results show that the rate of protein synthesis increased and the rate of protein degradation decreased with addition of porcine insulin to the cultures.

Insulin, an anabolic hormone, stimulates protein synthesis (Goldberg et al. 1980; Oddy, 1993). Thus, the increase in insulin in response to feed intake may explain the increased protein synthesis observed in the same period (Preedy and Garlick 1986). The response to insulin is rapid, which suggests that insulin stimulates the synthesis-activity of existing RNA (increased efficiency of translation) (Brown et al. 1981; Garlick et al. 1983), and that the role of insulin is to maintain a "normal" translational efficiency (Reeds and Davis 1992). However, infusion of insulin to post absorptive rats was not able to restore the same protein synthesis or the same protein synthesis activity per RNA as in fed rats, although the insulin level exceeded physiological values (Garlick et al. 1983). This fact was later explained via results produced by Preedy and Garlick (1986) and Garlick and Grant (1988), showing that insulin and amino acids seem to act together in stimulating muscle protein synthesis, and that amino acids change the insulin threshold concentration necessary for muscle protein synthesis. Insulin infusion has also been shown to have a depressing effect on muscle protein degradation in lambs (Douglas et al. 1991; Oddy 1993), which is also supported by major reduction in protein degradation (both myofibrillar and total) in L6 myotubes, human skeletal muscle cells, and porcine satellite cell culture upon insulin administration (Fernandez and Sainz 1997; Crown et al. 2000; Oksbjerg, unpublished data) (Figure 6.2). However, in cultured mammalian cells, it was shown that the major anabolic response to insulin is on protein synthesis (Ballard 1982), but the effect of insulin might differ between ruminant and non-ruminant tissues (Lobley 1992).

Growth hormone and IGF-I are also anabolic hormones that stimulate muscle protein synthesis in rats (Goldberg 1969a), pigs (Séve et al. 1993), lambs (Douglas et al. 1991), steers (Eisemann et al. 1989), and porcine satellite cell culture (Oksbjerg, unpublished data; Figure 6.1). In addition, Douglas et al. (1991) saw a decreased loss of protein from the whole body upon IGF-I administration to lambs, and Oddy (1993) saw a decreased rate of muscle protein degradation in response to IGF-I infusion in lambs, supported by similar results obtained in L6 myotubes and human skeletal myotubes upon IGF-I administration (Fernandez and Sainz 1997; Crown et al. 2000,), whereas we found very little response of IGF-I administration on degradation in porcine satellite cell culture (Oksbjerg, unpublished data, see Figure 6.1). Treatment with GH *in vivo* increases the DNA content of muscle tissue (Beermann et al, 1990; Maltin et al. 1990). This suggests increased satellite cell proliferation after GH treatment. Treatment of pigs with GH also increases the FBR (Sevé et al. 1993).

Regarding protein accretion, glucocorticoids are generally catabolic hormones (Goldberg et al. 1980; Spencer 1985) and thus suppress protein synthesis and increase protein degradation (Goldberg 1969b; Reeds 1989; Reeds and Davis 1992). The site of action is suggested to be both at the transcriptional and the translational level (Reeds and Davis 1992).

The thyroid hormones are generally known to increase whole body protein turnover; however, at thyroid levels above normal, the protein degradation is stimulated without an additional increase in synthesis (Goldberg et al. 1980; Spencer 1985; Jepson et al. 1988; Skjaerlund et al. 1988). Zeman et al. (1986) have shown that the proteolytic effect of triiodothyronine (T_3) might be linked to the proteolytic calpains, as the stimulating effect of T_3 on muscle protein degradation was depressed if an inhibitor against calpain (leupeptin) was added. Thyroid hormones also seem to have a regulatory effect on the activity of the cathepsins (Goldberg et al. 1980). The thyroid hormones are suggested to act at both the transcriptional and translational level but they are probably not involved in the acute regulation of translation, as is insulin (Brown et al. 1981).

The sex-hormones testosterone and estradiol stimulate muscle growth, and especially high levels of testosterone may explain the larger muscle mass in male animals. Testosterone stimulates protein synthesis (rats) (Martinez et al. 1984) and suppresses protein degradation (lambs) (Lobley et al. 1987). In addition, the effect of testosterone seems to depend on animal maturity, as no effect was found on protein synthesis and degradation in intact and castrated pigs from birth to 4 weeks of age (Skjaerlund et al. 1994). The anabolic effect of the sex hormones has been suggested to be indirect via the GH-IGF-axis because the IGF-I concentration increased when steers were implanted with estradiol or a mixture of trenbolone acetate and estradiol (Breier et al. 1988; Johnson et al. 1998). Lack of effect on the protein turnover when a skeletal muscle cell line was treated with either testosterone or estradiol (Desler et al. 1996) supports this idea.

6.4 DEPENDENCY OF AGE

Total daily protein synthesis and degradation usually increase with increasing live weight as found in pigs (Reeds et al. 1980), bulls (Morgan et al. 1993a; Van Eenaeme et al. 1994, 1998), and heifers (Harris and Milne 1981). However, both FSR and FBR decrease with age in rats (Millward et al. 1975; Garlick et al. 1989), lambs (Oddy and Lindsay 1985), and in cattle (Harris and Milne 1981; McCarthy et al. 1983; Lobley et al. 2000). A linear relationship between the fractional growth rate (growth rate in percentage of total weight) and the FBR has been found in well-fed rats (Millward et al. 1975). Similarly, Gopinath and Kitts (1984) observed that fluctuations in FBR corresponded to changes in growth rate of steers. Accordingly, a high growth rate was accompanied by high rates of muscle protein breakdown. Consequently, in these situations, the FSR may be elevated as well.

The changes in the protein turnover with age can be related to changes in the activity of enzymes involved. Thus, Ou et al. (1991), Northcutt et al. (1998), and Veiset et al. (2004) found a decreased activity of the calpain system with increased age. Ou et al. (1991) observed decreased activity of both μ- and m-calpain and calpastatin in lambs from 1 day of age to 3 months, where after there were no further changes up until 6 months of age. In contrast, Veiset et al. (2004) observed changes in calpastatin activity from 2 to 10 months of age in lambs, no change in μ-calpain, and a decreased activity in m-calpain from 6 to 10 months of age. In turkeys, Northcutt et al. (1998) found decreased activity of μ- and m-calpain and calpastatin from 5 to 9 weeks of age, where after there were no changes through to 17 weeks of age.

6.5 DEPENDENCY ON FIBER TYPES

The fiber type seems to influence the protein turnover rate. El Haj et al. (1986) and Garlick et al. (1989) have found higher protein synthesis and protein degradation rates in rat muscles rich in slow-twitch fibers (Type I) compared with muscles rich in fast-twitch fibers (Type II). Garlick et al. (1989) found a general positive relationship between fractional synthesis rate and percent area of slow oxidative fibers. As there was no relationship between the ratio of fast glycolytic:fast oxidative and glycolytic (Type IIB:Type IIA) and protein synthesis, they suggested that the speed of contraction is the most important factor determining protein turnover rates. Cattle possessing double muscling, as known from the Belgian white-blue breed, have a larger frequency of Type IIB fibers and less Type IIA fibers compared with normal cattle (Batjoens et al. 1989) and, in addition, a lower protein synthesis. This could suggest a difference in the protein synthesis capacity between Type IIA and Type IIB fibers, which differ in their oxidative capacity. In studies on heifers (Therkildsen et al. 2002b) and pigs (Therkildsen et al. 2004), the RNA concentration was higher in muscles rich in Type I fibers — that is, *M. supraspinatus* (SS, heifers) and *M. semitendinosus*, red part (STr, pigs) — compared with *M. longissimus dorsi* (LD) rich in Type IIB fibers; this supports a larger synthesis capacity in Type I fibers, and is also coincident with a higher concentration of elongation factor 2 (eEF-2) in muscles rich in Type I fibers compared with muscles rich in Type IIB fibers. eEF-2 works distally to both transcription and translation, and may be an even better indicator of muscle protein synthesis (Therkildsen et al. 2004). In the study on heifers (Therkildsen et al. 2002b), LD responded more to an increased energy intake than SS with respect to percentage of weight change, increased protein synthesis capacity, and muscle fiber hypertrophy, which suggests that different muscle types and thus fiber types react differently to changes in the general protein turnover, whereas this difference in response between muscle types did not occur between pig LD and ST (Therkildsen et al. 2004).

Muscles containing mainly slow-twitch fibers show higher activities of μ-calpain than muscles containing mainly fast-twitch fibers (Ouali and Talmant 1990; Whipple and Koohmararie 1992; Therkildsen et al. 2002b) but at the same time there might also be increased calpastatin activity (Ouali and Talmant 1990; Therkildsen et al. 2002b); thus, the internal relationship between calpain and the inhibitor calpastatin will have a major influence if these differences lead to the differences seen in protein degradation between fiber types.

6.6 GENOTYPES

Various genotypes are often characterized by their different growth potentials, which often can be explained by differences in muscle protein turnover — either protein synthesis, protein degradation, or both.

Sheep selected over ten generations for high or low weaning weight differed with respect to protein synthesis and degradation. Thus, the lambs selected for high weaning weight had a higher protein synthesis and degradation in *M. vastus lateralis* (Speck et al. 1993). Likewise, Harbison et al. (1976) measured differences in the RNA concentration and the RNA:DNA ratio in pigs from a muscular line compared with an obese line, indicating that the muscular line had a higher protein synthesis capacity. These results were supported by Kretchmar et al. (1994), who observed higher activity of calpastatin in an obese pig strain compared with a lean pig strain, suggesting a general depression of protein turnover in the obese pig strain.

The callipyge gene in sheep causes extreme muscling in some muscles (Koohmaraie et al. 1995); and compared with muscles from normal lambs, the total content of protein, RNA, and DNA and the RNA:DNA-ratio are increased in callipyge lambs (Koohmaraie et al. 1995; Lorenzen et al. 2000). On the other hand, both the fractional synthesis rate and the fractional degradation rate are reduced in the callipyge lambs, suggesting that the callipyge-gene-induced hypertrophy is a result of a reduction in the rate of protein degradation rather than an increase in the rate of protein synthesis. This is supported by an increased calpastatin activity in the hypertrophied muscles (Koohmaraie et al. 1995; Lorenzen et al. 2000), indicating that proteolysis is decreased in these callipyge muscles.

Double muscling (DM) in cattle, especially known from Belgian White-blue cattle, is another inherited characteristic related to the myostatin gene and expressed as high daily gain and extreme muscling (Fiems et al. 1995; Georges et al. 1998). When protein turnover from DM-cattle is compared with normal (N) cattle, the DM-cattle have a lower whole body muscle protein turnover, and — measured *in vitro* on a muscle biopsy — a lower protein synthesis (Van Eenaeme et al. 1991). Moreover, Uytterhaegen et al. (1994b) have shown that μ-calpain and the μ-calpain:calpastatin-ratio at 1 hour postmortem are lower in the DM-cattle than in N-cattle, which indicates a lower proteolysis in the DM-cattle. If the levels of the enzymes at 1 hour postmortem resemble the level in the living muscles, this indicates lower protein degradation *in vivo* in DM-cattle, which thus contributes in combination with an increased number of Type IIB fibers to the higher muscle yield in these cattle. That this lower protein degradation does not result in a decreased tenderness postmortem (Wheeler et al. 2001; Ngapo et al. 2002) might be explained by a lower amount of connective tissue in the meat (Ngapo et al. 2002), caused by a larger number of Type IIB fibers, and characterized with a larger area.

The above-mentioned examples of the genetic effect on muscle growth clearly demonstrate that muscle growth can be regulated in different ways by changes in protein synthesis and degradation, although the final result is an increased growth of the muscle.

6.7 INFLUENCE OF EXERCISE ON MUSCLE PROTEIN TURNOVER

The effect of exercise on protein turnover in muscles is ambiguous, as it depends on the kind of exercise and the type of muscle studied (reviewed by Goldspink 1991). High-intensity resistance exercise training induces hypertrophy of the muscles involved in exercise by accumulation of myofibrillar proteins in the muscle fibers. Depending on the muscle, this accumulation has been associated with increased synthesis and either simultaneously increased or decreased degradation (Goldspink 1991). With endurance training, which more resembles the exercise production animals get when they are loose-housed or grazed compared with tie-stall housed, the main change in the muscles is enhancement of the fatigue resistance of the skeletal muscles, and only minor changes are seen in the muscle size. However, the muscle protein synthesis is decreased during the exercise but is increased at the completion of the exercise (Goldspink 1991). Lambs exercised up to 35 min a day, five times per week on a treadmill for 30 days had higher concentration of fast-twitch isoform myosin light chain-1 ($MLC1_f$) mRNA in *M. biceps femoris* compared with control lambs. The $MLC1_f$ mRNA concentration was used as an index for myofibrillar protein synthesis. After 60 days of exercise, the difference had disappeared (Garrett et al. 1999). Thus, the synthesis rate was increased with exercise but adapted to the exercise with time. This adaptation is supported by results on rats trained for 4 weeks, where no difference was found in either protein synthesis or degradation (Davis and Karl 1986). The pattern for skeletal muscle protein degradation during endurance training and in the recovery period are less well understood (Goldspink 1991). However, Kasperek et al. (1995) saw an increased degradation of total muscle protein but no change in the degradation of myofibrillar protein immediately after treadmill exercise or during the recovery period of rats. This agrees with Garrett et al. (1999), who used the activity of μ-calpain and calpastatin as an index for myofibrillar protein degradation, and found no effect from treadmill exercise of lambs on the activity of the enzyme system in *M. biceps femoris*. In a study with slaughter pigs, Ertbjerg et al. (1999) found an increased activity of μ-calpain and no change in calpastatin upon epinephrine injection and exercise on treadmill right before slaughter, suggesting an increased muscle protein degradation; however, the extreme energy depletion caused by the epinephrine injection would seldom be seen in a traditional production system.

6.8 INFLUENCE OF FEEDING LEVEL ON MUSCLE PROTEIN TURNOVER

The nutritional level has a marked influence on muscle protein turnover. Thus, when the energy level is restricted below maintenance for a period of time (in the short term), the protein synthesis decreases as expected but the protein degradation increases in order to supply the animal with sufficient nutrients (i.e., glucose and amino acids) (Millward and Waterlow 1978; McDonagh et al. 1999). In contrast, when both cattle and pigs are on a restricted diet (in the long term), both protein synthesis and degradation decreased, compared with cattle and pigs with free access to feed (Jones et al. 1990; Van Eenaeme et al. 1994, 1998; Therkildsen et al. 2004). In newborn lambs, a higher energy supply increased muscle protein synthesis as expected but had no effect on the protein degradation (Speck et al. 1993), so the developmental stage of the animal (maturity) seems to influence the effect of energy level on protein degradation.

Protein supply also has an effect on protein turnover rates. Thus, Reecy et al. (1996) showed that steers offered a protein-restricted diet had a decreased FSR and FBR compared with steers offered a diet containing 150% of the dietary nitrogen found in the protein-restricted diet. Likewise, Gerrits et al. (1998) observed increased fractional degradation rates upon increased protein intake in calves, and protein synthesis was highly correlated with protein intake in pigs (Reeds et al. 1980). In a study with pigs fed only 61.4% of the lysine requirement, the effect on protein turnover differed between muscles, thus in *M. longissimus dorsi* only protein synthesis was depressed, whereas in *M. masseter, M. adducter,* and *M. biceps femoris,* both protein synthesis and degradation were depressed upon lysine restriction (Chang and Wei 2005). Thus, although all muscles responded with a decreased fractional accretion rate, this is a result of various changes in protein synthesis and degradation, depending on muscle. In the same study, the depression in protein synthesis could be explained by a depression in RNA activity, and not in the RNA capacity; thus, regulation of the synthesis is in the translational step. In agreement with the effect of feeding sub-maintenance levels of energy, Fuller et al. (1987) and Van den Hemel-Grooten et al. (1995) found that protein restriction below maintenance or protein-free rations increased protein degradation in pigs. However, Van den Hemel-Grooten et al. (1995) did not see any change in activity of the calpain system, the cathepsins, or of the proteasome on the protein-free ration compared with controls; thus, these enzymes seemed not to be responsible for the increased protein degradation in the fasted state. Furthermore, the protein-free ration decreased the protein synthesis capacity at the transcription and the translation level (Van den Hemel-Grooten et al. 1995).

A model summarizing the expected relationship between muscle protein turnover — that is, muscle protein synthesis and degradation — and the level of feed intake is presented in Figure 6.3.

The conclusion is that by increasing the amount of a well-balanced diet to animals in growth, both protein synthesis and degradation can be expected to increase, although with different rates, and give rise to increased muscle protein accretion.

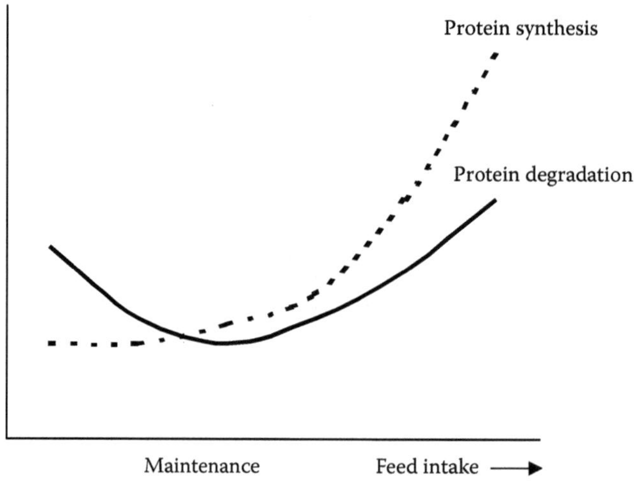

FIGURE 6.3 Model of the expected relationship between muscle protein synthesis and degradation, depending on level of feed intake in growing animals. (*Source:* Modified from Lobley 1998.)

6.9 COMPENSATORY GROWTH

Muscle protein turnover is not only dependent on the current energy intake for the time being, but also on the previous energy intake. Jones et al. (1990) and Therkildsen (2005) found that when previously restricted cattle were given free access to feed, both protein synthesis and degradation increased, and this to a significantly higher FSR and FBR than seen in cattle of the same age but on a continuously high feeding level. This pattern in protein turnover in response to energy level was also found in rats (Millward et al. 1975) and in lambs (Harris et al. 1992). The increase in protein turnover after re-alimentation seems to reach a maximum, which has been found to be between 42 and 70 days in slaughter pigs (Therkildsen et al. 2002c), after 66 days of re-alimentation in steer (Jones et al. 1990), and after 35 to 56 days in bull calves (Therkildsen 2005). The time of maximal protein turnover probably depends on the degree of previous restriction, length of restriction period, sex, and genotype. This increase in muscle protein turnover after re-alimentation has been suggested to explain the phenomenon of compensatory growth (Therkildsen et al. 2002c), which is observed in animals exposed to a period with restrictive access to feed, followed by a period with *ad libitum* access to feed, in which their growth rate will exceed the growth rate of animals offered feed *ad libitum* continuously (McMeekan 1940; Abdalla et al. 1988; Critser et al. 1995; Rossi et al. 2001; Oksbjerg and Therkildsen 2007).

6.10 LINK BETWEEN *IN VIVO* MUSCLE PROTEIN TURNOVER AND MEAT TENDERNESS

Tenderness development in meat postmortem is, to a large degree, an effect of proteolysis of the structural proteins in the muscle fiber, and the calpain system has been shown to be the rate-limiting enzyme system involved in this degradation. Likewise, the calpain system seems to be the rate-limiting enzyme system *in vivo* in muscle protein degradation. Thus, several studies have shown that manipulation with muscle protein turnover *in vivo* (e.g., through feeding level) (Aberle et al. 1981, McDonagh et al. 1999; Kristensen et al. 2002, 2004; Therkildsen et al. 2002a), growth path (Allingham et al. 1998; Kristensen et al. 2002, 2004; Therkildsen et al. 2008) and castration (Morgan et al. 1993b) have a marked influence on meat tenderness. However, recently it was also shown that different sexes and muscles may respond differently to these treatments (Kristensen et al. 2004; Therkildsen et al. 2008).

6.11 SUMMARY

In summary, muscle growth is a function of the difference between the rates of muscle protein synthesis and muscle protein degradation. Both the rate of muscle protein synthesis and degradation are higher than the rate of muscle growth. Consequently, small changes in either the rate of protein synthesis or degradation, or both, may cause profound changes in muscle growth. By age, both the fractional rates of protein synthesis and degradation decrease; and in mature animals, these rates are equal. Muscle protein turnover is regulated by hormones, of which insulin, GH/IGF/IGFBPs, and sex hormones are anabolic and glucocorticoids, which are catabolic. Feeding intensity (protein, energy, or global) is related with increasing rates of both protein synthesis and degradation, and the protein turnover is higher in slow-twitch muscles compared with fast-twitch muscles. Variations in growth by genotypes are also related to protein turnover, and also to the number of muscle fibers. Muscle protein turnover *in vivo* is linked to postmortem tenderness development, through the proteolytic enzymes responsible for protein degradation, as they are active both *in vivo* and postmortem. Thus, high protein degradation at slaughter will continue postmortem and affect meat tenderness in a positive way.

REFERENCES

Abdalla, H.O., D.G. Fox, and M.L. Thonney. 1988. Compensatory gain by Holstein calves after underfeeding protein. *J. Anim. Sci.* 66:2687–2695.

Aberle, E.D., E.S. Reeves, M.D. Judge, R.E. Hunsley, and T.W. Perry. 1981. Palatability and muscle characteristics of cattle with controlled weight gain: time on a high energy diet. *J. Anim. Sci.* 52:757–763.

Allingham, P.G., G.S. Harper, and R.A. Hunter. 1998. Effect of growth path on the tenderness of the *semitendinosus* muscle of Brahman-cross steers. *Meat Sci.* 48:65–73.

Asgar, A. and A.R. Bhatti. 1987. Endogenous proteolytic enzymes in skeletal muscle: their significance in muscle physiology and during postmortem aging events in carcasses. *Adv. Food Res.* 31:343–451.

Ballard, F.J. 1982. Regulation of protein accumulation in cultured cells. *Biochem. J.* 208:275–287.

Batjoens, P., J. Van Hoof, and D. Vereecke. 1989. The influence of muscle fiber composition on some meat quality characteristics in young bulls. *Proc. 35th ICoMST*, Copenhagen, Denmark 1082–1087.

Beermann, D.H., V.K. Fishell, K. Roneker, R.D. Boyd, G. Armbruster, and L. Souza. 1990. Dose-response relationships between porcine somatotropin, muscle composition, muscle fiber characteristics and pork quality. *J. Anim. Sci.* 68:2690–2697.

Breier, B.H., P.D. Gluckman, and J.J. Bass. 1988. The somatotrophic axis in young steers: influence of nutritional status and oestradiol-17β on hepatic high- and low-affinity somatotrophic binding sites. *J. Endocr.* 116:169–177.

Brown, J.G., P.C. Bates, M.A. Holliday, and D.J. Millward. 1981. Thyroid hormones and muscle protein turnover. *Biochem. J.* 194:771–782.

Calkins, C.R. and S.C. Seideman. 1988. Relationships among calcium-dependent protease, cathepsins B and H, meat tenderness and the response of muscle to aging. *J. Anim. Sci.* 66:1186–1193.

Chang, Y.-M. and H.-W. Wei. 2005. The effect of dietary lysine deficiency on muscle protein turnover in postweanling pigs. *Asian-Austr. J. Anim. Sci.* 18:1326–1335.

Critser, D.J., P.S. Miller, and A.J. Lewis. 1995. The effects of dietary protein concentration on compensatory growth in barrows and gilts. *J. Anim. Sci.* 73:3376–3383.

Crown, A.L., X.L. He, J.M.P. Holly, S.L. Lightman, and C.E.H. Stewart. 2000. Characterisation of the IGF system in a primary adult human skeletal muscle cell model, and comparison of the effects of insulin and IGF-I on protein metabolism. *J. Endocrinol.* 167:403–415.

Davis, T.A. and I.E. Karl. 1986. Response of muscle protein turnover to insulin after acute exercise and training. *Biochem. J.* 240:651–657.

Desler, M.M., S.J. Jones, C.W. Smith, and T.L. Woods. 1996. Effects of dexamethasone and anabolic agents on proliferation and protein synthesis and degradation in C2C12 myogenic cells. *J. Anim. Sci.* 74:1265–1273.

Douglas, R.G., P.D. Gluckman, K. Ball, B. Breier, and J.H.F. Shaw. 1991. The effects of infusion of insulinlike growth factor (IGF) I, IGF-II, and insulin and glucose and protein metabolism in fasted lambs. *J. Clin. Invest.* 88:614–622.

Eisemann, J.H., A.C. Hammond, and T.S. Rumsey. 1989. Tissue protein synthesis and nucleic acid concentration in steers treated with somatotropin. *Br. J. Nutr.* 62:657–671.

El Haj, A.J., S.E.M. Lewis, D.F. Goldspink, B.J. Merry, and A.M. Holehan. 1986. The effect of chronic and acute dietary restriction on the growth and protein turnover of fast and slow types of rat skeletal muscle. *Comp. Biochem. Physiol.* 85A:281–287.

Ertbjerg, P., P. Henckel, A. Karlsson, L.M. Larsen, and A.J. Møller. 1999. Combined effect of epinephrine and exercise on calpain/calpastatin and cathepsin B and L activity in porcine longissimus muscle. *J. Anim. Sci.* 77:2428–2436.

Fernandez, C. and R.D. Sainz. 1997. Pathways of protein degradation in L6 myotubes (44092). *Proc. Soc. Exp. Biol. Med.* 214:242–247.

Fiems, L.O., J. Van Hoof, L. Uytterhaegen, C. v. Boucqué, and D.I. Demeyer. 1995. Comparative quality of meat from double-muscled and normal beef cattle. In *Expression of Tissue Proteinases and Regulation of Protein Degradation as Related to Meat Quality*. A. Ouali, D.I. Demeyer, and F.J.M. Smulders, Eds., ECCEAMST, Utrecht, The Netherlands, p. 381–393.

Fuller, M.F., P.J. Reeds, A. Cadenhead, B. Seve, and T. Preston. 1987. Effects of the amount and quality of dietary protein on nitrogen metabolism and protein turnover of pigs. *Br. J. Nutr.* 58:287–300.

Garlick, P.J., M. Fern, and V.R. Preedy. 1983. The effect of insulin infusion and food intake on muscle protein synthesis in postabsorptive rats. *Biochem. J.* 210:669–676.

Garlick, P.J. and I. Grant. 1988. Amino acid infusion increases the sensitivity of muscle protein synthesis *in vivo* to insulin. *Biochem. J.* 254:579–584.

Garlick, P.J., C.A. Maltin, A.G.S. Baillie, M.I. Delday, and D.A. Grubb. 1989. Fiber-type composition of nine rat muscles. II. Relationship to protein turnover. *Am. J. Physiol.* 257:828–832.

Garrett, R.P., K.G. Britain, J.W. Savell, J.W. Edwards, and S.B. Smith. 1999. Body composition of lambs receiving 30 or 60 days of exercise training and (or) fenoterol treatment. *Meat Sci.* 52:235–246.

Georges, M., L. Grobet, D. Poncelet, L.J. Royo, D. Pirottin, and B. Brouwers. Positional candidate cloning of the bovine *MH* locus identifies an allelic series of mutations disrupting the myostatin function and causing double-muscling in cattle. *Proc. 6th World Congr. Genetica Applied to Livestock Production,* Armidale, Australia, January 11–16, 1998, p. 195–202.

Gerrits, W.J.J., J.W. Schrama, S. Tamminga, and M.W.A. Verstegen. 1998. Effect of protein and protein-free energy intake on myofibrillar protein degradation in preruminant calves of 120 and 200 kilograms of live weight. *J. Anim. Sci.* 76:1370.

Goldberg, A.L. 1969a. Protein turnover in skeletal muscle. I. Protein catabolism during work-induced hypertrophy and growth induced with growth hormone. *J. Biol. Chem.* 244:3217–3222.

Goldberg, A.L. 1969b. Protein turnover in skeletal muscle. II. Effects of denervation and cortisone on protein catabolism in skeletal muscle. *J. Biol. Chem.* 244:3223–3229.

Goldberg, A.L. and J.F. Dice. 1974. Intracellular protein degradation in mammalian and bacterial cells. *Annu. Rev. Biochem.* 43:835–869.

Goldberg, A.L., M. Tishler, G. DeMartino, and G. Griffin. 1980. Hormonal regulation of protein degradation and synthesis in skeletal muscle. *Federation Proc.* 39:31–36.

Goldspink, D.F. 1991. Exercise-related changes in protein turnover in mammalian striated muscle. *J. Exp. Biol.* 160:127–148.

Goll, D.E., V.F. Thompson, R.G. Taylor, and A. Ouali. 1998. The calpain system and skeletal muscle growth. *Can. J. Anim. Sci.* 78:503–512.

Gopinath, R. and W.D. Kitts. 1984. Growth, N^{r}-methylhistidine excretion and muscle protein degradation in growing beef steers. *J. Anim. Sci.* 59:1262–1269.

Harbison, S.A., D.E. Goll, F.C. Parrish, Jr., V. Wang, and E.A. Kline. 1976. Muscle growth in two genetically different lines of swine. *Growth* 40:253–283.

Harris, C.I. and G. Milne. 1981. The urinary excretion of N^{r}-methyl histidin by cattle: validation as an index of muscle protein breakdown. *Br. J. Nutr.* 45:411–422.

Harris, P.M., P.A. Skene, V. Buchan, E. Milne, A.G. Calder, S.E. Anderson, A. Connell, and G.E. Lobley. 1992. Effect of food intake on hind-limb and whole body protein metabolism in young growing sheep: chronic studies based on arterio-venous techniques. *Br. J. Nutr.* 68:389–407.

Houbak, M.B., P. Ertbjerg, and M. Therkildsen. 2008. *In vitro* study to evaluate the degradation of bovine muscle proteins *post-mortem* by proteasome and μ-calpain. *Meat Sci.* 79:77–85.

Jepson, M.M., P.C. Bates, and D.J. Millward. 1988. The role of insulin and thyroid hormones in the regulation of muscle growth and protein turnover in response to dietary protein in the rat. *Br. J. Nutr.* 59:397–415.

Johnson, B.J., N. Halstead, M.E. White, M.R. Hathaway, A. DiCostanzo, and W.R. Dayton. 1998. Activation state of muscle satellite cells isolated from steers implanted with a combined trenbolone acetate and estradiol implant. *J. Anim. Sci.* 76:2779–2786.

Jones, S.J., D.L. Starkey, C.R. Calkins, and J.D. Crouse. 1990. Myofibrillar protein turnover in feed-restricted and realimented beef cattle. *J. Anim. Sci.* 68:2707–2715.

Kasperek, G.J. 1995. Effect of exercise during dietary restriction on skeletal muscle protein degradation in the rat. *Nutr. Res.* 15:517–520.

Koohmaraie, M., M.P. Kent, S.D. Shackelford, E. Veiseth, and T.L. Wheeler. 2002. Meat tenderness and muscle growth: is there any relationship? *Meat Sci.* 62:345–352.

Koohmaraie, M., S.D. Shackelford, T. L. Wheeler, S.M. Lonergan, and M.E. Doumit. 1995. A muscle hypertrophy condition in lamb (Callipyge): characterization of effects on muscle growth and meat quality traits. *J. Anim. Sci.* 73:3596–3607.

Kretchmar, D.H., M. Koohmaraie, and H.J. Mersmann. 1994. Comparison of proteolytic variables in a lean and obese strain of pig at the ages of 2.5 and 7 months. *Lab. Anim. Sci.* 44:38–41.

Kristensen, L., M. Therkildsen, B. Riis, M.T. Sørensen, N. Oksbjerg, P.P. Purslow, and P. Ertbjerg. 2002. Dietary-induced changes of muscle growth rate in pigs: effect on *in vivo* and postmortem muscle proteolysis and meat quality. *J. Anim. Sci.* 80:2862–2871.

Kristensen, L., M. Therkildsen, M. Dall Aaslyng, N. Oksbjerg, and P. Ertbjerg. 2004. Compensatory growth improves meat tenderness in gilts but not in barrows. *J. Anim. Sci.* 82:3617–3624.

Lobley, G.E. 1992. Control of the metabolic fate of amino acids in ruminants: a review. *J. Anim. Sci.* 70:3264–3275.

Lobley, G.E. 1998. Nutritional and hormonal control of muscle and peripheral tissue metabolism in farm species. *Livest. Prod. Sci.* 56:91–114.

Lobley, G.E., A. Connell, V. Buchan, P.A. Skene, and J.M. Fletcher. 1987. Administration of testosterone to wether lambs: effect on protein and energy metabolism and growth hormone status. *J. Endocrinol.* 115:439–445.

Lobley, G.E., V. Milne, M.J.M. Lovie, P.J. Reeds, and K. Pennie. 1980. Whole body and tissue protein synthesis in cattle. *Br. J. Nutr.* 43:491–502.

Lobley, G.E., K.D. Sinclair, C.M. Grant, L. Miller, D. Mantle, A.G. Calder, C.C. Warkup, and C.A. Maltin. 2000. The effects of breed and level of nutrition on whole-body and muscle protein metabolism in pure-bred Aberdeen Angus and Charolais beef steers. *Br. J. Nutr.* 84:275–284.

Lorenzen, C.L., M. Koohmaraie, S.D. Shackelford, F. Jahoor, H.C. Freetly, T.L. Wheeler, J.W. Savell, and M.L. Fiorotto. 2000. Protein kinetics in callipyge lambs. *J. Anim. Sci.* 78:78–87.

Maltin, C.A., M.I. Delday, S.M. Hay, G.M. Innes, and P.E.V. Williams. 1990. Effects of bovine pituitary growth hormone alone or in combination with the β-agonist clenbuterol on muscle growth and composition in veal calves. *Br. J. Nutr.* 63:535–545.

Martinez, J.A., P.J. Buttery, and J.T. Pearson. 1984. The mode of action of anabolic agents: the effect of testosterone on muscle protein metabolism in the female rat. *Br. J. Nutr.* 52:515–521.

McCarthy, F.D., W.G. Bergen, and D.R. Hawkins. 1983. Muscle protein turnover in cattle of differing genetic background as measured by urinary N^{r}-methylhistidin excretion. *Br. J. Nutr.* 69:743–755.

McDonagh, M.B., C. Fernandez, and V.H. Oddy. 1999. Hind-limb protein metabolism and calpain system activity influence post-mortem change in meat quality in lamb. *Meat Sci.* 52:9–18.

McMeekan, C. 1940. Growth and development in the pig, with special references to carcass quality characters. III. Effects of plane of nutrition on the form and composition of the bacon pig. *J. Agric. Sci.* 30:511–569.

Millward, D.J., P.J. Garlick, W.P.T. James, D.O. Nnanyelugo, and J.S. Ryatt. 1973. Relationship between protein synthesis and RNA content in skeletal muscle. *Nature* 241:204–205.

Millward, D.J., P.J. Garlick, R.J.C. Stewart, D.O. Nnanyelugo, and J.C. Waterlow. 1975. Skeletal-muscle growth and protein turnover. *Biochem. J.* 150:235–243.

Millward, D.J. and J.C. Waterlow. 1978. Effect of nutrition on protein turnover in skeletal muscle. *Federation Proc.* 37:2283–2290.

Morgan, J.B., T.L. Wheeler, M. Koohmaraie, J.D. Crouse, and J.W. Savell. 1993a. Effect of castration on myofibrillar protein turnover, endogenous proteinases activities, and muscle growth in bovine skeletal muscle. *J. Anim. Sci.* 71:408–414.

Morgan, J.B., T.L. Wheeler, M. Koohmaraie, J.W. Savell, and J.D. Crouse. 1993b. Meat tenderness and the calpain proteolytic system in longissimus muscle of young bulls and steers. *J. Anim. Sci.* 71:1471–1476.

Ngapo, T.M., P. Berge, J. Culioli, E. Dransfield, S. De Smet, and E. Claeys. 2002. Perimysial collagen crosslinking and meat tenderness in Belgian Blue double-muscled cattle. *Meat Sci.* 61:91–102.

Northcutt, J.K., T.D. Pringle, J.A. Dickens, R.J. Buhr, and L.L. Young. 1998. Effects of age and tissue type on the calpain proteolytic system in turkey skeletal muscle. *Poultry Sci.* 77:367–372.

Oddy, H.V. 1993. Regulation in muscle protein metabolism in sheep and lambs: nutritional, endocrine and genetic aspects. *Austr. J. Agric. Res.* 44:901–913.

Oddy, H.V. and D.B. Lindsay. 1985. Determination of rates of protein synthesis, gain and degradation in intact hind-limb muscle of lambs. *Biochem. J.* 233(2):417–425

Oddy, V.H., D.B. Lindsay, P.J. Barker, and A.J. Northrop. 1987. Effect of insulin on hind-limb and whole-body leucine and protein metabolism in fed and fasted lambs. *Br. J. Nutr.* 58:437–452.

Oksbjerg, N. and M. Therkildsen. 2007. Compensatory growth in pigs: effects on performance, protein turnover and meat quality. In *Paradigms in Pig Science*. J. Wiseman, M.A. Varley, S. McOrist and B. Kemp, Eds. Nottingham University Press, p. 417–426.

Ou, B., H.H. Meyer, and N.E. Forsberg. 1991. Effects of age and castration on activities of calpains and calpastatin in sheep skeletal muscle. *J. Anim. Sci.* 69:1919–1924.

Ouali, A. and A. Talmant. 1990. Calpains and calpastatin distribution in bovine, porcine and ovine skeletal muscles. *Meat Sci.* 28:331–348.

Pedersen, P.H., Oksbjerg, N., Karlsson, A.H., Busk, H., Bendixen, E., Henckel, P. 2001. A within litter comparison of muscle fibre characteristics and growth of halothane carrier and haltahan free crossbreed pigs. *Livest. Prod. Sci.* 73:15–24

Pontremoli, S. and E. Melloni. 1986. Extralysosomal protein degradation. *Annu. Rev. Biochem.* 55:455–481.

Preedy, V.R. and P.J. Garlick. 1986. The response of muscle protein synthesis to nutrient intake in postabsorptive rats: the role of insulin and amino acids. *Biosci. Rep.* 6:177–183.

Price, N.C. and L. Stevens. 1999. The cell and molecular biology of catalytic proteins. In *Fundamentals of Enzymology*. Oxford University Press.

Reecy, J.M., J.E. Williams, M.S. Kerley, R.S. MacDonald, W.H. Thornton, Jr.,, and J.L. Davis. 1996. The effect of postruminal amino acid flow on muscle cell proliferation and protein turnover. *J. Anim. Sci.* 74:2158–2169.

Reeds, P.J. 1989. Regulation of protein turnover. In *Animal Growth Regulation*. D.R. Campion, G.J. Hausman, and R.J. Martin, Eds. Plenum Press, New York, p. 183–210.

Reeds, P.J., A. Cadenhead, M.F. Fuller, G.E. Lobley, and J.D. McDonald. 1980. Protein turnover in growing pigs. Effects of age and food intake. *Br. J. Nutr.* 43:445–455.

Reeds, P.J. and T.A. Davis. 1992. Hormonal regulation of muscle protein synthesis and degradation. *In The Control of Fat and Lean Deposition*. K.N. Boorman, P.J. Buttery, and D.B. Lindsay, Eds. Oxford, U.K., Butterworth-Heinemann Ltd. Proc. 51st Easter School, University of Nottingham, Sutton Bonnington, p. 1–26.

Robert, N., M. Briand, R. Taylor, and Y. Briand. 1999. The effect of proteasome on myofibrillar structures in bovine skeletal muscle. *Meat Sci.* 51:149–153.

Rossi, J.E., S.C. Loerch, H.L. Keller, and L.B. Willett. 2001. Effects of dietary crude protein concentration during periods of feed restriction on performance, carcass characteristics, and skeletal muscle protein turnover in feedlot steers. *J. Anim. Sci.* 79:3157.

Séve, B., O. Ballévre, P. Ganier, J. Noblet, J. Prugnaud, and C. Obled. 1993. Recombinant porcine somatotropin and dietary protein enhance protein synthesis in growing pigs. *J. Nutr.* 123:529–540.

Skjaerlund, D.M., D.R. Mulvaney, W.G. Bergen, and R.A. Merkel. 1994. Skeletal muscle growth and protein turnover in neonatal boars and barrows. *J. Anim. Sci.* 72:315–321.

Skjaerlund, D.M., D.R. Mulvaney, R.H. Mars, A.L. Schroeder, M.A. Stachiw, W.G. Bergen, and R.A. Merkel. 1988. Measurement of protein turnover in skeletal muscle strips. *J. Anim. Sci.* 66:687–698.

Speck, P.A., B.C. Thomson, K.M. Collingwood, R.S. Gilmour, R.D. Sainz, R.G. Bardsley, P.J. Buttery, and H.V. Oddy. 1993. Effect of genotype and nutrition on calpastatin inhibitory activity and mRNA abundance in milk-fed lambs. *Biochimie* 75:925–929.

Spencer, G.S.G. 1985. Hormonal systems regulating growth. A review. *Livest. Prod. Sci.* 12:31–46.

Sugden, P.H. and S.J. Fuller. 1991. Regulation of protein turnover in skeletal and cardiac muscle. *Biochem. J.* 273:21–37.

Taylor, R.G., D.E. Goll, and A. Ouali. 1995. Enzyme localization during postmortem muscle tenderization. In *Expression of Tissue Proteinases and Regulation of Protein Degradation as Related to Meat Quality*. A. Ouali, D.I. Demeyer, and F.J.M. Smulders, Eds. ECCEAMST, Utrecht, The Netherlands, p. 347–358.

Therkildsen, M. 2005. Muscle protein degradation in bull calves with compensatory growth. *Livest. Prod. Sci.* 98:205–218.

Therkildsen, M., M. B. Houbak, and D. V. Byrne. 2008. Feeding strategy for improving tenderness has opposite effects in two different muscles. *Meat Sci.* 80:1037–1045.

Therkildsen, M., L. Melchior Larsen, H.G. Bang, and M. Vestergaard. 2002a. Effect of growth rate on tenderness development and final tenderness of meat from Friesian calves. *Anim. Sci.* 74:253–264.

Therkildsen, M., L. Melchior Larsen, and M. Vestergaard. 2002b. Influence of growth rate and muscle type on muscle fibre type characteristics, protein synthesis capacity and activity of the calpain system in Friesian calves. *Anim. Sci.* 74:243–251.

Therkildsen, M., B. Riis, A. Karlsson, L. Kristensen, P. Ertbjerg, P.P. Purslow, M. Dall Aaslyng, and N. Oksbjerg. 2002c. Compensatory growth response in pigs, muscle protein turn-over and meat texture: effects of restriction/realimentation period. *Anim. Sci.* 75:367–377.

Therkildsen, M., M. Vestergaard, H. Busk, M.T. Jensen, B. Riis, A.H. Karlsson, L. Kristensen, P. Ertbjerg, and N. Oksbjerg. 2004. Compensatory growth in slaughter pigs — *in vitro* muscle protein turnover at slaughter, circulating IGF-I, performance and carcass quality. *Livest. Prod. Sci.* 88:63–75.

Uytterhaegen, L., E. Claeys, D. Demeyer, M. Lippens, L.O. Fiems, C.Y. Boucqué, G. Van de Voorde, and A. Bastiaens. 1994. Effects of double-muscling on carcass quality, beef tenderness and myofibrillar protein degradation in Belgian Blue White bulls. *Meat Sci.* 38:255–267.

Van den Hemel-Grooten, H.N.A., M. Koohmaraie, J.T. Yen, J.R. Arbona, J.A. Rathmacher, S.L. Nissen, M.L. Fiorotto, G.J. Garssen, and M.W.A. Verstegen. 1995. Comparison between 3-methylhistidine production and proteinase activity as measures of skeletal muscle breakdown in protein-deficient growing barrows. *J. Anim. Sci.* 73:2273–2281.

Van Eenaeme, C., A. Clinquart, P. Baldwin, V. Hollo, L. Istasse, and J.M. Bienfait. 1991. Muscle protein turnover at whole animal and cellular level in young beef bulls of different breed and type. *Proc. 6th Int. Symp. Protein Metabolism and Nutrition*, 2, 179–180, Herning Denmark.

Van Eenaeme, C., A. Clinquart, L. Uytterhaegen, J. Hornick, D. Demeyer, and L. Istasse. 1994. Post mortem proteases activity in relation to muscle protein turnover in Belgian Blue bulls with different growth rates. *Science des Aliments* 14:475–483.

Van Eenaeme, C., M. Evrard, J.L. Hornick, P. Baldwin, M. Diez, and L. Istasse. 1998. Nitrogen balance and myofibrillar protein turnover in double muscled Belgian Blue Bulls in relation to compensatory growth after different periods of restricted feeding. *Can. J. Anim. Sci.* 78:549–559.

Veiseth, E., S.D. Shackelford, T.L. Wheeler, and M. Koohmaraie. 2004. Factors regulating lamb longissimus tenderness are affected by age at slaughter. *Meat Sci.* 68:635–640.

Wheeler, T.L., S.D. Shackelford, E. Casas, L.V. Cundiff, and M. Koohmaraie. 2001. The effects of Piedmontese inheritance and myostatin genotype on the palatability of longissimus thoracis, gluteus medius, semimembranosus, and biceps femoris. *J. Anim. Sci.* 79:3069–3074.

Whipple, G. and M. Koohmaraie. 1992. Effects of lamb age, muscle type, and 24-hour activity of endogenous proteinases on postmortem proteolysis. *J. Anim. Sci.* 70:798–804.

Zeman, R.J., P.L. Bernstein, R. Ludemann, and J.D. Etlinger. 1986. Regulation of Ca^{2+}-dependant protein turnover in skeletal muscle by thyroxine. *Biochem. J.* 240:269–272.

7 Collagen

Richard J. McCormick

CONTENTS

7.1 INTRODUCTION

Collagens form the scaffold that is responsible for maintaining the structural integrity of skeletal muscle; they play a vital role in muscle growth and development and contribute to muscle diversity, functionality, as well as the quality characteristics of meat. From a quantity/volume standpoint, collagens are by far the largest component of what is generally referred to as the extracellular matrix (ECM).

However, if not as abundant as collagen, other ECM components such as proteoglycans, glycoproteins, receptors, growth factors, and proteases are recognized as contributing essential structural and functional properties to the ECM and to the muscle as a whole.

Traditionally, a distinction has been made between the collagen or ECM "compartment" of muscle and the cell itself. One thesis of this chapter is that the concept of compartmentalization of intracellular and ECM requires modification. This is particularly true if the role of ECM components in muscle growth and development is considered. A second theme of this chapter is the remarkable plasticity of the skeletal ECM; clearly, if an animal is to grow, muscle tissue including the ECM requires the ability to change and adapt. This is obvious in the fetal, neonate, and immature animal, but is an equally valid concept for mature and senescent animals.

This chapter focuses mainly on the collagen and proteoglycan components of muscle, with emphasis on the most current data and their relevance to skeletal muscle growth and development, and meat palatability characteristics.

7.2 ECM: COMPOSITION AND ORGANIZATION

The ECM is comprised of several components, including collagens (by far the most abundant), elastin, proteoglycans, glycoproteins, and transmembrane complexes that anchor the ECM to the cytoskelelton of the myofiber as well as providing receptors for matrix- and cell-associated signaling (Sanes 2003; Jenniskens et al. 2006).

The ECM is arranged hierarchically into several domains (Figure 7.1). The basement membrane is composed of two layers: the basal lamina that directly overlays the sarcolemma and an outer layer, the fibrillar reticular lamina.

The basement membrane is composed of collagenous and non-collagenous proteins, proteoglycans, and glycoproteins. The basal lamina (50 to 100 nm thick) lies close and completely surrounds the muscle cell membrane (sarcolemma). It is composed of, in addition to predominantly Type IV collagen, various protein, proteoglycan, and glycoprotein domains, some of which connect the ECM with muscle cell structures, others with cell surface receptors, and others that respond to multiple growth factors. Jenniskens et al. (2006) list more than 40 components and their isoforms of the basal lamina and the cell surface of muscle cells. Overlying the basement membrane and surrounding each muscle fiber is a relatively thin collagenous layer called the endomysium; it provides a direct physical connection between adjacent myofiber, both head-to-tail and laterally. A considerably thicker

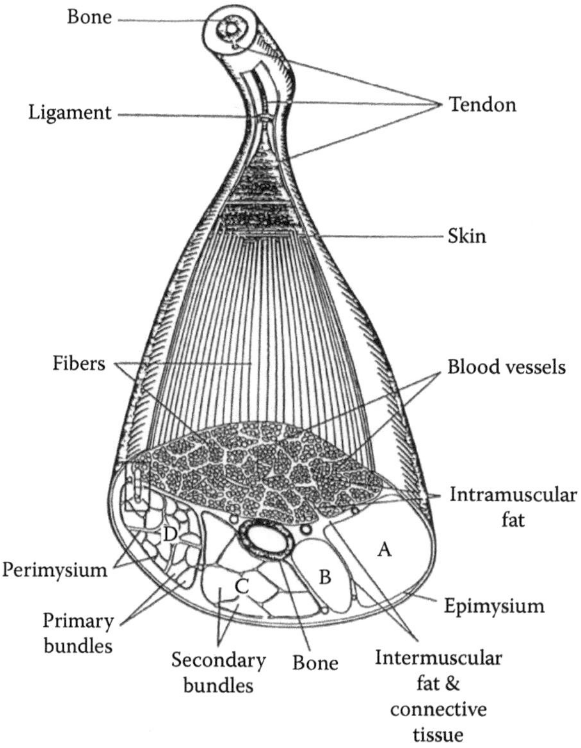

FIGURE 7.1 Gross composition of muscle indicating epimysium and perimysium. (*Source:* from McCormick, 1994b)

collagenous layer, the perimysium, forms bundles or fasciculi of endomysium-ensheathed myofibers. The whole muscle may be surrounded by a thick collagenous sheath, the epimysium, which may be contiguous with tendon, as in a muscle-to-bone connection or form aponeuroses in the case of muscle-to-muscle junctions. By convention, ICMT usually refers specifically to the endo- and-perimysial connective tissues and excludes basement membrane and epimysial levels of the ECM.

7.3 THE COLLAGEN SUPERFAMILY

Barely four decades ago, collagen was thought to be a single molecule; currently, at least 29 distinct collagen types, the products of more than 30 genes, have been identified (Myllyharju and Kivirikko 2004; Veit et al. 2006; Soderhall et al. 2007). The superfamily includes structurally diverse molecules with wide-ranging functions and tissue distributions. By way of standardization, all collagen types are identified by a Roman numeral, (e.g., I to XXIX). Collagen genes are named after the collagen type and the α-chain that is encoded (e.g., COL3A1 for the α-chain for Type III collagen). Collagens are divided into groups: (a) the fibrillar collagens, (b) the network collagens, (c) the filamentous collagens, and (d) the fibril-associated collagens or FACITs and several subgroups including membrane collagens (Hulmes 1992; Boot-Handford et al. 2003; Ricard-Blum et al. 2005). To date in IMCT, collagen Types I, III, V (fibrillar), IV (network), VI (filamentous), XII and XIV (FACITs), and XIII (membrane) and XVIII (multiplexin) have been identified (Bailey and Sims 1977; Listrat et al. 1999; Listrat and Hocquette 2004; Jenniskens et al. 2006). In mammalian IMCT, the most abundant collagens are Types I and III. Across a wide range of mature bovine muscles, in the perimysium Type I collagen predominates with Type III collagen comprising about 24% (L. dorsi) to 43% (P. major) of the combined Type I and Type III total. In contrast, in the endomysium, the proportion of Type III exceeds Type I collagen (53% to 58%, depending on muscle type) (Light et al. 1985). Type IV collagen is the major connective tissue element of the basal lamina and is also found in the endomysium although it occurs in small amounts compared to the fibrillar collagens (Mayne and Sanderson 1985). In the fish myocommata, collagen Types I, III, IV, V, and VI have been identified, with considerably more Type V and much less Type III present than in mammalian muscle (Sato et al. 1998; Bruggemann and Lawson 2005). With regard to both skeletal muscle growth and functionality, in mammals produced for food, research has focused almost exclusively on the most abundant collagens of the IMCT, Types I and III and in fish meat on Types I and V collagens (Sato et al. 1994; Sato et al. 1997). Unlike other tissues, notably cartilage and bone (Eyre and Wu 2005), the functions of the network, filamentous and fibril-associated collagens of skeletal muscle remain largely uninvestigated (Figure 7.2).

7.4 COLLAGEN BIOSYNTHESIS AND ASSEMBLY

Many comprehensive reviews of biosynthesis of collagen exist. The current reviews focus on the more recently described collagens, while somewhat older works provide exhaustive detail on the biosynthesis of the fibrillar collagen Types I, II, III, and V (Nimni and Harkness 1988; Ricard-Blum et al. 2005; Hulmes 2008) (Figure 7.3).

By way of nomenclature, the collagenous domain of each collagen molecule consists of three α-chains with a [Gly-X-Y] repeating motif. The α-chains may be identical (homotrimer) or different (heterotrimer). Type III collagen is a homotrimer; it consists of three identical α-chains. Type I is a heterotrimer with two α-chains alike and one different. The convention for describing different collagens involves assigning an Arabic number and a Roman numeral to each α-chain to differentiate them. For example, $[\alpha\ \mathrm{III}]_3$ is the designation for Type III collagen (three identical α-chains constituting Type III collagen). The designation for Type I collagen is $\alpha 1(\mathrm{I})_2\ \alpha 2(\mathrm{II})$.

In skeletal muscle tissue, collagen is produced primarily by fibroblasts and myocytes. Collagen is synthesized intracellularly in the lumen of the endoplasmic reticulum and extensively modified after translation, both intracellularly and extracellularly.

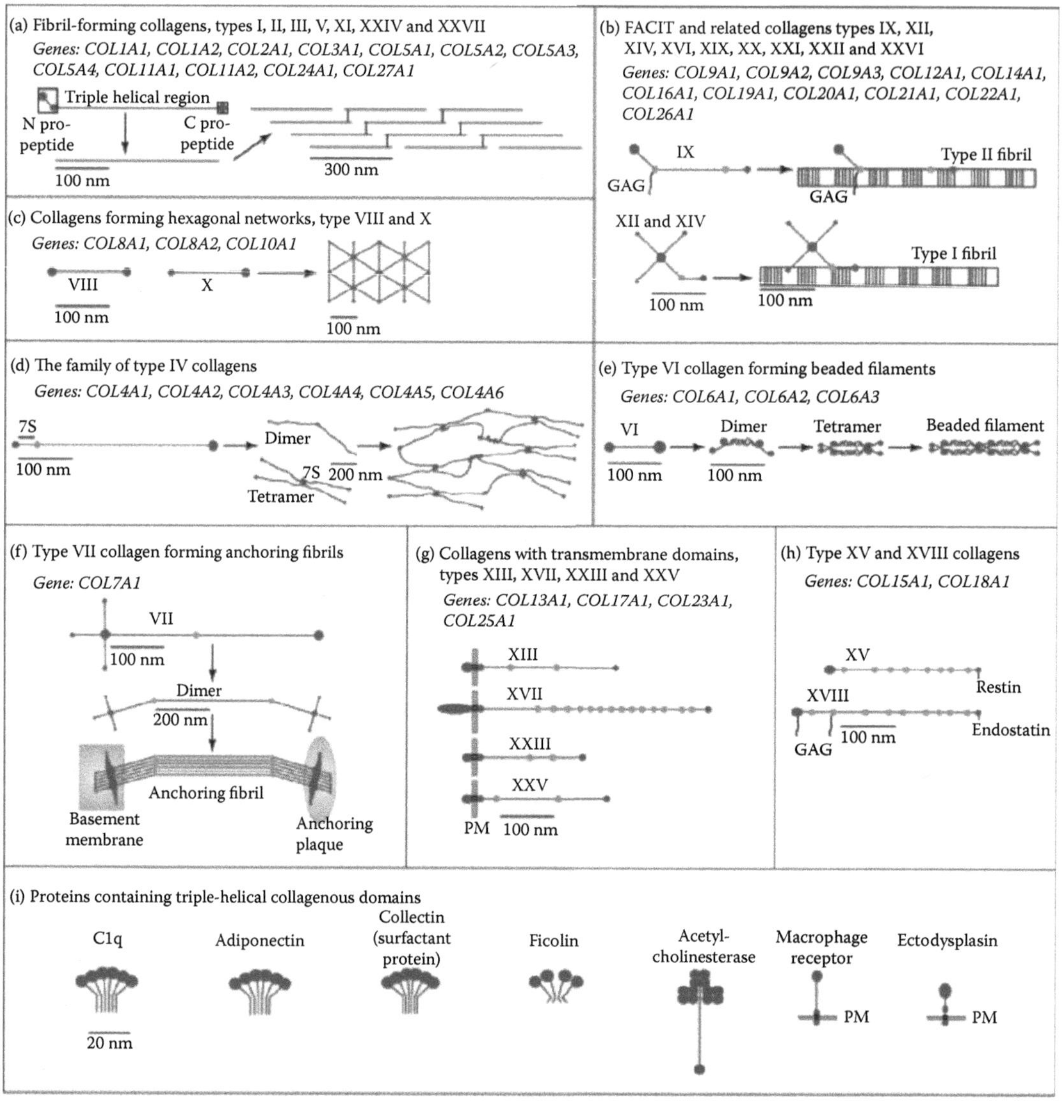

FIGURE 7.2 Members of the collagen superfamily, their supermolecular assemblies, and the genes that encode them. Proteins containing collagenous domains but not considered collagens are also included. Helical domains in purple and non-helical domains in pink. (*Source:* from Myllyharju and Kivirikko 2004.)

Collagen synthesis begins with the mRNAs specific for each individual α-chain, followed by message translation on the rough endoplasmic reticulum (ER). At this point, posttranslational modifications occur beginning with the hydroxylation of specific prolines by 3-proline or 4-proline hydroxylase (the predominant form in most tissues is 4-hydroxyproline). Specific lysine residues are then hydroxylated by lysyl hydroxylase and some hydroxylysines are glycosylated by a galactosyltransferase; some galactosyl residues are then modified by the addition of a glucose via glycosyl transferase. As cofactors, both prolyl and lysyl hydroxylase have requirements for oxygen, ferrous iron, a-ketoglutarate, and ascorbic acid (Nimni and Harkness 1988). Both hydroxylases exist in multiple isoforms. The prolyl 4-hydroxylases exist in several forms, three of which hydroxylate vertebrate collagens. One may be involved in the synthesis of collagen in embryos (Myllyharju 2003).

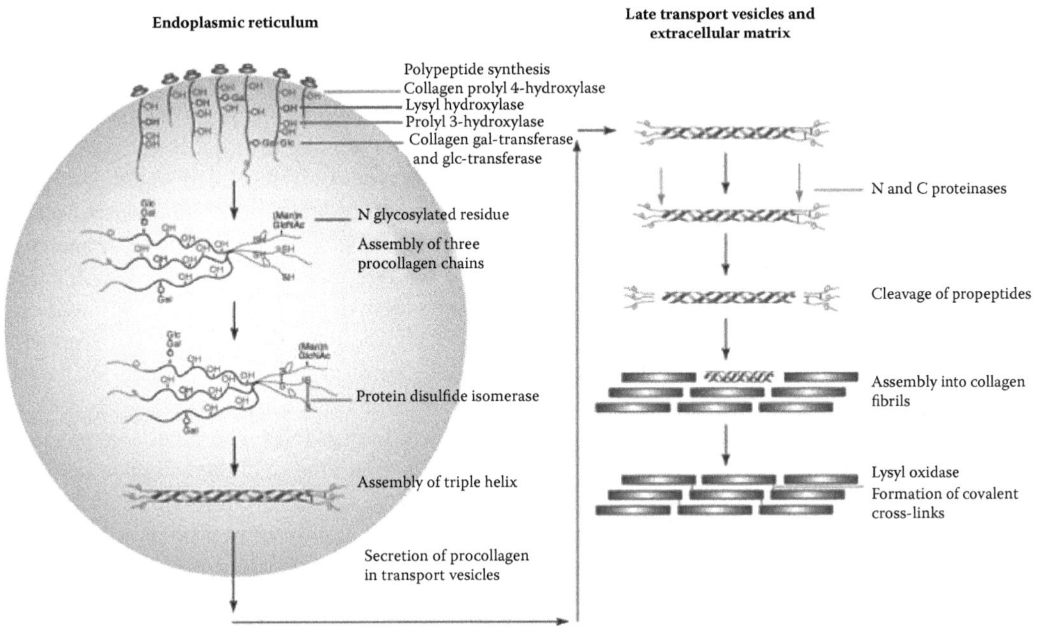

FIGURE 7.3 Scheme outlining the intracellular and extracellular steps in the biosynthesis of fibrillar collagen. (*Source:* from Myllyharju and Kivirikko 2004.)

Lysyl hydroxylases exist as three isoforms, and the expression of different isoforms in vertebrate tissue can have profound effects on collagen crosslinking, as discussed below (Mercer et al. 2003; Van der Slot et al. 2005; Takaluoma et al. 2007a; Takaluoma et al. 2007b).

The procollagen α-chains are synthesized with large globular domains, telopeptides, at the N- and C-termini. As the α-chains are released from the ER they assemble themselves in a triple helix. Recognition and trimerization of the individual α-chains followed by triple helix formation is a complex process involving the C-terminal propeptide. Triple helix formation begins at the C-terminal end of the molecules and proceeds toward the N-terminus (Snellman et al. 2007). Triple helix formation is assisted by a number of proteins that act as chaperones (Morimoto et al. 2008). The newly formed collagen propeptides are transported by the Golgi apparatus and secreted from the cell. As they exit the extracellular space, portions of the N- and C-telopeptides are excised by procollagen peptidases N and C.

Once in the extracellular space, the collagen molecules align themselves into microfibrils. The lateral displacement of each molecule relative to the other is about a quarter of its length. Thus, the degree of displacement is measured in D-periods (0–4). Microfibrils undergo lateral and end-to-end aggregation to form fibrils. This pattern of displacement and aggregation gives rise to the striated appearance of collagen fibrils. Crosslinking begins once the microfibril is formed (Nimni and Harkness 1988).

7.4.1 Crosslink Formation

Collagen fibrils owe much of their stability (including thermal stability) to the process of intermolecular crosslinking. Detailed reviews of the crosslinking process have been published (Eyre et al. 1987; Reiser et al. 1992; Eyre and Wu 2005).

Crosslinking sites on collagen molecules are highly conserved; there are only four per molecule: two at the N-terminus (9N, 87) and two at the C-terminus (6C, 930). The crosslinking site on one molecule is close to the crosslinking site of another when molecules align head-to-tail and adjacent molecules are staggered by 4 D-periods as in the microfibril.

Crosslinking is initiated by the oxidative deamination via the enzyme lysyl oxidase of specific lysines or hydroxylysines, which produces peptidyl aldehydes, termed allysine or hydroxyallysine, respectively. The head-to-tail lateral alignment of collagen molecules in a quarter-stagger array allows the aldehyde functions to react with other peptidyl aldehydes or unmodified lysine or hydroxylysine residues on adjacent α-chains. The initial condensation products form reducible crosslinks that are divalent. Reducible crosslinks contain Schiff base double bonds that can be reductively labeled and can join only two collagen molecules. There are two major pathways by which crosslinks form: the first, the allysine pathway is based on lysine aldehydes and produces aldimine crosslinks; the second, the hydroxylallysine pathway, produces crosslinks rising from hydroxylysine aldehydes. Amadori rearrangement of the initial aldimine crosslinks formed between lysine and hydroxylysine aldehydes can produce ketoamine derivatives. In muscle tissue, crosslinking follows the hydroxyallysine pathway. The reducible crosslinks vary in their stability, with ketoamine crosslinks being heat stable and aldimine crosslinks heat labile. Crosslinking of collagen is a progressive process, and the reducible crosslinks undergo further condensation reactions and are replaced with mature nonreducible crosslinks. In the hydroxyallysine pathway, there are two mature, nonreducible, trivalent crosslinks: hydroxylsylpyridinium (HP) and a pyrrole-derived crosslink, sometimes called Ehrlich chromogen (EC) (Kuypers et al. 1992; Kuypers et al. 1994). Both HP and EC are mature end products in the hydroxyallysine pathway. Two dehydroxylated forms of these crosslinks, lysyl pyridinium (LP) and lysyl pyrrole, occur; LP is found in negligible amounts in muscle tissue (McCormick 1994a; Bosselmann et al. 1995). Both HP and LP are heat stable and their occurrence in muscle is also highly correlated with the thermal stability of collagen (Horgan, 1991). The pyridinium crosslinks have been the subject of substantially more investigation than the pyrrolic crosslinks (Eyre et al. 1984) (Figure 7.4).

Crosslinking patterns are regulated by the action of the lysyl hydroxylases (LH), with hydroxylation of specific lysines determining the crosslink formed (Eyre and Wu 2005). Three LH isoforms have been shown to hydroxylate collagen sequences: LH1, LH2, and LH3. LH2 has been identified as a specific telopeptide hydroxylase. As more telopeptide lysines are hydroxylated, the degree of pyridininum crosslinking increases (Van der Slot et al. 2005). LH isoforms have been demonstrated in skeletal muscle (Mercer et al. 2003), although patterns of expression have not been investigated.

7.4.2 Turnover and Degradation

Collagen is usually considered a protein with a long half-life; once deposited, it remains *in situ* indefinitely. Collagen turnover studies, however, suggest otherwise, with extremely high and more modest turnover rates reported (McAnulty and Laurent 1987; Rucklidge et al. 1992). In normal growing animals, serum hydroxylproline levels attest to the degradation of collagen, which is absolutely necessary for growth (Miller et al. 1989).

Collagen and other EMC components are degraded by the action of matrix metalloproteinases (MMPs). The activity of MMPs is regulated at several levels, transcriptionally by zymogen activation and by tissue inhibitors (TIMPs) (Visse and Nagase 2003; Carmeli et al. 2004). Measurement of proteinases and their inhibitors in bovine skeletal muscle revealed the presence of two collagenases (MMP-1, MMP-2), one gelatinase (MMP-2 or MMP-9), elements of an MMP activation system, and TIMPs-1, -2, -3 (Balcerzak et al. 2001). In lambs, MMP-2 activity was detected at slaughter and at 21 days postmortem; animals with high MMP-2 activity had increased amounts of soluble collagen and increased hydroxyproline, indicating that collagen could be degraded to free amino acids (Sylvestre et al. 2002). There is ample evidence that the ECM, including the collagen network, is degraded postmortem (see below).

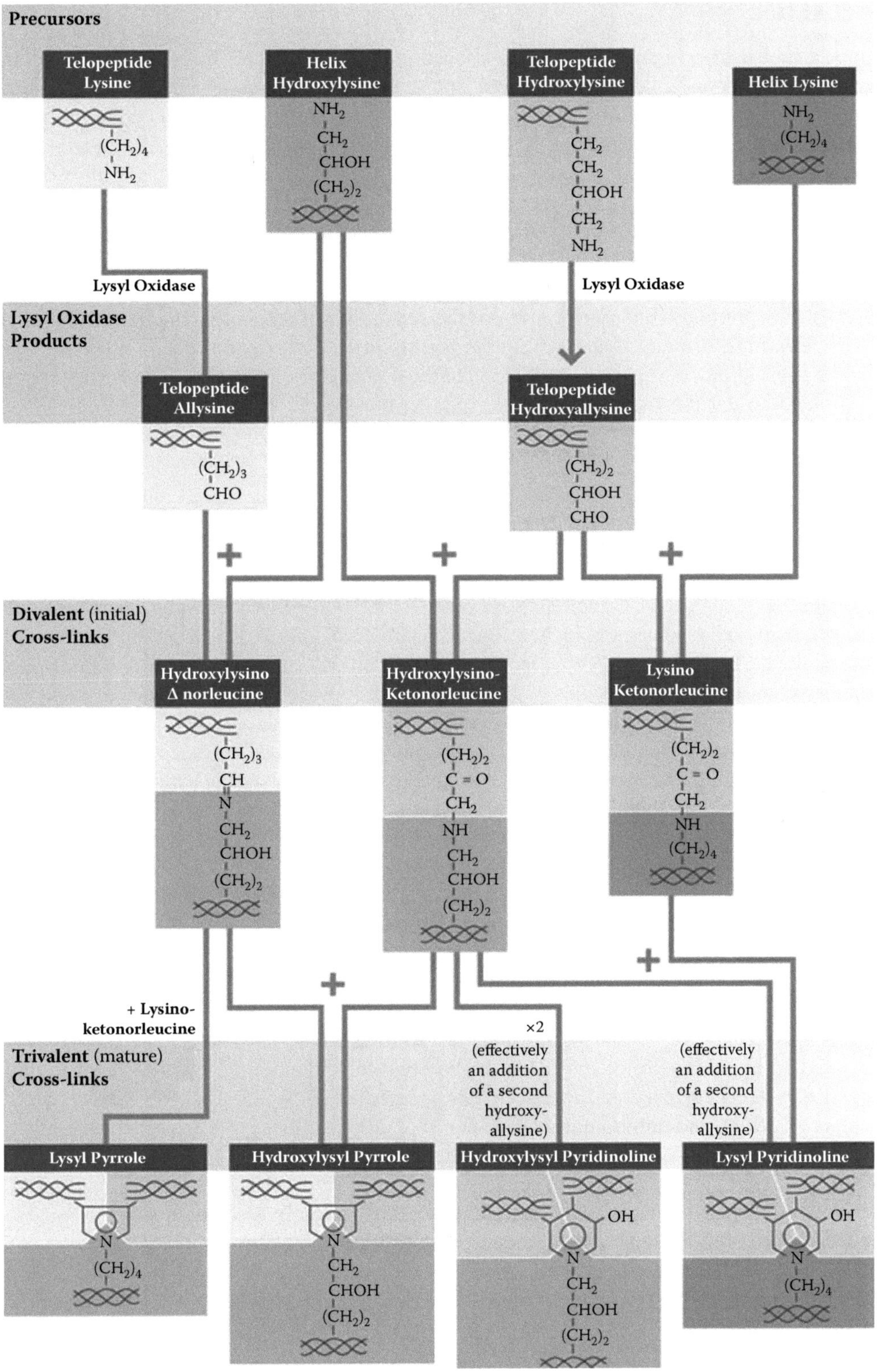

FIGURE 7.4 Precursors, divalent crosslinks, and mature trivalent crosslinks formed on the hydroxyallysine pathway. (*Source:* From Eyre and Wu 2005.)

7.5 ELASTIN

Elastin is a rubber-like, highly insoluble component of skeletal muscle. Its properties arise largely from its tetravalent crosslinks desmosine and isodesmosine (Reiser et al. 1992). Elastin comprises only a small fraction of the total collagenous tissue in muscle. Bendall (1967) conducted the most comprehensive study of the content of elastin in a variety of bovine muscles. Elastin as percent of dry weight ranged from a high of 2.0 in latissimus dorsi to a low of .07 in longissimus dorsi and psoas major. A muscle of some commercial importance, semitendinosus contains 1.82% elastin. Elastin as a percent of the collagen in muscle is generally quite low, 2% to 3%, the exceptions being latissimus dorsi and semitendinosus, 33% and 37% respectively.

Histochemical studies have identified the locations of elastin, which is generally observed throughout the epimysium and perimysium of the semitendinosus muscle. In contrast, in longissimus dorsi, elastin is virtually absent within the perimysium (Bailey and Light 1989). While elastin may make substantial contributions to the mechanical behavior of muscles high in elastin content (e.g., semitendinosus), it is of less importance in other muscles.

7.6 PROTEOGLYCANS

Proteoglycans represent some of the most diverse, complex, and multifunctional molecules in the biological world. Their contributions to muscle development are extremely significant and complex. The rapidly expanding biological world of collagens is exceeded in some measure by developments in the realm of proteoglycans. Comprehensive reviews on the role that proteoglycans play in skeletal muscle structure and development include Buckingham et al. (1994), Brandan and Larrain (1998), Carrino et al. (1999), Velleman (2002), Jenniskens et al. (2006), and Brandan et al. (2008).

Proteoglycans are macromolecules with a protein core to which at least one and as many as 100 glycosylaminoglycan (GAG) chains are covalently bonded (Wight et al. 1991; Carrino et al. 1999). The highly variable core proteins range in size from about 40 to 400 kDa (Iozzo and Murdoch 1996; Velleman 2003). GAGs are repeating dissaccharide units of highly variable length (see Carrino et al. 1999 for a description of their composition). GAGs are distributed among several major classes, including hyaluronic acid (HA) chondroitin sulfate, dermatan sulfate (DS), keratan sulfate (KS), heparan sulfate (HS), and heparin (Iozzo and Murdoch 1996; Carrino et al. 1999; Velleman 2002). All GAGs are highly sulfated and bear a strong negative charge, the exception being HA, which is neither sulfated nor attached to a core protein (Iozzo and Murdoch 1996; Carrino et al. 1999); it exists as a free GAG in the ECM and as such is not a proteoglycan.

Proteoglycans, being extremely complex, seem to defy a single means of classification. Iozzo and Murdoch (1996) suggested a binary system based on the nature of the core proteins: short leucine-rich proteoglycans (SLRPs) and modular proteoglycans. Alternatively, proteoglycans are often classified by the nature of their GAGs. Proteoglycans identified in skeletal muscle thus far are the SLRPs decorin (Iozzo and Murdoch 1996), biglycan (Bowe et al. 2000; Casar et al. 2004; Brandan et al. 2008), and fibromodulin (Pedersen et al. 2001). Other proteoglycans, generally bearing HS GAGS, in muscle, include dematan sulfate proteoglycan (Eggen et al. 1994), syndecans 1–4 (Stringer and Gallagher 1997; Velleman 2002; Zhang et al. 2007), perlecan, versican , and HS-containing collagen Type XVIII (Halfter et al. 1998). Closely associated with proteoglycans, particularly those of the basement membrane, are several glycoproteins. In muscle, these molecules include primarily the integrins, laminins, tenascins, nidogen, and the dystrophin-glycoprotein complex (Bosman and Stamenkovic 2003).

7.7 MORPHOLOGY OF THE IMCT SCAFFOLD

Recent developments in IMCT preparation techniques, the use of scanning electron microscopy and X-angle diffraction, as well as the ability to make sensitive mechanical measurements on stretched,

FIGURE 7.5 Collagen fibers in epimysium of bovine sternomandibularis muscle. The edges of individual muscle fibers are visible as horizontal lines; the overlying cross-ply of collagen fibers is approximately ±54° to the direction of muscle fiber. (*Source:* From Purslow 2005.)

passive, and contracted muscle tissue have allowed a heightened understanding of the structure and function of IMCT. The significance of morphological and mechanical observations of IMCT on muscle function *in vivo* is not to be discussed here. There are several comprehensive reviews on the subject available (Purslow 2002; Huijing and Jaspers 2005; Purslow et al. 2008).

Earlier microscopic studies (Rowe 1974; Borg and Caulfield 1980; Rowe 1981) describe the endomysium as a dense, felt-like network of collagen fibers overlaid by a less dense, thicker connective tissue layer, the perimysium. Rowe (1974, 1981) characterized the perimysium as composed of three distinct regions: (1) coarse, crimped fibers with a well-ordered cross-ply pattern; (2) a looser felt-like network; and (3) fine, randomly organized fibers. Rowe (1974) had previously reported that the cross-ply array of coarse, well-ordered, crimped fibers lay at an angle to the muscle fiber direction. Rowe (1974) further observed that the angle of the perimysial collagen fiber relative to the axis of the muscle fiber varied in the range of 45° to 60°. Purslow (1989) refined these measurements by plotting collagen fiber angle relative to sarcomere length. Over the sarcomere range (1.1 to 3.9 μm), the collagen fiber range angle varied from 20° (longest sarcomere length) to 80° (shortest). An at-rest sarcomere length (2.0 μm) produced a collagen fiber angle of about 55° (Figure 7.5).

Recent studies have expanded our understanding of the structural organization of the perimysium (Passerieux et al. 2006; Passerieux et al. 2007). They describe collagenous attachments between the perimysium and the myofiber, called perimysial junction plates (PJP), and a network of tubular structures uniting two tendons. It was noted that at the PJP there was a distinct concentration of myonuclei and mitochondria beneath the sarcolemma. These findings suggest that the perimysium is involved in the transmission of contraction forces and that these forces are relayed to the muscle cell at the PJP, perhaps acting as a mechanosensing mechanism (Figure 7.6).

7.8 ECM: ROLE IN GROWTH AND DEVELOPMENT

Collagens, proteoglycans, and glycoproteins play a critical role in muscle growth and development from myogenesis to fibrillogenesis. A detailed discussion of the complex interplay between non-collagenous ECM components, including proteoglycans and glycoproteins, and transmembrane structures, growth factors, and cells is beyond the scope of this chapter; however, some recent reports concerning meat animals will be reviewed.

It has been accepted for decades that collagen is part of the signaling process in myogenesis (Hauschka and Konigsberg 1966). Myoblast adhesion to collagen is necessary for fusion and further differentiation (Ruoslahti and Pierschbacher 1987), and collagen synthesis by myoblasts is a requirement for myogenesis. It was observed that addition or absence of non-collagenous molecules could inhibit collagen synthesis (Cates et al. 1984; Mayne and Sanderson 1985).

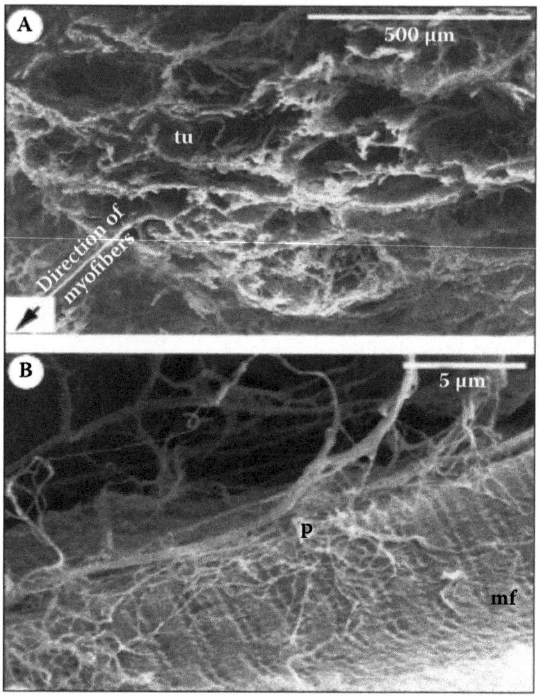

FIGURE 7.6 (A) Tubular arrangement (tu) of perimysium; and (B) perimysial collagen fiber bundles (p) connecting the perimysium to myofiber (mf). (*Source:* From Passerieux et al. 2007.)

As a class of molecules that regulate myogenesis and further growth and development, proteoglycans have been intensively studied. With regard to muscle development, the SLRP decorin, so named because it "decorates" collagen fibrils, has probably received the most attention. Decorin is a multifunctional molecule; it binds to collagen fibrils and also at the basal lamina. It is a regulator of fibrillogenesis and arrangement of collagen fibrils in the ICMT (Nishimura et al. 2002; Nishimura et al. 2003; Ferdous et al. 2007). It also binds transforming growth factor β, a potent regulator of muscle cell proliferation (Brandan et al. 2008). It has a role in determining collagen architecture as well as modulating muscle cell growth by interacting, with growth factors (Li et al. 2006).

The heparan sulfate proteoglycans, including syndecans and glypicans, and their influence on myogenic satellite cell proliferation and differentiation during different stages of muscle development have been rather intensively studied (Velleman et al. 2007; Zhang et al. 2007; Zhang et al. 2008). The developmental expression of selected proteoglycans has been the subject of numerous studies (Velleman et al. 1999; Carrino et al. 1999; Nishimura et al. 2002; Nishimura et al. 2003; Li et al. 2006; Liu et al. 2006).

Alterations in collagen composition and architecture during the course of muscle development from the fetal stage through the postnatal period into maturity have also been reported. In the bovine there was a shift in location (between perimysium and endomysium) of collagen Types I, III, V, and VI with increasing fetal age and Type XII and XIV collagens colocated in the perimysium. Type XII diminished after birth while Type XIV increased. As the fetus aged, collagen content decreased (Listrat et al. 1999; Listrat et al. 2000).

Similar studies have been undertaken with the goat (Shiba et al. 2007). Likewise in the fowl, developmental changes in the collagen characteristics of several muscle have been documented (Lawson and Purslow 2001; Nakamura et al. 2004).

7.9 COLLAGEN AND MEAT QUALITY

Meat is a complex, composite substance. It consists of myofibers, connective tissue, and lipid. It is not surprising that coming to definitive conclusions about the causes of toughening in meat has been difficult. A good deal of work and ink has been expended in an attempt to understand the basis of tenderness and toughness in meat. The role that ECM plays in meat tenderness was recently reviewed by Purslow (2005), with emphasis placed on structural and morphological aspects of connective tissue and their relationship to meat tenderness. The focus here is on biochemical qualities of EMC that influence tenderness.

7.9.1 The Roles that Collagen Crosslinking and Type Play in Meat Texture

Generally, greater fractions of both Type III collagen and heat-soluble collagen are indicative of more youthful, labile, less crosslinked collagen. Indeed, Type III collagen is considered the embryonic form of fibrillar collagen, with greatest amounts occurring in the fetus and neonate. Type III collagen also tends to form smaller fibers than Type I collagen (Light et al. 1985). With maturation, Type III collagen is replaced by both Type I collagen and a greater complement of mature crosslinks (Kovanen and Suominen 1989). Efforts to correlate changes in proportions of Type I to III collagen to differences in texture have proved inconclusive. A positive correlation was reported (Burson and Hunt 1986), but other studies found no relationship between type and tenderness (Light et al. 1985).

Before the widespread quantitation of specific crosslinks, indeed before crosslinks were identified, an estimation of their presence was made using the heat solubility/insolubility procedure of Hill (1966). However, the correlation of collagen solubility with either meat tenderness or crosslinking is often not good (McCormick 1994a; Lepetit 2008). Powell et al. (2000) describe an enzymatic assay using pronase digestion to assess collagen solubility that may be more discriminating than the heat solubility technique.

Some of the very earliest reports of crosslinking in muscle linked it to mechanical properties, notably toughening (Shimokomaki et al. 1972). The earliest researchers were first able to measure the reducible crosslinks but were handicapped by their lack of knowledge of the mature end-products, although they were predicted (Bailey and Light 1989). With the recognition that pyridinium crosslinks and later Ehrlich chromogen (EC) were mature, heat-stable, endpoint crosslinks on the crosslink pathway, which predominates in muscle, two good candidates for assessing collagen properties in meat were available. Most interest and effort have focused on the pyridinoline crosslink (HP), perhaps because it is relatively easily to measure.

Lepetit (2007) recently analyzed ten studies in which collagen crosslinks in muscle tissue were measured (Maiorano et al. 1993; Bosselmann et al. 1995; Field et al. 1996; Berge et al. 1997; McCormick 1999; Boutten et al. 2000; Shiba et al. 2000; Ngapo et al. 2002a; Ngapo et al. 2002b; Coro et al. 2002). He concluded that measurement of crosslinks, particularly pyridinoline, is a reasonable predictor of tenderness. A few studies found little or no correlation between HP levels and tenderness (Young et al. 1994; Avery et al. 1996); whether this is due to a lack of proportionality between HP and EC or other factors is not clear.

The caveat is that meat often varies in parameters other than connective tissue such as muscle fiber resistance, sarcomere length, pH, and collagen morphology (Lepetit 2007). An example is the comparison between normal and callipyge lambs (Field et al. 1996). Callipyge lambs produce higher shear force values yet have fewer HP crosslinks per molecule of collagen. The lack of myofibrillar protein proteolysis postmortem in the callipyge genotype provides an explanation for the failure of collagen crosslinking to correlate with shear force.

The collagen characteristics of fish muscle present a different issue than that of red meat animals. The soft texture of fish is deemed undesirable and texture is improved as crosslinking increases. Increased reducible and HP crosslinks contribute to a firmer, more desirable texture in Atlantic salmon (Li et al. 2005; Johnston et al. 2006).

7.9.2 Age and Crosslinking Patterns

It is well known that with chronological age crosslinks progress from immature to more mature forms. Furthermore, the concentration of mature crosslink increases with age (McCormick 1994a). Immature reducible crosslinks occur transiently and in many tissues can be considered as intermediate products. In skeletal muscle, reducible crosslinks are rapidly replaced with more mature forms. Shiba et al. (2007) noted that in goats, HP crosslink was present in muscle from fetuses 2 weeks prior to parturition. In steers at 1 year of age , the concentration of HP in IMCT was more than twofold greater than its reducible precursor, the ketoamine crosslink, dihydroxylysinonorleucine (McCormick 1994a). With regard to HP, a steady increase in this crosslink over time has been reported in both poultry (Coro et al. 2002) and cattle (Bosselmann et al. 1995). Interestingly, the progression of crosslinking occurs significantly faster in avian skeletal muscle than in the mammals examined. For example, in fowl (Leghorn chicken) pectoral muscle, HP values were in excess of 1.0 mol HP/mol collagen by 1 year of age, although this high level of crosslinking occurred concomitant with an overexpression of decorin (Velleman et al. 1996).

7.9.3 Muscle Type

The concentration of lysine aldehyde-derived crosslinks varies with muscle type and with a host of conditions related to time, growth, and adaptation. Data on muscle HP concentrations from various species, including cattle, sheep, pigs, and rats, suggest a relatively wide range of values for different muscles. This variability reflects muscle function and the extreme range of perimysial collagen amounts in different muscles (Light et al. 1985). When collagen concentration and crosslinking in five different bovine muscles that span a wide range of texture variation were compared, less tender muscles, such as biceps femoris (BF) and semimembranosus (SM), collagen concentrations were high, approaching 3%. Levels of HP were also high, at nearly 0.5 mol HP/mol collagen. Conversely, a muscle such as LD, the tenderness of which was much more acceptable than that of either BF or SM, possessed significantly less collagen (1.86%) and less crosslinked collagen (0.36 mol HP/mol collagen). Gluteus medius (GM) and psoas major (PM) muscles present an interesting contrast. The GM possessed a relatively high concentration of collagen (2.77%) but low concentrations of the HP crosslink (0.35 mol HP/mol collagen). The PM had little collagen (1.41%) but elevated crosslink levels (0.45 mol PM/mol collagen). Nevertheless, both muscles possessed excellent textural characteristics, with PM acknowledged as the most tender muscle (McCormick 1999). In young bulls, flexor digitorum and infraspinatus muscles had high collagen contents while PM had the lowest content. The insoluble collagen contents were the highest in triceps brachii and infraspinatus (Torrescano et al. 2003). In another study, three porcine muscles (LD, semimembranosus, and infraspinatus) were compared for collagen content and solubility and shrinkage during cooking; collagen solubility was highest in infraspinatus and lowest in semimembranosus (Voutila et al. 2007).

7.9.4 Animal Sex

Animal sex affects collagen crosslinking in cattle muscle (Huff and Parrish 1993; Bosselmann et al. 1995). Using testosterone-mediated growth of lambs as the model, we examined collagen concentration, collagen heat solubility, collagen phenotype, crosslinking, and shear force of the LD of both rams and wethers (Maiorano et al. 1993). Meat from intact males is tougher than that from similar-age castrates. The IMCT from rams clearly possessed greater mature crosslinking (about 40%) than IMCT from wethers, but at the same time had proportionately more Type III collagen and heat-labile collagen. Because rams under the influence of testosterone synthesized more collagen than wethers (Miller et al. 1989), the youthful nature of their IMCT as typified by phenotype and solubility is understandable. These data suggest that textural differences are due largely to increased mature crosslinking, the effect of which may be exacerbated by elevated amounts of collagen in the muscle of rams compared to wethers. Previous studies have documented significantly increased metabolic activity of collagen (synthesis and turnover) in the presence of testosterone, which probably accounts for the somewhat elevated muscle collagen concentration in rams. Harder to explain is the association of these attributes with rapid and extensive crosslinking of the IMCT. Certainly our inability to relate some physical and biochemical properties of muscle tissue to crosslinking patterns stems from a lack of knowledge of the mechanisms that regulate crosslinking in muscle. Reports that show that the expression of lysyl oxidase is affected by sex may shed light on the gender effort on collagen crosslinking (Werman et al. 1995).

7.9.5 Growth

Treatment or conditions that result in connective tissue turnover, and usually muscle growth, also markedly influence collagen characteristics. Although variations in collagen characteristics between muscle types is reasonably predictable, a more complex issue is the change that occurs in connective tissue of a muscle with the multiple factors that affect muscle growth and adaptation over time. Within a muscle type, numerous treatments that apparently result in remodeling of the muscle connective tissue matrix will influence crosslinking patterns. Examples include compensatory growth, somatotropin-mediated growth (McCormick 1994a), nutrition-enhanced growth rate (Maiorano et al. 2007), and exercise (Zimmerman et al. 1993).

7.9.6 Meat Tenderness: Raw or Cooked?

With regard to shear and fracture of samples of meat, raw meat behaves differently than cooked meat. There is often a very good correlation between shear force values and collagen concentration in a wide range of muscles across species (Liu et al. 1996; Dransfield et al. 2003; Torrescano et al. 2003). In cooked meat, however, the correlations between collagen content and shear force are generally lower (Dransfield et al. 2003; Serra et al. 2008). It has been noted that proteoglycans can also influence the mechanical properties of raw meat (Nishimura et al. 1995, 1996) but they are solubilized with heat (Bailey and Light 1989).

7.9.7 Shortening and Tension Development

Collagen is a unique molecule because, when assembled into a fibril, thermal denaturation results in shrinkage of the fibril accompanied by force or tension development. The unraveling of the collagen helix when the thermal shrinkage temperature (Ts) is reached produces a spring-like effect in the unfolded collagen molecules. The degree of tension or shrinkage developed is a function of how heavily the collagen is crosslinked with mature, heat-stable crosslinks (Bailey and Light 1985). As the temperature increases from 80 to 90°C and above, the temperature of denaturation (Td) is achieved and the collagen will gelatinize.

Step-wise, temperature-dependent increases in tension development and shrinkage occurring in both single muscle fibers (endomysium) and in muscle fiber bundles were demonstrated using mechanical and histochemical techniques (Bendall and Restall 1983). The initial shrinkage, which is minimal and occurs around 58°C, probably corresponds to shrinkage of Type IV basement membrane collagen surrounding the myofibers. However, when temperature exceeds about 64°C, the critical shrink temperature of the perimysial collagen (Types I and III), tension development, shrinkage, and fluid expression rapidly increase (Bouton et al. 1976) (Figure 7.7).

As the temperature is raised incrementally past the critical temperature (>64°C), tension also increases stepwise. The tension development corresponds to the unraveling and contraction of the collagen triple helix. When the temperature decreases to 38°C, the collagen molecules assume their native, rigid rod-like conformation and tension is relieved. In the case of beef sternomandibularis muscle, tension development continues to increase up to 94°C. Significantly, when PM is subjected

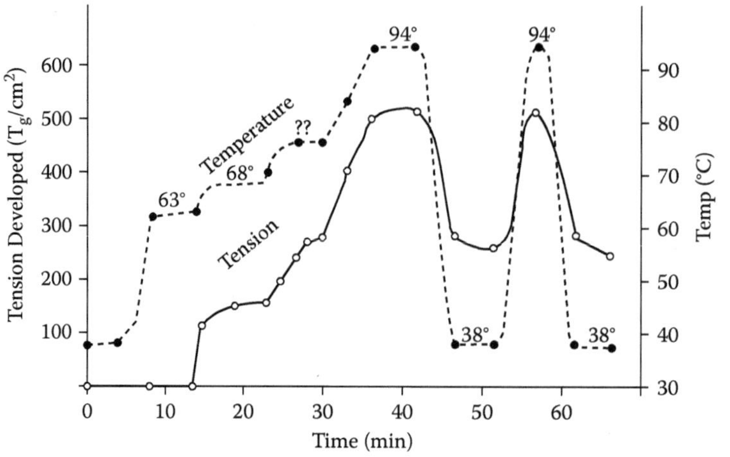

FIGURE 7.7 Plot of tension developed in muscle strips over time as temperature increases from 30 to 94°C. Note the relaxation of tension when strips are cooled. (*Source:* From Bendall and Restall 1983.)

to a similar heating regimen, it can develop only about a third of the force of the sternomandibularis muscle. The difference in tension development between the two muscles is a function of collagen content and how heavily crosslinked the collagen is (Bailey and Light 1985; Lepetit 2007).

7.9.8 Cooking and Toughening

As meat is cooked, the shear force (toughening) increases in a biphasic manner. The first phase of toughening from 20 to 50°C has been attributed to the perimysial fraction of the meat (Bouton et al. 1976; Bouton et al. 1981; Lewis and Purslow 1989; Christensen et al. 2000). Above 50°C, the breaking strength of the perimysium declines steadily (Christensen et al. 2000); there is a second phase of toughening beginning at 60 to 65°C. The toughening of meat above this point is attributed to the denaturation of the myofibrillar component, which hardens and perhaps dehydrates. One explanation for the increase in perimysial breaking strength at 20 to 50°C is the loss of crimp in the perimysial collagen fibers. Straightening of the crimp would effectively increase the concentration of collagen fibers in a given area and increase breaking strength (Christensen et al. 2000).

As the IMCT reaches and exceeds Ts, the shrinkage of collagen and tension development expresses fluid from the myofibers. The contribution of fluid expression and other factors such as denaturation of the cytoskeletal proteins to toughening has not been adequately investigated (Purslow 2005).

7.9.9 pH of Postmortem Muscle

The temperature required for collagen gelatinization is highest in the pH range 5 to 6, the isoelectric point of collagen. Further decreases in pH lead to the hydration of collagen fibers, which decreases the temperature needed for thermal transition, that is, Ts and Td (Miles et al. 2005). The decrease in thermal transition temperature is also associated with the breakage of acid-labile crosslinks (Offer et al. 1989; Lewis et al. 1991; Usha and Ramasami 2000). Due to enhanced gelatinization, the tension developed by collagen shrinking is reduced and the meat retains more water on cooking (Gault 1985; Lewis et al. 1991). However, the low pH of PSE pork (ultimate pH 5.43) did not affect the thermal characteristics of connective tissue when compared with normal pork (ultimate pH 5.64) (Voutila et al. 2008). A larger decrease in pH may be needed to affect transition temperature.

7.9.10 Postmortem Conditioning

Postmortem aging of carcasses is a very effective method of increasing meat tenderness. Such tenderization is mainly due to the hydrolysis of myofibrillar proteins. Numerous studies indicate a degradation of the collagen network with increasing time postmortem (Nishimura et al. 1995). Aging also decreases the breaking strength of the perimysial connective tissue in raw meat (Lewis et al. 1991). In addition, connective tissue resistance to shear force is reduced for meat cooked to 50°C but not for meat cooked to 60°C or above, the usual temperature for meat (Lewis et al. 1991). During postmortem aging, proteoglycans are likewise degraded, resulting in the weakening of perimyisum and the separation of collagen fibrils and fibers from endomysium and perimysium (Nishimura et al., 1996). The data clearly indicate substantial degradation of connective tissue during postmortem ageing, although its contribution to the reduction in toughness of cooked meat is minor (Lewis et al. 1991).

7.10 SUMMARY

The ECM is critical to the growth, development, and maintenance of skeletal muscle. There is an elaborate set of signaling mechanisms that permit communication between the ECM and the myofiber. Exactly how these signals might affect cellular metabolism under various conditions is not fully elucidated. A fuller understanding of the relationship between the myofiber and the surrounding ECM has the potential to allow manipulation of skeletal muscle growth and characteristics.

Much progress has been made toward understanding the chemical nature of collagen crosslinks and the pathways that result in their formation. Furthermore, an increasing body of evidence indicates that crosslinking is important to meat texture and is influenced by a myriad of practices and circumstances that accompany animal production. Future research should focus on elucidating mechanisms governing crosslinking formation in skeletal muscle.

REFERENCES

Avery, N.C., T.J. Sims, C. Warkup, and A.J. Bailey. 1996. Collagen cross-linking in porcine M longissimus lumborum: absence of a relationship with variation in texture at pork weight. *Meat Sci.* 42:355–369.

Bailey, A.J. and N.D. Light. 1985. Intermolecular cross-linking in fibrotic collagen. *Ciba Foundation Symposia* 114:80–96.

Bailey, A.J. and N.D. Light. 1989. *Connective Tissue in Meat and Meat Products.* Elsevier Applied Science, New York.

Bailey, A.J. and T.J. Sims. 1977. Meat tenderness — distribution of molecular-species of collagen in bovine muscle. *J. Sci. Food Agric.* 28:565–570.

Balcerzak, D., L. Querengesser, W.T. Dixon, and V.E. Baracos. 2001. Coordinate expression of matrix-degrading proteinases and their activators and inhibitors in bovine skeletal muscle. *J. Anim. Sci.* 79:94-107.

Bendall, J.R. 1967. Elastin content of various muscles of beef animals. *J. Sci. Food Agric.* 18:553–558.

Bendall, J.R. and D.J. Restall. 1983. The cooking of single myofibers, small myofiber bundles and muscle strips from beef m-psoas and M-sternomandibularis muscles at varying heating rates and temperatures. *Meat Sci.* 8:93–117.

Berge, P., J. Lepetit, M. Renerre, and C. Touraille. 1997. Meat quality traits in the emu (*Dromaius novaehollandiae*) as affected by muscle type and animal age. *Meat Sci.* 45:209–221.

Boot-Handford, R.P., D.S. Tuckwell, D.A. Plumb, C.F. Rock, and R. Poulsom. 2003. A novel and highly conserved collagen (pro alpha 1(XXVII)) with a unique expression pattern and unusual molecular characteristics establishes a new clade within the vertebrate fibrillar collagen family. *J. Biol. Chem.* 278:31067–31077.

Borg, T.K. and J.B. Caulfield. 1980. Morphology of connective tissue in skeletal muscle. *Tissue Cell* 12:197–207.

Bosman, F.T. and I. Stamenkovic. 2003. Functional structure and composition of the extracellular matrix. *J. Pathol.* 200:423–428.

Bosselmann, A., C. Moller, H. Steinhart, M. Kirchgessner, and F.J. Schwarz. 1995. Pyridinoline cross-links in bovine muscle collagen. *J. Food Sci.* 60:953–958.

Bouton, P.E., P.V. Harris, and D. Ratcliff. 1981. Effect of cooking temperature and time on the shear properties of meat. *J. Food Sci.* 46:1082–1087.

Bouton, P.E., P.V. Harris, and W.R. Shorthose. 1976. Factors influencing cooking losses from meat. *J. Food Sci.* 41:1092–1095.

Boutten, B., M. Brazier, N. Morche, A. Morel, and J.L. Vendeuvre. 2000. Effects of animal and muscle characteristics on collagen and consequences for ham production. *Meat Sci.* 55:233–238.

Bowe, M.A., D.B. Mendis, and J.R. Fallon. 2000. The small leucine-rich repeat proteoglycan biglycan binds to alpha-dystroglycan and is upregulated in dystrophic muscle. *J. Cell Biol.* 148:801–810.

Brandan, E., C. Cabello-Verrugio, and C. Vial. 2008. Novel regulatory mechanisms for the proteoglycans decorin and biglycan during muscle formation and muscular dystrophy. *Matrix Biology* 10.1016/j.matbio.2008.07.004.

Brandan, E. and J. Larrain. 1998. Heparan sulfate proteoglycans during terminal skeletal muscle cell differentiation: possible functions and regulation of their expression. *Basic Appl. Myol.* 8:107–113.

Bruggemann, D.A. and M.A. Lawson. 2005. The extracellular matrix of Gadus morhua muscle contains types III, V, VI and IV collagens in addition to type I. *J. Fish Biol.* 66:810–821.

Buckingham, B., K. Reiser, and R. Shalwitz. 1994. Vitamin-E decreases collagen associated fluorescence in diabetic rats. *Diabetes* 43:A96–A96.

Burson, D.E. and M.C. Hunt. 1986. Proportion of collagen type-I and type-III in 4 bovine muscles differing in tenderness. *J. Food Sci.* 51:51–53.

Carmeli, E., M. Moas, A.Z. Reznick, and R. Coleman. 2004. Matrix metalloproteinases and skeletal muscle: a brief review. *Muscle & Nerve* 29:191–197.

Carrino, D.A., J.M. Sorrell, and A.I. Caplan. 1999. Dynamic expression of proteoglycans during chicken skeletal muscle development and maturation. *Poult. Sci.* 78:769–777.

Casar, J.C., B.A. McKechnie, J.R. Fallon, M.F. Young, and E. Brandan. 2004. Transient up-regulation of biglycan during skeletal muscle regeneration: delayed fiber growth along with decorin increase in biglycan-deficient mice. *Dev. Biol.* 268:358–371.

Cates, G.A., A.M. Brickenden, and B.D. Sanwal. 1984. Possible involvement of a cell-surface glycoprotein in the differentiation of skeletal myoblasts. *J. Biol. Chem.* 259:2646–2650.

Christensen, M., P.P. Purslow, and L.M. Larsen. 2000. The effect of cooking temperature on mechanical properties of whole meat, single muscle fibres and perimysial connective tissue. *Meat Sci.* 55:301–307.

Coro, F.A.G., E.Y. Youssef, and M. Shimokomaki. 2002. Age related changes in poultry breast meat collagen pyridinoline and texture. *J. Food Biochem.* 26:533–541.

Dransfield, E., J.F. Martin, D. Bauchart, S. Abouelkaram, J. Lepetit, J. Culioli, C. Jurie, and B. Picard. 2003. Meat quality and composition of three muscles from French cull cows and young bulls. *Anim. Sci.* 76:387–399.

Eggen, K.H., A. Malmstrom, and S.O. Kolset. 1994. Decorin and a large dermatan sulfate proteoglycan in bovine striated-muscle. *Biochim. Biophys. Acta – Protein Struct. Mol. Enzymol.* 1204:287–297.

Eyre, D.R., S. Apon, J.J. Wu, L.H. Ericsson, and K.A. Walsh. 1987. Collagen type-IX — Evidence for covalent linkages to type-II collagen in cartilage. *FEBS Lett.* 220:337–341.

Eyre, D.R., M.A. Paz, and P.M. Gallop. 1984. Cross-Linking in collagen and elastin. *Annu. Rev. Biochem.* 53:717–748.

Eyre, D.R. and J.J. Wu. 2005. Collagen cross-links. *Top. Curr. Chem.* 247:207–229.

Ferdous, Z., V.M. Wei, R. Iozzo, M. Hook, and K.J. Grande-Allen. 2007. Decorin-transforming growth factor — interaction regulates matrix organization and mechanical characteristics of three-dimensional collagen matrices. *J. Biol. Chem.* 282:35887–35898.

Field, R.A., R.J. McCormick, D.R. Brown, F.C. Hinds, and G.D. Snowder. 1996. Collagen crosslinks in longissimus muscle from lambs expressing the callipyge gene. *J. Anim. Sci.* 74:2943–2947.

Gault, N.F.S. 1985. The relationship between water-holding capacity and cooked meat tenderness in some beef muscles as influenced by acidic conditions below the ultimate pH. *Meat Sci.* 15:15–30.

Halfter, W., S. Dong, B. Schurer, and G.J. Cole. 1998. Collagen XVIII is a basement membrane heparan sulfate proteoglycan. *J. Biol. Chem.* 273:25404–25412.

Hauschka, S.D. and I.R. Konigsberg. 1966. The influence of collagen on the development of muscle clones. *Proc. Natl. Acad. Sci. U.S.A.* 55:119–126.

Hill, F. 1966. Solubility of intramuscular collagen in meat animals of various ages. *J. Food Sci.* 31:161–166.

Horgan, D.J. 1991. The estimation of the age of cattle by the measurement of thermal-stability of tendon collagen. *Meat Sci.* 29:243–249.

Huff, E.J. and F.C. Parrish. 1993. Bovine longissimus muscle tenderness as affected by postmortem aging time, *Anim. Age Sex. J. Food Sci.* 58:713–716.

Huijing, P.A. and R.T. Jaspers. 2005. Adaptation of muscle size and myofascial force transmission: a review and some new experimental results. *Scand. J. Med. Sci. Sports* 15:349–380.

Hulmes, D.J.S. 1992. The collagen superfamily — diverse structures and assemblies. *Essays in Biochem* 27:49–67.

Hulmes, D.J.S. 2008. Collagen diversity, synthesis and assembly. In *Collagen: Structure and Mechanics,* P. Fratzl, Ed. Springer, New York, p. 15–47.

Iozzo, R.V. and A.D. Murdoch. 1996. Proteoglycans of the extracellular environment: clues from the gene and protein side offer novel perspectives in molecular diversity and function. *FASEB J.* 10:598–614.

Jenniskens, Y.M., W. Koevoet, A.C.W. de Bart, H. Weinans, H. Jahr, J.A.N. Verhaar, J. DeGroot, and G.J.V.M. van Osch. 2006. Biochemical and functional modulation of the cartilage collagen network by IGF1, TGF beta 2 and FGF2. *Osteoarthritis and Cartilage* 14:1136–1146.

Johnston, I.A., X.J. Li, V.L. A. Vieira, D. Nickell, A. Dingwall, R. Alderson, P. Campbell, and R. Bickerdike. 2006. Muscle and flesh quality traits in wild and fanned Atlantic salmon. *Aquaculture* 256:323–336.

Kovanen, V. and H. Suominen. 1989. Age-related and training-related changes in the collagen-metabolism of rat skeletal-muscle. *Eur. J. Appl. Physiol.* 58:765–771.

Kuypers, R., M. Tyler, L.B. Kurth, and D.J. Horgan. 1994. The molecular location of Ehrlich chromogen and pyridinoline cross-links in bovine perimysial collagen. *Meat Sci.* 37:67–89.

Kuypers, R., M. Tyler, L.B. Kurth, I.D. Jenkins, and D.J. Horgan. 1992. Identification of the loci of the collagen-associated Ehrlich chromogen in type-I collagen confirms its role as a trivalent cross-link. *Biochem. J.* 283:129–136.

Lawson, M.A. and P.P. Purslow. 2001. Development of components of the extracellular matrix, basal lamina and sarcomere in chick quadriceps and pectoralis muscles. *Brit. Poult. Sci.* 42:315–320.

Lepetit, J. 2007. A theoretical approach of the relationships between collagen content, collagen cross-links and meat tenderness. *Meat Sci.* 76:147–159.

Lepetit, J. 2008. Collagen contribution to meat toughness: theoretical aspects. *Meat Science* 10.1016/j.meatsci.2008.06.016.

Lewis, G.J. and P.P. Purslow. 1989. The strength and stiffness of perimysial connective-tissue isolated from cooked beef muscle. *Meat Sci.* 26:255–269.

Lewis, G.J., P.P. Purslow, and A.E. Rice. 1991. The effect of conditioning on the strength of perimysial connective-tissue dissected from cooked meat. *Meat Sci.* 30:1–12.

Li, X., D.C. McFarland, and S.G. Velleman. 2006. Effect of transforming growth factor-beta on decorin and beta1 integrin expression during muscle development in chickens. *Poult. Sci.* 85:326–332.

Li, X.J., R. Bickerdike, E. Lindsay, P. Campbell, D. Nickell, A. Dingwall, and I.A. Johnston. 2005. Hydroxylysyl pyridinoline cross-link concentration affects the textural properties of fresh and smoked Atlantic salmon (*Salmo salar* L.) flesh. *J. Agric. Food Chem.* 53:6844–6850.

Light, N., A.E. Champion, C. Voyle, and A.J. Bailey. 1985. The role of epimysial, perimysial and endomysial collagen in determining texture in 6 bovine muscles. *Meat Sci.* 13:137–149.

Listrat, A. and J.F. Hocquette. 2004. Analytical limits of total and insoluble collagen content measurements and of type I and III collagen analysis by electrophoresis in bovine muscles. *Meat Sci.* 68:127–136.

Listrat, A., C. Lethias, J.F. Hocquette, G. Renand, F. Menissier, Y. Geay, and B. Picard. 2000. Age-related changes and location of types I, III, XII and XIV collagen during development of skeletal muscles from genetically different animals. *Histochem. J.* 32:349–356.

Listrat, A., B. Picard, and Y. Geay. 1999. Age-related changes and location of type I, III, IV, V and VI collagens during development of four foetal skeletal muscles of double-muscled and normal bovine animals. *Tissue & Cell* 31:17–27.

Liu, A., T. Nishimura, and K. Takahashi. 1996. Relationship between structural properties of intramuscular connective tissue and toughness of various chicken skeletal muscles. *Meat Sci.* 43:43–49.

Liu, C., D.C. McFarland, K.E. Nestor, and S.G. Velleman. 2006. Differential expression of membrane-associated heparan sulfate proteoglycans in the skeletal muscle of turkeys with different growth rates. *Poult. Sci.* 85:422–428.

Maiorano, G., C. Cavone, R.J. McCormick, A. Ciarlariello, M. Gambacorta, and A. Manchisi. 2007. The effect of dietary energy and vitamin E administration on performance and intramuscular collagen properties of lambs. *Meat Sci.* 76:182–188.

Maiorano, G., R.J. McCormick, R.A. Field, and G. Snowder. 1993. Intramuscular collagen characteristics of ram, wether, and zeranol-implanted ram lambs. *J. Anim. Sci.* 71:1817–1822.

Mayne, R. and R.D. Sanderson. 1985. The extracellular matrix of skeletal muscle. *Coll. Relat. Res.* 5:449–468.

McAnulty, R.J. and G.J. Laurent. 1987. Collagen — synthesis and degradation *in vivo* — evidence for rapid rates of collagen turnover with extensive degradation of newly synthesized collagen in tissues of the adult-rat. *Collagen Related Res.* 7:93–104.

McCormick, R.J. 1994a. The flexibility of the collagen compartment of muscle. *Meat Sci.* 36:79–91.
McCormick, R.J. 1994b. Structure and properties of tissues. In *Muscle Foods,* D.M. Kinsman, A.W. Kotula, B.B. Breidenstein, Eds., Chapman & Hall Publishers, pp. 25–62.
McCormick, R.J. 1999. Extracellular modifications to muscle collagen: implications for meat quality. *Poult. Sci.* 78:785–791.
Mercer, D.K., P.F. Nicol, C. Kimbembe, and S.P. Robins. 2003. Identification, expression, and tissue distribution of the three rat lysyl hydroxylase isoforms. *Biochem. Biophys. Res. Commun.* 307:803–809.
Miles, C.A., N.C. Avery, V.V. Rodin, and A.J. Bailey. 2005. The increase in denaturation temperature following cross-linking of collagen is caused by dehydration of the fibres. *J. Mol. Biol.* 346:551–556.
Miller, L.F., M.D. Judge, M.A. Diekman, R.E. Hudgens, and E.D. Aberle. 1989. Relationships among intramuscular collagen, serum hydroxyproline and serum testosterone in growing rams and wethers. *J. Anim. Sci.* 67:698–703.
Morimoto, H., J. Wada, B. Font, J.D. Mott, D.J.S. Hulmes, T. Ookoshi, H. Naiki, A. Yasuhara, A. Nakatsuka, K. Fukuoka, Y. Takatori, H. Ichikawa, S. Akagi, K. Nakao, and H. Makino. 2008. Procollagen C-proteinase enhancer-1 (POPE-1) interacts with beta 2-microglobulin (beta 2-m) and may help initiate beta 2-m amyloid fibril formation in connective tissues. *Matrix Biol.* 27:211–219.
Myllyharju, J. 2003. Prolyl 4-hydroxylases, the key enzymes of collagen biosynthesis. *Matrix Biol.* 22:15–24.
Myllyharju, J. and K.I. Kivirikko. 2004. Collagens, modifying enzymes and their mutations in humans, flies and worms. *Trends Genet.* 20:33–43.
Nakamura, Y.N., H. Iwamoto, N. Shiba, H. Miyachi, S. Tabata, and S. Nishimura. 2004. Developmental states of the collagen content, distribution and architecture in the pectoralis, iliotibialis lateralis and puboischiofemoralis muscles of male Red Cornish x New Hampshire and normal broilers. *Brit. Poult. Sci.* 45:31–40.
Ngapo, T.M., P. Berge, J. Culioli, and S. De Smet. 2002a. Perimysial collagen crosslinking in Belgian Blue double-muscled cattle. *Food Chem.* 77:15–26.
Ngapo, T.M., P. Berge, J. Culioli, E. Dransfield, S. De Smet, and E. Claeys. 2002b. Perimysial collagen crosslinking and meat tenderness in Belgian Blue double-muscled cattle. *Meat Sci.* 61:91–102.
Nimni, M.E. and R.D. Harkness. 1988. Molecular structures and functions of collagen. In *Collagen, Vol. I, Biochemistry,* M.E. Nimni, Ed. CRC Press, Boca Raton, FL, p. 1–77.
Nishimura, T., E. Futami, A. Taneichi, T. Mori, and A. Hattori. 2002. Decorin expression during development of bovine skeletal muscle and its role in morphogenesis of the intramuscular connective tissue. *Cells Tissues Organs* 171:199–214.
Nishimura, T., A. Hattori, and K. Takahashi. 1995. Structural weakening of intramuscular connective-tissue during conditioning of beef. *Meat Sci.* 39:127–133.
Nishimura, T., A. Hattori, and K. Takahashi. 1996. Relationship between degradation of proteoglycans and weakening of the intramuscular connective tissue during post-mortem ageing of beef. *Meat Sci.* 42:251–260.
Nishimura, T., A. Taneichi, J. Wakamatsu, and A. Hattori. 2003. Effect of skeletal muscle decorin on collagen fibrillogenesis *in vitro*. *Anim. Sci. J.* 74:399–405.
Offer, G., P. Knight, R. Jeacocke, R. Almond, T. Cousins, J. Elsey, N. Parsons, A. Sharp, R. Starr, and P. Purslow. 1989. The structural basis of the water-holding, appearance and toughness of meat and meat-products. *Food Microstructure* 8:151–170.
Passerieux, E., R. Rossignol, A. Chopard, A. Carnino, J.F. Marini, T. Letellier, and J.P. Delage. 2006. Structural organization of the perimysium in bovine skeletal muscle: junctional plates and associated intracellular subdomains. *J. Struct. Biol.* 154:206–216.
Passerieux, E., R. Rossignol, T. Letellier, and J.P. Delage. 2007. Physical continuity of the perimysium from myofibers to tendons: involvement in lateral force transmission in skeletal muscle. *J. Struct. Biol.* 159:19–28.
Pedersen, M.E., M.A. Kulseth, S.O. Kolset, S. Velleman, and K.H. Eggen. 2001. Decorin and fibromodulin expression in two bovine muscles (M-semitendinosus and M-psoas major) differing in texture. *J. Muscle Foods* 12:1–17.
Powell, T.H., M.C. Hunt, and M.E. Dikeman. 2000. Enzymatic assay to determine collagen thermal denaturation and solubilization. *Meat Sci.* 54:307–311.
Purslow, P.P. 1989. Strain-induced reorientation of an intramuscular connective-tissue network — implications for passive muscle elasticity. *J. Biomechanics* 22:21–31.
Purslow, P.P. 2002. The structure and functional significance of variations in the connective tissue within muscle. *Compar. Biochem. Physiol., A— Mol. Integrat. Physiol.* 133:947–966.
Purslow, P.P. 2005. Intramuscular connective tissue and its role in meat quality. *Meat Sci.* 70:435–447.

Purslow, P.P., I.B. Mandell, T.M. Widowski, J. Brown, C.F.M. Delange, J.A.B. Robinson, E.J. Squires, M.C. Cha, and G. VanderVoort. 2008. Modelling quality variations in commercial Ontario pork production. *Meat Sci.* 80:123–131.

Reiser, K., R.J. McCormick, and R.B. Rucker. 1992. Enzymatic and nonenzymatic cross-linking of collagen and elastin. *FASEB J.* 6:2439–2449.

Ricard-Blum, S., F. Ruggiero, and M. Van der Rest. 2005. The collagen superfarmily. *Collagen* 247:35–84.

Rowe, R. 1974. Collagen fibre arrangement in intramuscular connective tissue. Changes associated with muscle shortening and their possible relevance to raw meat toughness measurements. *J. Food Technol.* 9:501–508.

Rowe, R.W. 1981. Morphology of perimysial and endomysial connective tissue in skeletal muscle. *Tissue Cell* 13:681–690.

Rucklidge, G.J., G. Milne, B.A. Mcgaw, E. Milne, and S.P. Robins. 1992. Turnover rates of different collagen types measured by isotope ratio mass-spectrometry. *Biochim. Biophys. Acta* 1156:57–61.

Ruoslahti, E. and M.D. Pierschbacher. 1987. New perspectives in cell-adhesion — Rgd and integrins. *Science* 238:491–497.

Sanes, J.R. 2003. The basement membrane/basal lamina of skeletal muscle. *J. Biol. Chem.* 278:12601–12604.

Sato, K., M. Ando, S. Kubota, K. Origasa, H. Kawase, H. Toyohara, M. Sakaguchi, T. Nakagawa, Y. Makinodan, K. Ohtsuki, and M. Kawabata. 1997. Involvement of type V collagen in softening of fish muscle during short-term chilled storage. *J. Agric. Food Chem.* 45:343–348.

Sato, K., C. Ohashi, M. Muraki, H. Itsuda, Y. Yokoyama, M. Kanamori, K. Ohtsuki, and M. Kawabata. 1998. Isolation of intact type V collagen from fish intramuscular connective tissue. *J. Food Biochem.* 22:213–225.

Sato, K., A. Sakuma, K. Ohtsuki, and M. Kawabata. 1994. Subunit composition of eel (*Anguilla-Japonica*) Type-V collagen — evidence for existence of a novel 4Th alpha-4(V) chain. *J. Agric. Food Chem.* 42:675–678.

Serra, X., L. Guerrero, M.D. Guardia, M. Gil, C. Sanudo, B. Panea, M.M. Campo, J.L. Olleta, M.D. Garcia-Cachan, J. Piedrafita, and M.A. Oliver. 2008. Eating quality of young bulls from three Spanish beef breed-production systems and its relationships with chemical and instrumental meat quality. *Meat Sci.* 79:98–104.

Shiba, N., M. Matsuzaki, and E. Tsuneishi. 2000. Effect of endurance exercise on intramuscular collagen characteristics in goats. In *Proc. 46th Int. Congr. Meat Sci. Technol.*, Argentina. p. 430–431.

Shiba, N., M. Matsuzaki, E. Tsuneishi, and H. Iwamoto. 2007. Changes in the pyridinoline concentration of the gastrocnemius and soleus muscle in goats from 2 weeks prenatal to 24 weeks of age. *Small Ruminant Res.* 71:144–149.

Shimokomaki, M., D.F. Elsden, and A.J. Bailey. 1972. Meat tenderness — age related changes in bovine intramuscular collagen. *J. Food Sci.* 37:892–896.

Snellman, A., A. Tuomisto, A. Koski, A. Latvanlehto, and T. Pihlajaniemi. 2007. The role of disulfide bonds and alpha-helical coiled-coils in the biosynthesis of type XIII collagen and other collagenous transmembrane proteins. *J. Biol. Chem.* 282:14898–14905.

Soderhall, C., I. Marenholz, T. Kerscher, F. Ruschendorf, J. Esparza-Gordillo, M. Worm, C. Gruber, G. Mayr, M. Albrecht, K. Rohde, H. Schulz, U. Wahn, N. Hubner, and Y.A. Lee. 2007. Variants in a novel epidermal collagen gene (COL29A1) are associated with atopic dermatitis. *PLoS Biol.* 5:e242.

Stringer, S.E. and J.T. Gallagher. 1997. Heparan sulphate. *Int. J. Biochem. Cell Biol.* 29:709–714.

Sylvestre, M.N., D. Balcerzak, C. Feidt, V.E. Baracos, and J.B. Bellut. 2002. Elevated rate of collagen solubilization and postmortem degradation in muscles of lambs with high growth rates: possible relationship with activity of matrix metalloproteinases. *J. Anim. Sci.* 80:1871–1878.

Takaluoma, K., M. Hyry, J. Lantto, R. Sormunen, R.A. Bank, K.I. Kivirikko, J. Myllyharju, and R. Soininen. 2007a. Tissue-specific changes in the hydroxylysine content and cross-links of collagens and alterations in fibril morphology in lysyl hydroxylase 1 knock-out mice. *J. Biol. Chem.* 282:6588–6596.

Takaluoma, K., J. Lantto, and J. Myllyharju. 2007b. Lysyl hydroxylase 2 is a specific telopeptide hydroxylase, while all three isoenzymes hydroxylate collagenous sequences. *Matrix Biol.* 26:396–403.

Torrescano, G., A. Sanchez-Escalante, B. Gimenez, P. Roncales, and J.A. Beltran. 2003. Shear values of raw samples of 14 bovine muscles and their relation to muscle collagen characteristics. *Meat Sci.* 64:85–91.

Usha, R. and T. Ramasami. 2000. Effect of crosslinking agents (basic chromium sulfate and formaldehyde) on the thermal and thermomechanical stability of rat tail tendon collagen fibre. *Thermochim. Acta* 356:59–66.

Van der Slot, A.J., E.A. van Dura, E.C. de Wit, J. De Groot, T.W. Huizinga, R.A. Bank, and A.M. Zuurmond. 2005. Elevated formation of pyridinoline cross-links by profibrotic cytokines is associated with enhanced lysyl hydroxylase 2b levels. *Biochim. Biophys. Acta* 1741:95–102.

Veit, G., B. Kobbe, D.R. Keene, M. Paulsson, M. Koch, and R. Wagener. 2006. Collagen XXVIII, a novel von Willebrand factor A domain-containing protein with many imperfections in the collagenous domain. *J. Biol. Chem.* 281:3494–3504.

Velleman, S. 2003. Extracellular matrix and growth. In *Biology of Growth of Domestic Animals*. C.G. Scanes, Ed. Blackwell Publishing, Oxford, U.K., p. 134–146.

Velleman, S.G. 2002. Role of the extracellular matrix in muscle growth and development. *J. Anim. Sci.* 80:E8-E13.

Velleman, S.G., C.S. Coy, and D.C. McFarland. 2007. Effect of syndecan-1, syndecan-4, and glypican-1 on turkey muscle satellite cell proliferation, differentiation, and responsiveness to fibroblast growth factor 2. *Poult. Sci.* 86:1406–1413.

Velleman, S.G., X. Liu, K.H. Eggen, and K.E. Nestor. 1999. Developmental regulation of proteoglycan synthesis and decorin expression during turkey embryonic skeletal muscle formation. *Poult. Sci.* 78:1619–1626.

Velleman, S.G., J.D. Yeager, H. Krider, D.A. Carrino, S.D. Zimmerman, and R.J. McCormick. 1996. The avian low score normal muscle weakness alters decorin expression and collagen crosslinking. *Connective Tissue Res.* 34:33–39.

Visse, R. and H. Nagase. 2003. Matrix metalloproteinases and tissue inhibitors of metalloproteinases — structure, function, and biochemistry. *Circulation Res.* 92:827–839.

Voutila, L., A.M. Mullen, M. Ruusunen, D. Troy, and E. Puolanne. 2007. Thermal stability of connective tissue from porcine muscles. *Meat Sci.* 76:474–480.

Voutila, L., M. Ruusunen, and E. Puolanne. 2008. Comparison of the thermal characteristics of connective tissue in loose structured and normal structured porcine M. semimembranosus. *Meat Sci.* 80:1024–1030.

Werman, M.J., E. Barat, and S.J. Bhathena. 1995. Gender, dietary copper and carbohydrate source influence cardiac collagen and lysyl oxidase in weanling rats. *J. Nutr.* 125:857–863.

Wight, T.N., D.K. Heinegard, and V.C. Hascall. 1991. Proteoglycans: structure and function. In *Cell Biology of Extracellular Matrix, 2nd ed.* Plenum Press, New York.

Young, O.A., T.J. Braggins, and G.J. Barker. 1994. Pyridinoline in ovine intramuscular collagen. *Meat Sci.* 37:297–303.

Zhang, X., C. Liu, K.E. Nestor, D.C. McFarland, and S.G. Velleman. 2007. The role of glypican-1 glycosaminoglycan chains in myogenic satellite cell proliferation, differentiation, and fibroblast growth factor 2 responsiveness. *J. Anim. Sci.* 85:389–389.

Zhang, X., K.E. Nestor, D.C. McFarland, and S.G. Velleman. 2008. The role of syndecan-4 and attached glycosaminoglycan chains on myogenic satellite cell growth. *Matrix Biol.*

Zimmerman, S.D., R.J. Mccormick, R.K. Vadlamudi, and D.P. Thomas. 1993. Age and training alter collagen characteristics in fast-twitch and slow-twitch rat limb muscle *J. Appl. Physiol.* 75:1670–1674.

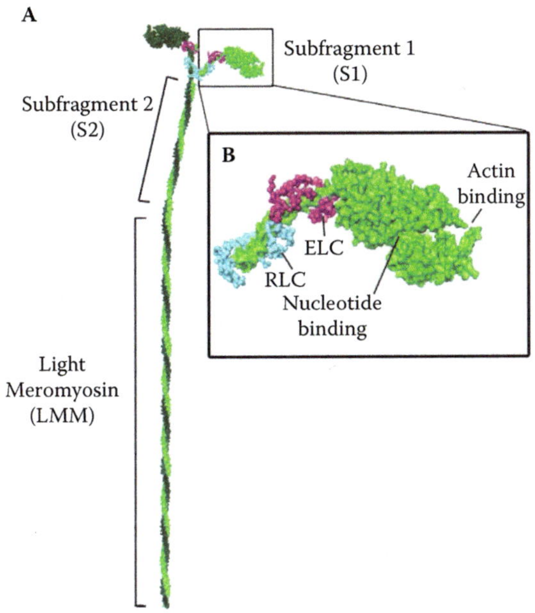

FIGURE 1.5 Myosin structure in vertebrate striated muscle. The myosin molecule (A) includes two heavy chains (rendered in light and dark green) and two light chains (magenta and cyan), and can be divided into three subfragments as shown. The crystal structure of S1 (B) shows locations of light chains and binding sites for actin and nucleotides. Images were rendered with MacPyMol using pdb file 2MYS for S1 and 2FXO for S2.

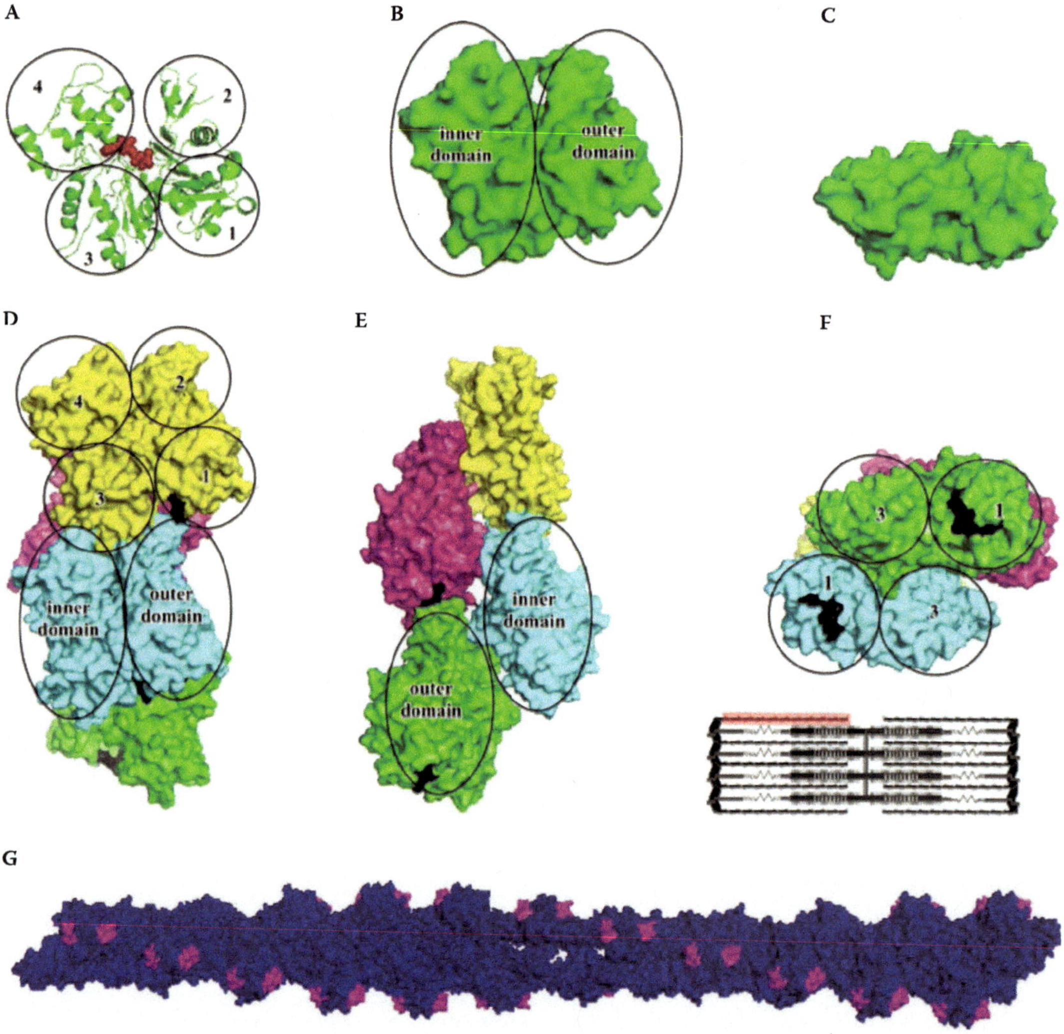

FIGURE 1.6 Actin monomer and polymer structure. The crystal structure of G-actin is shown in α-carbon cartoon view (A) and surface rendered views (B and C). The subdomains are noted, as well as the inner and outer domains. Tilting of the structure 90° into the page shows the bottom (subdomains 1 and 3) and that the molecule is mostly flat (C). The crystal structure of an F-actin filament containing four monomers is shown (D through F). Each subunit is colored differently and the subdomains and domains are noted. The C-terminal three residues for all monomers is colored in black for reference. View (D) shows the filament from the front, (E) from the side, and (F) from the barbed (+) end of the filament. A smaller view of a longer filament is shown in (G) with the α-actinin binding sites colored magenta. The barbed (+) end is to the left and pointed (-) end is to the right. Note the spiral of the binding sites resulting from the helical nature of the filament. Images were adapted from Pirani et al. (2006) supplemental files.

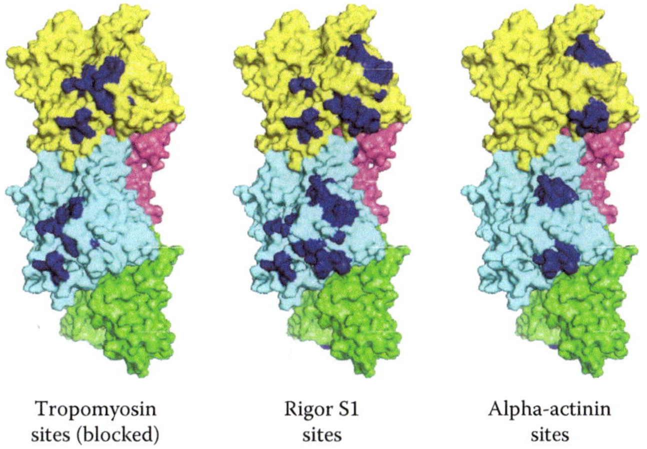

FIGURE 1.7 Binding sites on the actin filament. The F-actin model from Figure 1.6 was modified by coloring the amino acid residues (dark blue) thought to be involved in the binding of tropomyosin in the blocked state, S1 in rigor, and α-actinin. Note that similar residues are involved for all three proteins. The residues involved are from McGough (1998) and images were rendered as in Figure 1.5.

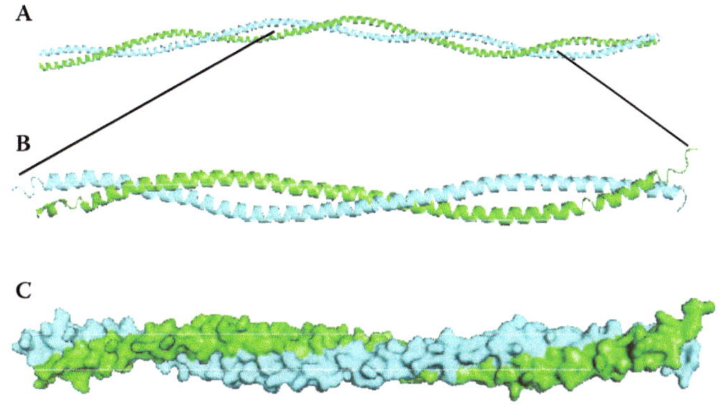

FIGURE 1.8 **Tropomyosin structure.** A low-resolution structure of tropomyosin in α-carbon cartoon view is shown in (A), and a higher-resolution structure of part of the molecule (B). The high-resolution structure is also shown in surface view in (C). One monomer is colored cyan and the other green. Note the coiling of the monomers to form the coiled-coil dimer.

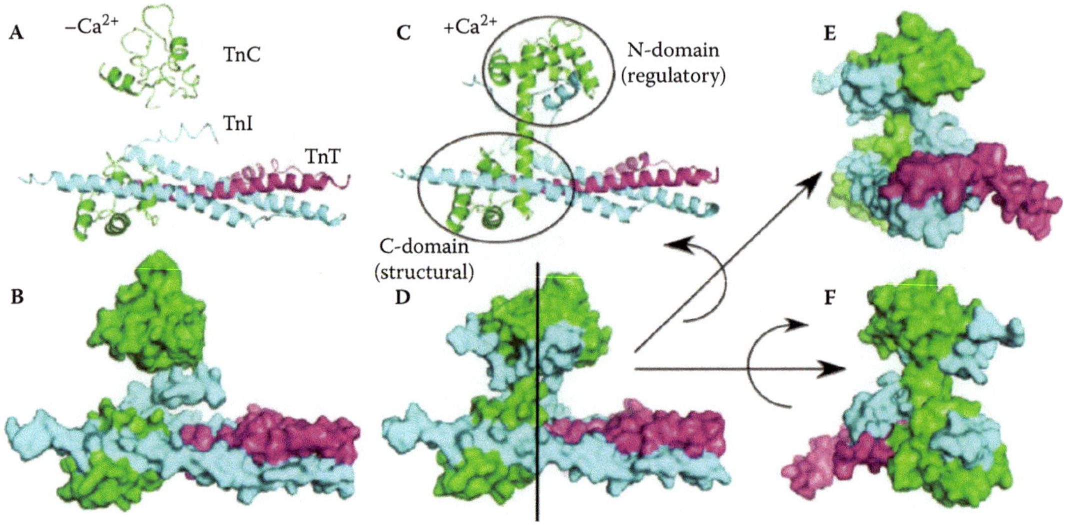

FIGURE 1.9 Structure of the troponin core domain. Most of TnC (green) and part of TnI (cyan) and TnT (magenta) are shown as α-carbon cartoon structures (A and C) or surface views (B, D, E, and F) The core domain without regulatory calcium bound is shown (A and B) and with regulatory calcium bound (C through F). Views (E) and (F) are viewed rotated 90° either out of (E) or into (F) the page. Note in (A) and (C) the greater amount of structure detected in the calcium-bound state as a result of calcium-dependent regulatory interactions between TnC and TnI.

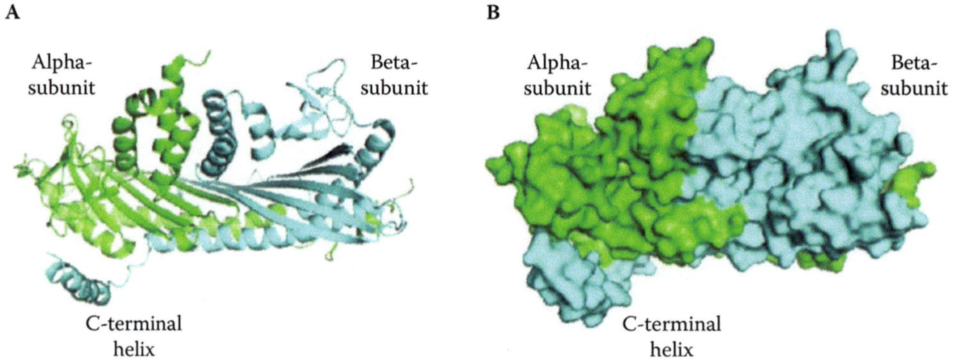

FIGURE 1.10 Structure of CapZ. The α-subunit is colored green and β-subunit cyan in carbon cartoon (A) and surface (B) views. Note the similarity in general structure between the subunits and the C-terminal extension in the β-subunit that is thought to be involved in binding to the barbed (+) end of actin filaments.

A

B

C

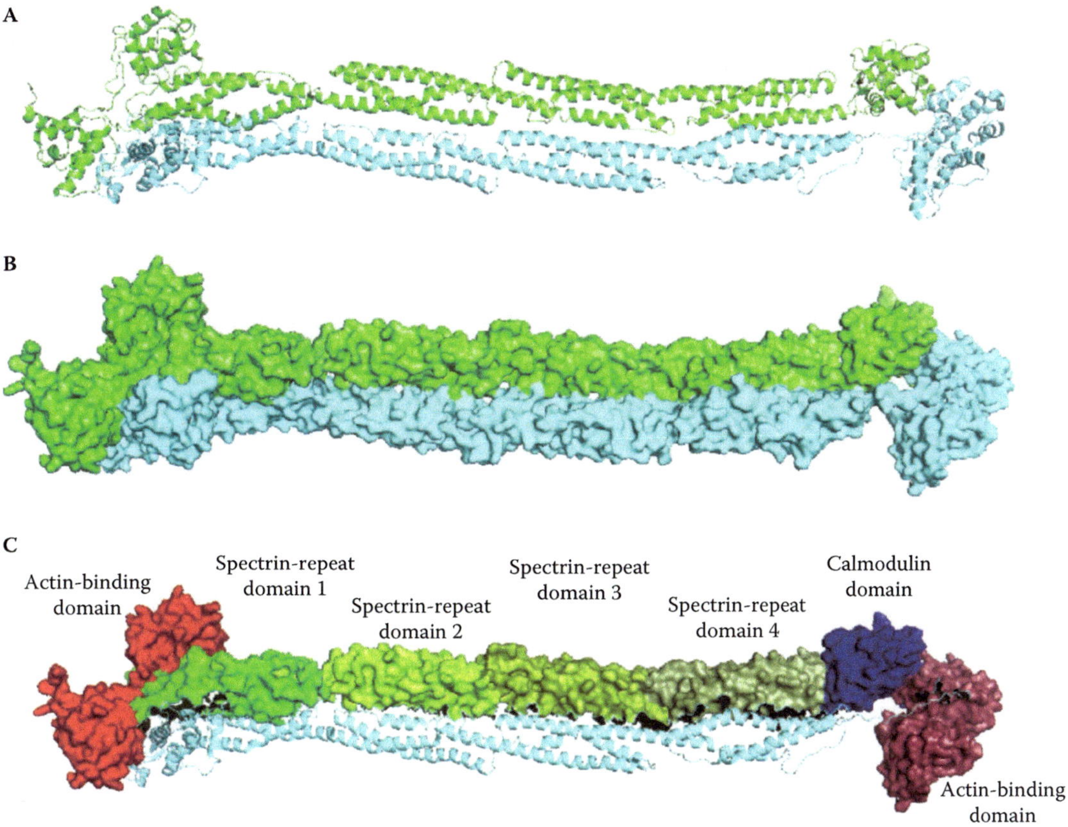

FIGURE 1.11 Model structure and the domain organization of α-actinin dimer. One monomer is colored green while the other is colored cyan in the α-carbon cartoon (A) and surface (B) views. The complete domain structure of one of the monomers is shown (C) with the actin-binding domain in red, spectrin repeats in shades of green, and the calmodulin-like domain in blue. The actin-binding domain of the other monomer is shown in dark salmon to demonstrate its location opposite the calmodulin domain.

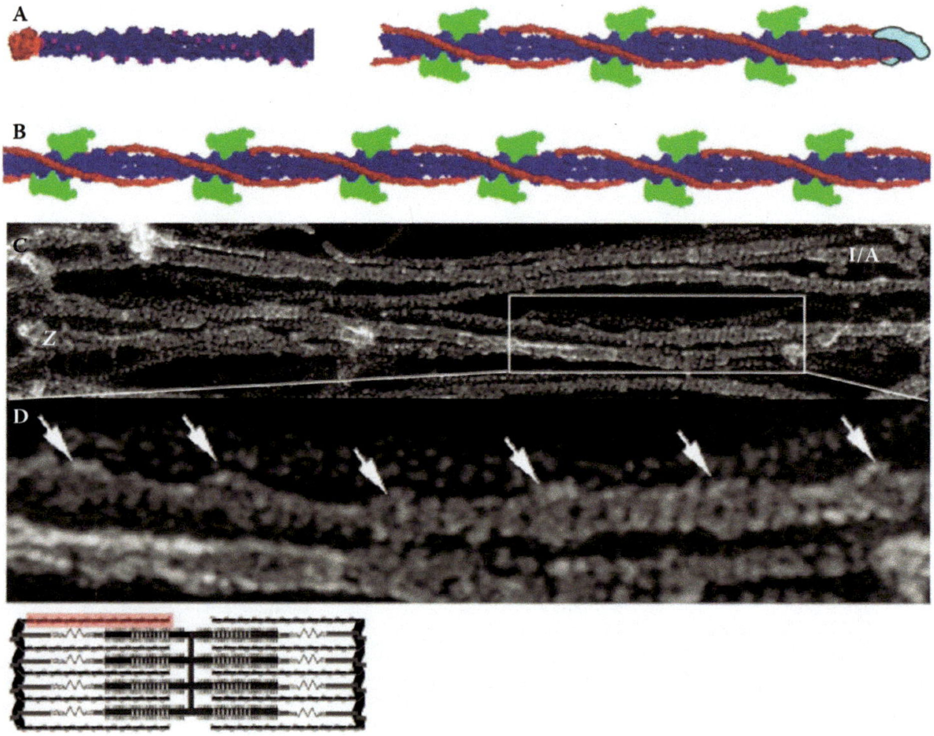

FIGURE 1.15 Thin filament structure. Cartoons of the thin filament (A and B) and transmission electron micrographs (C and D) of thin filaments in the non-overlap region from plaice fin muscle after freeze fracture show the protein organization along the filament. Cartoons of the ends (Z-line and tip in the overlap region) of the thin filaments (A) show the capping of the filament. For the plus or Z-line end (leftmost), actin is colored blue and CapZ is colored red. For the minus or tropomodulin end, actin is blue, Tm is red, Tn is green, and tropomodulin is cyan. Arrows in D denote the troponin bulges. The rotation of each successive troponin pair around the filament axis has been omitted to simplify the rendering.

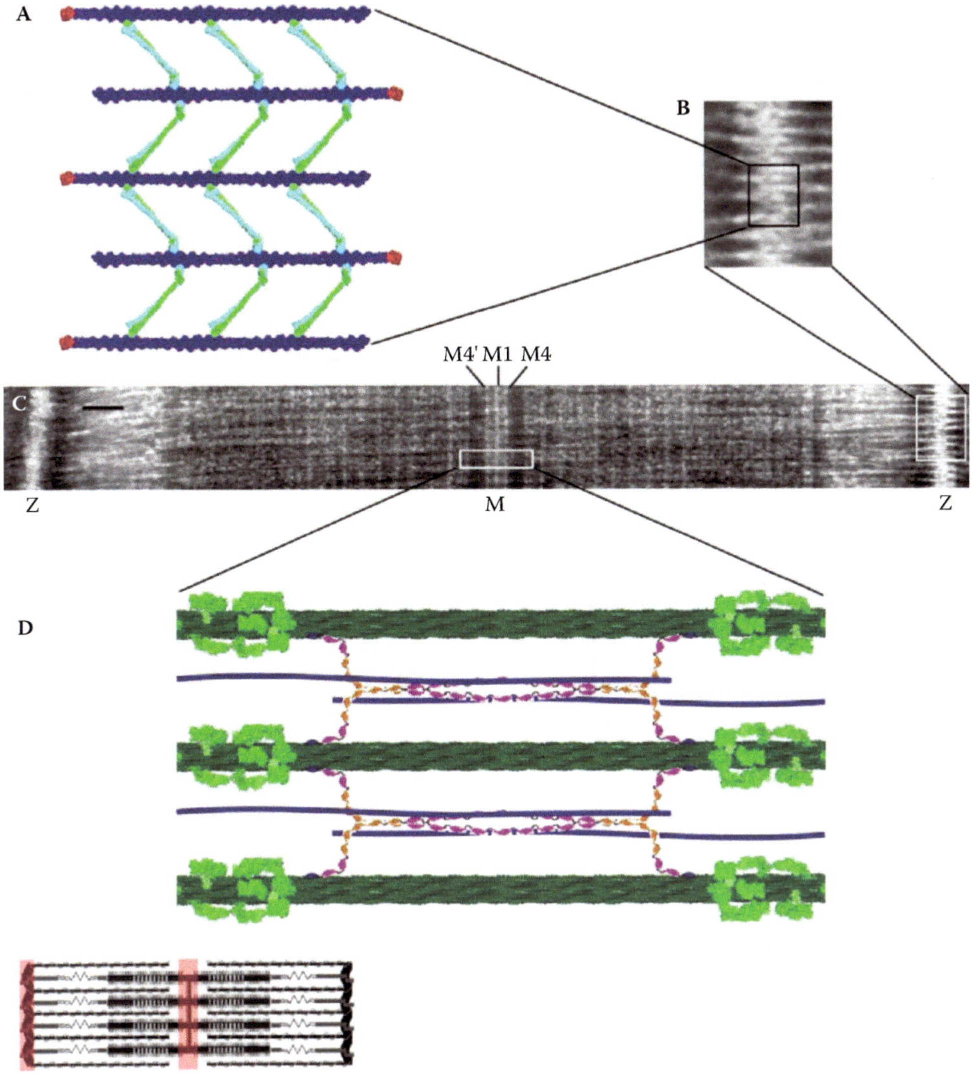

FIGURE 1.16 Structure of the Z-line/Z-disc and M-band. A cartoon of the Z-line (A) was made using F-actin structures from Figure 1.8 and the α-actinin structure in Figure 1.13. F-actin is in blue, CapZ in red, and α-actinin in cyan and green. This zigzag appearance is evident in electron micrographs from negatively stained sections of frog semitendinosus muscle (B and C). The M-band as envisioned in cartoon form (D) was made using Fn and Ig domain models for myomesin and the thick filament cartoons in Figure 1.14. The general organization was interpreted from Lange et al. (2002). Myosin is colored green, titin blue, Ig domains magenta, Fn domains orange, and myosin binding domain black. Myomesin bridges are shown more widely spaced than expected for clarity. The M-lines (M4′, M1, and M4) are noted. Bar in (C) = 100 nm.

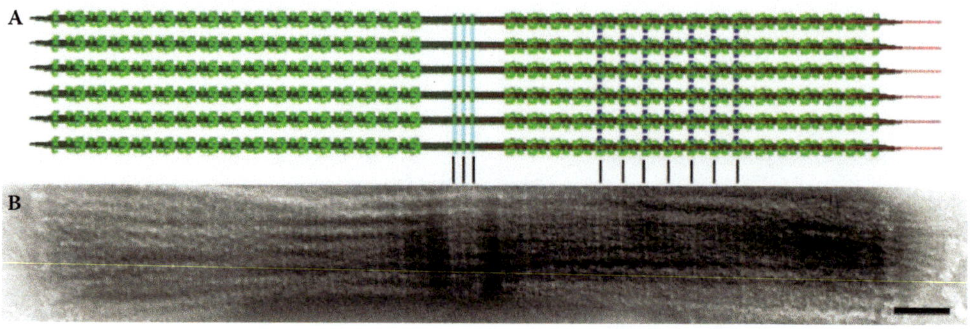

FIGURE 1.17 Structure of the A-band. A cartoon of the A-band is shown (A) with thick filaments in green (from Figure 1.14C), M-line forming the M-band in cyan, MyBP-C in dark blue, and titin in red. A transmission electron micrograph of a negatively stained mechanically isolated A-band from goldfish bodywall shows the A-band structures noted in the cartoon (B). Bar = 100 nm.

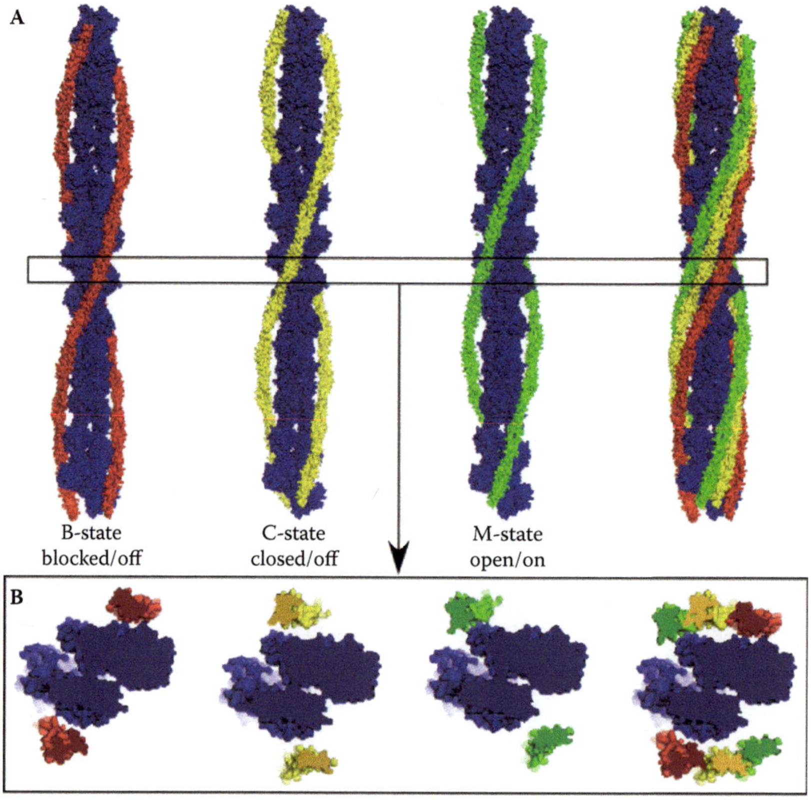

FIGURE 1.22 Tropomyosin position on actin filaments associated with thin filament regulation. Longitudinal (A) and cross-sectional (B) views of F-actin with bound tropomyosin are shown with actin in blue and tropomyosin in red, yellow, or green for the B-, C-, and M-states, respectively. Note that tropomyosin is mostly on the inner domain of actin for the B-state and on the outer domain for the M-state. (The cartoons were made as in Figure 1.5 using the supplemental files in Pirani et al., 2006.)

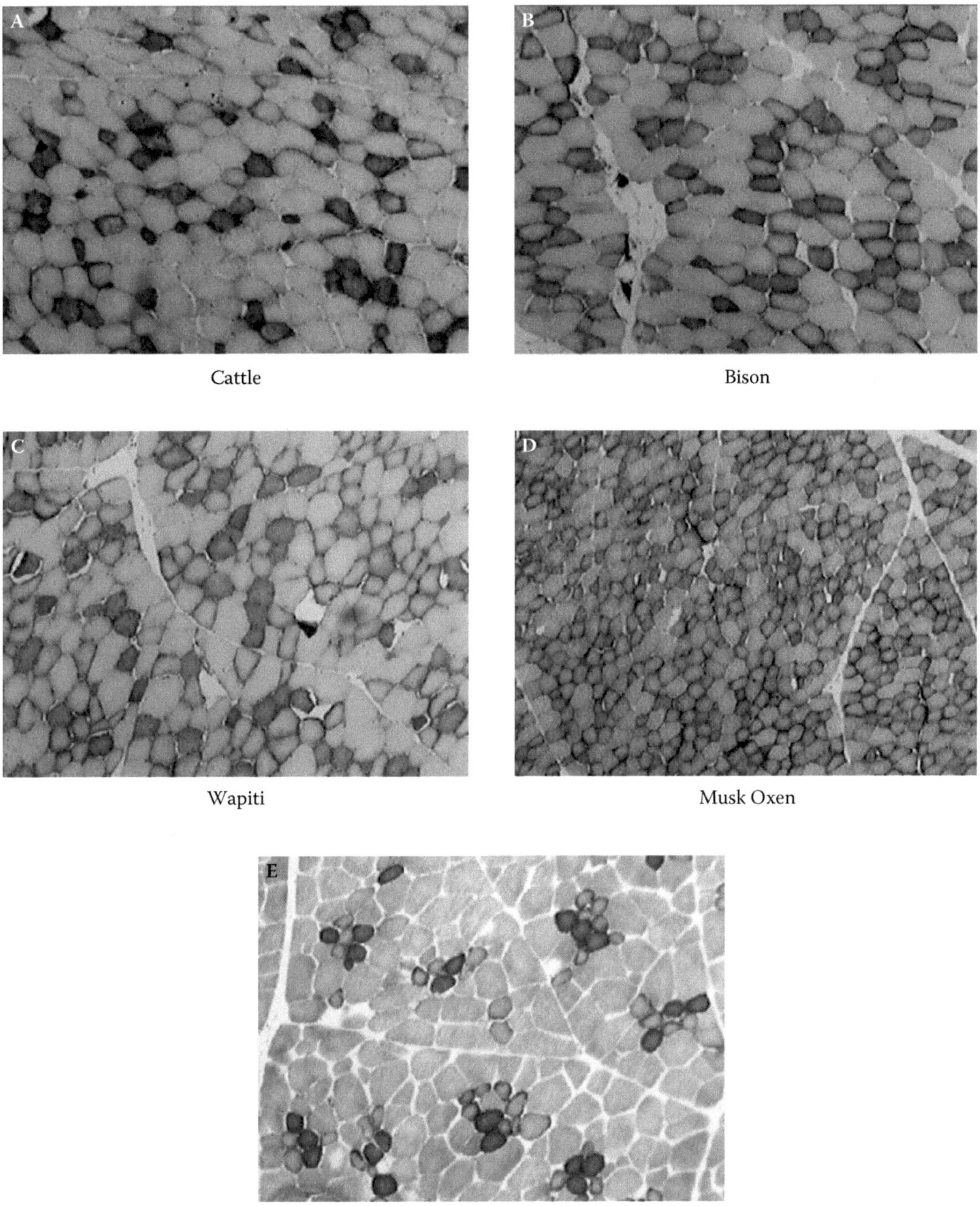

FIGURE 5.1 Combined ATPase and SDH stain used to differentiate β R (brown), α R (blue), and α W (light) fibers in the *longissimus lumborum* muscle (80X magnification) of five different species.

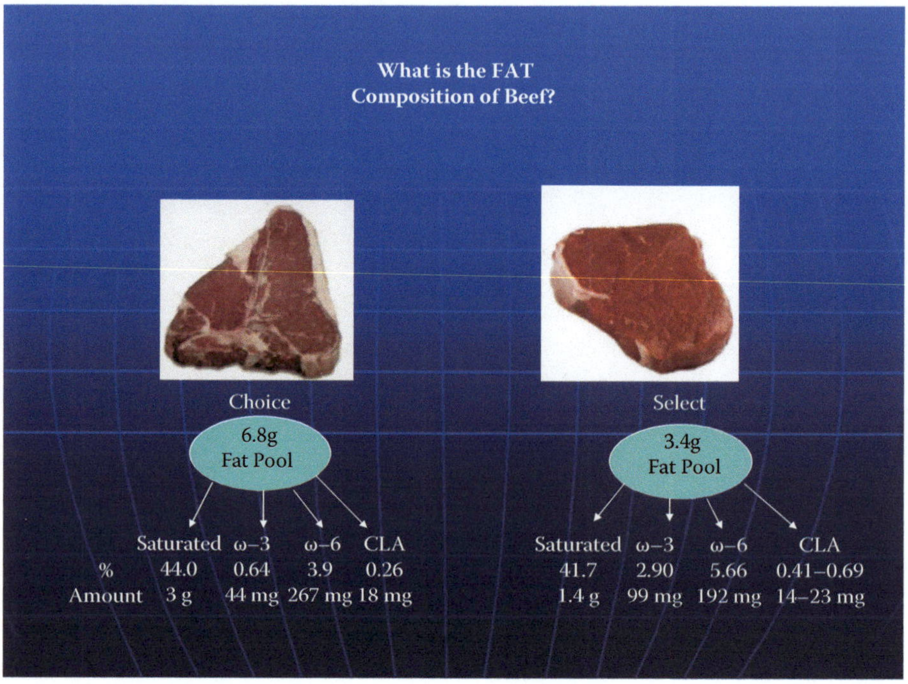

FIGURE 13.18 Fatty acid composition of choice and select beef.

FIGURE 15.5 Microarray. This image represents a two-color bovine oligo microarray. RNA samples were obtained from skeletal muscle from two different steers. One sample was labeled with Cy3 and the second was labeled with Cy5 and co-hybridized to the array. Genes that were expressed at a greater level in the first animal are pseudo-colored green, and those expressed higher in the second animal are red. Genes expressed at equivalent levels are yellow. (*Source:* Image provided by R. N. Vaughn and P.K. Riggs.)

8 Protein Degradation Postmortem and Tenderization

D.L. Hopkins and G.H. Geesink

CONTENTS

8.1 INTRODUCTION

Multiple factors — including palatability, water-holding capacity, color, nutritional value, and food safety — determine meat quality. The importance of these factors varies, depending on both the product and the end user of the product. Flavor, juiciness, and tenderness influence the palatability of meat. Among these traits, tenderness is ranked as most important for beef meat (Thompson 2002), but has lesser influence in sheep meat (Thompson et al. 2005). The three factors that determine meat tenderness are "background toughness," the toughening phase, and the tenderization phase. While the toughening and tenderization phases take place during the postmortem storage, or ageing period, background toughness exists at the time of slaughter and does not change during the storage period. The effect of postmortem storage on tenderness is illustrated in Figure 8.1.

Variation in the background toughness is due to the connective tissue component of muscle. In particular, the organization of the perimysium appears to affect the background toughness, because a general correlation between characteristics of the perimysium and tenderness of muscles has been found for both beef and chicken (Strandine et al. 1949). The toughening phase is caused by muscle shortening during rigor development (Wheeler and Koohmaraie 1994; Koohmaraie et al. 1996). The known effect of the interaction of the rate of postmortem glycolysis and temperature decline on shortening have led to industry recommendations on optimal processing conditions and procedures aimed at minimizing shortening for beef (Thompson 2002) and sheep (Thompson et al. 2005).

While the toughening phase is similar for carcasses under similar processing conditions, there is a large variation in both the rate and extent of tenderization. It has been known for a long time that meat tenderness improves during refrigerated storage, and it was suggested almost a century

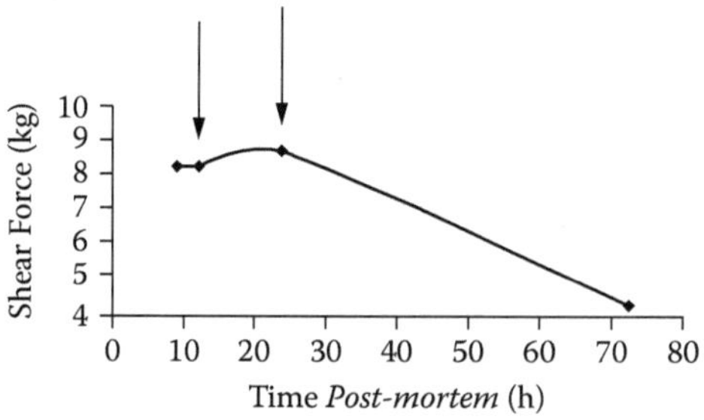

FIGURE 8.1 Proposed time course changes in shear force of ovine longissimus during postmortem storage. Maximum contraction occurs somewhere between the arrows as the muscle enters *rigor*. (*Source:* Adapted from Wheeler and Koohmaraie. With permission of the American Society of Animal Science.)

ago that this is due to enzymatic activity (Hoagland et al. 1917). It is now well-established that postmortem proteolysis of myofibrillar and associated proteins is responsible for this process, but it is a matter of debate about which proteases are responsible for tenderization and which muscle proteins are degraded (Hopkins and Thompson 2002b; Koohmaraie and Geesink 2006; Ouali et al. 2006).

The protease systems that have been studied in relation to postmortem tenderization include the calpain system, cathepsins, the multicatalytic proteinase complex, and, to a lesser extent, caspases. The purpose of this chapter is to discuss the involvement of these protease systems in meat tenderization and to refer the reader to key publications concerning this issue. However, for a proper understanding of factors determining meat tenderness, some basic knowledge of the organization of muscle tissue is a prerequisite. A brief overview of muscle structure will be presented in the following text.

Skeletal muscle is a highly organized tissue, built up of individual cells known as fibers, that are held together by connective tissue. The order of component breakdown is (1) muscle (2) muscle fasciculus (3) muscle fiber/muscle cell, and (4) myofibril. Each muscle fiber consists of a large number of single strands called myofibrils, and these are comprised of myofilaments that are based on thin and thick filaments, predominantly made up of actin and myosin, respectively (Rawn 1989). Myosin is a large protein consisting of 2 heavy chains, each with a molecular weight of 220 kDa and 2 pairs of small subunits referred to as light chains (Bandman 1999). Two regulatory proteins (tropomyosin and troponin) are associated with actin. Troponin is a complex of 3 subunits (Wong 1991). Myofibrils are laterally linked to the sarcolemma and to other myofibrils in the Z-disk region via transverse intermediate filaments, termed costameres, which include such proteins as desmin, filamin, talin, and vinculin. Desmin also connects the Z-disk proteins between myofibrils (Taylor et al. 1995a) and contributes to the maintenance of sarcomere structure. Integrins are cell adhesion proteins that link the extracellular matrix to the cytoskeleton, which is also important for signaling pathways. Titin (or connectin) is a large protein (up to 3700 kDa; Labeit and Kolmerer 1995) that is crucial to the orderly organization of the sarcomere. The sarcomere and cytoskeletal structure have been reviewed by Greaser (1991) and Small et al. (1992). This protein provides, along its length, binding sites for other sarcomere proteins, and the carboxyl-terminal region of titin is reported to bind to myosin (Maruyama 1997). Taylor et al. (1995a) provides a schematic diagram of how titin interacts with actin and myosin, and also illustrates where other proteins such as desmin and vinculin are located in the sarcomere and in Figure 8.2 a proposed arrangement for proteins in the Z-disk is provided.

After death, muscle filaments are in a continuous state of contraction and relaxation, and as glycolysis proceeds, glycogen levels drop, ATP is depleted and the filaments enter rigor and a contracted state, although this process does not occur uniformly throughout a sarcomere (Hwang et al. 2003). Myosin molecules devoid of ATP become permanently bound to actin filaments to form the actomyosin complex (Rawn 1989), causing muscles to become inextensible. Protein degradation

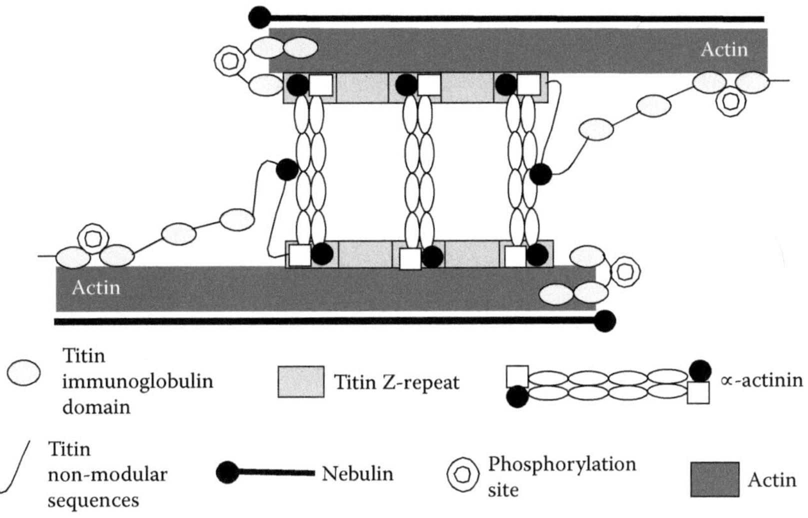

FIGURE 8.2 Molecular structure of the Z-disk adapted from that proposed by Young et al. (1998). This shows the proposed interaction of titin with α-actinin by the symbol • and illustrates the close proximity of part of the protein nebulin with a section of titin.

(proteolysis) of the myofibrillar proteins can precede (i.e., before rigor) tenderization (Hopkins and Thompson 2001c), and the degree of disruption of the proteins (such as titin and nebulin) will influence the final tenderness of meat.

8.2 MECHANISMS OF TENDERIZATION

Although there is good evidence that specific myofibrillar muscle proteins are degraded during the immediate postmortem period (Bandman and Zdanis 1988), the total proportion of protein degraded is considered very small (Davey and Gilbert 1966), while there are also substantial reductions in toughness. There is also debate as to whether the cysteine proteases, in particular the calpains (Dransfield 1993; Uytterhaegen et al. 1994; Koohmaraie 1996; Koohmaraie and Geesink 2006), are largely responsible for this degradation, or whether the cathepsins and other enzyme systems (Goll et al. 1992; Ouali 1992; Zeece et al. 1992; O'Halloran et al. 1997; Roncales et al. 1995; Quali et al. 2006) have a role. These issues will be discussed in this chapter in relation to general principles drawn from studies on bovine and ovine muscle. The mechanism that drives tenderization post-rigor has demanded significant attention, and several explanations for this process have been proposed; but given the requirement for remodeling of protein in living muscle (Goll et al. 1998), it is apparent that such enzymatic pathways must also be central to postmortem proteolysis. There are a growing number of reviews that summarize the biochemical changes occurring in meat during the postmortem period (Ouali 1992; Koohmaraie 1996; Geesink et al. 2000; Hopkins and Thompson 2002b, Sentandreu et al. 2002; Hopkins and Taylor 2004; Ouali et al. 2006), and we will draw on the overview provided by Hopkins and Taylor (2004). Quantification of the contribution of individual proteins to tenderization remains a challenge and, although the rate of disappearance of individual proteins may not always match changes in toughness (Dransfield 1997), this does not diminish the strategic and cumulative contribution that such degradation may have.

Most of the reduction in toughness is attributed to changes in the myofibrillar proteins as discussed below. In general, connective tissue is stable postmortem (as reviewed: McCormick 1994; Purslow 1994) although after extended periods of storage, intramuscular connective tissue (IMCT) does show signs of structural changes (Nishimura et al. 1995). In fact, Nishimura et al. (1998) have described a method to examine the mechanical strength of IMCT and have shown that after 10 days of ageing there was a decrease in strength of this tissue in beef meat. A curve illustrating the reduction during storage of toughness post-rigor (i.e., tenderization) is shown in Figure 8.1. There are two important features of this curve: the post-rigor slope and the maximum toughness (measured as shear force) obtained at rigor. It is accepted that individual curves for the same muscle from different animals will show considerable variation (Koohmaraie et al. 2002), but the generalized curve serves as a useful model for discussing mechanisms. The absolute values obtained at maximum contraction (Figure 8.2) are of great importance, for these will influence the rate of ageing post-rigor (Herring et al. 1967) and also because it will reflect the degree of ageing required to reach acceptable tenderness levels. Temperature during pre-rigor storage has been found to have a large impact on the extent of shortening (Locker and Hagyard 1963), and this varies considerably between different muscles in the carcass (Cena et al. 1992). As such, there is opportunity to alter the maximum toughness achieved (measured as shear force) by managing chilling conditions pre-rigor within the restraints imposed by food safety regulations or by the use of pre-rigor stretching.

The degradation of myofibrillar muscle proteins during the postmortem period has also been studied using a number of techniques, including the determination of free amino acids, fragmentation of myofibrils, measurement of protein solubility, measurement of non-protein nitrogen, gel electrophoresis, and Western blotting. Recent developments in the proteomics field have provided more detailed information on the degradation of proteins postmortem (Bendixen 2005). In their review, Hopkins and Thompson (2002a) provided a detailed overview of the degradation of various myofibrillar proteins as assessed by conventional electrophoresis and some relevant facts will be reiterated here.

8.3 DEGRADATION OF MYOFIBRILLAR PROTEINS

In discussing the patterns of degradation of individual skeletal proteins, it is unrealistic to expect that the disappearance of a single myofibrillar protein could be used to explain tenderization. The evidence indicates that a number of proteins associated with the thin and thick filaments and others found at the periphery of the Z-disk are degraded under normal postmortem chiller conditions. Contrary to earlier suggestions that Z-disk degradation was significant during postmortem storage of muscle (Davey and Gilbert 1968), the evidence from the work of Taylor et al. (1995a) shows that degradation of proteins near the Z-disk is what actually occurs. This conclusion can now extend at least for titin to other regions of the sarcomere (Greaser et al. 2000). In fact, several ultrastructural studies from the 1970s clearly show that the myofiber breaks in the I-band region and that the Z-disk is stable (Davey and Graafhuis 1976; Abbot et al. 1977; Gann and Merkel 1978).

The large protein titin (up to 3700 kDa; Labeit and Kolmerer 1995), crucial to the orderly organization of the sarcomere, has been clearly shown to degrade into fragments during the postmortem period (Robson et al. 1991), as has another large protein nebulin (Huff-Lonergan et al. 1996a) that is associated with actin. Degradation of both titin and nebulin were the suggested reasons for the increased fragility of myofibrils in the I-band region (Taylor et al. 1995a) and were linked to the major decrease in toughness that occurs between 24 to 72 hours postmortem (Dransfield et al. 1992a; Wheeler and Koohmaraie 1994), but the extent of changes will be mediated by muscle types (Taylor et al. 1995a). Proteins associated with muscle contraction, such as myosin, actin, and α-actinin, have been reported to be resistant to degradation under "normal" chiller conditions, and purified calpain does not degrade actin or α-actinin (Goll et al. 1991). The apparent resistance of these proteins to degradation is used to discount the role of cathepsins in tenderization (Koohmaraie 1996; Roncales et al. 1995) as these proteins are degraded in the presence of cathepsins at elevated temperatures and pH. However, the recent report of Hwang et al. (2005), using 2-dimensional electrophoresis and mass spectrometry (proteomics), did show some degradation of these proteins under "normal" chiller conditions. Hwang et al. (2005) did contend, however, that the extent of the degradation was unlikely to contribute to tenderization, but proteomics will continue to shed light on protein changes that previously could not be detected.

Troponin-T is associated with the contractile proteins myosin and actin and has been clearly shown to degrade postmortem (Penny and Dransfield 1979), giving rise to a 30-kDa protein fragment (Ho et al. 1994). This fragment has been associated with a decrease in toughness (Huff-Lonergan et al. 1995; Huff-Lonergan et al. 1996a; Steen et al. 1997). However, given that Troponin-T does not have a structural role in myofibrils, its degradation likely contributes little to tenderization. Its degradation can be interpreted as an indicator of postmortem proteolysis, but not as causative of tenderization.

Other proteins that are important in maintaining the functional capability of muscle are those of the costameres and the cytoskeleton. Costameres connect Z-disks to the sarcolemma and are made up of proteins such as vinculin, talin, desmin, dystrophin, and β-spectrin. A summary of the characteristics of these proteins was given by Robson et al. (1997). Work with human skeletal muscle using confocal microscopy and immunofluorescence (Mondello et al. 1996) suggests that vinculin, desmin, and talin are located at both the Z-disk and the M-line. If this is also true for bovine/ovine muscle, then their degradation will be of more significance for detachment of myofibrils from the sarcolemma and thus tenderization. Vinculin was shown by Taylor et al. (1995a) to be readily degraded under normal postmortem chiller conditions, and a decrease in shear force and an increase in myofibrillar fragmentation index (MFI) in ovine muscle paralleled its disappearance. This degradation, in part, explains the detachment of the sarcolemma from the myofibrils, which was observed in electron micrographs of aged versus fresh muscle (Taylor et al. 1995a). Related to this, the degradation of vinculin and filamin has been implicated in the detachment of adjacent myofibrils (Taylor et al. 1995a; Taylor and Koohmaraie 1998). Wheeler and Koohmaraie (1999) showed that desmin, an intermediate filament protein, was degraded at the same rate in shortened as in control muscle over 10 days of ageing, and that the reduction in shear force was similar between shortened and control muscle. They also reported that the disappearance rate of desmin was highly correlated with the disappearance of troponin-T. Huff-Lonergan et al. (1996a) found that the disappearance of desmin and filamin was faster in meat that had a faster rate of tenderization. There has been extensive work to discover which enzyme groups are responsible for protein degradation in postmortem muscle, and the main contenders are discussed in subsequent sections.

8.4 CATHEPSINS

The calpains and a number of the cathepsins contain a cysteine residue that binds covalently to a substrate (Ertbjerg 1996). Goll et al. (1989) provided a classification of the cathepsins with indicative pH levels for optimal activity, which are all in the acidic range; and Zeece et al. (1992) have reviewed their potential role in meat ageing. Much of the focus on cathepsins and their role in meat tenderization has been directed toward B (EC 3.4.22.1), L (EC 3.4.22.15) and D (EC 3.4.23.5), which are endopeptidases. Cathepsins B and L (cysteine proteases; EC 3.4.22) have also been shown to exhibit exopeptidase activity. The endogenous family of cystatins located in the sarcoplasm inhibits their activity, and a review of these inhibitors was provided by Turk and Bode (1991). A short summary of their characteristics was provided by Ouali et al. (2006).

Cathepsins are located in the lysosomes (Goll et al. 1983) and thus to play a part in myofibril degradation, they must be released from the lysosomes, which is feasible given the failure of ion pumps in membranes during the development of rigor. Pommier et al. (1987) showed that the amount of free cathepsin D (an aspartic protease) increased during ageing, which they attributed to a fall in pH and lysosomal rupture, but concluded this did not have an impact on tenderization; whereas in ostrich meat, Thomas et al. (2004) reported that this enzyme remained active over 12 days of ageing after extraction from meat, but there was no relationship with tenderness. Others have suggested that there is no lysosomal rupture during ageing (Whipple et al. 1990). Attempts to study the structural integrity of lysosomes during ageing have used the electron microscope (Chambers et al. 1994) and, more recently in a preliminarily report, histochemistry was used (Mobark et al. 1999). Both studies suggested that leakage from the lysosomes does occur during ageing, although the activity of cathepsins D and B was minimal in the latter study. The low activity was attributed to either a lack of sensitivity of the technique or the action of cystatins. In addition, lysosomal enzyme activity is often associated with lipid droplets in muscle and is also detected in endoplasmic reticulum after 14 days of postmortem storage (Taylor et al. 1995b). It is perhaps the reticulum staining, which was reported by Dutson and Lawrie (1974) as diffuse staining near the Z-line within 5 days of postmortem storage. It is noteworthy that Chambers et al. (1994), Dutson and Lawrie (1974), and Taylor et al. (1995b) all found intact lysosomes during postmortem storage for up to 14 days, indicating that lysosomes are stable structures. However, O'Halloran et al. (1997) did find a shift in cathepsin B activity from the lysosomal to the soluble fraction during ageing. Interestingly, they also found a reduction in the level of activity of cathepsin B in the soluble fraction between 24 hr and 6 days postmortem in fast glycolysing muscle and an increase in the lysosomal fraction, although when considered together this did not occur for cathepsins B and L. Adding further support for the stability of lysosomes postmortem, Jung et al. (2000) showed that beef muscle lysosomes are stable in beef aged for 2 days and subjected to pressures up to 600 MPa.

Careful scrutiny of the studies on cathepsins show little support for the conclusion that these enzymes have a role in postmortem proteolysis and thus tenderization, as invariably they were conducted under conditions atypical for the myofibrillar environment during the postmortem period. Matsukura et al. (1981) and Mikami et al. (1987) demonstrated that of the cathepsins, type L degraded the largest number of different myofibrillar proteins. In both studies, degradation was studied under high-temperature, low-pH conditions and was evidenced throughout the sarcomere. Robbins and Cohen (1976) also reported that incubation of bovine muscle with cathepsins caused loss of Z-lines, a structural change that is not normally observed postmortem. In some cases, Z-disk structure was lost in addition to changes in the I-band. The fact that actin and myosin are not generally observed to degrade during normal chilled storage of meat is a major hurdle for those who propose a role for cathepsins in tenderization, because as evidenced in a number of studies (Schwartz and Bird 1977; Matsukura et al. 1981; Okitani et al. 1981), these proteins are commonly degraded when myofibrils are incubated in the presence of cathepsins. Having said this, it should be noted, however, that proteomics may reveal subtle changes that previously could not be detected (Hwang et al. 2005).

In a study by Koohmaraie et al. (1991), it was shown that although lamb and beef had similar cathepsin B and B + L activity measured at death, the rates of tenderization were very different. Whipple et al. (1990), working in the same laboratory, found that the activity of cathepsins B and B + L did not explain the differences in tenderness of muscle from *Bos indicus* and *Bos taurus* cattle. Ertbjerg et al. (1999) made very similar conclusions, finding that cathepsin B and L activity in beef is closely related to actin and myosin degradation but not to the appearance of degradation products normally found in meat, such as the 30-kDa protein. O'Halloran et al. (1997) suggested that the higher activity of cathepsins B and B + L in the soluble fraction from fast glycolysing muscle was, in part, responsible for a lower shear force measured at 2 and 6 days postmortem, compared with slow glycolysing muscle. By contrast, the report of Thomas et al. (2004) demonstrated that the activity of extracted cathepsins (B, L, H, and D) was still detectable 12 days after death but there was no reduction in shear force, and thus no tenderization. Such conflicting conclusions may well reflect variations in extraction techniques, methods of assaying enzyme activity, and, in general, a lack of rigorous statistical techniques to quantitatively link variation in cathepsin activity with measures of toughness such as shear force.

In a review of the mechanisms controlling tenderization, Ouali (1992) suggested that cathepsins do contribute to degradation, citing results from experiments in which both rabbit and beef muscles were studied; but in a more recent review, Ouali et al. (2006) provided no new substantive evidence to support the earlier claim. The accumulated evidence makes it difficult to conclude that cathepsins are a major contributor to tenderization, particularly in the early postmortem period, given the following points: (1) the requirement for these enzymes to leak from lysosomes and into the cytosol; (2) the acidic conditions that these enzymes require for optimum activity; (3) the fact that under normal chilling conditions, proteins such as myosin and actin are not degraded, yet these are readily degraded when myofibrils are incubated with cathepsins; and (4) the lack of significant studies that have demonstrated a link between tenderization and the activity of these enzymes. This conclusion is reinforced by the results of studies in which inhibitors to these enzymes have been used as recognized in the review of Sentandreu et al. (2002).

8.5 CALPAINS

Busch et al. (1972) showed that an endogenous protease activated by calcium could degrade myofibrillar proteins, resulting in a significant decrease in isometric tension. Optimal activity of the unidentified enzyme was at pH 7.0 and required levels above 0.1 mM Ca^{2+} *in vitro*. Subsequent data showed that the action of the Ca^{2+} ions was toward the enzyme itself and not on the myofibril substrates (Dayton et al. 1976). Additionally, it was suggested that cysteine side chains on the enzyme were important for its proteolytic activity, but that at 37°C, the enzyme underwent autolysis in the presence of Ca^{2+} ions. The first report of calpain activity related to meat tenderness was by Olson et al. (1977) and is discussed further below.

Murachi et al. (1981) described the enzyme (EC 3.4.22.17) as having two forms, which were given the names calpain I and calpain II, the latter form requiring much higher levels of Ca^{2+} ions for activation. These authors indicated that the enzymes had been found in a range of tissues, with the earliest report in brain tissue (Guroff 1964). An endogenous inhibitor of the enzymes was named calpastatin (Murachi et al. 1981). It was reported that less of this inhibitor was required to reduce calpain II activity and that inhibition occurred by the binding of Ca^{2+} ions to the calpains (I and II), resulting in a conformational change, thereby allowing interaction with calpastatin and inhibition. A detailed review by Wendt et al. (2004) described the inhibitory mode of calpastatin and discussed the importance of Ca^{2+} for calpastatin effectiveness.

More recently, another member of the calpain family has been identified, usually referred to as p94 or calpain 3. Calpain 3 was originally reported to be specific to muscle tissue (Sorimachi et al. 1989), but this has now expanded to other tissues (Goll et al. 2003). This calpain has been reported to bind to 2 distinct sites on titin (Labeit et al. 1997). Other ubiquitous forms in skeletal muscle have also been characterized. These include isoforms such as calpain 10, which has been reported to be associated with the sarcolemma and does not possess domain IV, which binds calcium (Ma et al. 2001). The review by Goll et al. (2003) provided an exhaustive overview of the presence of the different forms of calpains across a range of species.

8.5.1 The Role of Calcium

Etherington (1984) suggested that calpain II required a Ca^{2+} ion concentration of 1 to 2 mM and calpain I, 50 to 100 μM Ca^{2+} for maximal activity. Subsequent studies suggested that calpain II, now commonly called m-calpain, required somewhere between 300 and 800 μM Ca^{2+} for half-maximal activity (Boehm et al. 1998), a level well above that found in living muscle cells (Etherington 1984); and more recent studies on Ca^{2+} binding have supported these higher requirements of m-calpain (Goll et al. 2003).

In terms of postmortem tenderization, the calcium requirement of the calpains presents some important constraints. Jeacocke (1993) stated that the intracellular concentration of free Ca^{2+} must rise to more than 100 μM at rigor, although the maximum level was not specified and the use of the dye arsenazo may have inflated the estimate as this dye can cleave Ca^{2+} bound to proteins and anions. Hopkins and Thompson (2001c) showed that the free calcium concentration in ovine muscle reached a plateau post-rigor of approximately 110 μM when measured with a Ca^{2+}-sensitive electrode, a level too low to activate m-calpain, in support of the conclusion of Geesink and Koohmaraie (1999a). Similar results were reported by Geesink et al. (2001), who found a free calcium level of 106 μM in ovine longissimus muscle at 24 hr postmortem when determined by atomic absorption (AA) and 60 μM when determined using the dye arsenazo III.

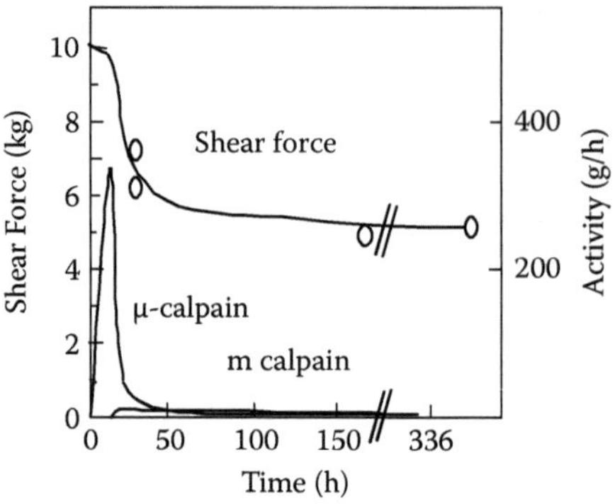

FIGURE 8.3 Change in calpain activity in relation to change in shear force postmortem. (*Source:* Adapted from Dransfield (1993) and based on the data of Koohmaraie et al. (1991) and Whipple et al. (1990). With permission.)

Protein degradation reported in some initial studies was at very high Ca^{2+} ion concentrations (Busch et al. 1972; Dayton et al. 1976), beyond the levels expected under intracellular conditions, even those after rigor. Further to this, Ducastaing et al. (1985) showed that the activity of calpain I (μ-calpain) decreased significantly during the onset of rigor mortis. By comparison, the activity of m-calpain was largely unaltered for up to 3 days postmortem and in some cases this has extended to 56 days postmortem (Geesink and Koohmaraie 1999a). Insufficient free calcium to activate the enzyme is the suggested reason for m-calpain stability (Veiseth et al. 2001).

It is notable that several studies by Ducastaing et al. (1985), Koohmaraie et al. (1987), and Boehm et al. (1998) have considered the activity of extractable μ-calpain, m-calpain, and calpastatin maximal "at death" with a subsequent decline postmortem under *in vitro* conditions (at very different rates). By contrast, Dransfield (1993,) using modeling, predicted minimal activity of these enzymes "at death" under *in situ* conditions. According to his model, μ-calpain reached a maximum activity approximately 10 hr after death (Figure 8.3) and he proposed that this enzyme was not active at death, probably because the Ca^{2+} ion concentration is too low (Dransfield 1992). There is reason to question this conclusion, given new knowledge about "free" Ca^{2+} ion concentrations after death (Hopkins and Thompson 2001c; Geesink et al. 2001) and the fact that calpains have been implicated in protein turnover in living muscle (Goll et al. 2003) under presumably much lower "free" Ca^{2+} ion concentrations.

The autolysis (self-proteolysis) of the calpains has been shown to lower the Ca^{2+} ion concentration required for activation of both μ and m autolyzed types (Edmunds et al. 1991), but the requirement is still beyond the level that is reported to exist in living skeletal muscle cells (Etherington 1984). However, in kidney tissue taken from rats, Yoshimura et al. (1983) reported that μ-calpain only required a Ca^{2+} ion concentration of 10 μM for full activity, and it is apparent that electrical stimulation of living cells will cause an elevation in Ca^{2+} ion levels intracellularly. Kerr et al. (2000), using optical imaging techniques and proteins that fluoresce in the presence of calcium, have recently verified this in "real time." An electrostatic switch mechanism, whereby m-calpain may be activated at physiological calcium concentrations *in vivo,* was proposed by Strobl et al. (2000) and it may be that under postmortem conditions, this mechanism is ineffective, explaining the significantly increased calcium requirement. Related to this, Lawson and Schäfer (2002) in a preliminary paper reported that m-calpain co-localizes with the protein integrin on muscle cell membranes and exhibits activity; but the mechanism of activation was not discussed. This question was reviewed by Goll et al. (2003), who cited studies in a range of tissues where proteins that activate calpains were identified (e.g., in rat skeletal muscle, Michetti et al. 1991) that lower the Ca^{2+} requirement for activity, but the role and significance of such proteins remains to be clarified.

It should be acknowledged that a number of reports (Hattori and Takahashi 1979; Takahashi et al. 1992; Takahashi et al. 1995) endorse the notion that Ca^{2+} ions result in a non-enzymatic weakening of the myofibrillar structure. It is of note that maximum degradation of titin (Takahashi et al. 1992) and nebulin (Tatsumi and Takahashi 1992) occurred at a pH of 7.0 with decreasing fragmentation as the pH declined, which is consistent with a reduction in the activity of the calpains. The fact that Triton X-100 (which causes a release of Ca^{2+} ions) does not reduce toughness when combined with a calpain inhibitor (Uytterhaegen et al. 1994) provides no support for the calcium theory of tenderization (Takahashi 1996). It has been shown, using autoradiography, that Ca^{2+} ions bind to titin (Takahashi et al. 1992) and, more specifically, using fluorescence detection, to the major subfragment of this degraded protein where the 1200-kDa fragment is cleaved (Tatsumi et al. 1999). These results are consistent with the report of Ouali et al. (1991). However, because calpains bind Ca^{2+} ions for activation, and titin has been shown to be a good substrate for these enzymes (Robson et al. 1997), these results are not inconsistent with the proposed role of the calpains. Christensen et al. (2003), studying muscle fibers, concluded that Ca^{2+} ions can lead to conformational changes in some proteins but this does not result in decreasing the breaking strength of fibers, which was attributed to calpain activity with a 70% decrease in the forces needed to fracture fibers when they were incubated in μ-calpain.

8.5.2 Optimal Conditions for Calpain Activity

Early research showed that the optimal conditions for *in vitro* calpain activity were pH 7.5 at 25°C, and this has been confirmed repeatedly. However, this is a combination that does not generally occur in postmortem muscle, but which nevertheless indicates a temperature/pH dependency for the activity of the calpains *in vitro.* Minimal activity has been reported at 5°C with a pH of 5.5, conditions that exist approximately 24 hours postmortem; and Boehm et al. (1998) suggested that the level of activity after 24 hours would be less than 10%, a conclusion supported by Camou et al. (2007a). In contrast, Koohmaraie et al. (1986) suggested that μ-calpain retained 24% to 28% of its maximum activity at pH 5.5 to 5.8 and 5°C (as measured against pH 7.5 and 25°C) when myofibrils were used as the substrate. However, this was challenged by Kanawa et al. (2002), who reported no activity under these conditions when using myofibrils. Activity was reduced to less than 15% when casein was used as the substrate (Koohmaraie et al. 1986), which the authors suggested could have been due to reduced solubility of casein at the low pH and temperature. It should be stressed that the type of assay used will determine the ability to detect μ-calpain activity; and when ^{14}C-labeled casein was used by Geesink and Koohmaraie (1999a), activity could still be detected at 56 days postmortem in ovine muscle. Another aspect that may have impacted on previous estimates of calpain activity was the pH of the extraction buffer used during homogenization (Veiseth and Koohmaraie 2001). It was shown that unless the homogenate pH was kept above 6.5, there was a significant reduction of calpain activity due, it was suggested, to precipitation of these enzymes. Published results indicate large variability in the loss of calpastatin activity postmortem (Ducastaing et al. 1985; Koohmaraie et al. 1987; Boehm et al. 1998); but in all cases, the rate of loss of μ-calpain activity was the greatest, as confirmed by Veiseth et al. (2004). There are differing opinions about the influence of declining pH on the ability of calpastatin to inhibit the calpains. The model proposed by Dransfield (1993) relies to some degree on the claim that the ability of calpastatin to inhibit calpain activity is reduced below a pH of 6.5. Goll et al. (1997), in a conference review paper, questioned this conclusion and suggested there is evidence, albeit from cardiac muscle, that there was no loss of inhibitory activity until the pH drops below 6.2. At this pH, the calpains start to precipitate from aqueous solutions, which makes it difficult to assess the inhibitory activity of calpastatin below this level (Otsuka and Goll 1987). Geesink and Koohmaraie (1999b) suggested that pH has little effect on the inhibition of μ-calpain by calpastatin under *in vitro* conditions of ideal temperature and Ca^{2+} levels, a conclusion supported by Maddock et al. (2005).

Camou et al. (2007a) measured the change in μ-calpain activity using small sections of diaphragm muscle held for up to 120 hours postmortem at a range of pHs (7.5, 7.0, 6.5, or 5.8) under conditions designed to lessen the impact of increases in ionic strength on activity. Activity was assessed at 22 to 24°C using zymography, and this showed that there was minimal effect on activity over the pH range 6.5 to 7.5, but that at pH 5.8 activity was significantly reduced, and was zero after 24 hours postmortem. This latter result probably reflects the difficulty of extracting calpains at low pHs, however, and this must be taken into account when accounting for these outcomes. More recently, Camou et al. (2007b) conducted an investigation of calpain activity in muscle aged for 11 to 13 days postmortem, and the difficulties of extracting calpains and calpastatin were discussed. Using a microplate assay, μ-calpain activity was detected after 11 days of ageing, whereas there was no evidence using zymography. Unfortunately, there was no comparison of activity levels with shear force measures. These results highlight the impact that methodologies can have on interpretation.

A further consideration with respect to the activity of the calpains is the effect of ionic strength. As muscle enters rigor, there is a rapid rise in the ionic strength of the sarcoplasmic fluid, with a peak and subsequent plateau (Ouali 1992); this has been challenged recently by Veiseth et al. (2004), however, who reported results that showed that osmolality increased over 15 days of postmortem storage. It was demonstrated by Geesink and Koohmaraie (1999b) that the activity of partly autolyzed μ-calpain in the absence of calcium was reduced by 50% when incubated in a buffer containing 0.3 M NaCl, after 30 min of incubation. This salt concentration was assumed to be similar to the ionic strength found in post-rigor muscle (Ouali 1992). Therefore, Geesink and Koohmaraie (1999b) indicated that part of the loss of μ-calpain activity in postmortem muscle can be explained by the instability of the partially autolyzed enzyme under high ionic conditions. This conclusion was used to explain why minimal proteolysis of myofibrillar proteins was observed beyond 7 days of incubation in a solution mimicking postmortem conditions, despite the fact that the μ-calpain present was only partially autolyzed, and therefore in a seemingly active state. In a report by Goll et al. (2000), the authors concluded that the activity of autolyzed μ- and m-calpain was significantly reduced when the ionic strength was increased above 0.3 M, and this effect was attributed to a rise in ionic strength and not NaCl (Geesink and Koohmaraie 2000). Veiseth et al. (2004) suggested that if salt solutions higher than 0.1 M were used for assaying activity, this did not equate to conditions representative of postmortem muscle and thus extraction of calpains under higher ionic strengths would lead to an artificial reduction in activity, a possible explanation for the results presented by Maddock et al. (2005). Related to this, it has been reported by Goll et al. (2000) that autolyzed μ-calpain binds irreversibly to the chromatography columns used for separation of the calpains and calpastatin, which would explain the reported decrease in activity, due to low recovery. A comparison by Veiseth et al. (2004) of elution methods again illustrates the impact that different methods can have on conclusions about activity levels. For example, using a gradient elution, μ-calpain and m-calpain activity was shown to decrease during the ageing of meat, whereas with a two-step elution no decrease in m-calpain activity was recorded. It is apparent from the results of these studies that methods of determining calpain activity *in vitro* have real limitations and there is conflict about whether the calpains can be separated from calpastatin using elution with differing concentrations of NaCl (e.g., McDonagh 1998; Geesink and Koohmaraie 1999c). An alternative approach based on determination of mRNA levels may prove useful (Ilian et al. 1998), but this approach also has some limitations. Other alternatives that have been used include ^{14}C-labeled casein and casein zymography but as already discussed, zymography lacks the sensitivity to detect all activity. There is a complication in extrapolating from different assay conditions where variables such as pH, temperature, or calcium concentration are often held constant compared to the dynamic postmortem muscle environment where pH and temperature both decline as muscle passes through rigor and the "free" calcium concentration rises.

8.5.3 Protein Degradation due to Calpain Activity

Early studies found that the action of the calpains led to the release of α-actinin from the Z-disk and degradation of troponins-T and -I, and C-protein (Etherington 1984). While a number of the structural and cytoskeletal proteins are degraded by the calpains (Goll et al. 1989), actin and myosin are generally not considered to be degraded (Dayton et al. 1976) and this mirrors the resilience of these proteins to postmortem degradation as previously discussed. This adds further support to the argument for the role of the calpains in postmortem tenderization.

After incubating myofibrils with m-calpain for up to 30 min at temperatures ranging from 5°C to 25°C and at pHs from 6.5 to 7.5, Zeece et al. (1986b) found that titin and troponin-T were degraded, but this was reduced as the temperature and pH decreased. Degradation of titin was still detected at 5°C and pH 7.5; but at 5°C and pH < 6.5, there was no significant decrease in the number of sarcomeres per myofibril (a measure of myofibril fragmentation) after a 30-min incubation. In general, the degradation of troponin-T to the 30-kDa fragment by calpain extracts mimics the pattern observed for meat during postmortem ageing. Huff-Lonergan et al. (1996b) followed the degradation of specific muscle proteins, using Western blotting techniques, after myofibrils were incubated with μ-calpain for varying times at 5°C and pH 5.6 with 100-μM Ca^{2+}. This approach revealed rapid degradation of titin and nebulin, and a slower degradation of filamin, desmin, and troponin-T, with the production of the same proteolytic fragments found in studies of postmortem meat degradation (Huff-Lonergan et al. 1995; Taylor et al. 1995a). It has been argued that the degradation of myofibrillar proteins such as titin and desmin is minimal during the first 24 hours after death (Boehm et al. 1998) and hence a role for μ-calpain in tenderization was questioned. However, Koohmaraie et al. (1988) states that "about 50% of the ageing response is completed by 24 hours of postmortem aging" in beef, and Wheeler and Koohmaraie (1994) have also demonstrated significant proteolysis in lamb within the first 24 hours. Related to this, Hopkins and Thompson (2001b, 2001c) challenged the claim of minimal proteolysis in the 24-hour period after death and presented evidence using a range of indicators that proteolysis was very rapid during this period. This outcome was consistent with activation of μ-calpain and the significant tenderization of muscle post-rigor. Indeed, autolysis of μ-calpain as early as 3 hours postmortem in ovine muscle has been shown by Veiseth et al. (2001) as indicative of proteolytic activity before the onset of rigor. More recently, Veiseth et al. (2004) showed significant degradation of desmin at 9 hours postmortem and a significant increase in MFI values at 6 hours postmortem compared to "at death" values; however, measuring MFI in pre-rigor muscle is of dubious worth given that the muscle can contract during testing. Using mice in which the gene coding for μ-calpain was deleted (knockout), Geesink et al. (2006) demonstrated, using Western blotting, that specific proteins such as metavinculin and troponin-T were not degraded during postmortem storage. Degradation of other proteins such as dystrophin and desmin was not completely inhibited in μ-calpain knockout mice. However, contrary to what is observed for ovine and bovine muscle, m-calpain in murine muscle appears to be active in postmortem muscle (Kent et al. 2004; Geesink et al. 2006). Thus, the limited postmortem proteolysis in μ-calpain knockout muscles could possibly be attributed to the action of m-calpain, and this suggests that the knockout mice "model" may not be completely appropriate for extrapolation to ovine or bovine muscle.

As mentioned earlier, the use of proteomics has provided a more complete understanding of the degradation of muscle proteins during the postmortem period, and several reports have now emerged showing the degradation of proteins such as actin and myosin (Lametsch et al. 2004; Hwang et al. 2005). Despite this, desmin was a much better substrate for μ-calpain-mediated degradation (Lametsch et al. 2004), and it should be noted that the methodology would have increased the effective concentration of μ-calpain over that present in the isolated myofibrils and this may explain the degradation of actin. Related to this, there was no decrease in actin detected by conventional two-dimensional electrophoresis, illustrating that only a small amount of the actin was degraded.

It has been shown using autoradiography that Ca^{2+} ions bind to titin (Takahashi et al. 1992) and more specifically using fluorescence detection to the major sub-fragment of this degraded protein where the 1200-kDa fragment is cleaved (Tatsumi et al. 1999). Because calpains bind Ca^{2+} ions for activation and titin is a good substrate for calpain (Robson et al. 1997), these results are not inconsistent with the proposed role of the calpains and the suggested degradation sites of titin (Greaser et al. 2000). Additionally, calpain 3 has been reported to bind to titin (Sorimachi et al. 1995; Robson et al. 1997). Despite the emphasis on Ca^{2+} ions, there have been few studies that have measured the levels of these ions in meat throughout the rigor process as outlined earlier. The Ca^{2+} ion has a central role in tenderization, and Hopkins and Thompson (2002c) showed that the increase in postmortem levels of the ion was related to indices of proteolysis. Further studies that quantify the rate of accumulation of "free" ions during the postmortem period with a focus on the activation of the calpains are considered important.

8.5.4 The Contribution of the Calpains to Tenderization

Considering the circumstantial evidence linking the activity of calpains to proteolysis, there have been few attempts to model enzyme activity with the rate of tenderization, or rate of proteolysis. Koohmaraie et al. (1987) demonstrated that MFI values increased for 6 days postmortem when measured initially at death (plateaued after 6 days); and Taylor et al. (1995a) also showed an increase from death through the postmortem period in MFI values, although the rate varied between different muscles. As MFI increased, the activity of extractable μ-calpain decreased in the work of Koohmaraie et al. (1987) to almost a plateau by 1 day postmortem, and a similar effect was reported by Veiseth et al. (2004), depending on how the activity of μ-calpain was determined. Adopting the Ouali (1992) approach of basing comparisons on an index of the enzyme to its inhibitor, the relationship of this index to MFI for the data of Koohmaraie et al. (1987) appeared highly correlated up to 6 days postmortem. It was debatable, however, whether the at-death measures of MFI are meaningful, as pre-rigor muscle will shorten when excised (Parrish et al. 1973) and this likely explained the low MFI values reported at death by both Koohmaraie et al. (1987) and Veiseth et al. (2004).

Taking a slightly different approach, Dransfield et al. (1992b) predicted that tenderization (defined as a decrease in shear force) commenced at pH 6.1, which even in fast glycolysing muscle would be at least 3 hours after death according to their data. Subsequently, it was suggested that μ-calpain was not fully activated until this pH (6.1) was reached (Dransfield 1993) and that this coincided with the start of tenderization. However, this does not preclude proteolysis occurring before this pH as the prediction by Dransfield (1993) sees an increase in μ-calpain activity up to a maximum at 10 hours (Figure 9.3). Thus, proteolysis could proceed before tenderization commences, as suggested by the results of Hopkins and Thompson (2001c). It could be speculated that because single fibers do not enter rigor simultaneously (Jeacocke 1984), then neither would the activation of μ-calpain occur uniformly throughout fiber bundles.

The interaction between enzyme activity, pH, and temperature cannot be ignored when modeling tenderization. Dransfield et al. (1992a) indicated that μ-calpain activity was unaffected by temperature until the pH dropped to 6.2 and was reduced thereafter at a faster rate as the temperature increased. However, it should be noted that this estimate was based on limited data and without consideration for the change in activity of calpastatin. This latter point is very important. Dransfield (1992) stated that the *in vivo* activity of μ-calpain below pH 6.1 was likely to be similar to the *in vitro* (extractable by chromatography) activity because the calpastatin activity was largely removed. However, this conclusion was questioned by Goll et al. (1997). Clearly, this must have an impact on the result of Dransfield (1992), who indicated that 68% of the variation in shear values could be accounted for by μ-calpain activity. Hwang and Thompson (2001), using an experimental design aimed at unravelling the impact of pH and temperature on calpain activity, reported an interaction between temperature and pH at 1.5 hours postmortem. As a result, when the chilling was slow, the activity of μ-calpain and calpastatin decreased when pH decline was rapid; whereas when chilling was rapid, the activity of μ-calpain was largely unaffected by pH decline. In the former case, this led to a reduction in ageing potential due to an exhaustion of μ-calpain.

Another consideration for this modeling of shear force and calpain activity was the well-documented reduction in ageing rate for meat with an intermediate ultimate pH (Watanabe et al. 1996). Additionally, although MFI values were lower for high ultimate pH meat (>6.3) in the data of Watanabe et al. (1996) (as would be predicted from the modeling of Dransfield 1992), indicative of reduced proteolysis, shear force values were not the highest (as would be expected). The effect on shear force was contrary to the prediction from the modeling but was consistent with other reports for high pH meat (Silva et al. 1999). Steen et al. (1997) discuss some of the contrasting results reported between ultimate pH and shear force, and it should not be forgotten that changes to the water-holding capacity of meat are also likely to have an impact on shear force values. Thus, meat with a high pH will have an increased water-holding capacity (Purchas 1990). Also of consideration from the data of McDonagh et al. (1999) is the fact that indices of calpain/calpastatin activity are likely to be more strongly related to MFI than to shear force.

Regulation of calpain activity postmortem will not only be influenced by the pH decline and temperature interaction, but also by the activity of calpastatin. A correlation between ultimate toughness and calpastatin activity at 24 hours postmortem has been shown (Steen et al. 1997). O'Halloran et al. (1997) reported a lower activity of calpastatin at 3 hours postmortem in fast glycolysing muscle than in slow glycolysing muscle, but not at subsequent times postmortem. Interestingly, Jones et al. (1999) has reported fiber type specific differences in calpain 3 expression; but until this enzyme is purified and its activity characterized, we do not know its importance in meat aging (see below). In a conference review paper, Geesink et al. (2000) discussed the apparent paradox of the activity of μ-calpain in the presence of excess calpastatin. They suggested that it is likely that calpastatin's ability to bind and therefore inhibit calpain is modified as they interact, or that the degradation of calpastatin by the calpains enables the enzyme group to maintain activity. Another explanation is that calpastatin distribution in cells is altered by rises and falls in intracellular "free" calcium levels and this regulates activity of the calpains (Averna et al. 2001).

Zamora et al. (1996), using a measure of the resistance (measured by compression) of raw meat, showed a positive correlation ($r = 0.65$) between calpastatin activity measured at 1 hour postmortem and resistance measured at 14 days. Additionally, the rate of ageing was strongly correlated ($r = 0.90$) with the rate of decline of calpastatin activity. The work of Zamora et al. (1996) was more informative than most reports because of the repeated measurements of calpain and calpastatin activity throughout the postmortem period. In a follow-up study, Zamora et al. (2005) used a modeling approach to ascribe variance to the prediction of toughness of the longissimus muscle and concluded that loss of μ-calpain activity explained only 9% of the variance, with serine protease inhibitors explaining 40% of the variation in toughness. This modeling suggested that when μ-calpain activity loss was fast, the meat was more tender — an outcome that was not inconsistent with the findings of Hwang and Thompson (2001). However, Zamora et al. (2005) presented data also for the rectus abdominis, which showed that this muscle was tougher but had higher residual levels of μ-calpain activity and then concluded that "the higher the activity of μ-calpain the tougher the muscle." In a letter to the editor, Hopkins (2006) pointed out the need to be careful when comparing results between muscles as the conclusion was potentially flawed. Having said this, most other studies have relied on one or two measurements and correlations have been used to examine relationships, whereas Zamora et al. (2005) used multiple measures during the postmortem period, thus providing the opportunity to model relationships.

Given the slow decrease in m-calpain activity (Ducastaing et al. 1985; Koohmaraie et al. 1987; Boehm et al. 1998), most attention has focused on the activity of μ-calpain. This activity can only be measured for the μ-calpain that can be extracted. Geesink and Goll (1995), in a preliminary report, suggested that μ-calpain becomes associated with an unextractable fraction in postmortem muscle during storage (i.e., is myofibrillar bound) and that, as a consequence, the proteolytic activity of the enzyme was underestimated. This would have important ramifications for attempts to model enzyme activity and changes in toughness. However, in follow-up work, Boehm et al. (1998) presented evidence that suggested that the bound μ-calpain was virtually inactive under *in vitro* conditions. However, based on the results of Delgado et al. (2001), it could be postulated that under *in vivo* conditions, the bound μ-calpain is protected from the action of calpastatin. The limitation with the studies to date has been the inability to measure the *in situ* activity of both the calpains and their inhibitor calpastatin, although recent evidence from work using rat satellite cells has implicated these enzymes in degradation and remodeling of muscle cytoskeletal proteins during cell fusion (Temm-Grove et al. 1999). In this work, rat satellite cells were microinjected with calpastatin, which stopped fusion of myoblasts, whereas injection with m-calpain increased the rate of fusion. This does not mean, however, that the full mode of action of the calpains is understood in living muscle (Goll et al. 2003).

Olson et al. (1977) demonstrated that the inherently tender psoas muscle had a significantly lower "calpain" activity than either semitendinosus or longissimus muscle and that by 24 hours postmortem, the level of activity was close to zero. This lower "calpain" activity for the m. psoas was later verified by Koohmaraie et al. (1988), who reported muscle-specific levels, but it should be remembered that this muscle is stretched in the carcass and has a low connective tissue content that also contributes to the low shear force of the muscle. Indeed, the relative contribution of these three variables (i.e., enzyme activity, sarcomere length, and connective tissue content) varies between muscles and thus impacts on the extent of tenderization (Wheeler et al. 2000), and this was one of the points made by Hopkins (2006) when commenting on the work of Zamora et al. (2005).

The discovery of calpain 3 or p94 has raised new challenges and possibilities for the development of a tenderization model. Mutations in the gene coding for this calpain are responsible for limb-girdle muscular dystrophy type 2A. Calpain 3 was observed to undergo very rapid autolysis in muscle cells when expressed under *in vitro* conditions (Branca et al. 1999) although it is expressed abundantly at the mRNA level (Ono et al. 1998). This enzyme has been found to degrade a number of different substrate types, including proteins degraded by μ-calpain and m-calpain (Ono et al. 2007) and to be located in the M- and Z-line of the sarcomere (Ilian et al. 2003). There have been limited studies undertaken to examine whether the amount of calpain 3 is related to variation in toughness. This is because the unstable nature of the enzyme prevents the proteolytic activity from being defined and hinders elucidation of its structure. However, by raising antibodies to a protein sequence based on a specific region of calpain 3, Parr et al. (1999) presented data for porcine muscle that showed no association between the level of this enzyme and toughness assessed by shear force after 8 days of ageing. Probably of more importance was the finding that the stability of the enzyme during the first 24 hours postmortem did not vary between samples with very different rates of tenderization. In contrast, Ilian et al. (2001) claimed, using ovine and bovine muscle, that there was a significant correlation between the abundance of mRNA for calpain 3 (determined by use of a DNA probe) and toughness. However, the determination of mRNA abundance does not necessarily reflect the activity of an enzyme and, unfortunately, the analysis conducted by Ilian et al. (2001) was restricted to correlation so it was impossible to draw conclusions about causation. But, in follow-up work, Ilian et al. (2004b) presented evidence to support the specificity of the antibodies they used to calpain 3 and they modeled tenderization over 7 days postmortem in longissimus muscle so as to relate changes in toughness to enzyme levels, but not activity levels. In this work, a reduction in desmin and nebulin was measured during ageing, which was mirrored by an increase in MFI values and an improvement in tenderness. Degradation of desmin has been shown by others to also relate to tenderness measures (Martin et al. 2006), and this protein is a good substrate for calpain degradation (Huff-Lonergan et al. 1996b). In the study of Ilian et al. (2004b), the disappearance of calpains 1 and 3 mirrored increases in MFI and tenderness and the disappearance of nebulin. This resulted in high correlations between the measures (all above 0.9), but lower correlations with the disappearance of desmin. As stressed by the results of Ilian et al. (2004c), an association does not mean causation. Using Western blotting, it has been established that calpain 3 does autolyze in postmortem muscle (Anderson et al. 1998; Ilian et al. 2004b; Parr et al. 1999). In contrast with μ- and m-calpain, calpain 3 is not inhibited by calpastatin (Sorimachi et al. 1993). This observation appears to exclude a major role of calpain 3 on postmortem proteolysis and tenderization, given the great influence of calpastatin on these events. This conclusion was corroborated by results of a more recent study showing that postmortem proteolysis of a series of muscle proteins is not affected in calpain 3 knockout mice (Geesink et al. 2005). Clearly, the role of calpain 3 in tenderization has not been studied as intensively as the role of μ- and m-calpain. However, it is worth remembering that calpain 3 was identified 25 years after μ- and m-calpains I and II. Thus, research focusing on the isolation and identification of muscle-specific enzymes will continue to fill knowledge gaps in meat science and provide a more complete understanding of the processes that drive tenderization. Research on other calpain isoforms, such as calpains 3 and 10, falls into this category, especially with the finding that calpain 10 is present in skeletal muscle and may have some association with tenderization (Ilian et al. 2004a).

8.6 INHIBITION OF CATHEPSINS AND CALPAINS

There are few studies in which inhibitors of the cysteine proteases have been used *in situ* to quantify their contribution to postmortem tenderization. Sugita et al. (1980) showed that in an *in vitro* system E-64 (*trans*-epoxysuccinyl-L-leucylamido(4-guanidino)-butane) inactivated the calpains; Barrett et al. (1982) reported that this compound also inhibited the cathepsins B, H, and L. In contrast to this, Tatsumi et al. (1998) questioned the action of protease inhibitors, suggesting that a number of these inhibitors bound to proteins such as titin and nebulin causing a conformation change and, as a result, suppressing their ability to bind calcium. This result was advanced as evidence in support of the calcium theory of tenderization. In this theory, a role for enzymes in tenderization was dismissed (Takahashi 1996), and it was claimed that tenderization was due to the binding of calcium to specific sites on proteins. However, the claim of the binding of protease inhibitors to sarcomere proteins appears in conflict with the reported action of inhibitors such as E-64 (Fujishima et al. 1997). In the literature, several researchers have attempted to use infusion techniques to introduce enzyme inhibitors such as E-64 into a carcass immediately postmortem, but the use of low concentrations and problems with the permeability and diffusion of the inhibitors to muscle cells resulted in disparate outcomes.

To study the process of tenderization Uytterhaegen et al. (1994) utilized a range of enzyme inhibitors injected into muscle at 24 hours postmortem. It should be stressed that Uytterhaegen et al. (1994) used small cores of muscle to ensure the various inhibitors were effectively dispersed in the treated muscle. This *in situ* approach was an attempt to overcome the limitations of *in vitro* studies, where often the environment is atypical of that within muscle during ageing. The results of these workers clearly showed that catheptic enzymes (B, L, D, and H) could not be implicated in protein degradation in postmortem muscle when held at chiller temperatures at least up to 8 days postmortem. By contrast however, the data of Alarcon-Rojo and Dransfield (1995) suggested that cathepsin D and cathepsins B + L may contribute to tenderization late in the ageing period. At 6 days postmortem, muscle strips soaked at 10°C from day 1 in a number of cathepsin inhibitors, were tougher than control samples. However enzymes such as cathepsin D have a reduced activity at chiller temperatures (Zeece et al. 1986a). In addition, when Uytterhaegen et al. (1994) injected muscle with either pepstatin (60 μM) or a peptidyl-diazomethane (0.2 mM) at 1 day postmortem, these inhibitors of cathepsins D and B + L, respectively, did not inhibit ageing.

Uytterhaegen et al. (1994) also found that a range of calpain inhibitors, including leupeptin and E-64 injected at concentrations of 400 μM and 1.4 mM, respectively, at 24 hours postmortem, effectively prevented tenderization. This approach provided strong evidence for the role of calpains in meat tenderization. The use of other inhibitors selective for the cathepsins B, L, D, and H ruled out their contribution to postmortem tenderization as measured by Uytterhaegen et al. (1994), and this was supported by the work of Hopkins and Thompson (2001b, 2001c). When muscle was injected with E-64, indices of proteolysis such as protein solubility and myofibrillar fragmentation were reduced, and there was no reduction of shear force with ageing. By comparison, an inhibitor of cathepsins B and L (peptidyl-diazomethane) had no such effect, and injected muscle behaved like control muscle exhibiting extensive proteolysis and tenderization. Like Uytterhaegen et al. (1994), Hopkins and Thompson (2001b, c) used a method of injection that they had previously demonstrated was able to deliver an inhibitor throughout muscle (Hopkins and Thompson 2001a). It was also verified that the peptidyl-diazomethane inhibitor was soluble at the temperature and pH characteristic of muscle soon after death (Hopkins 2000). The one cautionary note is the possibility that the inhibitor was in some way blocked from accessing the active site of the target enzymes. Despite this possibility, the evidence adds further weight to the conclusion that cathepsins play little or no role in early postmortem ageing, and that the evidence points to the calpains as being responsible for tenderization.

The conclusion by Dransfield (1999) that studies based on the use of inhibitors have not clarified the role of enzyme groups in tenderization must be reviewed based on the results of studies published by Uytterhaegen et al. (1994) and Hopkins and Thompson (2001b, 2001c). An interesting observation in the results of Uytterhaegen et al. (1994) and Hopkins and Thompson (2001c) was the partial degradation of the myofibrillar protein troponin-T in the presence of the inhibitor E-64, despite a lack of tenderization. This may indicate that the degradation of particular proteins does not relate to the toughness of the cooked meat, which could be associated with the labile nature of some proteins after heating. Alternatively, such degradation in the presence of effective synthetic calpain inhibitors suggests that other enzymes may also have a role in tenderization.

In spite of the current limitations in extracting cathepsins and calpains from muscle, the evidence does suggest that the latter group of enzymes have a pivotal role in proteolysis and postmortem tenderization. The contrast in degradation patterns for myofibrils incubated with the respective enzymes shows that the calpains can mimic the pattern observed when meat is chilled. It is considered that research, which focuses on developing new approaches to the isolation and characterization of muscle enzymes, will provide significant benefits to meat science.

8.7 SERINE PROTEASES AND PROTEASOMES

The evidence suggests that the calpains can be implicated in the tenderization of meat through proteolytic action, and that although their exact activity pattern is yet to be decisively modeled, the changes they elicit are likely to be significant. Thus, they may only be required to be active for relatively short periods of time as initial proteolysis has been suggested as potentially more important than subsequent proteolysis (Dransfield 1998).

Despite these conclusions, it is interesting to consider what drives tenderization after long storage periods such as 60 days (Simmons et al. 2000). Such findings suggest that as meat science gleans new insight from medical and biochemical research about the activity of different enzyme groups, other candidate groups may well emerge that can explain apparent anomalies. The serine proteases such as thrombin fall into this category as this enzyme has been found in skeletal muscle (Citron et al. 1997), and recently a serine protease (M) was identified in mice that could bind to myofibrils and cause degradation (Sangorrín et al. 2002). The evidence for the involvement of these enzymes in myofibrillar degradation is, however, not compelling at the moment. For example, when Uytterhaegen et al. (1994) injected meat with the serine protease inhibitor PMSF, the meat aged and exhibited proteolysis, suggesting no involvement of this enzyme group in myofibrillar degradation. Countering this, Zamora et al. (2005) suggested that, at death, endogenous serine peptidase inhibitors explained 40% of the variation in the rate of tenderization, but which specific inhibitors were acting was not established.

Doubt regarding the contribution of serine proteases also applies to the multicatalytic protease complex (MCP) named by some as proteasome. According to the summary provided by Koohmaraie (1992), it is latent upon extraction from muscle but a range of conditions can activate the enzyme once extracted from muscle, such as heating or incubation with sodium dodecyl sulfate. Given the fact that MCP was incubated under conditions atypical to postmortem muscle before it became active, there appeared little evidence to implicate this protease in the process of tenderization. However, more recently using crude extracts, the activity of proteasome was measured under conditions that mimicked those in postmortem muscle with related changes in pH (Lamare et al. 2002). This showed that proteasomes maintained activity during the development of rigor and post-rigor; and it has been suggested that under high pH conditions, these enzymes may play a role in myofibrillar degradation. When myofibrils are incubated with purified proteasome, structural changes are induced after 24 hours at 37°C (Taylor et al. 1995c), whereas under similar conditions calpain will cause extensive structural changes within 10 min. These results suggest, again, that calpains are the major system causing structural change postmortem.

8.8 CASPASES

Caspases are a group of cysteine peptidases involved in programmed cell death or apoptosis (Fuentes-Prior and Salvesen 2004). Caspases are produced as inactive zymogens, possessing a large and a small subunit. Upstream (or initiator) caspases can autolytically activate themselves. Downstream (or effector) caspases can sequentially be activated by the initiator caspases. Based on their role in apoptosis, the following classes of caspases can be distinguished: caspases involved in initiating apoptosis (caspases 8, 9, 10); effector caspases (caspases 3, 6, 7); and caspases involved in inflammatory processes (caspases 1, 4, 5). For a detailed description of the characteristics of the caspases, the reader is referred to some recent reviews (Fuentes-Prior and Salvesen 2004; Turk and Stoka 2007). In meat science, the caspases have received little attention pertaining to their possible role in meat tenderization. Some recent reports have suggested a role for the caspases in postmortem proteolysis (Ouali et al. 2006; Camou et al. 2007b). However, these suggestions were not supported by data from actual experiments. Results from targeted experiments to elucidate the role of caspases in postmortem tenderization were published by Kemp et al. (2006a, b). For porcine muscle, evidence was presented that the activated form of caspase 3 can be detected in early postmortem muscle and that caspase activity changes early postmortem in porcine muscle. Moreover, a negative relationship between shear force (e.g., toughness) and the degradation of a putative caspase substrate, alpha II spectrin, was observed, suggesting that caspase activity may contribute to tenderization. In contrast, Underwood et al. (2008) failed to detect the activated form of caspase 3 in bovine muscle using Western blotting. Clearly, at present, the numbers of studies on the involvement of caspases in postmortem proteolysis are too few to evaluate their role in postmortem tenderization.

8.9 SUMMARY

Connective tissue provides support at a number of levels via the endomysium, perimysium, or epimysium and maintains the integrity of the contractile apparatus made up of myofibrillar proteins such as actin and myosin and important associated proteins such as titin and nebulin. Proteins such as desmin, talin, and α-actinin also help to maintain the framework within which the contractile proteins function. Degradation of proteins such as titin, nebulin, and desmin will lead to myofibril disruption and contribute to tenderization.

The available evidence does not support a role for the cathepsins B and L in early postmortem proteolysis and points to the major candidate as being the calpain system, despite the fact that the mode of action of the calpains is not yet fully defined. This is consistent with protein turnover in living muscle; and studies have shown that factors that lead to a reduction in protein degradation (i.e., β-agonists) also lead to lower calpain activity. The rise in ionic strength as muscle enters rigor must be considered complementary to biochemical processes, which lead to tenderization post-rigor. The development of rigor and the shortening of fibers counter early proteolysis, but post-rigor the cumulative effect of proteolysis (largely driven by the calpains) reverses the rise in toughness and tenderization will occur. This proposed model is far from complete and this reflects, in part, our incomplete knowledge of the action of endogenous muscle enzymes.

There are several issues that remain to be resolved with respect to the calpains: (1) elucidation of the controlling mechanism for activation in the *in vivo* environment, (2) the role of m-calpain and whether the activation of this enzyme can be manipulated postmortem, (3) the contribution of calpains 3 and 10 to tenderization, and (4) the contribution of myofibril bound μ-calpain to proteolysis and tenderization. Additionally, the observation that the degradation of myofibrillar proteins occurs in the presence of effective synthetic and natural calpain inhibitors suggests that other enzymes may also have a role in tenderization such as the caspases. Inevitably, the accumulated evidence points to a complex system likely to involve interacting proteases and ions, and only through open-minded investigation with reliance on developments in the medical and biochemical fields will a more complete model of tenderization be developed.

REFERENCES

Abbot M.T., A.M. Pearson, J.F. Price, and G.R. Hooper. 1977. Ultrastructural changes during autolysis of red and white porcine muscle. *J. Food Sci.* 42:1185–1188.

Alarcon-Rojo, A.D. and E. Dransfield. 1995. Alteration of post-mortem ageing in beef by the addition of enzyme inhibitors and activators. *Meat Sci.* 41:163–178.

Anderson, L.V.B., K. Davison, J.A.Moss, I. Richard, M. Fardeau, F.M.S. Tomé, C. Hubner, A. Lasa, J. Colomer, and J.S. Beckmann. 1998. Characterization of monoclonal antibodies to calpain 3 and protein expression in muscle from patients with limb-girdle muscular dystrophy type 2 A. *Am. J. Pathol.* 153:1169–1179.

Averna, M., R. De Tullio, M. Passalacquia, F. Salamino, S. Pontremoli, and E. Melloni. 2001. Changes in intracellular calpastatin localization are mediated by reversible phosphorylation. *Biochem. J.* 354:25–30.

Bandman, E. 1999. Solubility of myosin and the binding quality of meat products. In *Proc. 45th Int. Congr. Meat Scie. Technol.*, Yokohama, Japan, p. 236–244.

Bandman, E. and D. Zdanis. 1988. An immunological method to assess protein degradation in post-mortem muscle. *Meat Sci.* 22:1–19.

Barrett, A.J., A.A. Kembhavi, M.A. Brown, H. Kirschke, C.G. Knight, M. Tamai, and K. Hanada. 1982. L-*trans*-epoxysuccinyl-leucylamido (4-guanidino) butane (E-64) and its analogues as inhibitors of cysteine proteinases including cathepsins B, H and L. *Biochem. J.* 201:189–198.

Bendixen, E. 2005. The use of proteomics in meat science. *Meat Sci.* 71:138–149.

Boehm, M.L., T.L. Kendall, V.F. Thompson, and D.E. Goll. 1998. Changes in the calpains and calpastatin during postmortem storage of bovine muscle. *J. Anim. Sci.* 76:2415–2434.

Branca, D., A. Gugliucci, D. Bano, M. Brini, and E. Carafoli. 1999. Expression, partial purification and functional properties of the muscle specific calpain isoform p94. *Eur. J. Biochem.* 265:835–846.

Busch, W.A., D.E. Goll, and F.C. Parrish, Jr. 1972. Molecular properties of postmortem muscle. Isometric tension development and decline in bovine, porcine and rabbit muscle. *J. Food Sci.* 37:289–299.

Camou, J.P., S.W. Mares, J.A. Marchello, R. Vazquez, M. Taylor, V.F. Thompson, and D.E. Goll. 2007a. Isolation and characterization of μ-calpain, m-calpain, and calpastatin from postmortem muscle. I. Initial steps. *J. Anim. Sci.* 85:3400–3414.

Camou, J.P., J.A. Marchello, V.F. Thompson, S.W. Mares, and D.E. Goll. 2007b. Effect of post-mortem storage on activity of μ- and m-calpain in five bovine muscles. *J, Anim. Sci.* 85:670–2681.

Cena, P., I. Jaime, J.A. Beltran, and P. Roncales. 1992. Postmortem shortening of lamb longissimus oxidative and glycolytic fibers. *J. Muscle Foods* 3:253–260.

Chambers, J.J., W.J. Reville, and M.G. Zeece. 1994. Lysosomal integrity in post mortem bovine skeletal muscle. *Sciences des Aliments* 14:441–457.

Christensen, M., R.D. Young, M.A. Lawson, L.M. Larsen, and P.P. Purslow. 2003. Effect of added μ-calpain and post-mortem storage on the mechanical properties of bovine single muscle fibers extended to fracture. *Meat Sci.* 66:105–112.

Citron, B.A., I.V. Smirnova, M.N. Zoubine, and B.W. Festoff. 1997. Quantitative PCR analysis reveals novel expression of prothrombin mRNA and regulation of its levels in developing mouse muscle. *Thrombosis Res.* 87:303–313.

Davey, C.L. and K.V. Gilbert. 1966. Studies in meat tenderness. II. Proteolysis and the aging of beef. *J. Food Sci.* 31:135–140.

Davey, C.L. and K.V. Gilbert. 1968. Studies in meat tenderness. 6. The nature of myofibrillar proteins extracted from meat during aging. *J. Food Sci.* 33:343–348.

Davey C.L. and A.E. Graafhuis. 1976. Structural changes in beef muscle during ageing. *J. Sci. Food Agric.* 27:301–306.

Dayton, W.R., W.J. Reville, D.E. Goll, and M.H.A. Stromer. 1976. A Ca^{2+}-activated protease possibly involved in myofibrillar protein turnover. Partial characterization of the purified enzyme. *Biochemistry* 15:2159–2167.

Delagado, E.F., G.H. Geesink, J.A. Marchello, D.E. Goll, and M. Koohmaraie. 2001. Properties of myofibril-bound calpain activity in longissimus muscle of callipyge and normal sheep. *J. Anim. Sci.* 79:2097–2107.

Dransfield, E. 1992. Modelling *post-mortem* tenderisation-III: role of calpain 1 in conditioning. *Meat Sci.* 31:85–94.

Dransfield, E. 1993. Modelling *post-mortem* tenderisation-IV: role of calpains and calpastatin in conditioning. *Meat Sci.* 34:217–234.

Dransfield, E. 1997. When the glue comes unstuck. In *Proc. 43rd Int. Congr. Meat Sci. Technol.* Auckland, New Zealand, p. 52–63.

Dransfield, E. 1998. Letter to the editor: initial toughness of meat. *Meat Sci.* 48:319–321.

Dransfield, E. 1999. Meat tenderness — the μ-calpain hypothesis. In: *Proc. 45th Int. Congr. Meat Sci. Technol.* Yokohama, Japan, p. 220–228.

Dransfield, E., D.J. Etherington, and M.A.J. Taylor. 1992a. Modelling post-mortem tenderisation. II. Enzyme changes during storage of electrically stimulated and non-stimulated beef. *Meat Sci.* 31:75–84.

Dransfield, E., D.K. Wakefield, and I.D. Parkham. 1992b. Modelling post-mortem tenderisation. I. Texture of electrically stimulated and non-stimulated beef. *Meat Sci.* 31:57–73.

Ducastaing, A., C. Valin, J. Schollmeye, and R. Cross. 1985. Effects of electrical stimulation on postmortem changes in the activities of two Ca dependent neutral proteinases and their inhibitor in beef muscle. *Meat Sci.* 15:193–202.

Dutson T.R. and R.A. Lawrie. 1974. Release of lysosomal enzymes during post mortem conditioning and their relationship to tenderness. *J. Food Technol.* 9:43–50.

Edmunds, T., P.A. Nagainis, S.K. Sathe, V.F. Thompson, and D.E. Goll. 1991. Comparison of the autolyzed and unautolyzed forms of μ- and m-calpain from bovine skeletal muscle. *Biochim. Biophys. Acta* 1077:197–208.

Ertbjerg, P. 1996. The Potential of Lysosomal Enzymes on Post-mortem Tenderisation of Meat. PhD. thesis, The Royal Veterinary and Agricultural University Copenhagen, Denmark.

Ertbjerg P., M.M. Mielche, L.M. Larsen, and A.J. Moller. 1999. Relationship between proteolytic changes and tenderness in prerigor lactic acid marinated beef. *J. Sci. Food Agric.* 79:970–978.

Etherington E.J. 1984. The contribution of proteolytic enzymes to postmortem changes in muscle. *J. Anim. Sci.* 59:1644–1650.

Fuentes-Prior, P. and G.S. Salvesen. 2004. The protein structures that shape caspase-activity, specificity and inhibition. *Biochem. J.* 384:201–232.

Fujishima, A., Y. Imai, T. Nomura, Y. Fujisawa, Y. Yamamoto, and T. Sugawara. 1997. The crystal structure of human cathepsin L complexed with E-64. *FEBS Lett.* 40:747–750.

Gann G.L. and R.A. Merkel 1978. Ultrastructural changes in bovine longissimus muscle during postmortem ageing. *Meat Sci.* 2:129–144.

Geesink, G.H. and D.E. Goll. 1995. Measurement of calpain activity in postmortem muscle extracts underestimates levels of μ-calpain. In *Proc. 41st Int. Congr. Meat Sci. Technol.,* San Antonio, TX, p. 547–549.

Geesink, G.H., M.A. Ilian, J.D. Morton, and R. Bickerstaffe. 2000. Involvement of calpains in post-mortem tenderisation. A review of recent research. In *Proc. 60th New Zealand Soc. Anim. Production Conf.*, Hamilton: New Zealand, p. 99–102.

Geesink, G.H. and M. Koohmaraie. 1999a. Postmortem proteolysis and calpain/calpastatin activity in callipyge and normal lamb biceps femoris during extended postmortem storage. *J. Anim. Sci.* 77:1490–1501.

Geesink, G.H. and M. Koohmaraie. 1999b. Effect of calpastatin on degradation of myofibrillar proteins by μ-calpain under postmortem conditions. *J. Anim. Sci.* 77:2685–2692.

Geesink, G.H. and M. Koohmaraie. 1999c. Technical note: a rapid method for quantification of calpain and calpastatin activities in muscle. *J. Anim. Sci.* 77:3225–3229.

Geesink, G.H. and M. Koohmaraie. 2000. Ionic strength-induced inactivation of μ-calpain in postmortem muscle. *J. Anim. Sci.* 78:2336–2343.

Geesink, G.H., S. Kuchay, A.H. Chishti, and M. Koohmaraie. 2006. μ-Calpain is essential for post-mortem proteolysis of muscle proteins. *J. Anim. Sci.* 84:2834–2840.

Geesink, G.H., R.G. Taylor, A.E.D. Bekhit, and R. Bickerstaffe. 2001. Evidence against the non-enzymatic calcium theory of tenderisation. *Meat Sci.* 59:417–422.

Geesink, G.H., R.G. Taylor, and M. Koohmaraie. 2005. Calpain 3/p94 is not involved in post-mortem proteolysis. *J. Anim. Sci.* 83:1646–1652.

Goll, D.E., M.L. Boehm, G.H. Geesink, and V.F. Thompson. 1997. What causes postmortem tenderisation? In *Proc. 50th Reciprocal Meat Conf.,* American Meat Sci. Association, Chicago, IL, p. 60–67.

Goll, D.E., W.R. Dayton, I. Singh, and R.M. Robson. 1991. Studies of the α-actinin/actin interaction in the Z-disk by using calpain. *J. Biol. Chem.* 266:8501–8510.

Goll, D.E., W.C. Kleese, and A. Szpacenko. 1989. Skeletal muscle proteases and protein turnover. In *Animal Growth Regulation*. Campion, D.R., Hausman, G.J., and Martin R.J. Eds., Plenum Publishing Corporation, New York, p. 141–182.

Goll, D.E., Y. Otsuka, P.A. Nagainis, J.D. Shannon, S.K. Sathe, and M. Mugurma. 1983. Role of muscle proteinases in maintenance of muscle integrity and mass. *J. Food Biochem.* 7:137–177.

Goll D.E., R.G. Taylor, J.A. Christiansen, and V.F. Thompson. 1992. Role of proteinases and protein turnover in muscle growth and meat quality. In *Proc. 45th Reciprocal Meat Conf.,* American Meat Science Association, Chicago, IL, p. 25–36.

Goll, D.E., V.F. Thompson, L. Honqi, G.H. Geesink, and M.L. Boehm. 2000. Effects of autolysis on properties of the calpains. In *Proc. Int. Conf. on Cysteine Proteinases and Their Inhibitors: The New Millennium,* Portorož: Slovenia, p. 21.

Goll, D.E., V.F. Thompson, H. Li, W. Wei, and J. Cong. 2003. The calpain system. *Physiol. Rev.* 83:731–801.

Goll, D.E., V.F. Thompson, R.G. Taylor, and A. Ouali. 1998. The calpain system and skeletal muscle growth. *Can. J. Anim. Sci.* 78:503–512.

Greaser M.L. 1991. An overview of the muscle cell cytoskeleton. In *Proc. 44th Reciprocal Meat Conf.,* American Meat Science Association, Chicago, IL, p. 1–5.

Greaser, M.L., C. Boyer-Berri, Y. Kumazawa, M. Szalata, and E. Pospiech. 2000. Titin and tenderness. In *Proc. 46th Int. Congr. Meat Sci. and Technol.,* Buenos Aires, Argentina, p. 420–423.

Guroff, G. 1964. A neutral, calcium-activated proteinase from the soluble fraction of rat brain. *J. Biol. Chem.* 239:149–155.

Hattori, A. and K. Takahashi. 1979. Studies on the post-mortem fragmentation of myofibrils. *J. Biochem.* 85:47–56.

Herring, H.K., R.G. Cassens, G.G. Suess, V.H. Brungardt, and E.J. Briskey. 1967. Tenderness and associated characteristics of stretched and contracted bovine muscles. *J. Food Sci.* 32:317–323.

Ho, C.Y., M.H. Stromer, and R.M. Robson. 1994. Identification of the 30 kDa polypeptide in post mortem skeletal muscle as a degradation product of troponin-T. *Biochimie* 76:369–375.

Hoagland, R., C.N. McBryde, and W.C. Powick. 1917. Changes in fresh beef during cold storage above freezing. *U. S. Dept. Agric. Bull.* 433:1–100.

Hopkins, D.L. 2000. The Relationship between Actomyosin, Proteolysis and Tenderisation Examined using Protease Inhibitors. PhD thesis, University of New England, Australia.

Hopkins D.L. 2006. Reply: Paper by Zamora et al. (2005). Serine peptidase inhibitors, the best predictor of beef ageing amongst a large set of quantitative variables. *Meat Sci.* 72:185.

Hopkins, D.L. and R.G. Taylor. 2004. Post-mortem muscle proteolysis and meat tenderness. In *Muscle Development of Livestock Animals: Physiology, Genetics, and Meat Quality*, Everts, M, te Pas, M., and Haagsman, H. (Eds.), CAB International, U.K., p. 363–388.

Hopkins, D.L. and J.M. Thompson. 2001a. The relationship between tenderness, proteolysis, muscle contraction and dissociation of actomyosin. *Meat Sci.* 57:1–12.

Hopkins, D.L. and J.M. Thompson. 2001b. Inhibition of protease activity. 1. The effect on tenderness and indicators of proteolysis in ovine muscle. *Meat Sci.* 59:175–185.

Hopkins, D.L. and J.M. Thompson. 2001c. Inhibition of protease activity. 2. Degradation of myofibrillar proteins, myofibril examination and determination of free calcium levels. *Meat Sci.* 59:199–209.

Hopkins, D.L. and J.M. Thompson. 2002a. The degradation of myofibrillar proteins in beef and lamb meat using denaturing electrophoresis — an overview. *J. Muscle Foods* 13:81–102.

Hopkins, D.L. and J.M. Thompson. 2002b. Factors contributing to proteolysis and disruption of myofibrillar proteins and the impact on tenderisation in beef and sheep meat. *Austr. J. Agric. Res.* 53:149–166.

Hopkins, D.L. and J.M. Thompson. 2002c. The relationship between post-mortem calcium concentration or pH and indicators of proteolysis in ovine muscle. *Meat Sci.* 61:411–414.

Huff-Lonergan, E., T. Mitsuhashi, D.D. Beekman, F.C. Parrish, Jr., D.G. Olson, and R.M. Robson. 1996b. Proteolysis of specific muscle structural proteins by mu-calpain at low pH and temperature is similar to degradation in postmortem bovine muscle. *J. Anim. Sci.* 74:993–1008.

Huff-Lonergan, E., T. Mitsuhashi, F.C. Parrish Jr., and R.M. Robson. 1996a. Sodium dodecyl sulfate-polyacrylamide gel electrophoresis and Western blotting comparisons of purified myofibrils and whole muscle preparations for evaluating titin and nebulin in postmortem muscle. *J. Anim. Sci.* 74:779–785.

Huff-Lonergan, E., F.C. Parrish, Jr., and R.M. Robson. 1995. Effects of postmortem aging time, animal age, and sex on degradation of titin and nebulin in bovine longissimus muscle. *J. Anim. Sci.* 73:1064–1073.

Hwang, I.H., C.E. Devine, and D.L. Hopkins. 2003. The biochemical and physical effects of electrical stimulation on beef and sheep meat tenderness — a review. *Meat Sci.* 65:677–691.

Hwang, I.H., B.Y. Park, J.H. Kim, S.H. Cho, and J.M. Lee. 2005. Assessment of postmortem proteolysis by gel-based proteome analysis and its relationship to meat quality traits in pig longissimus. *Meat Sci.* 69:79–91.

Hwang, I.H. and J.M. Thompson. 2001. The interaction between pH and temperature decline early postmortem on the calpain system and objective tenderness in electrically stimulated beef longissimus dorsi muscle. *Meat Sci.* 58:167–174.

Ilian, M.A., A. Bekhit, and R. Bickerstaffe. 2004a. Does the newly discovered calpain 10 play a role in meat tenderization during post-mortem storage? *Meat Sci.* 66:317–327.

Ilian, M.A., A. Bekhit, and R. Bickerstaffe. 2004b. The relationship between meat tenderization, myofibril fragmentation and autolysis of calpain 3 during post-mortem aging. *Meat Sci.* 66:387–397.

Ilian, M.A., A. Bekhit, B. Stevenson, J.D. Morton, P. Isherwood, and R. Bickerstaffe. 2004. Up- and down-regulation of longissimus tenderness parallels changes in the myofibril-bound calpain 3 protein. *Meat Sci.* 67:433–445.

Ilian, M.A., R. Bickerstaffe, and M.L. Greaser. 2003. Postmortem changes in myofibrillar-bound calpain 3 revealed by immunofluorescence microscopy. *Meat Sci.* 66:231–240.

Ilian, M.A., R.S. Gilmour, and R. Bickerstaffe. 1998. Quantification of ovine and bovine calpain I, calpain II, and calpastatin mRNA by ribonuclease protection assay. *J. Anim. Sci.* 76:853–864.

Ilian, M.A., J.D. Morton, M.P. Kent, C.E. Le Couteur, J. Hickford, R. Cowley, and R. Bickerstaffe. 2001. Intermuscular variation in tenderness: association with the ubiquitous and muscle-specific calpains. *J. Anim. Sci.* 79:122–132.

Jeacocke, R.E. 1984. The kinetics of *rigor* onset in beef muscle fibres. *Meat Sci.* 11:237–241.

Jeacocke, R.E. 1993. The concentration of free magnesium and free calcium ions both increase in skeletal muscle fibres entering *rigor mortis*. *Meat Sci.* 35:27–45.

Jones S.W., T. Parr, P.L. Sensky, G.P. Scothern, Bardsley R.G., and P.J. Buttery. 1999. Fibre type-specific expression of p94, a skeletal muscle-specific calpain. *J. Muscle Res. Cell Motil.* 20:417–424.

Jung S., M. Lamballerie-Anton, R.G. Taylor, and M. Ghoul. 2000. High pressure effects on lysosome integrity and lysosomal enzyme activity in bovine muscle. *J. Agric. Food Chem.* 48:2467–2471.

Kanawa, R., J.R. Ji, and K. Takahashi. 2002. Inactivity of m-calpain throughout postmortem aging of meat. *J. Food Sci.* 67:635–638.

Kemp, C.M., R.G. Bardsley, and T. Parr. 2006a. Changes in caspase activity during the post-mortem conditioning period and its relationship to shear force in porcine longissimus muscle. *J. Anim. Sci.* 84:2841–2846.

Kemp, C.M., T. Parr, R.G. Bardsley, and P.J. Buttery. 2006b. Comparison of the relative expression of caspase isofroms in different porcine skeletal muscles. *Meat Sci.* 73:426–431.

Kent, M.P. M.J. Spencer, and M. Koohmaraie. 2004. Postmortem proteolysis is reduced in transgenic mice overexpressing calpastatin. *J. Anim. Sci.* 82:794–801.

Kerr, R., V. Lev-Ram, G. Baird, P. Vincent, R.Y. Tsien, and W.R. Schafer. 2000. Optical imaging of calcium transients in neurons and pharyngeal muscle of *C. elegans*. *Neuron* 26:583–594.

Koohmaraie, M. 1992. Ovine skeletal muscle multicatalytic proteinase complex (proteasome): purification, characterization, and comparison of its effects on myofibrils with μ-calpains. *J. Anim. Sci.* 70:3697–3708.

Koohmaraie, M. 1996. Biochemical factors regulating the toughening and tenderisation processes of meat. *Meat Sci.* 43:S193–S201.

Koohmaraie, M., M. Doumit, and T.L. Wheeler. 1996. Meat toughening does not occur when rigor shortening is prevented. *J. Anim. Sci.* 74:293–2942.

Koohmaraie, M. and G.H. Geesink. 2006. Contribution of postmortem muscle biochemistry to the delivery of consistent meat quality with particular focus on the calpain system. *Meat Sci.* 74:34–43.

Koohmaraie, M., M.P. Kent, S.D. Shackelford, E. Veiseth, and T.L. Wheeler. 2002. Meat tenderness and muscle growth: is there a relationship? *Meat Sci.* 62:345–352.

Koohmaraie, M., J.E. Schollmeyer, and T.R. Dutson. 1986. Effect of low-calcium-requiring calcium activated factor on myofibrils under varying pH and temperature conditions. *J. Food Sci.* 51:28–32/65.

Koohmaraie, M., S.C. Seideman, J.E. Schollmeyer, T.R. Dutson, and A.S. Babiker. 1988. Factors associated with the tenderness of three bovine muscles. *J. Food Sci.* 53:407–410.

Koohmaraie, M., S.C. Seideman, J.E. Schollmeyer, T.R. Dutson, and J.D. Crouse. 1987. Effect of post-mortem storage on Ca^{2+}-dependent proteases, their inhibitor and myofibril fragmentation. *Meat Sci.* 19:187–196.

Koohmaraie, M., G. Whipple, D.H. Kretchmar, J.D. Crouse, and H.J. Mersmann. 1991. Postmortem proteolysis in longissimus muscle from beef, lamb and pork carcasses. *J. Anim. Sci.* 69:617–624.

Labeit, S. and B. Kolmerer. 1995. Titins: giant proteins in charge of muscle ultrastructure and elasticity. *Science* 270:293–296.

Labeit, S., B. Kolmerer, and W.A. Linke. 1997. The giant protein titin. Emerging roles in physiology and pathophysiology. *Circulation Res.* 80:290–294.

Lamare, M., R.G. Taylor, L. Farout, Y. Briand, and M. Briand. 2002. Changes in proteasome activity during postmortem aging of bovine muscle. *Meat Sci.* 61:199–204.

Lametsch, R., P. Roepstorff, H.S. Møller, and E. Bendixen. 2004. Identification of myofibrillar substrates for μ-calpain. *Meat Sci.* 86:515–521.

Lawson, M.A. and A. Schäfer. 2002. The relationship between integrins, calpains and drip channel formation in pork. In *Proc. 48th Int. Congr. Meat Sci. Technol.* Rome, Italy, p. 572–573.

Locker, R.H. and C.J. Hagyard. 1963. A cold shortening effect in beef muscles. *J. Sci. Food Agric.* 14:787–793.

Ma, H., C. Fukiage, Y.H. Kim, M.K. Duncan, N.A. Reed, M. Shih, M. Azuma, and T.R. Shearer. 2001. Characterization and expression of calpain 10. *J. Biol. Chem.* 276:28525–28531.

Maddock, K.R., E. Huff-Lonergan, L.J. Rowe, and S.M. Lonergan. 2005. Effect of pH and ionic strength on μ- and m-calpain inhibition by calpastatin. *J. Anim. Sci.* 83:1370–1376.

Martin, K.M., D.L. Hopkins, G.E. Gardner, and J.M. Thompson. 2006. Effects of stimulation on tenderness of lamb with a focus on protein degradation. *Austr. J. Exp. Agric.* 46:891–896.

Maruyama, K. 1997. Connectin/titin, giant elastic protein of muscle. *FASEB J.* 11:341–345.

Matsukura, U., A. Okitani, T. Nishimuro, and H. Kato. 1981. Mode of degradation of myofibrillar proteins by an endogenous protease, cathepsin L. *Biochim. Biophys. Acta* 66:241–247.

McCormick R.J. 1994. The flexibility of the collagen compartment of muscle. *Meat Sci.* 36:79–91.

McDonagh, M.B. 1998. Calpain Activity Affects Muscle Protein Turnover and Meat Tenderisation. PhD thesis, The University of New England, Australia.

McDonagh, M.B., C. Fernandez, and V.H. Oddy. 1999. Hind-limb protein metabolism and calpain system activity influence post-mortem change in meat quality in lamb. *Meat Sci.* 52:9–18.

Michetti, M., P.L. Viotti, E. Melloni, and S. Pontremoli. 1991. Mechanism of action of the calpain activator protein in rat skeletal muscle. *Eur. J. Biochem.* 202:1177–1180.

Mikami, M., A.H. Whiting, M.A. Tallor, R.A. Maciewicz, and D.J. Etherington. 1987. Degradation of myofibrils from rabbit, chicken and beef by cathepsin L and lysosomal lysates. *Meat Sci.* 21:81–97.

Mobark, M., S. Connell, W. Reville, and M. Zeece. 1999. Localisation of lysosomal cathepsin B during the post-mortem storage of bovine muscle at 4°C and 25°C using enzyme fluorescence histochemistry. In *Proc. 45th Int. Congr. Meat Sci. Technol.* Yokohama, Japan, p. 284–285.

Mondello, M.R., P. Bramanti, G. Cutroneo, G. Santoro, D.D. Mauro, and G. Anastasi. 1996. Immunolocalization of the costameres in human skeletal muscle fibers: confocal microscope investigations. *Anatomical Record* 245:3–9.

Murachi, T., K. Tanaka, M. Hatanaka, and T. Murakami. 1981. Intracellular Ca^{2+}-dependent protease (calpain) and its high-molecular-weight endogenous inhibitor (calpastatin). In *Advances in Enzyme Regulation.* G. Weber, Ed., Pergamon Press, New York, Vol. 19, p. 407–424.

Nishimura, T., A. Hattori, and K. Takahashi. 1995. Structural weakening of intramuscular connective tissue during conditioning beef. *Meat Sci.* 39:127–133.

Nishimura, T., A. Hattori, and K. Takahashi. 1998. Changes in mechanical strength of intramuscular connective tissue during postmortem aging of beef. *J. Anim. Sci.* 76:528–532.

O'Halloran, G.R., D.J. Troy, D.J. Buckley, and W.J. Reville. 1997. The role of endogenous proteases in the tenderisation of fast glycolysing muscle. *Meat Sci.* 47:187–210.

Okitani, A., T. Matsumoto, Y. Kitamura, and H. Kato. 1981. Purification of cathepsin D from rabbit skeletal muscle and its action towards myofibrils. *Biochem. Biophys. Acta* 662:202–209.

Olson, D.G., F.C. Parrish, Jr., W.R. Dayton, and D.E. Goll. 1977. Effect of postmortem storage and calcium activated factor on myofibrillar proteins of bovine skeletal muscle. *J. Food Sci.* 42:117–124.

Ono, Y., C. Hayashi, N. Doi, F. Kitamura, M. Shindo, K. Kudo, T. Tsubata, M. Yanagida, and H. Sorimachi. 2007. Comprehensive survey of p94/calpain 3 substrates by comparive proteomics — possible regulation of protein synthesis by p94. *Biotechnol. J.* 2:565–576.

Ono, Y., H. Shimada, H. Sorimachi, I. Richard, T.C. Saido, J.S. Beckmann, S. Ishiura, and K. Suzuki. 1998. Functional defects of a muscle-specific calpain, p94, caused by mutations associated with limb-girdle muscular dystrophy type 2A. *J. Biol. Chem.* 273:17073–17078.

Otsuka, Y. and D.E. Goll. 1987. Purification of the Ca^{2+}-dependent proteinase inhibitor from bovine cardiac muscle and its interaction with the millimolar Ca^{2+}-dependent proteinase. *J. Biol. Chem.* 262:5839–5851.

Ouali, A. 1992. Proteolytic and physicochemical mechanisms involved in meat texture development. *Biochimie* 74:251–265.

Ouali, A., C.H. Hernan-Mendez, G. Coulis, S. Beclia, A. Boudjellal, L. Aubury, and M.A. Sentandreu. 2006. Revisting the conversion of muscle into meat and the underlying mechanisms. *Meat Sci.* 74:44–58.

Ouali, A., X.Vigon, and M. Bonnet. 1991. Osmotic pressure changes in postmortem bovine muscles:factors of variation and possible causative agents. In *Proc. 37th Int. Congr. Meat Sci. Technol.,* Kulmbach, Germany, p. 452–455.

Parr, T., P.L. Sensky, G.P. Scothern, R.G. Bardsley, P.J. Buttery, J.D. Wood, and C. Warkup. 1999. Relationship between skeletal muscle-specific calpain and tenderness of conditioned porcine longissimus muscle. *J. Anim. Sci.* 77:661–668.

Parrish, Jr., F.C. R.B.Young, B.E. Miner, and L.D. Andrew. 1973. Effect of postmortem conditions on certain chemical, morphological and organoleptic properties of bovine muscle. *J. Food Sci.* 38:690–695.

Penny, I.F. and E. Dransfield. 1979. Relationship between toughness and troponin T in conditioned beef. *Meat Sci.* 3:135–141.

Pommier, S.A., L.M. Postes, and G. Butler. 1987. Effect of low voltage electrical stimulation on the distribution of cathepsin D and the palatability of the longissimus dorsi from Holstein veal calves fed a corn or barley diet. *Meat Sci.* 21:203–218.

Purchas, R.W. 1990. An assessment of the role of pH differences in determining the relative tenderness of meat from bulls and steers. *Meat Sci.* 27:129–140.

Purslow P.P. 1994. The structural basis of meat toughness:what role does the collagenous component play? In *Proc. 40th Int. Congr. Meat Sci. Technol.*, The Hague, Netherlands, p. 27–34.

Rawn, D.J. 1989. *Biochemistry*. Burlington, NC: Neil Patterson Publishers, p. 1085.

Rhee, M.S., Y.C. Ryu, J.Y. Imm, and B.C. Kim. 2000. Combination of low voltage electrical stimulation and early postmortem temperature conditioning on degradation of myofibrillar proteins in Korean native cattle (Hanwoo). *Meat Sci.* 55:391–396.

Robbins, F.M. and S.H. Cohen. 1976. Effects of catheptic enzymes from spleen on the microstructure of bovine semimembranosus muscle. *J. Texture Studies* 7:137–142.

Robson, R.M., E. Huff-Lonergan, F.C. Parrish, Jr., C.H. Ho, M.H. Stromer, T.W. Huiatt, R.M. Bellin, and S.W. Sernett. 1997. Postmortem changes in the myofibrillar and other cytoskeletal proteins in muscle. In *Proc. 50th Reciprocal Meat Conf. American Meat,* Chicago, IL, p. 43–52.

Robson, R.M., T.W. Huiatt, and F.C. Parrish, Jr. (1991) Biochemical and structural properties of titin, nebulin and intermediate filaments in muscle. In *Proc. 50th Reciprocal Meat Conf. American Meat Sci. Assoc.,* Chicago, IL, p. 7–20.

Roncales, P., G.H. Geesink, R.L. van Laack, I. Jaile, J.A. Beltran, V.M.H. Barnier, and F.J.M. Smulders. 1995. Meat tenderisation: enzymatic mechanisms. In *Expression of Tissue Proteinases and Regulation of Protein Degradation as Related to Meat Quality*, A. Ouali, D.I. Demeyer, and F.J.M. Smulders, Eds. ECCEAMST, Utrecht, The Netherlands, p. 311–330.

Sangorrín, M.P., C.B. Martone, and J.J. Sánchez. 2002. Myofibril-bound serine protease and its endogenous inhibitor in mouse: extraction, partial characterization and effect on myofibrils. *Compar. Biochem. Physiol., Part B* 131:713–723.

Schwartz, W. and J.W. Bird. 1977. Degradation of myofibrillar proteins by cathepsins B and D. *Biochem. J.* 167:811–820.

Sentandreu, M.A., G. Coulis, and A. Ouali. 2002. Role of muscle endopeptidases and their inhibitors in meat tenderness. *Trends Food Sci. Technol.* 13:398–419.

Silva, J.A., L. Patarata, and C. Martins. 1999. Influence of ultimate pH on bovine meat tenderness during ageing. *Meat Sci.* 52:453–459.

Simmons, N.J., M.M. Auld, B. Thomson, J.M. Cairney, and C.C. Daly. 2000. Relationship between intermediate pH and toughness in the striploin and other muscles of the beef carcass. In *Proc. 60th New Zealand Soc. Anim. Production Conf.*, Hamilton, New Zealand, p. 117–119.

Small, J.V., D.O. Fürst, and L.E. Thornell. 1992. The cytoskeletal lattice of muscle cells. *Eur. J. Biochem.* 208:559–572.

Sorimachi, H., S. Imajoh-Ohmi, Y. Emori, H. Kawasaki, S. Ohno, Y. Minami, and K. Suzuki. 1989. Molecular cloning of a novel mammalian calcium-dependent protease distinct from both m- and mμ-types. Specific expression of the mRNA in skeletal muscle. *J. Biol. Chem.* 264:20106–20111.

Sorimachi, H., K. Kinbara, S. Kimura, M. Takahashi, S. Ishiura, N. Sasagawa, N. Sorimachi, K. Tagawa, K. Maruyama, and K. Suzuki. 1995. Muscle-specific calpain, p94, responsible for limb girdle muscular dystrophy type 2A, associates with connectin through IS2, a p94 specific sequence. *J. Biol. Chem.* 270:31158–31162.

Sorimachi, H., N. Toyama-Sorimach, T.C. Saido, H. Kawasaki, H. Sugita, M. Miyasaka, K. Arahata, S. Ishiura, and K. Suzuki. 1993. Muscle-specific calpain, p94, is degraded by autolysis immediately after translation, resulting in disappearance from muscle. *J. Biol. Chem.* 268:10593–10605.

Steen, D., E. Claeys, L. Uytterhaegen, D. De Smet, and D. Demeyer. 1997. Early post-mortem conditions and the calpain/calpastatin system in relation to tenderness of double-muscled beef. *Meat Sci.* 45:307–319.

Strandine, E.J., C.H. Koonz, and J.M. Ramsbottom. 1949. A study of variations in muscle of beef and chicken. *J. Anim. Sci.* 8:483–494.

Strobl S., C. Fernandez-Catalan, M. Braun, R. Huber, H. Masumoto, K. Nakagawa, A. Irie, H. Sorimachi, G. Bourenkow, H. Bartunik, K. Suzuki, and W. Bode. 2000. The crystal structure of calcium-free human m-calpain suggests an electrostatic switch mechanism for activation by calcium. *Proc. Natl. Acad. Sci. U.S.A.* 97:588–592.

Sugita, H., S. Ishiura, K. Suzuki, and K. Imahori. 1980. Inhibition of epoxide derivates on chicken calcium-activated neutral protease (CANP) *in vitro* and *in vivo*. *J. Biochem.* 87:339–341.

Takahashi, K. 1996. Structural weakening of skeletal muscle tissue during post-mortem ageing of meat:the non-enzymatic mechanism of meat tenderisation. *Meat Sci.* 43:s67–s80.

Takahashi K, A. Hattori, and H. Kuroyanagi. 1995. Relationship between the translocation of paratropomyosin and the restoration of *rigor*-shortened sarcomeres during *post-mortem* ageing of meat — a molecular mechanism of meat tenderization. *Meat Sci.* 40:415–423.

Takahashi, K., A. Hattori, R. Tatsumi, and K. Takai. 1992. Calcium-induced splitting of connectin filaments into β-connectin and a 1,200-kDa subfragment. *J. Biochem.* 111:778–782.

Tatsumi, R., A. Hattori, and K. Takahashi. 1998. Deterioration of connectin/titin and nebulin filaments by an excess of protease inhibitors. *Biosci. Biotechnol. Biochem.* 62:927–934.

Tatsumi, R., K. Maeda, and K. Takahashi. 1999. Calcium-induced splitting of connectin/titin filaments: localization and characterization of calcium-binding sites. In *Proc. 45th Int. Congr. Meat Sci. Technol.*, Yokohama, Japan, p. 270–271.

Tatsumi, R. and K. Takahashi. 1992. Calcium-induced fragmentation of skeletal muscle nebulin filaments. *J. Biochem.* 112:775–779.

Taylor, R.G., G.H. Geesink, V.F. Thompson, M. Koohmaraie, and D.E. Goll. 1995a. Is Z-disk degradation responsible for postmortem tenderisation? *J. Anim. Sci.* 73:1351–1367.

Taylor, R.G., A. Ouali, and D.E. Goll. 1995b. Enzyme localization during post mortem muscle tenderisation. In *Expression of Tissue Proteinases and Regulation of Protein Degradation as Related to Meat Quality,* A. Ouali, D. Demeyer, and F.J.M. Smulders, Eds., ECCEAMST, Utrecht, Netherlands, Session 6:347–358.

Taylor, R.G., C. Tassy, M. Briand, N. Robert, Y. Briand, and A. Ouali. 1995c. Proteolytic activity of proteasome on myofibrillar structures. *Mol. Biol. Rep.* 21:71–73.

Taylor, R.G. and M. Koohmaraie. 1998. Effects of postmortem storage on the ultrastructure of the endomysium and myofibrils in normal and callipyge longissimus. *J. Anim. Sci.* 76:2811–2817.

Temm-Grove, C.J., D. Wert, V.F. Thompson, R.E. Allen, and D.E. Goll. 1999. Microinjection of calpastatin inhibits fusion in myoblasts. *Exp. Cell Res.* 247:293–303.

Thomas, A.R., H. Gondoza, L.C. Hoffman, V. Oosthuizen, and R.J. Naudé. 2004. The role of the proteasome, and cathepsins B, L, H and D, in ostrich meat tenderisation. *Meat Sci.* 67:113–120.

Thompson, J.M. 2002. Managing meat tenderness. *Meat Sci.* 62:295–308.

Thompson, J.M., D.L. Hopkins, D. D'Sousa, P.J. Walker, S.R. Baud, and D.W. Pethick. 2005. Sheep meat eating quality: the impact of processing on sensory and objective measurements of sheep meat eating quality. *Austr. J. Exp. Agric.* 45:561–573.

Turk, B. and V. Stoka. 2007. Protease signaling in cell death: caspases versus cysteine cathepsins. *FEBS Lett.* 581:2761–2767.

Turk, V. and W. Bode. 1991. The cystatins: protein inhibitors of cysteine proteinases. *FEBS Lett.* 285:213–219.

Underwood, K.R., W.J. Means, and M. Du. 2008. Caspase 3 is not likely involved in the postmortem tenderization of beef muscle. *J. Anim. Sci.* 86:960–966.

Uytterhaegen, L., E. Claeys, and D. Demeyer. 1994. Effects of exogenous protease effectors on beef tenderness development and myofibrillar degradation and solubility. *J. Anim. Sci.* 72:1209–1223.

Veiseth, E. and M. Koohmaraie. 2001. Effect of extraction buffer on estimating calpain and calpastatin activity in postmortem ovine muscle. *Meat Sci.* 57:325–329.

Veiseth, E., S.D. Shackelford, T.L. Wheeler, and M. Koohmaraie. 2001. Effect of postmortem storage on μ-calpain and m-calpain in ovine muscle. *J. Anim. Sci.* 79:1502–1508.

Veiseth, E., S.D. Shackelford, T.L. Wheeler, and M. Koohmaraie. 2004. Indicators of tenderization are detectable by 12 h post-mortem in ovine longissimus. *J. Anim. Sci.* 82:1428–1436.

Watanabe, A., C.C. Daly, and C.E. Devine. 1996. The effects of the ultimate pH of meat on tenderness changes during ageing. *Meat Sci.* 42:67–78.

Wendt, A., V.F. Thompson, and D.E. Goll. 2004. Interaction of calpastatin with calpain: a review. *Biol. Chem.* 385:465–472.

Wheeler, T.L. and M. Koohmaraie. 1994. *Prerigor* and *postrigor* changes in tenderness of ovine longissimus muscle. *J. Anim. Sci.* 72:1232–1238.

Wheeler, T.L. and M. Koohmaraie. 1999. The extent of proteolysis is independent of sarcomere length in lamb longissimus and psoas major. *J. Anim. Sci.* 77:2444–2451.

Wheeler, T.L., S.D. Shackelford, and M. Koohmaraie. 2000. Variation in proteolysis, sarcomere length, collagen content, and tenderness among major pork muscles. *J. Anim. Sci.* 78:958–965.

Whipple, G., M. Koohmaraie, M.E. Dikeman, J.D. Crouse, M.C. Hunt, and R.D. Klemm. 1990. Evaluation of attributes that affect longissimus muscle tenderness in *Bos taurus* and *Bos indicus* cattle. *J. Anim. Sci.* 68:2716–2728.

Wong, S.S. 1991. *Chemistry of Protein Conjugation and Cross-Linking*. CRC Press, Boca Raton, FL, p. 226.

Young, P., C. Ferguson, S. Bañuelos, and M. Gautel. 1998. Molecular structure of the sarcomeric Z-disk: two types of titin interactions lead to an asymmetrical sorting af α-actinin. *Embo J.* 17:1614–1624.

Yoshimura, N., T. Kikuchi, T. Sasaki, M. Hatanaka, and T. Murachi. 1983. Two distinct Ca^{2+} proteases (calpain I and calpain II) purified concurrently by the same method from rat kidney. *J. Biol. Chem.* 258:8883–8889.

Zamora, F., L. Aubury, T. Sayd, J. Lepetit, A. Lebert, M.A. Sentandreu, and A. Ouali. 2005. Serine peptidase inhibitors, the best predictor of beef ageing amongst a large set of quantitative variables. *Meat Sci.* 71:730–742.

Zamora, F., E. Debiton, J. Lepetit, A. Lebert, E. Dransfield, and A. Ouali. 1996. Predicting variability of ageing and toughness in beef *M. longissimus lumborum et thoracis*. *Meat Sci.* 43:321–333.

Zeece, M.G., K. Katoh, R.M. Robson, and F.C. Parrish, Jr. 1986a. Effect of cathepsin D on bovine myofibrils under different conditions of pH and temperature. *J. Food Sci.* 51:769–773.

Zeece, M.G., R.M. Robson, M.L. Lusby, and F.C. Parrish, Jr. 1986b. Effect of calcium activated protease (CAF) on bovine myofibrils under different conditions of pH and temperature. *J. Food Sci.* 51:797–803.

Zeece, M.G., T.L. Woods, M.A. Keen, and W.J. Reville. 1992. Role of proteinases and inhibitors in postmortem muscle protein degradation. *Proc. 45th Reciprocal Meat Conf.,* American Meat Science Association, Chicago, IL, p. 51–61.

9 Regulation of Postmortem Glycolysis and Meat Quality

Qingwu W. Shen, Min Du, and Warrie J. Means

CONTENTS

9.1 INTRODUCTION

Postmortem glycolysis is one of the most important factors determining the quality of meat. Rapid and/or excessive glycolysis leads to pale, soft, and exudative (PSE) meat, while insufficient glycolysis results in dark, firm, and dry (DFD) meat. Both PSE and DFD meat induce significant losses to the meat industry.

9.2 PALE, SOFT, AND EXUDATIVE (PSE) MEAT

9.2.1 PSE Meat and Mechanisms Leading to PSE Meat

The PSE condition in the pig was first described by Ludvigsen in 1953 as being muscle degeneration. PSE is characterized by low pH, a pale and exudative appearance, and a soft texture. PSE problems in meat have been recognized for decades and remain unsolved. Intensive selection for lean muscle and against fat deposition in pigs has contributed to the increased incidence of PSE meat (Lee and Choi 1999; Solomon et al. 1998). PSE problems are even more serious for the turkey and chicken industry (Barbut 1998a; Woelfel et al. 2002). Fast and/or excessive glycolysis in postmortem muscle causes an early build-up of lactic acid (a rapid pH drop) in early-stage postmortem muscle, leading to PSE syndrome in the meat of swine, turkeys, and chickens, and is also related to the incidence of "acid meat" in Hampshire pigs (Enfalt et al. 1997; Hamilton et al. 2000; Moeller et al. 2003). Rapid glycolysis also generates significant amounts of heat, which increases the carcass temperature. In combination with low pH during early-stage postmortem, this temperature rise denatures muscle proteins, resulting in an undesirable pale color and low water-holding capacity of meat (Freise et al. 2005; Owens et al. 2000a; Owens et al. 2000b; Sams and Alvarado 2004b). Under the above conditions, myosin denatures and shrinks, causing a reduction of the filament spacing. Thus, water is expelled from the cells and lost in the form of purge or drip (Offer et al. 1989). In pig muscle that becomes PSE meat, the pH declined much faster than that in normal muscle, with a speed of 1.04 versus 0.65 units/hr at 37°C (Bendall et al. 1963). Furthermore, in some extreme instances, the ultimate pH can be reached in 15 min postmortem (Solomon et al. 1998). The rate of glycolysis in pale turkey muscle was two times faster than in normal-colored turkey muscle (Sosnicki and Greaser 1996). Therefore, slowing down glycolysis in early-stage postmortem muscle is the key to solving the PSE problem.

In addition, PSE conditions cause additional quality problems for the meat processing industry due to the inferior quality of resultant products (Fernandez et al. 2002b; Owens et al. 2000a; Owens et al. 2000b). PSE meat has a high drip loss, a low cooking yield, and a tough texture after cooking. Furthermore, denaturation of protein results in reduced extractability of muscle proteins. Consequently, products processed from PSE meat not only have lower water-binding capacity (low yields), but also exhibit a decreased cohesiveness (poor sliceability).

Intensive selection for rapid lean growth results in an increase in PSE meat due to the increase in the number of fast-twitch glycolytic (FG) muscle fibers and a concomitant reduction in slow-twitch oxidative (SO) muscle fibers. Because FG muscle fibers are characterized for their glycolytic metabolism, postmortem glycolysis in FG fibers is rapid (Vanhoof 1979).

Despite common occurrence in pork and poultry, PSE is highly unlikely to occur and is not considered a serious problem in beef (Hunt and Hedrick 1977). This is due to the relatively low proportion of FG fibers in beef compared to pork musculature. In instances where PSE occurs in beef, it is often a localized phenomenon in the psoas major muscle (Fischer et al. 1977a; Fischer et al. 1977b) or in hip muscles such as the semimembranosus and adductor (Aalhus et al. 1998). In these muscles, PSE develops because slow heat loss causes the muscles to be exposed to relatively high temperatures during the postmortem period.

9.2.2 PSE Meat Induces Huge Losses to the Meat Industry

The high incidence of PSE in pork, turkey, and chicken causes a huge economic loss to the animal slaughter and meat processing industry (Kauffman et al. 1992; Woelfel et al. 2002). For the animal slaughter industry, the Pork Chain Quality Audit suggests that "10.2% of the carcasses were classified as having PSE muscle" (Cannon et al. 1996). The normal incidence of PSE in pork has been reported to range from 10% to 30%, but in some isolated instances is up to 60% (Lee and Choi 1999). The direct economic loss to the pork industry alone is $32 million per year (Li and Wick 2001). For turkey, observations in the slaughterhouse (especially in the United States) have shown that PSE meat could represent 5% to 30% of the slaughter turkeys (Barbut 1998b). In chicken, using color L* value of 3 and 24 hr postmortem fillets as an indicator, approximately 47% of the 3554 fillets tested were pale and could potentially exhibit poor water-holding capacity (Woelfel et al. 2002). When it comes to the meat processing industry, the PSE problem becomes more apparent. Meat products, such as hams, made of PSE meat have low cooking yield, increased purge, poor binding ability, and tough and coarse texture, resulting in additional economic losses (Fernandez et al. 2002a; Fernandez et al. 2002b; Fernandez et al. 2002d; Owens et al. 2000a; Owens and Sams 2000). Cannon et al. (1996) estimated that the PSE and reddish-pink, soft, and exudative (RSE) meats lead to an approximately $100 million annual loss in the United States.

9.2.3 AMPK, Glycolysis, and PSE Meat

AMP-activated protein kinase (AMPK) is a serine/threonine kinase composed of three subunits: a catalytic subunit (α) and two regulatory subunits (β and γ). The key function of AMPK is to regulate the energy balance within the cell. Thus, AMPK is usually referred to as the intracellular energy sensor and the master energy metabolic switch. AMPK is activated in response to ATP depletion, or more accurately, an increase in the AMP/ATP ratio (Corton et al. 1994). Once activated, AMPK then phosphorylates downstream substrate and the biological effects of AMPK are pleiotropic: switching off ATP-consuming processes such as biosynthetic pathways and switching on catabolic processes to generate ATP, including glycolysis (Carling 2004; Hardie et al. 2006; Hardie et al. 2003). AMPK is activated in ischemic heart and hypoxic skeletal muscle. Activated AMPK increases glycolysis *in vivo* by two signaling pathways. First, AMPK phosphorylates and activates phosphorylase kinase, which then phosphorylates and activates glycogen phosphorylase, an enzyme controlling glycogenolysis and catalyzing the production of substrate for glycolysis (Fraser et al. 1999; Russell et al. 1999; Young et al. 1997). This is supported by the observation that N-cyclohexaneadenosine, an adenosine agonist, inhibits both ischemia-induced AMPK activation and glycogen phosphorylase (Fraser et al. 1999). Second, AMPK regulates glycolysis by phosphorylating and activating phosphofructokinase-2 (PFK-2) (Marsin et al. 2000). PFK-2 catalyzes the production of fructose-2,6-biphosphate, which is a potent allosteric activator of PFK-1. PFK-1 is the most important rate-controlling enzyme in glycolysis. *In vitro* phosphorylation of PFK-2 by purified AMPK further confirmed a direct involvement of AMPK in increasing glycolysis (Marsin et al. 2000).

Involvement of AMPK in postmortem glycolysis regulation was first investigated by Shen and Du (2005a,b). Experiments carried out on mice showed that activation of AMPK by pre-euthanasia forced swim or intraperitoneal injection of AMPK activator (5-amino-4-imidazolecarboxamide riboside, AICAR) led to faster pH decline and lower ultimate pH values in postmortem muscle (Shen and Du 2005b). Conversely, inhibition of AMPK by compound C resulted in lower AMPK activity, higher pH value, and less lactic acid accumulation in postmortem muscle when compared to control experiments (Shen et al. 2008). In agreement with these data, mice (whether exercised or not) overexpressing dominant-negative mutant AMPK had the lowest postmortem AMPK activity, the lowest glycolytic rate, and the highest pH values in muscle, compared to other treatments (Shen and Du 2005b). Furthermore, dietary α-lipoic acid supplementation not only inhibits postmortem AMPK activation, but also resulted in less postmortem glycolysis and higher pH values in mouse longissimus dorsi muscle (Shen and Du 2005a; Shen et al. 2005). All these results have clearly demonstrated that AMPK regulates postmortem glycolysis in mouse muscle and suggest that PSE meat might be prevented or at least reduced by inhibiting postmortem AMPK activation.

Obviously, mice and pigs are genetically different. Non-genetic conditions whereby mouse and pig muscle undergo postmortem metabolism are also quite different. To this end, Shen et al. (2006) measured postmortem AMPK activity from muscle that became both normal and PSE pork. AMPK was activated earlier and AMPK activity was much higher early postmortem, especially within half an hour after exsanguination in the muscle that became PSE meat (Shen et al. 2006b), suggesting the involvement of AMPK in PSE pork occurrence. This is supported by the observation that preslaughter stress increased the risk of PSE meat as well as AMPK activity in pork loin early postmortem (Shen et al. 2006a). Based on the data obtained, it is proposed that pre-slaughter transport, as well as halothane gene, accelerates energy depletion shown by higher IMP concentration and an increase in [AMP+IMP]/ATP ratio in postmortem muscle (Shen et al. 2007a). The lower energy level activates AMPK. Activated AMPK then enhances glycolysis in postmortem muscle during the very early postmortem period, which increases the risk of PSE meat. Pre-slaughter rest after transport restored energy status in muscle before animals were killed. The restored energy status then prevents early and rapid AMPK activation in postmortem muscle and lessens the negative effects of pre-slaughter transport on meat quality. These data suggest that AMPK might regulate postmortem glycolysis by up-regulating PFK, in agreement with a previous study that demonstrated an involvement of PFK in postmortem glycolysis regulation (Kastenschmidt et al. 1968). However, further detailed studies are still needed.

9.3 DARK, FIRM, AND DRY (DFD) MEAT AND GLYCOGEN ACCUMULATION IN MUSCLE

DFD meat mainly exists in beef, and is more commonly referred to as dark cutting beef. Dark cutting beef was recognized more than 50 years ago (Munns and Burrell 1965) but remains an unsolved problem and causes significant losses to the meat industry. The costs of DFD to the beef industry are estimated at $172 million annually, resulting in an average loss of $6.08 per head harvested in the United States (Apple et al. 2005a; Scanga et al. 1998; Wulf et al. 2002). In the 2000 National Beef Quality Audit, 2.3% of beef carcasses were dark cutters (McKenna et al, 2002); and in the 1995 Audit, 2.7% were classified as dark cutters (Boleman et al. 1998). In another survey, the DFD incidence was 4.7% for steers and 5.6% for heifers (Lorenzen et al. 1993). Among 8000 heifers slaughtered at a commercial plant, 1.7% were dark cutters (Kreikemeier et al. 1998). DFD caused a 1.4% decline in value for steers and a 1.7% decline in value for heifers, according to a survey of beef carcasses at three packing plants in Illinois, Texas, and Ohio (Page et al. 2001). Consumers are not likely to purchase the dark-colored meat associated with DFD, which then causes decreased profits (Killinger et al. 2004). Furthermore, the undesirable redness in cooked DFD beef causes additional losses for the meat industry (Gasperlin et al. 2000; Moiseev and Cornforth 1999).

DFD results from a high pH in postmortem muscle. At high pH, the respiration of mitochondria is maintained, which consumes oxygen and causes the formation of deoxymyoglobin. This alters light absorption in the muscle surface and results in a dark red color (Ashmore et al. 1973; Ashmore et al. 1971). A pH value of 5.87 in the longissimus muscle is the approximate cut-off between normal and DFD (Page et al. 2001). The high pH in the postmortem muscle of DFD is due to glycogen deficiency in the muscle (Mcveigh et al. 1982; Warriss et al. 1984). It appears that a glycolytic potential at 100 μmol/g muscle is the threshold for the occurrence of DFD (Wulf et al. 2002). Factors influencing glycogen content in muscle at slaughter directly affect the incidence of DFD (Bartos et al. 1993; Kenny and Tarrant 1988; Kreikemeier et al. 1998; Scanga et al. 1998). Therefore, to solve the DFD problem, it is necessary to increase the glycogen content in muscle.

Studies in rodents show that GSK-3 and AMPK play regulatory roles in glycogen accumulation (see Chapter 10 for a detailed discussion of AMPK). AMP-activated protein kinase is recognized as a critical regulator of energy metabolism (Hardie 2004; Kim et al. 2004). Once activated, AMPK enhances glucose uptake from blood, which increases glycogen content in muscle (Aschenbach et al. 2002). AMPK activation also promotes glycogen mobilization through enhancing glycogenolysis and glycolysis (Du et al. 2005). In addition, AMPK phosphorylates muscle isoform of glycogen synthase (GS), which decreases its activity (Aschenbach et al. 2002; Miyamoto et al. 2007). However, *in vivo*, such glycolysis and inhibition of glycogen synthesis are rapidly down-regulated when cellular energy is high (Aschenbach et al. 2002), resulting in a net increase in glycogen content in muscle following AMPK activation. The role of AMPK in glycogen accumulation is well demonstrated in Hampshire pigs carrying a point mutation, Arg^{200} to Gln in the AMPK γ subunit. This mutation increases the AMPK activity and enhances muscle glucose uptake, leading to enhanced glycogen accumulation (Andersson 2003; Enfalt et al. 1997; Hardie and Hawley 2001; Milan et al. 2000b; Yu et al. 2006).

Glycogen synthase kinase-3 (GSK-3) is a serine/threonine kinase. GSK-3 regulates glycogen synthase activity by phosphorylation. Because GSK-3 is constitutively active in unstimulated muscle cells, glycogen synthase mainly exists in a phosphorylated and inactive state (MacAulay et al. 2005). However, in response to the stimulation of insulin, GSK-3 is itself phosphorylated and inactivated by Akt (another name for PKB, protein kinase B) (MacAulay et al. 2005; Van der Velden et al. 2006). Thus, glycogen synthase is dephosphorylated by protein phosphatase and becomes active (MacAulay et al. 2005). Incubation of rat epitrochlearis and soleus muscles with CT-98014, a specific GSK-3 inhibitor, increased glycogen synthase activity by 49% and 50%, respectively (Henriksen et al. 2003). In L6 myotubes, inhibition of GSK-3 was associated with a marked dephosphorylation and activation of glycogen synthase by four- to sevenfold (MacAulay et al. 2003).

9.4 FACTORS LEADING TO ABNORMAL POSTMORTEM GLYCOLYSIS

Past and current research is focused on the postmortem glycolysis and its association with the incidence of PSE meat. Although DFD meat results in significant losses to the beef industry, such losses are relatively small compared to PSE meat. Also, the negative effect of DFD meat on quality is not as severe as PSE meat. In addition, the increasing incidence of PSE meat in recent decades makes it urgent to solve this problem. Therefore, the following discussion mainly focuses on PSE meat.

9.4.1 Pre-slaughter Factors, Glycolysis, and PSE Meat

Numerous studies have been conducted to identify factors leading to the incidence of PSE meat, and tremendous progress has been made by showing that pre-slaughter stresses, post-slaughter handling, and genetics are the main factors contributing to the increased incidence of PSE meat (Leheska et al. 2002b; Rosenvold and Andersen 2003).

Pre-slaughter handling includes loading, transportation, lairage, mixing of strange animals, stunning methods, and slaughter conditions (Ekkel et al. 1997; Gade and Christensen 1998; Velarde et al. 2000). These handling practices can all induce stresses, either psychologically or physically, and consequently can adversely affect meat quality. Poor farm handling increases the susceptibility to pre-slaughter stress and results in a higher incidence of PSE meat compared with pigs that are handled correctly on the farm (D'Souza et al. 1998b). Long-term stress that leads to the depletion of glycogen in muscle is mainly associated with DFD meat, while short-term stress-activated glycolytic enzymes mainly lead to PSE meat. However, these two types of stresses are not definitively separated.

9.4.1.1 Mixing Unfamiliar Pigs

Pre-slaughter mixing or regrouping animals during collection, transport, and lairage is practically unavoidable. Mixing pigs from different groups causes stress to the animals, increases their excitability, and leads unfamiliar individuals to fight (Bradshaw et al. 1996b; Faucitano 1998; Warriss and Brown 1985; Warriss et al. 1998b). The most evident consequence of fighting is the damage to the skin of animals (Bradshaw et al. 1996b; Faucitano 1998, 2001; Karlsson and Lundstrom 1992). When compared to the unmixed groups, mixing pigs before slaughter decreases meat quality, increases weight loss and the incidence of both PSE and DFD meat, as well as increases skin lacerations (Karlsson and Lundstrom 1992). It has been reported that stress from mixing unfamiliar pigs increases cortisol and creatine phosphokinase (CPK) levels in blood (Bradshaw et al. 1996b; Oliver et al. 1996). This physiological response of pigs to mixing and fighting could be responsible for the inferior meat quality and increased occurrence of both PSE and DFD meat. It is suggested that, if unavoidable, mixing should be done at loading, as pigs fight less in the novel environment of the lorry interior and during transportation (Lambooij 1988). In lairage, the practice of mixing should be avoided. If mixing in lairage is unavoidable, the lairage time should be kept as short as possible because fighting in lairage increases carcass surface damage and eliminates the benefits of careful handling in previous procedures (Faucitano 1998).

9.4.1.2 Loading

Loading at the farm is usually believed to be the most stressful part of pre-slaughter handling because animals suffer several stressors in the same procedure: the first time in their life to leave the familiar finishing pen and move to an unknown environment, mixing of unfamiliar individuals, being forced to walk ahead, and others (Faucitano 1998). The widely used loading system on the farm, the loading ramp, is believed to be stressful (Faucitano 1998; Lee and Choi 1999). The conditions of the loading facilities, such as the floor type and the ramp angle, are of importance in determining the level of induced stress. Lee and Choi (1999) recommended that the maximal angle for a permanent pig ramp should be 20°, and the loading ramp should be wide enough to allow two or three pigs to climb together. Livestock move more easily up and down stair-step ramps (Lee and Choi 1999). In the range of 0.33 to 3.0 hr, Guardia et al. (2004) observed that increasing the loading time decreased the risk of PSE. They explained that short loading times might involve worse handling practices than longer ones or, alternatively, the loading ramp used was not good enough to allow rapid loading of pigs without inducing additional stress. When compared to the loading ramp, the so-called stress-free loading system, a hydraulic lift capable of holding eight to ten pigs in a group, is much better and significantly decreases the incidence of PSE carcasses and the hemorrhage in the shoulder (Guardia et al. 2004; Nanni Costa et al. 1996).

9.4.1.3 Transport

A negative relationship between transport time and the risk of developing PSE (2.9% per hour) has been observed (Guardia et al. 2004). A large body of literature has reported the effect of transport duration on animal performance and meat quality. Most data reported show that short transport duration increases PSE occurrence and longer transport times are associated with DFD meat. Guardia et al. (2004) also suggested that short transport times increase PSE risk because during the first 3 hr of transport, pigs tend to remain stressed and standing up; especially when the loading density is low, pigs can move around and are ready to be involved in fighting. After a 3-hour transport, animals start to adapt to the new environment. With space available that allows pigs to lie down and rest, the risk of PSE decreases (Guardia et al. 2004). However, transport duration decreases carcass yield (Guardia et al. 2004; Lee and Choi 1999; Mota-Rojas et al. 2006). More seriously, too long a transport time is believed to increase the transport mortality rate, and procedures need to be implemented to reduce losses due to death (Gregory 1994; Lee and Choi 1999).

Sufficient space in transport vehicles should be provided for individual animals during transport — especially in hot, humid weather — to reduce stress-induced poor-quality meat and the incidence of PSE, as well as death loss (Lee and Choi 1999). High loading density has been reported to have detrimental effects on PSE (Guise and Penny 1989; Von Wenzlawowicz et al. 1996). However, the risk of PSE increases with loading density only when the transit is longer than 3 hours, while for short transit, the availability of space increases PSE occurrence (Guardia et al. 2004). In their study, the researchers found that there is an interaction between transport time and stocking density: the higher the stocking density, the lower the effect of the transport time on the risk of PSE. When the stocking density is as high as 0.25 m^2/100-kg pig, the risk of PSE hardly changes between transits of 1 to 7 hours but, after doubling the space availability (to 0.5 m^2/100-kg pig), the transport time can explain up to 6% of PSE meat (Guardia et al. 2004).

In addition to transport duration and loading density, other factors such as transport vehicle conditions, road conditions, noise, and driving style are of importance for the stress level in animals induced during transportation (Bradshaw et al. 1996a; Grandin 1994; Rosenvold and Andersen 2003). Guardia et al. (2004) detected, for the first time, a significant influence of the lorry floor surface on pork PSE risk. Their data show that a polyester surface decreased the risk of PSE meat by about 1.5% compared with the aluminum and iron surface because polyester provides a more comfortable environment for pigs during transportation, particularly because it is less slippery, less noisy, and a better thermal insulator (Guardia et al. 2004).

9.4.1.4 Lairage

Holding pigs in lairage for a certain time is supposed to provide animals with an opportunity for rest and recovery from stressful experiences during transport, thus improving pork quality and reducing the occurrence of PSE meat. Studies using stressed pigs have confirmed the benefits of resting animals in lairage before slaughter. Fortin (2002) reported that a 50-minute transport before slaughter was associated with a high occurrence of PSE pork loins, and a 3-hour rest period in lairage significantly improved meat quality by reducing the incidence of PSE loins (Fortin 2002). Another study also reported that a 2-hour rest was sufficient for pigs to recover from transport-induced stress and significantly reduced the prevalence of PSE pork (Owen et al. 2000).

A body of literature has reported the relationship between lairage time and meat quality. Although some opposite results were obtained (Guardia et al. 2004), most researchers have reported that the frequency of PSE progressively decreases with the extension of lairage time (Costa et al. 2002; Milligan et al. 1998; Van der Wal et al. 1997; Warriss et al. 1998a). During prolonged lairage, however, the percentage of DFD carcasses and skin damage increases due to glycogen depletion. Also, long periods of lairage decrease carcass yield due to long-time fasting (Costa et al. 2002; Milligan et al. 1998; Van der Wal et al. 1997; Warriss et al. 1998a). Based on the above studies, it is recommended that the optimal lairage time is 1 to 3 hours for most slaughter pigs in terms of both meat quality and consideration for the pigs' welfare (Faucitano 1998; Lee and Choi 1999; Warriss 2003). Too long a lairage time (such as overnight) may increase skin damage several-fold (Warriss et al. 1998a).

Lairage does not always result in decreased PSE occurrence. On the contrary, improper handling of pigs in this stage may result in additional stress and physical pain, leading to further deterioration in meat quality (Fraqueza et al. 1998; Geverink et al. 1998; Roseiro et al. 1996). The effects of lairage on meat quality and animal welfare vary, depending on a variety of factors, including the intensity of stress experienced previously, the lairage temperature and humidity, lairage conditions, susceptibility (genotype) of animals to stress, group size, and mixing animals in lairage.

The stresses pigs experienced previously influence the effect of lairage on meat quality and the optimal lairage time. Usually, the higher the stress the pigs suffered during transport, the longer the lairage time required for animals to recover from the stress (Faucitano 1998; Grandin 1994). If animals are not particularly stressed before lairage, lairage is not necessary in terms of meat quality and welfare. In pigs subjected to low-stress pre-slaughter handling, the lairage time had no effect on meat quality and lairage could be skipped (Aaslyng and Gade 2001).

The influence of temperature and humidity on the effects of lairage time and meat quality was investigated by Fraquenza et al. (1998). In their study, 23 groups of about 30 pigs each were kept in an environmentally controlled room in lairage, at temperatures of 20°C or 35°C for periods of 0.5 or 3 hours. Meat-quality analyses showed that there is an interaction between lairage temperature and lairage time: a longer lairage time (3 hours) reduces PSE occurrence at low lairage temperature (20°C, also low humidity); while at higher lairage temperature (also high humidity), longer lairage times increase PSE prevalence from 31.1% to 35.5%. Others (Santos et al. 1997) also reported that a high lairage temperature (35°C) led to a higher incidence of PSE meat. These results show that stressful conditions during lairage further deteriorate meat quality. Thus, the authors suggested that at high temperature and humidity, it was better to slaughter pigs within 30 min after they arrived at lairage (Fraqueza et al. 1998; Santos et al. 1997).

Fighting is usually the main cause of skin damage and hematoma in the underlying tissue of the carcass, which decreases meat quality and might cause significant economic losses (Faucitano, 2001). According to previous studies, the fighting of pigs usually peaks at about 40 to 60 min after arriving at lairage. After 1.5 hours, pigs start to lie down for a rest (Brown et al. 1999). Mixing of unfamiliar animals increases fighting and carcass surface damage (D'Souza et al. 1999a; Geverink et al. 1998). Moreover, fighting at the same time accelerates glycogen depletion within muscle and increases the incidence of DFD meat; thus, it is recommended that familiar pigs should be kept together throughout all pre-slaughter handling to improve meat quality and animal welfare, as well as decrease economic losses (Karlsson and Lundstrom 1992).

Group size is also associated with lairage effects on meat quality. If mixing of unfamiliar pigs is unavoidable in commercial abattoirs, it is better to mix in groups of up to 200 pigs rather than in small groups of 6 to 50 pigs. In a big group, the pigs being attacked have more chance to escape from the aggressive ones (Grandin 1990).

Some other factors associated with the effects of lairage time on meat quality and pig welfare include ventilation, noise, and even fasting and feeding (Warriss 2003). Showering pigs before slaughter with cool water (9°C) helps to remove excess body heat, especially during the summer, and reduces the incidence of PSE carcasses (Knowles et al. 1998; Long and Tarrant 1990; Weeding et al. 1993). However, Weeding et al. (1993) recommended that pigs should not be showered if the temperature inside or outside the lairage falls below 5°C, to avoid shivering (Knowles et al. 1998).

9.4.1.5 Fasting

The main incentive of fasting before transport is to lower mortality rates during transport, particularly in hot weather, and to reduce bacterial contamination from gut contents during evisceration (Faucitano 1998; Lee and Choi 1999). It was also found that feed withdrawal on the farm increased meat pH and decreased the incidence of PSE pork (Eikelenboom et al. 1991). Eikelenboom et al. (1991) observed that the PSE occurrence decreased after on-farm fasting intervals of 24 hours and the ultimate pH in loin and ham was higher than in the non-fasted group due to the depletion of muscle glycogen. In general, it is accepted that on-farm fasting of 12 to 18 hours is a good practice that decreases the death rate during transport and lowers physiological responses in pigs (Lee and Choi 1999; Oliver et al. 1996). A prolonged on-farm fasting period is not desirable because it increases the pigs' aggressiveness (Murray et al. 2001), increases DFD meat due to glycogen depletion, and reduces carcass yield (Eikelenboom et al. 1991). On the other hand, too short a fasting period (such as less than 12 hours) increases the risk of PSE meat (Guardia et al. 2004).

9.4. 1.6 Stunning

The stunning method affects meat quality. For pigs, there are two main stunning methods used: (1) carbon dioxide (CO_2) and (2) electrical stunning. Muscle contraction induced by electrical stunning has a negative effect on pork quality, increasing the risk of PSE meat in both halothane-positive and -negative pigs (Channon et al. 2000, 2002; Van der Wal et al. 1997). Carbon dioxide stunning reduces the incidence of PSE meat.

Proper handling of pigs right before stunning or during stunning is crucial for the control of meat quality. Rough handling when moving pigs forward to the stunning point and the stunning process itself at an abattoir can eliminate all the benefits of previous careful handling, causing the pig to stress enough to increase PSE (Lee and Choi 1999; Rosenvold and Andersen 2003). Short-term stress immediately before slaughter, even the last 5 minutes in stunning, increases cortisol content, lactic acid content, and creatine phosphate kinase activity in blood, decreases meat water-holding capacity, and increases the percentage of PSE carcasses (Brown et al. 1998; D'Souza et al. 1998a, b; Van der Wal et al. 1999; Van der Wal et al. 1999). The detrimental effects of this stress are more apparent in muscles with a high glycolytic fiber ratio (Hambrecht et al. 2005). When moving pigs forward for stunning, electric goad should not be used and handlers need to be trained to move pigs quietly (D'Souza et al. 1999a; D'Souza et al. 1998a; Lee and Choi 1999).

9.4.2 Post-slaughter Handling

9.4.2.1 Chilling

Because pH/temperature in early postmortem is critical for the formation of PSE meat, rapid cooling is expected to reduce the PSE problem. Borchert and Briskey (1964) showed that liquid nitrogen chilling prevented the incidence of PSE meat. Afterward, many studies confirmed that accelerated chilling improves meat quality (Kerth et al. 2001; Long and Tarrant 1990; Park et al. 2007; Rathgeber et al. 1999; Sams and Alvarado 2004a; Taylor and Dant 1971). The effects of chilling on pork and poultry quality have been extensively investigated (James et al. 2006; Savell et al. 2005). One problem with accelerated cooling is that it may cause cold shortening, which may occur in beef and lamb if the temperature of muscle decreases too rapidly while plenty of ATP still exists in the muscle. This is evidenced by the tougher texture of rapidly chilled meat compared to conventionally chilled meat (Jones et al. 1993; Van der Wal et al. 1995).

9.4.2.2 Electrical Stimulation

Electrical stimulation is usually used to prevent cold shortening and improve meat quality from beef and lamb carcass as electrical stimulation accelerates postmortem metabolism in muscle (Rosenvold and Andersen 2003). In pigs, electrical stimulation is also used to improve pork tenderness (Rosenvold and Andersen 2003). However, it has been observed that electrical stimulation accelerates pH decline and stimulates PSE pork development (Bowker et al. 1999; Hallund and Bendall 1965; Hammelman et al. 2003; Taylor et al. 1995a, b; Warriss et al. 1995). Some studies show that electrical stimulation applied later than 20 min postmortem neither decreases the water-holding capacity of muscle nor increases the frequency of PSE meat (Hammelman et al. 2003; Taylor and Martoccia 1995; Taylor and Tantikov 1992). Because PSE development is not a concern to beef and lamb, electrical stimulation is commonly used in beef to improve tenderness and other quality characteristics.

9.4.3 Halothane Gene, Rendement Napole Gene, and Glycolysis in Postmortem Muscle

Genetic factors contribute to the incidence of PSE syndrome in pigs. Genetic selection for lean growth dramatically increased the incidence of PSE meat (Lee and Choi 1999). Pigs that carry either the halothane gene or the Rendement Napole (*RN*⁻) gene are prone to develop PSE or "acid meat" syndrome. The so-called halothane gene results from a point mutation in the ryanodine receptor, a Ca^{2+} channel, which leads to hypersensitive Ca^{2+}-channel gating and animals that elicit malignant hyperthermia (MH) under stress or halothane treatment (Fujii et al. 1991). High cellular Ca^{2+} levels can enhance glycolysis directly and also through activation of AMPK by CaMKK (see Chapter 10 for detailed discussion) (Shen et al. 2007b). In addition, Ca^{2+} stimulates muscle contraction, accelerating energy consumption, and activating AMPK and glycolysis (Shen et al. 2007a). However, a halothane carrier grows faster, has higher carcass yield and fat-free lean content, but has a higher incidence of PSE meat (Apple et al. 2002a; Depreux et al. 2002; Leach et al. 1996; Weaver et al. 2000). It is expected that halothane-positive pigs should have a much lower PSE incidence if they are subjected to enough rest before slaughter (Warriss et al. 1987). DeSmet et al. (1996) demonstrated the different responses of pig genotypes (NN, Nn, and nn, where N stands for one allele of halothane gene) to lairage time. The data obtained showed that resting pigs in lairage for 4 to 5 hours compared to no resting before slaughter resulted in a more pronounced improvement in the quality of meat from nn pigs than from NN pigs: increased pH (40 minutes postmortem), decreased lightness, and drip loss within nn genotype, and only slightly decreased drip loss within the NN genotype (DeSmet et al. 1996). What needs to be mentioned is that this only showed that lairage time has greater effect on the quality of meat from nn than NN pigs. As reported by others (Rosenvold and Andersen 2003), the halothane gene in this study was still the predominant factor influencing meat quality (DeSmet et al. 1996). Stress-susceptible pigs need relatively long holding for them to recover from stresses experienced previously (Lee and Choi 1999).

"Acid meat" is characterized by a normal pH decline but a low ultimate pH (Hamilton et al. 2000). The RN^- gene is associated with "acid meat" in Hampshire pigs (Josell et al. 2004; Miller et al, 2000). It is reported that the RN^- gene encodes the AMPK $\gamma 3$ subunit (Milan et al. 2000b). A mutation in AMPK $\gamma 3$ (RN^-) results in an AMPK that is constitutively active (the degree of activation is low compared with AMPK activation stimulated by AMP) but insensitive to the cellular AMP level (Andersson 2003; Barnes et al. 2004; Ciobanu et al. 2001; Oliveira et al. 2003). The constant AMPK activity stimulates glucose uptake into skeletal muscle cells and thus enhances glycogen synthesis, resulting in excessive glycogen storage in muscle (Arad et al. 2002; Barnes et al. 2004; Fontanesi et al. 2003; Moeller et al. 2003). Thus, *RN*⁻ carriers have a high glycolytic potential, which is frequently used to identify the *RN*⁻ phenotype (Hamilton et al. 2002; Hamilton et al. 2003; Miller et al. 2000). The low ultimate pH in "acid meat" could be due to the sustained glycolysis stimulated by the constitutively activated AMPK, and fueled by the high level of stored glycogen, resulting in "acid meat" or red, soft, exudative (RSE) pork (Van Laack and Kauffman 1999; Warner et al. 1997).

9.5 PREVENTION OF PSE MEAT

Up to now, no single method exists for eliminating PSE pork from stress-susceptible pigs. Minimization of stress before or during slaughter reduces PSE occurrence. Short-term resting has long been used to reduce the frequency of PSE pork (Monin et al. 1981). A 3-hour resting after transport and before slaughter significantly improved meat quality by reducing the incidence of PSE loins (Fortin 2002).

Proper post-slaughter handling and/or other strategies if used appropriately reduce the frequency of PSE meat. It is reported that the carcass processing speed during slaughter has a significant influence on pork quality. Hanging carcasses on the slaughter line for 20 minutes before evisceration and entering the chillers increases the occurrence of PSE meat (D'Souza et al. 1998a). This is associated with the temperature early postmortem. Delayed post-slaughter processing rates increase the time that carcasses are subjected to high temperature (D'Souza et al. 1998a). That is why accelerated chilling reduces PSE and improves meat quality, as discussed above. Low temperature decreases glycolytic enzyme activity and glycolytic rate postmortem, as well as slows protein denaturation, leading to reduced PSE occurrence (Bendall and Swatland 1988; Briskey 1964; Lee and Choi 1999; Sayre et al. 1963; Solomon et al. 1998). More recently, some other strategies such as postmortem injection of sodium bicarbonate have been reported to retard postmortem pH decline, improve muscle color, and reduce drip loss (Kauffman et al. 1998).

Manipulation of feeding has been used for many years to reduce the occurrence of PSE meat. Dietary vitamin E supplementation has been successfully used to protect against lipid oxidation and improve the color stability of fresh meat (Rosenvold and Andersen 2003). Daily dietary supplementation of 500 mg vitamin E/kg diet for 46 days reduced drip loss in unfrozen longissimus thoracis by 45% and 54% in Nn and NN pigs, respectively (Cheah et al. 1995). In PSE-prone Landrace × Large White Nn pigs, daily supplementation of 1 g vitamin E/kg diet for the same period reduced excessive release of Ca^{2+} and prevented the formation of PSE carcasses (Buckley et al. 1995; Cheah et al. 1995). This improvement in meat quality by vitamin E is associated with its ability to stabilize cell membrane integrity by inhibiting phospholipid oxidation. Later, others also reported that vitamin E supplementation before slaughter improved pork water-holding capacity and color (Hamman et al. 2001), as well as decreased the percentage of PSE loins and hams (Kerth et al. 2001). Furthermore, vitamin supplementation increased the resistance of chickens to heat stress, inhibited poultry PSE, and improved meat functional properties (Olivo et al. 2001).

Several studies have shown that magnesium supplementation before slaughter reduces the glycolytic rate early postmortem and improves meat quality (Apple et al. 2005b; D'Souza et al. 1999b; D'Souza et al. 1998c, 2000). The beneficial effects of magnesium supplementation are associated with its reduction in neuromuscular stimulation due to the calcium antagonist effects of magnesium. Indeed, dietary magnesium supplementation led to higher plasma magnesium (D'Souza et al. 1999b) and lower plasma norepinephrine (D'Souza et al. 1998c) concentrations at slaughter, and reduced the detrimental effects of pre-slaughter stress on meat quality (Apple et al. 2005b; D'Souza et al. 2000); thus, dietary magnesium supplementation prior to slaughter reduces the incidence of PSE meat. Dietary magnesium supplementation has no effects on average daily gain (ADG), average daily feed intake (ADFI), or glycogen stores in muscle at slaughter (Apple et al. 2005b; Apple et al. 2002b; D'Souza et al. 1998c).

Tryptophan supplementation has also been reported to reduce PSE frequency and improve meat quality (Warner et al. 1998). Tryptophan is the precursor of the neurotransmitter serotonin. Serotonin was shown to decrease in pigs experiencing stress (Adeola and Ball 1992). Dietary overload of tryptophan in growing-finishing pigs increases serotonin synthesis in brain (Adeola and Ball 1992), and reduces stress response and the occurrence of PSE carcasses (Warner et al. 1998). Furthermore, supplementing pigs with an extra 0.5% tryptophan for 5 days before slaughter reduced the incidence of PSE carcass from 9% to 6% (Anon 1991).

Glycogen within muscle is the substrate for postmortem glycolysis; thus, it is expected to control postmortem glycolysis by manipulating muscle glycogen levels by feeding. Feeding pigs with diets high in fat (17% to 18%) and protein (22% to 24%) in combination with a low content of digestible carbohydrate (<5%) for 3 weeks right before slaughter has been shown to successfully decrease the glycogen content in muscle without compromising overall performance (Rosenvold et al. 2002; 2001a; Rosenvold et al. 2001b). This reduction in glycogen stores in the muscle of pigs at slaughter by strategic finishing feeding had no effect on ultimate (24 hr postmortem) values, but decreased glycometabolism early postmortem and improved the water-holding capacity of longissimus dorsi, biceps femoris, and semimembranosus muscle (Rosenvold et al. 2002; 2001a). Furthermore, strategic finishing feeding led to darker meat and significantly lower L* values (Rosenvold et al. 2001b). All these results seemed to suggest that glycogen stores in muscle at the time of slaughter are critical for meat quality (Briskey 1964), and manipulation of muscle glycogen level can be used to reduce PSE meat.

However, the reported results have been inconsistent. Leheska et al. (2002) fed swine an ultra-high-protein/low-carbohydrate (HIPRO) diet for up to 2 weeks. They found no effect of diet on meat quality, including postmortem pH decline rate and the ultimate pH. Thus, they concluded that feeding barrows the HIPRO diet for a time period prior to slaughter was not effective in improving pork quality (Leheska et al. 2002a). Furthermore, although a low-carbohydrate diet decreased glycolytic potential in postmortem muscle of barrows, neither the pH nor temperature of the LM measured from 30 to 330 min postmortem was affected by the diet, showing there was no difference in PSE risk between dietary treatments (Bee et al. 2006). This inconsistency between muscle glycogen storage and the PSE risk suggests that muscle glycogen levels may not be decisive for the occurrence of PSE meat. In fact, an abundance of evidence exists indicating that the activity of glycolytic enzymes plays a more important role in the control of PSE occurrence.

As mentioned above, it has long been reported that accelerated chilling by liquid nitrogen prevents PSE meat with no difference in initial muscle glycogen concentration (Borchert and Briskey 1964), showing that fast chilling decreases the PSE occurrence in a muscle glycogen concentration-independent manner. Subsequently, many studies have revealed that fast chilling reduced the PSE rate as a consequence of lower muscle temperature early postmortem: lower muscle temperature slows both glycolysis and protein denaturation (Savell et al. 2005). This lower glycolytic rate should be the result of lower enzyme activity at lower temperatures because the initial muscular glycogen concentrations are not or should not, theoretically, be different (Borchert and Briskey 1964; Kerth et al. 2001; Park et al. 2007). Furthermore, dietary magnesium supplementation reduces lactic acid accumulation in early postmortem muscle and the incidence of PSE meat independent of muscle glycogen concentration (D'Souza et al. 1998c), showing that the glycolytic enzymes — not the muscle glycogen levels — are responsible for the difference in early postmortem glycolytic rate. In this study, it also showed that dietary magnesium supplementation had no influence on ultimate pH values (D'Souza et al. 1998c).

No other examples are more convincing than *RN⁻* pigs in showing that it is enzyme activity — not glycogen concentration — that plays a more important role in the occurrence of PSE meat. The

RN⁻ phenotype is the result of a mutation in the AMPK $\gamma 3$ subunit (Milan et al. 2000a). *RN*⁻ pigs have 70% higher glycogen concentrations in skeletal muscle (Estrade et al. 1993) and, rather than PSE meat, *RN*⁻ pigs result in "acid meat" characterized by very low ultimate pH (Josell et al. 2003; Le Roy et al. 2000; Lebret et al. 1999; Milan et al. 1996; Van Laack and Kauffman 1999). Consistent with these reports, augmented glycolysis does not occur early postmortem in muscle from *RN*⁻ pigs (Copenhafer et al. 2006). This further shows that muscle glycogen storage has no decisive role in the occurrence of PSE meat. However, for the incidence of DFD meat, there exists a critical point in muscle glycogen concentration. Below this point, glycogen storage within muscle might play a decisive role in determining meat quality and is responsible for the development of DFD meat, which usually occurs in beef because beef has much lower glycogen storage levels than pork. Above this critical point, which means enough substrate available for glycolysis, enzyme activity controls the postmortem glycolytic rate, especially early postmortem when the carcass temperature is high. In fact, glycolysis may stop even if there is glycogen left in muscle, and it is speculated that some critical glycolytic enzyme plays an important role in this stoppage (Immonen et al. 2000; Lawrie 1955).

9.6 SUMMARY

A number of pre-slaughter factors, slaughter procedures, as well as genetic factors affect meat quality and the incidence of PSE meat. The frequency of PSE varies widely. It has been reported that the frequency of PSE pork was more than fivefold greater in some plants than in others (Lee and Choi 1999). However, it is well defined that PSE meat is caused by a too-fast pH decline, especially early postmortem while the muscle temperature is still high. All the factors discussed above influence meat quality and the frequency of PSE meat by increasing or inhibiting glycolysis, lactate accumulation, and thus the pH values in postmortem muscle. However, the molecular mechanism of how postmortem glycolysis is controlled, how all these pre-slaughter and postmortem factors regulate postmortem glycolysis remains unknown. To meet the quality demands of the meat industry in the future and for the control of PSE meat, it is necessary to elucidate the molecular mechanism of how postmortem glycolysis is regulated; this in turn, may lead to strategies to prevent PSE meat and the loss of millions of dollars every year in the meat industry.

REFERENCES

Aalhus, J.L., D.R. Best, A.C. Murray, and S.D. M. Jones. 1998. A comparison of the quality characteristics of pale, soft and exudative beef and pork. *J. Muscle Foods* 9:267–280.

Aaslyng, M.D. and P.B. Gade. 2001. Low stress pre-slaughter handling: effect of lairage time on the meat quality of pork. *Meat Sci.* 57:87–92.

Adeola, O. and R.O. Ball. 1992. Hypothalamic neurotransmitter concentrations and meat quality in stressed pigs offered excess dietary tryptophan and tyrosine. *J. Anim. Sci.* 70:1888–1894.

Andersson, L. 2003. Identification and characterization of AMPK gamma 3 mutations in the pig. *Biochem. Soc. Trans.* 31:232–235.

Anon. 1991. Pig meat processing R & D. *Food Austr.* 43:46–47.

Apple, J.K., E.B. Kegley, D.L. Galloway, T.J. Wistuba, and L.K. Rakes. 2005a. Duration of restraint and isolation stress as a model to study the dark-cutting condition in cattle. *J. Anim. Sci.* 83:1202–1214.

Apple, J.K., E.B. Kegley, C.V. Maxwell, L.K. Rakes, D. Galloway, and T.J. Wistuba. 2005b. Effects of dietary magnesium and short-duration transportation on stress response, postmortem muscle metabolism, and meat quality of finishing swine. *J. Anim. Sci.* 83:1633–1645.

Apple, J.K., C.V. Maxwell, M.R. Stivarius, L.K. Rakes, and Z.B. Johnson. 2002a. Effects of dietary magnesium and halothane genotype on performance and carcass traits of growing-finishing swine. *Livestock Production Sci.* 76:103–113.

Apple, J.K., C.V. Maxwell, M.R. Stivarius, L.K. Rakes, and Z.B. Johnson. 2002b. Effects of dietary magnesium and halothane genotype on performance and carcass traits of growing-finishing swine. *Livestock Production Sci.* 76:103–113.

Arad, M., D.W. Benson, A.R. Perez-Atayde, W.J. McKenna, E.A. Sparks, R.J. Kanter, K. McGarry, J.G. Seidman, and C.E. Seidman. 2002. Constitutively active AMP kinase mutations cause glycogen storage disease mimicking hypertrophic cardiomyopathy. *J. Clin. Invest.* 109:357–362.

Aschenbach, W.G., M.F. Hirshman, N. Fujii, K. Sakamoto, K.F. Howlett, and L.J. Goodyear. 2002. Effect of AICAR treatment on glycogen metabolism in skeletal muscle. *Diabetes* 51:567–573.

Ashmore, C.R., F. Carroll, L. Doerr, G. Tompkins, H. Stokes, and W. Parker. 1973. Experimental prevention of dark-cutting meat. *J. Anim. Sci.* 36:33–36.

Ashmore, C.R., L. Doerr, G. Foster, and F. Carroll. 1971. Respiration of mitochondria isolated from dark-cutting beef. *J. Anim. Sci.* 33:574–577.

Barbut, S. 1998a. Estimating the magnitude of the PSE problem in poultry. *J. Muscle Foods* 9:35.

Barbut, S. 1998b. Estimating the magnitude of the PSE problem in poultry. *J. Muscle Foods* 9:35–49.

Barnes, B.R., S. Marklund, T.L. Steiler, M. Walter, G. Hjalm, V. Amarger, M. Mahlapuu, Y. Leng, C. Johansson, D. Galuska, K. Lindgren, M. Abrink, D. Stapleton, J.R. Zierath, and L. Andersson. 2004. The 5'-AMP-activated protein kinase gamma3 isoform has a key role in carbohydrate and lipid metabolism in glycolytic skeletal muscle. *J. Biol. Chem.* 279:38441–38447.

Bartos, L., C. Franc, D. Rehak, and Stipkova. 1993. A practical method to prevent dark-cutting (DFD) in beef. *Meat Sci.* 34:275–282.

Bee, G., C. Biolley, G. Guex, W. Herzog, S.M. Lonergan, and E. Huff-Lonergan. 2006. Effects of available dietary carbohydrate and preslaughter treatment on glycolytic potential, protein degradation, and quality traits of pig muscles. *J. Anim. Sci.* 84:191–203.

Bendall, J.R., O. Hallund, and J. Wismer-Penersen. 1963. Postmortem changes in the muscles of Landrace pigs. *J. Food Sci.* 28:156–162.

Bendall, J.R. and H.J. Swatland. 1988. A review of the relationship of pH with physical aspects of pork quality. *Meat Sci.* 24:85–96.

Boleman, S.L., S.J. Boleman, W.W. Morgan, D.S. Hale, D.B. Griffin, J.W. Savell, R.P. Ames, M.T. Smith, J.D. Tatum, T.G. Field, G.C. Smith, B.A. Gardner, J.B. Morgan, S.L. Northcutt, H.G. Dolezal, D.R. Gill, and F.K. Ray. 1998. National Beef Quality Audit 1995: survey of producer-related defects and carcass quality and quantity attributes. *J. Anim. Sci.* 76:96–103.

Borchert, L.L. and E.J. Briskey. 1964. Prevention of pale soft exudative porcine muscle through partial freezing with liquid nitrogen post-mortem. *J. Food Sci.* 29:203.

Bowker, B.C., E.J. Wynveen, A.L. Grant, and D.E. Gerrard. 1999. Effects of electrical stimulation on early postmortem muscle. pH and temperature declines in pigs from different genetic lines and halothane genotypes. *Meat Sci.* 53:125–133.

Bradshaw, R.H., S.J. G. Hall, and D.M. Broom. 1996a. Behavioural and cortisol response of pigs and sheep during transport. *Vet. Rec.* 138:233–234.

Bradshaw, R.H., R.F. Parrott, J.A. Goode, D.M. Lloyd, R.G. Rodway, and D.M. Broom. 1996b. Behavioural and hormonal responses of pigs during transport: effect of mixing and duration of journey. *Anim. Sci.* 62:547–554.

Briskey, E.J. 1964. Etiological status and associated studies of pale, soft, exudative porcine musculature. *Adv. Food Res.* 13:89.

Brown, S.N., T.G. Knowles, J.E. Edwards, and P.D. Warriss. 1999. Behavioural and physiological responses of pigs to being transported for up to 24 hours followed by six hours recovery in lairage. *Vet. Rec.* 145:421–426.

Brown, S.N., P.D. Warriss, G.R. Nute, J.E. Edwards, and T.G. Knowles. 1998. Meat quality in pigs subjected to minimal preslaughter stress. *Meat Sci.* 49:257–265.

Buckley, D.J., P.A. Morrissey, and J.I. Gray. 1995. Influence of dietary vitamin-E on the oxidative stability and quality of pig meat. *J. Anim. Sci.* 73:3122–3130.

Cannon, J.E., J.B. Morgan, F.K. McKeith, G.C. Smith, S. Sonka, J. Heavner, and D.L. Meeker. 1996. Pork chain quality audit and survey: quantification of pork quality characteristics. *J. Muscle Foods* 7:29–44.

Carling, D. 2004. The AMP-activated protein kinase cascade — a unifying system for energy control. *Trends Biochem. Sci.* 29:18–24.

Channon, H.A., A.M. Payne, and R.D. Warner. 2000. Halothane genotype, pre-slaughter handling and stunning method all influence pork quality. *Meat Sci.* 56:291–299.

Channon, H.A., A.M. Payne, and R.D. Warner. 2002. Comparison of CO_2 stunning with manual electrical stunning (50 Hz) of pigs on carcass and meat quality. *Meat Sci.* 60:63–68.

Cheah, K.S., A.M. Cheah, and D.I. Krausgrill. 1995. Effect of dietary supplementation of vitamin-E on pig meat quality. *Meat Sci.* 39:255–264.

Ciobanu, D., J. Bastiaansen, M. Malek, J. Helm, J. Woollard, G. Plastow, and M. Rothschild. 2001. Evidence for new alleles in the protein kinase adenosine monophosphate-activated gamma(3)-subunit gene associated with low glycogen content in pig skeletal muscle and improved meat quality. *Genetics* 159:1151–1162.

Copenhafer, T.L., B.T. Richert, A.P. Schinckel, A.L. Grant, and D.E. Gerrard. 2006. Augmented postmortem glycolysis does not occur early postmortem in AMPK73-mutated porcine muscle of halothane positive pigs. *Meat Sci.* 73:590–599.

Corton, J.M., J.G. Gillespie, and D.G. Hardie. 1994. Role of the AMP-activated protein kinase in the cellular stress response. *Curr. Biol* 4:315–324.

Costa, L.N., D.P. Lo Fiego, S. Dall'Olio, R. Davoli, and V. Russo. 2002. Combined effects of pre-slaughter treatments and lairage time on carcass and meat quality in pigs of different halothane genotype. *Meat Sci.* 61:41–47.

D'Souza, D.N., F.R. Dunshea, B.J. Leury, and R.D. Warner. 1999a. Effect of mixing boars during lairage and pre-slaughter handling on pork quality. *Austr. J. Agric. Res.* 50:109–113.

D'Souza, D.N., F.R. Dunshea, R.D. Warner, and B.J. Leury. 1998a. The effect of handling pre-slaughter and carcass processing rate post-slaughter on pork quality. *Meat Sci.* 50:429–437.

D'Souza, D.N., R.D. Warner, F.R. Dunshea, and B.J. Leury. 1998b. Effect of on-farm and pre-slaughter handling of pigs on meat quality. *Austr. J. Agric. Res.* 49:1021–1025.

D'Souza, D.N., R.D. Warner, F.R. Dunshea, and B.J. Leury. 1999b. Comparison of different dietary magnesium supplements on pork quality. *Meat Sci.* 51:221–225.

D'Souza, D.N., R.D. Warner, B.J. Leury, and F.R. Dunshea. 1998c. The effect of dietary magnesium aspartate supplementation on pork quality. *J. Anim. Sci.* 76:104–109.

D'Souza, D.N., R.D. Warner, B.J. Leury, and F.R. Dunshea. 2000. The influence of dietary magnesium supplement type, and supplementation dose and duration, on pork quality and the incidence of PSE pork. *Austr. J. Agric. Res.* 51:185–189.

Depreux, F.F. S., A.L. Grant, and D.E. Gerrard. 2002. Influence of halothane genotype and body-weight on myosin heavy chain composition in pig muscle as related to meat quality. *Livestock Prod Sci.* 73:265–273.

DeSmet, S.M., H. Pauwels, S. DeBie, D.I. Demeyer, J. Callewier, and W. Eeckhout. 1996. Effect of halothane genotype, breed, feed withdrawal, and lairage on pork quality of Belgian slaughter pigs. *J. Anim. Sci.* 74:1854–1863.

Du, M., Q.W. Shen, and M.J. Zhu. 2005. Role of alpha-adrenoceptor signaling and AMP-activated protein kinase in glycolysis of postmortem skeletal muscle. *J. Agric. Food Chem.* 53:3235–3239.

Eikelenboom, G., A.H. Bolink, and W. Sybesma. 1991. Effects of feed withdrawal before delivery on pork quality and carcass yield. *Meat Sci.* 29:25–30.

Ekkel, E.D., B. Savenije, W.G.P. Schouten, V.M. Wiegant, and M.J.M. Tielen. 1997. The effects of mixing on behavior and circadian parameters of salivary cortisol in pigs. *Physiol. Behav.* 62:181–184.

Enfalt, A.C., K. Lundstrom, I. Hansson, S. Johansen, and P.E. Nystrom. 1997. Comparison of non-carriers and heterozygous carriers of the RN– allele for carcass composition, muscle distribution and technological meat quality in Hampshire-sired pigs. *Livestock Production Sci.* 47:221–229.

Estrade, M., X. Vignon, E. Rock, and G. Monin. 1993. Glycogen hyperaccumulation in white muscle-fibers of RN- carrier pigs — a biochemical and ultrastructural-study. *Compar. Biochem. Physiol. B– Biochem. Mol. Biol.* 104:321–326.

Faucitano, L. 1998. Preslaughter stressors effects on pork: a review. *J. Muscle Foods* 9:293–303.

Faucitano, L. 2001. Causes of skin damage to pig carcasses. *Can. J. Anim. Sci.* 81:39–45.

Fernandez, X., S. Gilbert, and J.L. Vendeuvre. 2002a. Effects of halothane genotype and pre-slaughter treatment on pig meat quality. 2. Physico-chemical traits of cured-cooked ham and sensory traits of cured-cooked and dry-cured hams. *Meat Sci.* 62:439–446.

Fernandez, X., E. Neyraud, T. Astruc, and V. Sante. 2002b. Effects of halothane genotype and pre-slaughter treatment on pig meat quality. 1. Post mortem metabolism, meat quality indicators and sensory traits of m. Longissimus lumborum. *Meat Sci.* 62:429–437.

Fernandez, X., V. Sante, E. Baeza, E. Lebihan-Duval, C. Berri, H. Remignon, R. Babile, G. Le Pottier, and T. Astruc. 2002d. Effects of the rate of muscle post mortem pH fall on the technological quality of turkey meat. *Brit. Poult. Sci.* 43:245–252.

Fischer, C., K. Hofmann, and E. Bluchel. 1977a. Electrophoretic picture of sarcoplasmic and myofibrillar proteins from pale and watery beef muscle. *Fleischwirtschaft* 57:2050–2051.

Fischer, C., J. Scheper, and R. Hamm. 1977b. Occurrence of pale, watery beef. *Fleischwirtschaft* 57:1826.

Fontanesi, L., R. Davoli, L. Nanni Costa, E. Scotti, and V. Russo. 2003. Study of candidate genes for glycolytic potential of porcine skeletal muscle: identification and analysis of mutations, linkage and physical mapping and association with meat quality traits in pigs. *Cytogenet. Genome Res.* 102:145–151.

Fortin, A. 2002. The effect of transport time from the assembly yard to the abattoir and resting time at the abattoir on pork quality. *Can. J. Anim. Sci.* 82:141–150.

Fraqueza, M.J., L.C. Roseiro, J. Almeida, E. Matias, C. Santos, and J.M. Randall. 1998. Effects of lairage temperature and holding time on pig behaviour and on carcass and meat quality. *Appl. Anim. Behav. Sci.* 60:317–330.

Fraser, H., G.D. Lopaschuk, and A.S. Clanachan. 1999. Alteration of glycogen and glucose metabolism in ischaemic and post-ischaemic working rat hearts by adenosine A(1) receptor stimulation. *Brit. J. Pharmacol.* 128:197–205.

Freise, K., S. Brewer, and J. Novakofski. 2005. Duplication of the pale, soft, and exudative condition starting with normal postmortem pork. *J. Anim. Sci.* 83:2843–2852.

Fujii, J., K. Otsu, F. Zorzato, S. de Leon, V.K. Khanna, J.E. Weiler, P.J. O'Brien, and D.H. MacLennan. 1991. Identification of a mutation in porcine ryanodine receptor associated with malignant hyperthermia. *Science* 253:448–451.

Gade, P.B. and L. Christensen. 1998. Effect of different stocking densities during transport on welfare and meat quality in Danish slaughter pigs. *Meat Sci.* 48:237–247.

Gasperlin, Z., B. Zlender, and V. Abram. 2000. Colour of normal and high pH beef heated to different temperatures as related to oxygenation. *Meat Sci.* 54:391–398.

Geverink, N.A., R.H. Bradshaw, E. Lambooij, V.R. Wiegant, and D.M. Broom. 1998. Effects of simulated lairage conditions on the physiology and behaviour of pigs. *Vet. Rec.* 143:241–244.

Grandin, T. 1990. Design of loading facilities and holding pens. *Appl. Anim. Behav. Sci.* 28:187–201.

Grandin, T. 1994. Solving livestock handling problems. *Vet. Med.* 89:989–998.

Gregory, N.G. 1994. Preslaughter handling, stunning and slaughter. *Meat Sci.* 36:45–56.

Guardia, M.D., J. Estany, S. Balasch, M.A. Oliver, M. Gispert, and A. Diestre. 2004. Risk assessment of PSE condition due to pre-slaughter conditions and RYR1 gene in pigs. *Meat Sci.* 67:471–478.

Guise, H.J. and R.H. Penny. 1989. Factors influencing the welfare and carcass meat quality of pigs. The effects of stocking density in transit and the use of electric goads. *Anim. Production* 49:511–515.

Hallund, O. and J.R. Bendall. 1965. Long-term effect of electrical stimulation on post-mortem fall of pH in muscles of Landrace pigs. *J. Food Sci.* 30:296–299.

Hambrecht, E., J.J. Eissen, D.J. Newman, C.H.M. Smits, M.W.A. Verstegen, and L.A. den Hartog. 2005. Preslaughter handling effects on pork quality and glycolytic potential in two muscles differing in fiber type composition. *J. Anim. Sci.* 83:900–907.

Hamilton, D.N., M. Ellis, M.D. Hemann, F.K. McKeith, K.D. Miller, and K.W. Purser. 2002. The impact of longissimus glycolytic potential and short-term feeding of magnesium sulfate heptahydrate prior to slaughter on carcass characteristics and pork quality. *J. Anim. Sci.* 80:1586–1592.

Hamilton, D.N., M. Ellis, K.D. Miller, F.K. McKeith, and D.F. Parrett. 2000. The effect of the Halothane and Rendement Napole genes on carcass and meat quality characteristics of pigs. *J. Anim. Sci.* 78:2862–2867.

Hamilton, D.N., K.D. Miller, M. Ellis, F.K. McKeith, and E.R. Wilson. 2003. Relationships between longissimus glycolytic potential and swine growth performance, carcass traits, and pork quality. *J. Anim. Sci.* 81:2206–2212.

Hamman, L.L., J.G. Gentry, C.B. Ramsey, J.J. Mcglone, and M.F. Miller. 2001. The effect of vitamin-mineral, nutritional modulation on the pork quality of halothane carriers. *J. Muscle Foods* 12:37–51.

Hammelman, J.E., B.C. Bowker, A.L. Grant, J.C. Forrest, A.P. Schinckel, and D.E. Gerrard. 2003. Early postmortem electrical stimulation simulates PSE pork development. *Meat Sci.* 63:69–77.

Hardie, D.G. 2004. AMP-activated protein kinase: a key system mediating metabolic responses to exercise. *Med. Sci. Sports Exercise* 36:28–34.

Hardie, D.G. and S.A. Hawley. 2001. AMP-activated protein kinase: the energy charge hypothesis revisited. *Bioessays* 23:1112–1119.

Hardie, D.G., S.A. Hawley, and J. Scott. 2006. AMP-activated protein kinase — development of the energy sensor concept. *J. Physiol. – London* 574:7–15.

Hardie, D.G., J.W. Scott, D.A. Pan, and E.R. Hudson. 2003. Management of cellular energy by the AMP-activated protein kinase system. *Febs Lett.* 546:113–120.

Henriksen, E.J., T.R. Kinnick, M.K. Teachey, M.P. O'Keefe, D. Ring, K.W. Johnson, and S.D. Harrison. 2003. Modulation of muscle insulin resistance by selective inhibition of GSK-3 in Zucker diabetic fatty rats. *Am. J. Physiol. – Endocrinol. Metab.* 284:E892–E900.

Hunt, M.C. and H.B. Hedrick. 1977. Chemical, physical and sensory characteristics of bovine muscle from four quality groups. *J. Food Sci.* 42:716–720.

Immonen, K., M. Ruusunen, K. Hissa, and E. Puolanne. 2000. Bovine muscle glycogen concentration in relation to finishing diet, slaughter and ultimate pH. *Meat Sci.* 55:25–31.

James, C., C. Vincent, T.I.D.A. Lima, and S.J. James. 2006. The primary chilling of poultry carcasses — a review. *Int. J. Refrigeration – Revue Internationale Du Froid* 29:847–862.

Jones, S.D.M., L.E. Jeremiah, and W.M. Robertson. 1993. The effects of spray and blast-chilling on carcass shrinkage and pork muscle quality. *Meat Sci.* 34:351–362.

Josell, A., A.C. Enfalt, G. Von Seth, G. Lindahl, I. Hedebro-Velander, L. Andersson, and K. Lundstrom. 2003. The influence of RN genotype, including the new V199I allele, on the eating quality of pork loin. *Meat Sci.* 65:1341–1351.

Josell, A., G. Von Seth, and E. Tornberg. 2004. Sensory and meat quality traits of pork in relation to post-slaughter treatment and RN genotype. *Meat Sci.* 66:113–124.

Karlsson, A. and K. Lundstrom. 1992. Meat quality in pigs reared in groups kept as a unit during the fattening period and slaughter. *Anim. Production* 54:421–426.

Kastenschmidt, L.L., W.G. Hoekstra, and E.J. Briskey. 1968. Glycolytic intermediates and co-factors in "fast-" and "slow-glycolyzing" muscles of the pig. *J. Food Sci.* 33:151–158.

Kauffman, R.G., R.G. Cassens, A. Scherer, and D.L. Meeker. 1992. Variations in pork quality. *National Pork Producers Council Bulletin.* Des Moines, IA.

Kauffman, R.G., R.L. Van Laack, R.L. Russell, E. Pospiech, C.A. Cornelius, C.E. Suckow, and M.L. Greaser. 1998. Can pale, soft, exudative pork be prevented by postmortem sodium bicarbonate injection? *J. Anim. Sci.* 76:3010–3015.

Kenny, F.J. and P.V. Tarrant. 1988. The effect of estrus behavior on muscle glycogen concentration and dark-cutting in beef heifers. *Meat Sci.* 22:21–31.

Kerth, C.R., M.A. Carr, C.B. Ramsey, J.C. Brooks, and R.C. Johnson. 2001. Vitamin-mineral supplementation and accelerated chilling effects on quality of pork from pigs that are monomutant or noncarriers of the Halothane gene. *J. Anim. Sci.* 79:2346–2355.

Killinger, K.M., C.R. Calkins, W.J. Umberger, D.M. Feuz, and K.M. Eskridge. 2004. Consumer visual preference and value for beef steaks differing in marbling level and color. *J. Anim. Sci.* 82:3288–3293.

Kim, J., R.S. Solis, E.B. Arias, and G.D. Cartee. 2004. Postcontraction insulin sensitivity: relationship with contraction protocol, glycogen concentration, and 5' AMP-activated protein kinase phosphorylation. *J. Appl. Physiol.* 96:575–583.

Knowles, T.G., S.N. Brown, J.E. Edwards, and P.D. Warriss. 1998. Ambient temperature below which pigs should not be continuously showered in lairage. *Vet. Rec.* 143:575–578.

Kreikemeier, K.K., J.A. Unruh, and T.P. Eck. 1998. Factors affecting the occurrence of dark-cutting beef and selected carcass traits in finished beef cattle. *J. Anim. Sci.* 76:388–395.

Lambooij, E. 1988. Road transport of pigs over a long distance: some aspects of behaviour, temperature and humidity during transport and some effects of the last two factors. *Anim. Production* 46:257–263.

Lawrie, R.A. 1955. Residual glycogen at high ultimate pH in horse muscle. *Biochim. Biophys. Acta* 17:282–282.

Le Roy, P., J.M. Elsen, J.C. Caritez, A. Talmant, H. Juin, P. Sellier, and G. Monin. 2000. Comparison between the three porcine RN genotypes for growth, carcass composition and meat quality traits. *Genet. Select. Evolut.* 32:165–186.

Leach, L.M., M. Ellis, D.S. Sutton, F.K. McKeith, and E.R. Wilson. 1996. The growth performance, carcass characteristics, and meat quality of halothane carrier and negative pigs. *J. Anim. Sci.* 74:934–943.

Lebret, B., L. Fefaucheur, and J. Mourot. 1999. Pig meat quality. Influence of rearing factors on skeletal muscle traits. *Productions Animales* 12:11–28.

Lee, Y.B. and Y.I. Choi. 1999. PSE (Pale, soft, exudative) pork: the causes and solutions — review. *Asian-Australasian J. Anim. Sci.* 12:244–252.

Leheska, J.M., D.M. Wulf, J.A. Clapper, R.C. Thaler, and R.J. Maddock. 2002a. Effects of high-protein/low-carbohydrate swine diets during the final finishing phase on pork muscle quality. *J. Anim. Sci.* 80:137–142.

Leheska, J.M., D.M. Wulf, and R.J. Maddock. 2002b. Effects of fasting and transportation on pork quality development and extent of postmortem metabolism. *J. Anim. Sci.* 80:3194–3202.

Li, C.T. and M. Wick. 2001. Improvement of the physicochemical properties of pale soft and exudative (PSE) pork meat products with an extract from mechanically deboned turkey meat (MDTM). *Meat Sci.* 58:189–195.

Long, V.P. and P.V. Tarrant. 1990. The effect of pre-slaughter showering and post-slaughter rapid chilling on meat quality in intact pork sides. *Meat Sci.* 27:181–195.

Lorenzen, C.L., D.S. Hale, D.B. Griffin, J.W. Savell, K.E. Belk, T.L. Frederick, M.F. Miller, T.H. Montgomery, and G.C. Smith. 1993. National Beef Quality Audit — survey of producer-related defects and carcass quality and quantity attributes. *J. Anim. Sci.* 71:1495–1502.

Ludvigsen, J. 1953. Muscular degeneration in pigs. *Int. Vet. Congr. 15th Congr.,* Stockholm, Sweden. 1:602.

MacAulay, K., A.S. Blair, E. Hajduch, T. Terashima, O. Baba, C. Sutherland, and H.S. Hundal. 2005. Constitutive activation of GSK3 down-regulates glycogen synthase abundance and glycogen deposition in rat skeletal muscle cells. *J. Biol. Chem.* 280:9509–9518.

MacAulay, K., E. Hajduch, A.S. Blair, M.P. Coghlan, S.A. Smith, and H.S. Hundal. 2003. Use of lithium and SB-415286 to explore the role of glycogen synthase kinase-3 in the regulation of glucose transport and glycogen synthase. *Eur. J. Biochem.* 270:3829–3838.

Marsin, A.S., L. Bertrand, M.H. Rider, J. Deprez, C. Beauloye, M.F. Vincent, G. Van den Berghe, D. Carling, and L. Hue. 2000. Phosphorylation and activation of heart PFK-2 by AMPK has a role in the stimulation of glycolysis during ischaemia. *Curr. Biol.* 10:1247–1255.

McKenna, D.R., D.L. Roeber, P.K. Bates, T.B. Schmidt, D.S. Hale, D.B. Griffin, J.W. Savell, J.C. Brooks, J.B. Morgan, T.H. Montgomery, K.E. Belk, and G.C. Smith. 2002. National Beef Quality Audit — 2000: survey of targeted cattle and carcass characteristics related to quality, quantity, and value of fed steers and heifers. *J. Anim. Sci.* 80:1212–1222.

Mcveigh, J.M., P.V. Tarrant, and M.G. Harrington. 1982. Behavioral stress and skeletal-muscle glycogen-metabolism in young bulls. *J. Anim. Sci.* 54:790–795.

Milan, D., J.T. Jeon, C. Looft, V. Amarger, A. Robic, M. Thelander, C. Rogel-Gaillard, S. Paul, N. Iannuccelli, L. Rask, H. Ronne, K. Lundstrom, N. Reinsch, J. Gellin, E. Kalm, P. Le Roy, P. Chardon, and L. Andersson. 2000a. A mutation in PRKAG3 associated with excess glycogen content in pig skeletal muscle. *Science* 288:1248–1251.

Milan, D., N. Woloszyn, M. Yerle, P. LeRoy, M. Bonnet, J. Riquet, Y. LahbibMansais, J.C. Caritez, A. Robic, P. Sellier, J.M. Elsen, and J. Gellin. 1996. Accurate mapping of the "acid meat" RN gene on genetic and physical maps of pig Chromosome 15. *Mammal. Genome* 7:47–51.

Miller, K.D., M. Ellis, F.K. McKeith, B.S. Bidner, and D.J. Meisinger. 2000. Frequency of the Rendement Napole RN– allele in a population of American Hampshire pigs. *J. Anim. Sci.* 78:1811–1815.

Milligan, S.D., C.B. Ramsey, M.F. Miller, C.S. Kaster, and L.D. Thompson. 1998. Resting of pigs and hot-fat trimming and accelerated chilling of carcasses to improve pork quality. *J. Anim. Sci.* 76:74–86.

Miyamoto, L., T. Toyoda, T. Hayashi, S. Yonemitsu, M. Nakano, S. Tanaka, K. Ebihara, H. Masuzaki, K. Hosoda, Y. Ogawa, G. Inoue, T. Fushiki, and K. Nakao. 2007. Effect of acute activation of 5'-AMP-activated protein kinase on glycogen regulation in isolated rat skeletal muscle. *J. Appl. Physiol.* 102:1007–1013.

Moeller, S.J., T.J. Baas, T.D. Leeds, R.S. Emnett, and K.M. Irvin. 2003. Rendement Napole gene effects and a comparison of glycolytic potential and DNA genotyping for classification of Rendement Napole status in Hampshire-sired pigs. *J. Anim. Sci.* 81:402–410.

Moiseev, I.V. and D.P. Cornforth. 1999. Treatments for prevention of persistent pinking in dark-cutting beef patties. *J. Food Sci.* 64:738–743.

Monin, G., P. Sellier, L. Ollivier, R. Goutefongea, and J.P. Girard. 1981. Carcass characteristics and meat quality of halothane negative and halothane positive Pietrain pigs. *Meat Sci.*5:413–423.

Mota-Rojas, D., M. Becerril, C. Lemus, P. Sanchez, M. Gonzalez, S.A. Olmos, R. Ramirez, and M. Alonso-Spilsbury. 2006. Effects of mid-summer transport duration on pre- and post-slaughter performance and pork quality in Mexico. *Meat Sci.* 73:404–412.

Munns, W.O. and D.E. Burrell. 1965. Use of rib-eye pH for detecting dark-cutting beef. *Food Technol.* 19:1432–1434.

Murray, A., W. Robertson, F. Nattress, and A. Fortin. 2001. Effect of pre-slaughter overnight feed withdrawal on pig carcass and muscle quality. *Can. J. Anim. Sci.* 81:89–97.

Nanni Costa, L., D.P. Lo Fiego, A. De Grossi, and V. Russo. 1996. Effect of loading method, stocking density and temperature on carcass and meat quality in heavy pigs. In *Proc. Eu-Seminar New Information on Welfare and Meat Quality of Pigs as Related to Handling, Transport and Lairage Conditions.* Mariensee, Germany, p. 83–93.

Offer, G., P. Knight, R. Jeacocke, R. Almond, T. Cousins, J. Elsey, N. Parsons, A. Sharp, R. Starr, and P. Purslow. 1989. The structural basis of the water-holding, appearance and toughness of meat and meat-products. *Food Microstruct.* 8:151–170.

Oliveira, S.M., J. Ehtisham, C.S. Redwood, I. Ostman-Smith, E.M. Blair, and H. Watkins. 2003. Mutation analysis of AMP-activated protein kinase subunits in inherited cardiomyopathies: implications for kinase function and disease pathogenesis. *J. Mol. Cell. Cardiol.* 35:1251–1255.

Oliver, M.A., C. Coll, L. Faucitano, M. Gispert, and A. Diestre. 1996. The effect of ante mortem treatment on bood parameters and carcass damage and its relationship with pig meat quality. In *Proc. 42nd Int. Congr. Meat Sci. and Tech.*, Lillehammer, Norway, p. 35–37.

Olivo, R., A.L. Soares, E.I. Ida, and M. Shimokomaki. 2001. Dietary vitamin E inhibits poultry PSE and improves meat functional properties. *J. Food Biochem.* 25:271–283.
Owen, B.L., J.L. Montgomery, C.B. Ramsey, and M.F. Miller. 2000. Preslaughter resting and hot-fat trimming effects on the incidence of pale, soft and exudative (PSE) pork and ham processing characteristics. *Meat Sci.* 54:221–229.
Owens, C.M., E.M. Hirschler, S.R. McKee, R. Martinez-Dawson, and A.R. Sams. 2000a. The characterization and incidence of pale, soft, exudative turkey meat in a commercial plant. *Poult. Sci.* 79:553–558.
Owens, C.M., E.M. Hirschler, S.R. McKee, R. Martinez-Dawson, and A.R. Sams. 2000b. The characterization and incidence of pale, soft, exudative turkey meat in a commercial plant. *Poult. Sci.* 79:553–558.
Owens, C.M., S.R. McKee, N.S. Matthews, and A.R. Sams. 2000b. The development of pale, exudative meat in two genetic lines of turkeys subjected to heat stress and its prediction by halothane screening. *Poult. Sci.* 79:430–435.
Owens, C.M. and A.R. Sams. 2000. The influence of transportation on turkey meat quality. *Poult. Sci.* 79:1204–1207.
Page, J.K., D.M. Wulf, and T.R. Schwotzer. 2001. A survey of beef muscle color and pH. *J. Anim. Sci.* 79:678–687.
Park, B.Y., J.H. Kim, S.H. Cho, K.H. Hah, S.H. Lee, C.H. Choi, D.H. Kim, J.M. Lee, Y.K. Kim, J.N. Ahn, and I.H. Hwang. 2007. Evidence of significant effects of stunning and chilling methods on PSE incidences. *Asian-Australasian J. Anim. Sci.* 20:257–262.
Rathgeber, B.M., J.A. Boles, and P.J. Shand. 1999. Rapid postmortem pH decline and delayed chilling reduce quality of turkey breast meat. *Poult. Sci.* 78:477–484.
Roseiro, L.C., M.J. Fraqueza, J. Almeida, E. Matias, C. Santos, and L. Sardinha. 1996. Influence of lairage temperature and time on postural attitude in pigs. In *Proc. 42nd Int. Congr. Meat Sci. and Technol.*, Lillehammer, Norway, p. 442–443.
Rosenvold, K. and H.J. Andersen. 2003. Factors of significance, for pork quality — a review. *Meat Sci.* 64:219–237.
Rosenvold, K., H.N. Laerke, S.K. Jensen, A.H. Karlsson, K. Lundstrom, and H.J. Andersen. 2001a. Strategic finishing feeding as a tool in the control of pork quality. *Meat Sci.* 59:397–406.
Rosenvold, K., H.N. Laerke, S.K. Jensen, A.H. Karlsson, K. Lundstrom, and H.J. Andersen. 2002. Manipulation of critical quality indicators and attributes in pork through vitamin E supplementation, muscle glycogen reducing finishing feeding and pre-slaughter stress. *Meat Sci.* 62:485–496.
Rosenvold, K., J.S. Petersen, H.N. Laerke, S.K. Jensen, M. Therkildsen, A.H. Karlsson, H.S. Moller, and H.J. Andersen. 2001b. Muscle glycogen stores and meat quality as affected by strategic finishing feeding of slaughter pigs. *J. Anim. Sci.* 79:382–391.
Russell, R.R., R. Bergeron, G.I. Shulman, and L.H. Young. 1999. Translocation of myocardial GLUT-4 and increased glucose uptake through activation of AMPK by AICAR. *Am. J. Physiol. – Heart Circulat. Physiol.* 277:H643–H649.
Sams, A.R. and C.Z. Alvarado. 2004. Turkey carcass chilling and protein denaturation in the development of pale, soft, and exudative meat. *Poultry Sci.* 83:1039–1046.
Santos, C., J.M. Almeida, E.C. Matias, M.J. Fraqueza, C. Roseiro, and L. Sardina. 1997. Influence of lairage environmental conditions and resting time on meat quality in pigs. *Meat Sci.* 45:253–262.
Savell, J.W., S.L. Mueller, and B.E. Baird. 2005. The chilling of carcasses. *Meat Sci.* 70:449–459.
Sayre, R.N., E.J. Briskey, and W.G. Hoekstra. 1963. Effect of excitement, fasting and sucrose feeding on porcine muscle phosphorylase and postmortem glycolysis. *J. Food Sci.* 28:472–477.
Scanga, J.A., K.E. Belk, J.D. Tatum, T. Grandin, and G.C. Smith. 1998. Factors contributing to the incidence of dark cutting beef. *J. Anim. Sci.* 76:2040–2047.
Shen, Q.W. and M. Du. 2005a. Effects of dietary alpha-lipoic acid on glycolysis of postmortem muscle. *Meat Sci.* 71:306–311.
Shen, Q.W. and M. Du. 2005b. Role of AMP-activated protein kinase in the glycolysis of postmortem muscle. *J. the Sci. Food Agric.* 85:2401–2406.
Shen, Q.W., D.E. Gerrard, and M. Du. 2008. Compound C, an inhibitor of AMP-activated protein kinase, inhibits glycolysis in mouse longissimus dorsi postmortem. *Meat Sci.* 78:323–330.
Shen, Q.W., C.S. Jones, N. Kalchayanand, M.J. Zhu, and M. Du. 2005. Effect of dietary alpha-lipoic acid on growth, body composition, muscle pH, and AMP-activated protein kinase phosphorylation in mice. *J. Anim. Sci.* 83:2611–2617.
Shen, Q.W., W.J. Means, S.A. Thompson, K.R. Underwood, M.J. Zhu, R.J. McCormick, S.P. Ford, and M. Du. 2006a. Pre-slaughter transport, AMP-activated protein kinase, glycolysis, and quality of pork loin. *Meat Sci.* 74:388–395.

Shen, Q.W., W.J. Means, K.R. Underwood, S.A. Thompson, M.J. Zhu, R.J. McCormick, S.P. Ford, M. Ellis, and M. Du. 2006b. Early post-mortem AMP-activated protein kinase (AMPK) activation leads to phosphofructokinase-2 and -1 (PFK-2 and PFK-1) phosphorylation and the development of pale, soft, and exudative (PSE) conditions in porcine longissimus muscle. *J. Agric. Food Chem.* 54:5583–5589.

Shen, Q.W., K.R. Underwood, W.J. Means, R.J. McCormick, and M. Du. 2007a. The halothane gene, energy metabolism, adenosine monophosphate-activated protein kinase, and glycolysis in postmortem pig longissimus dorsi muscle. *J. Anim Sci.* 85:1054–1061.

Shen, Q.W., M.J. Zhu, J. Tong, J. Ren, and M. Du. 2007b. Ca^{2+}/calmodulin-dependent protein kinase kinase is involved in AMP-activated protein kinase activation by alpha-lipoic acid in C2C12 myotubes. *American J. Physiol. – Cell Physiol.* 293:C1395–C1403.

Solomon, M.B., R.L.. Van Laack, and J.S. Eastridge. 1998. Biophysical basis of pale, soft, exudative (PSE) pork and poultry muscle: a review. *J. Muscle Foods* 9:1–11.

Sosnicki, A.A. and M.L. Greaser. 1996. Alterations of protein functionality in PSE-like poultry muscle. *1996 IFT Annual Meeting,* Abstract 2-3.

Taylor, A.A. and S.J. Dant. 1971. Influence of carcass cooling rate on drip loss in pigment. *J. Food Technol.* 6:131–139.

Taylor, A.A. and L. Martoccia. 1995. The effect of low-voltage and high-voltage electrical-stimulation on pork quality. *Meat Sci.* 39:319–326.

Taylor, A.A., G.R. Nute, and C.C. Warkup. 1995a. The effect of chilling, electrical-stimulation and conditioning on pork eating quality. *Meat Sci.* 39:339–347.

Taylor, A.A., A.M. Perry, and C.C. Warkup. 1995b. Improving pork quality by electrical-stimulation or pelvic suspension of carcasses. *Meat Sci.* 39:327–337.

Taylor, A.A. and M.Z. Tantikov. 1992. Effect of different electrical-stimulation and chilling treatments on pork quality. *Meat Sci.* 31:381–395.

Van der Velden, J.L., R.C. Langen, M.C. Kelders, E.F. Wouters, Y.M. Janssen-Heininger, and A.M. Schols. 2006. Inhibition of glycogen synthase kinase-3beta activity is sufficient to stimulate myogenic differentiation. *Am. J. Physiol. – Cell Physiol.* 290:C453–462.

Van der Wal, P.G., B. Engel, and B. Hulsegge. 1997. Causes for variation in pork quality. *Meat Sci.* 46:319–327.

Van der Wal, P.G., B. Engel, and H.G.M. Reimert. 1999. The effect of stress, applied immediately before stunning, on pork quality. *Meat Sci,* 53:101–106.

Van der Wal, P.G., B. Engel, G.Vanbeek, and C.H. Veerkamp. 1995. Chilling pig carcasses — effects on temperature, weight-loss and ultimate meat quality. *Meat Sci.* 40:193–202.

Van Laack, R.L. and R.G. Kauffman. 1999. Glycolytic potential of red, soft, exudative pork longissimus muscle. *J. Anim. Sci.* 77:2971–2973.

Vanhoof, J. 1979. Influence of ante-mortem and peri-mortem factors on biochemical and physical characteristics of turkey breast muscle. *Vet. Quart.* 1:29–36.

Velarde, A., M. Gispert, L. Faucitano, X. Manteca, and A. Diestre. 2000. The effect of stunning method on the incidence of PSE meat and haemorrhages in pork carcasses. *Meat Sci.* 55:309–314.

Von Wenzlawowicz, M., K. Von Holleben, and G. Von Mickwitz. 1996. Meat quality in pigs — comparative investigation in consideration of preliminary factors of strain, carcass weight and lean meat percentage. *Fleischwirtschaft* 76:301–307.

Warner, R.D., G.A. Eldridge, C.D. Hofmeyer, and J.L. Barnett. 1998. The effect of dietary tryptophan on pig behaviour and meat quality — preliminary results. *22nd Biennial Conference,* University of New England, Armidale, New South Wales, Australia, p. 22.

Warner, R.D., R.G. Kauffman, and M.L. Greaser. 1997. Muscle protein changes postmortem in relation to pork quality traits. *Meat Sci.* 45:339–352.

Warriss, P.D. 2003. Optimal lairage times and conditions for slaughter pigs: a review. *Vet. Rec.* 153:170–176.

Warriss, P.D. and S.N. Brown. 1985. The physiological-responses to fighting in pigs and the consequences for meat quality. *J. Sci. Food Agric.* 36:87–92.

Warriss, P.D., S.N. Brown, J.E. Edwards, and T.G. Knowles. 1998a. Effect of lairage time on levels of stress and meat quality in pigs. *Anim. Sci.* 66:255–261.

Warriss, P.D., S.N. Brown, M.A. Francome, and J.A. Higgins. 1987. The effect of pre-slaughter fasting on the characteristics of pig livers. *Int. J. Food Sci. and Technol.* 22:255–263.

Warriss, P.D., S.N. Brown, P.B. Gade, C. Santos, L.N. Costa, E. Lambooij, and R. Geers. 1998b. An analysis of data relating to pig carcass quality and indices of stress collected in the European Union. *Meat Sci.* 49:137–144.

Warriss, P.D., S.N. Brown, G.R. Nute, T.G. Knowles, J.E. Edwards, A.M. Perry, and S.P. Johnson. 1995. Potential interactions between the effects of preslaughter stress and postmortem electrical-stimulation of the carcasses on meat quality in pigs. *Meat Sci.* 41:55–68.

Warriss, P.D., S.C. Kestin, S.N. Brown, and L.J. Wilkins. 1984. The time required for recovery from mixing stress in young bulls and the prevention of dark cutting beef. *Meat Sci.* 10:53–68.

Weaver, S.A., W.T. Dixon, and A.L. Schaefer. 2000. The effects of mutated skeletal ryanodine receptors on hypothalamic-pituitary-adrenal axis function in boars. *J. Anim. Sci.* 78:1319–1330.

Weeding, C.M., H.J. Guise, and R.H.C. Penny. 1993. Factors influencing the welfare and carcass and meat quality of pigs - the use of water sprays in lairage. *Anim. Production* 56:393–397.

Woelfel, R.L., C.M. Owens, E.M. Hirschler, R. Martinez-Dawson, and A.R. Sams. 2002. The characterization and incidence of pale, soft, and exudative broiler meat in a commercial processing plant. *Poult. Sci.* 81:579–584.

Wulf, D.M., R.S. Emnett, J.M. Leheska, and S.J. Moeller. 2002. Relationships among glycolytic potential, dark cutting (dark, firm, and dry) beef, and cooked beef palatability. *J. Anim. Sci.* 80:1895–1903.

Young, L.H., Y. Renfu, R. Russell, X.Y. Hu, M. Caplan, J.M. Ren, G.I. Shulman, and A.J. Sinusas. 1997. Low-flow ischemia leads to translocation of canine heart GLUT-4 and GLUT-1 glucose transporters to the sarcolemma *in vivo*. *Circulation* 95:415–422.

Yu, H., M.F. Hirshman, N. Fujii, J.M. Pomerleau, L.E. Peter, and L.J. Goodyear. 2006. Muscle-specific over-expression of wild type and R225Q mutant AMP-activated protein kinase gamma3-subunit differentially regulates glycogen accumulation. *Am. J. Physiol. - Endocrinol. Metab.* 291:E557–E565.

10 AMP-Activated Protein Kinase in Muscle Growth, Fat Deposition, and Meat Quality

Min Du, Qingwu W. Shen, Keith R. Underwood, Jun F. Tong, and Meijun Zhu

CONTENTS

10.1 INTRODUCTION

AMP-activated protein kinase (AMPK) was first described in 1989 (Hardie et al. 1989). Since then, interest in AMPK has exploded due to its apparent association with obesity and diabetes (Hardie 2007). AMPK is a heterotrimeric complex comprising a catalytic subunit (α) and two regulatory subunits (β and γ). The approximate molecular masses are 63 and 38 kDa for the α- and β-subunits, respectively (Hardie and Carling 1997; Hardie et al. 1998; Winder and Hardie 1999). The γ-subunit molecular mass varies considerably among isoforms (Cheung et al. 2000). Each subunit is encoded by either two or three distinct genes (α1, α2, β1, β2, γ1, γ2, γ3) (Beri et al. 1994; Cheung et al. 2000; Hardie et al. 1998; Stapleton et al. 1997), which demonstrate tissue-specific distribution. In skeletal muscle, the α2-isoform is dominant, accounting for 80% of the total AMPK activity (Cheung et al. 2000; Stapleton et al. 1996, 1997; Thornton et al. 1998; Vavvas et al. 1997). Both β-subunit isoforms are present in skeletal muscle with the β2 isoform predominating (Stapleton et al. 1996; Thornton et al. 1998; Woods et al. 1996). The γ2-isoform mainly exists in cardiac muscle, while the γ3-isoform is skeletal muscle specific (Cheung et al. 2000). The AMPK γ3-isoform is especially abundant in white muscle fibers (Mahlapuu et al. 2004). Hampshire pigs carrying the *Rendement Napole* (*RN*) gene are characterized by superior lean growth and decreased fat deposition, but poor meat quality (Carr et al. 2006; Enfalt et al. 1997; Moeller et al. 2003; Monin and Sellier 1985). Recently, the discovery that *Rendement Napole* (*RN*) gene is due to a mutation in the AMPK γ-subunit has stimulated research interest in examining the role of AMPK in muscle growth, fat deposition, and meat quality.

10.2 MECHANISMS CONTROLLING AMPK ACTIVATION

10.2.1 Allosteric Activation by AMP

The γ-subunit of AMPK contains four cystathionine-β-synthase (CBS) domains, which are essential for AMP binding (Cheung et al. 2000). AMP allosterically activates AMPK through three separate mechanisms: (1) AMP directly activates phosphorylated AMPK (Hawley et al. 1995); (2) AMP binds to AMPK, changing its conformation and making it a better substrate for upstream AMPK kinase (LKB1), which phosphorylates the AMPK α-subunit at Thr 172 (Hawley et al. 1995); and (3) AMP binds to AMPK, making it a worse substrate for dephosphorylation by protein phosphatases (Davies et al. 1995). The coexistence of these three mechanisms make AMPK highly sensitive to the AMP concentration within cells (Hardie et al. 1999), all of which are antagonized by a high concentration of ATP (Davies et al. 1995; Hawley et al. 1996), suggesting that the allosteric sites bind AMP and ATP competitively. In striated muscle, AMPK is also regulated by the ratio of phosphocreatine to creatine (Fujii et al. 2000; Hardie and Carling 1997; Ponticos et al. 1998). Decreased phosphocreatine-to-creatine ratio can activate AMPK by two mechanisms, either (1) through direct activation or (2) through alteration of the AMP/ATP ratio by increasing the AMP concentration (Fujii et al. 2000; Ponticos et al. 1998; Tian et al. 1997).

10.2.2 AMPK Kinases, Phosphatase, and AMPK α-Subunit Phosphorylation at Thr 172

10.2.2.1 AMPK Phosphorylation at Thr 172 and Activation

The AMPK complex is inactive unless its α-subunit has been phosphorylated at Thr 172 by upstream AMPK kinase (AMPKK) (Hawley et al. 1996). Thr 172 lies within the activation loop of the kinase catalytic domain, a region conserved in many protein kinases requiring phosphorylation for their activity (Johnson et al. 1996). Phosphorylation of Thr 172 is essential for AMPK activity as site-directed mutagenesis of Thr 172 to alanine completely abolished kinase activity (Crute et al. 1998; Stein et al. 2000). In addition to Thr 172, two other phosphorylation sites — Thr 258 and Ser 485(α1)/491(α2) in the α-subunit of AMPK — have been identified recently (Woods et al. 2003b). Site-directed mutagenesis experiments indicated that phosphorylation at these new sites was not essential for AMPK activity (Woods et al. 2003b).

10.2.2.2 LKB1 as an Upstream Activator of AMPK

LKB1 is the first identified AMPKK (Hawley et al. 2003; Shaw et al. 2004; Woods et al. 2003a). Deficiency of LKB1 in skeletal muscle prevents contraction-stimulated AMPK activation in skeletal muscle as well as ischemia-mediated AMPK α2 activation in cardiac muscle (Sakamoto et al.

2004; 2005; 2006). Also, deletion of LKB1 in the liver of adult mice led to a nearly complete loss of AMPK activity (Shaw et al. 2005). This evidence emphasizes that LKB1 is an important upstream kinase essential for AMPK activation. LKB1 is required for activation of AMPK in response to an elevated AMP/ATP ratio or increased AMP analogues. However, the LKB1 complex itself is not regulated by AMP and is constitutively active (Lizcano et al. 2004; Sakamoto et al., 2004). AMP and its mimetic agents activate AMPK by binding to AMPK and making it a better substrate for LKB1 (Hawley et al. 2003; Hawley et al. 1995). Interestingly, LKB1 was identified as a tumor suppressor gene that when mutated results in a rare hereditary cancer termed Peutz-Jeghers syndrome (PJS) (Jenne et al. 1998), linking AMPK to cancer.

10.2.2.3 Ca^{2+}/Calmodulin-Dependent Protein Kinase Kinase (CaMKK) as AMPKK

CaMKK is activated in response to elevation of intracellular Ca^{2+} concentration, and the activation of AMPK by CaMKK requires intracellular Ca^{2+} mobilization. Ca^{2+}/calmodulin-dependent protein kinase has two isoforms: CaMKKα and CaMKKβ. While LKB1 is ubiquitously expressed, CaMKK, especially CaMKKβ, is mainly localized in neural tissues (Anderson et al. 1998). However, substantial CaMKKα expression is observed in skeletal muscle (Rose et al. 2006). Witczak et al. (2006) reported that CaMKKα is the only CaMKK isoform expressed in skeletal muscle. The AMPK activity in $LKB1^{-/-}$ mouse embryo fibroblasts increased multiple folds upon stimulation by Ca^{2+} ionophore, ionomycin, and/or A23187, indicating that CaMKK is an AMPKK along with LKB1 (Hawley et al. 2005; Woods et al. 2005).

Studies by three separate groups confirmed that CaMKKβ is an upstream AMPK kinase (Hawley et al. 2005; Hurley et al. 2005; Woods et al. 2005). In their studies, AMPK was activated by Ca^{2+} ionophores in LKB1 deficient cells, and this activation is prevented by either selective CaMKKβ inhibitor STO-609 or RNA interference (RNAi) targeting CaMKK. Also, AMPK activation in brain slices by depolarization (which increases intracellular Ca^{2+}) was blocked by STO-609 (Hawley et al. 2005). Less information is available regarding the role of CaMKKα as an upstream kinase of AMPK. In a previous study using C2C12 cells, CaMKKα mediated AMPK activation by α-lipoic acid (Shen et al. 2007b). In C2C12 cells, IL-6 activated AMPK through CaMKKα (Weigert et al. 2007). Also, caffeine induced calcium release, which activated AMPK in rodent soleus muscle, possibly mediated by CaMKKα (Jensen et al. 2007).

10.2.2.4 Other Proposed AMPKK

Transforming growth factor-β-activated kinase (TAK1) is a newly identified AMPKK (Momcilovic et al. 2006). TAK1 directly phosphorylates AMPK in HeLa cells (Momcilovic et al. 2006). However, TAK1 was placed upstream of LKB1 in another report (Xie et al. 2006). Apparently, more studies are needed to either support or repudiate TAK1 as an AMPKK.

10.2.2.5 Phosphatase

Protein phosphatase 2C (PP2C) dephosphorylates the AMPK α-subunit at Thr 172 (Marley et al. 1996; Sanders et al. 2007). Tumor necrosis factor α (TNFα) reduces AMPK activity in skeletal muscle through up-regulation of PP2C (Shen et al. 2007b; Steinberg et al. 2006b). High levels of TNFα in blood circulation are associated with insulin resistance, which is at least partially due to the enhancement of PP2C expression, inhibiting AMPK activity.

10.3 AMPK ACTIVITY IS REGULATED BY HORMONES AND CYTOKINES, AND DRUGS

10.3.1 Hormone and Cytokine Regulation of AMPK Activity

One of the most fascinating advancements in AMPK research is the demonstration that AMPK is regulated by hormones and cytokines. Leptin, a cytokine produced by adipocytes, stimulates fatty-acid oxidation by activating AMPK in skeletal muscle (Minokoshi et al. 2002). Chronic administration of leptin (2 weeks) increases the expression of AMPK α-subunit and its phosphorylation at Thr 172 in rodent muscle (Steinberg et al. 2003). In contrast to its effects in periphery tissues, leptin suppresses AMPK activity in the hypothalamus (Andersson et al. 2004; Minokoshi et al. 2004), which is necessary for the satiety control by leptin (Minokoshi et al. 2004).

AMPK is also activated by adiponectin (Yamauchi et al. 2002). Adiponectin is another peptide hormone secreted by adipocytes that stimulates glucose uptake and fatty acid oxidation and inhibits gluconeogenesis. Both globular and full-length adiponectin activate AMPK in skeletal muscle (Tomas et al. 2002; Yamauchi et al. 2002). In parallel with the activation of AMPK, adiponectin stimulates fatty acid oxidation, glucose uptake, and lactic acid accumulation in myocytes and adipocytes (Wu et al. 2003; Yamauchi et al. 2002). Blocking AMPK activation by a dominant-negative mutation (Yamauchi et al. 2002; Yoon et al. 2006) or by an AMPK inhibitor (Wu et al. 2003; Yoon et al. 2006) inhibits all these effects of adiponectin on lipid metabolism.

The adipocyte/macrophage-secreted hormone, resistin, induces insulin resistance and stimulates hepatic glucose production *in vivo*; this is at least partially mediated by down-regulation of AMPK activity (Banerjee et al. 2004; Muse et al. 2004; Satoh et al. 2004). *In vitro*, chronic incubation of skeletal muscle cells with resistin decreased fatty acid uptake and metabolism via a mechanism involving decreased cell surface fatty acid translocase FAT/CD36 and a decrease in the phosphorylation of AMPK and ACC. Because AMPK has a stimulatory effect on FAT/CD36, resistin-induced AMPK inhibition might mediate these effects (Palanivel and Sweeney 2005). Despite the above observations, mechanisms leading to AMPK activation by hormones and cytokines are far from clear and certainly warrant further studies.

10.3.2 Pharmacological Activation of AMPK

Metformin and rosiglitazone are two common anti-diabetic drugs. Metformin induces fatty acid oxidation, stimulates glucose uptake in muscle, and suppresses lipogenic enzyme expression in liver through activation of AMPK in both muscle and liver (Zhou et al. 2001). The mechanism by which metformin activates AMPK is controversial. Earlier literature tends to suggest that metformin activates AMPK without an increase in the AMP/ATP ratio (Fryer et al. 2002a; Hawley et al. 2002; Owen et al. 2000), indicating that LKB1 is not involved in the activation of AMPK by metformin. However, a more recent study shows that LKB1 is necessary for the activation of AMPK by metformin (Shaw et al. 2005). Rosiglitazone increases AMPK activity in muscle cells via an increase in the AMP/ATP ratio (Fryer et al. 2002a). Chronic treatment with rosiglitazone restores AMPK α_2 activity in the skeletal muscle of obese, insulin-resistant Zucker rats (Lessard et al. 2006).

To study the physiological function of AMPK *in vivo* requires the development of methods to manipulate AMPK activity other than cellular stresses that deplete ATP concentration. The widely used AMPK activator, 5-aminoimidazole-4-carboxamide ribonucleoside (AICAR), is phosphorylated to 5-amino-4-imidazolecarboxamide ribotide (ZMP) after entering into cell and mimics AMP to activate AMPK (Corton et al. 1995; Sabina et al. 1985).

α-Lipoic acid (ALA) naturally exists in a wide range of foods. It is a short-chain fatty acid with powerful antioxidant capacity (Packer et al. 2001), ALA has long been used as an anti-diabetes agent (Ziegler et al. 1997, 1999). ALA, like leptin, suppresses AMPK activity in hypothalamus, reduces food intake, and enhances whole-body energy expenditure (Kim et al. 2004b). In skeletal muscle, ALA improves insulin sensitivity, and increases glucose uptake and fatty acid oxidation by activating AMPK (Lee et al. 2005). We recently showed that ALA activates AMPK through CaMKKα (Shen et al. 2007b). In addition to ALA, certain compounds in plants also activate AMPK. Berberine is an alkaloid found in plants that activates AMPK (Cheng et al. 2006; Lee et al. 2006).

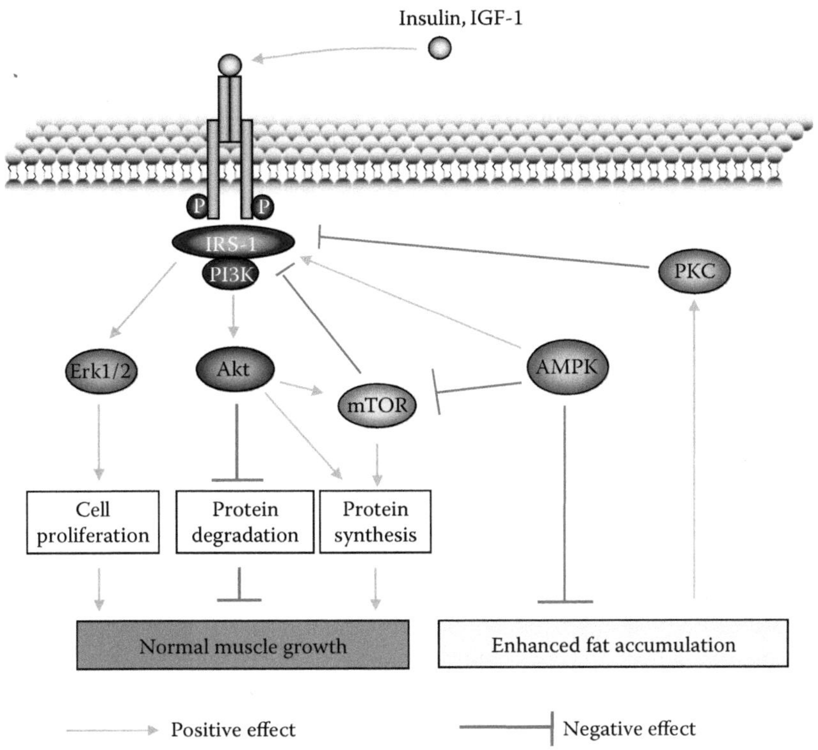

FIGURE 10.1 AMPK and insulin/IGF-1 signaling in skeletal muscle growth. AMPK: AMP-activated protein kinase; IGF-1: insulin-like growth factor-1; IRS-1: insulin receptor substrate-1; PI3K: phosphoinositide-3 kinase; PKC: protein kinase C; Erk1/2: extracellular signal-regulated kinase 1/2; Akt: protein kinase B; and mTOR: mammalian target of rapamycin.

10.4 AMPK IN MUSCLE GROWTH AND DEVELOPMENT

Skeletal muscle hypertrophy is defined as an increase in muscle mass and is a result of an increase in the size of preexisting skeletal muscle fibers, as opposed to an increase in number (hyperplasia). At the molecular level, skeletal muscle hypertrophy results from an increase in the rate of protein synthesis and an inhibition of protein degradation (Glass 2005). AMPK has a role in both protein synthesis and degradation in skeletal muscle.

10.4.1 AMPK and Muscle Protein Synthesis

It is well documented that insulin and IGF-1 stimulate skeletal muscle growth, mainly mediated by the phosphoinositide-3 kinase/protein kinase B (PI3K/Akt) signaling pathway (Figure 10.1) (Bush et al. 2003; Latres et al. 2005; Park et al. 2005; Song et al. 2005; Subramaniam et al. 2005; Vary 2006). First, Akt promotes muscle growth through mammalian target of rapamycin (mTOR) (Shaw and Cantley 2006; Wullschleger et al. 2006). The proteins TSC1 and TSC2 form a functional complex that negatively regulates mTOR signaling and cell growth (Inoki et al. 2003; McManus and Alessi 2002). Akt phosphorylates TSC2 and inhibits the function of the TSC1/TSC2 complex, leading to activation of mTOR (Figure 10.1) (Inoki et al. 2003). Akt also directly phosphorylates mTOR at Ser 2448 (Chiang and Abraham 2005). The downstream targets of mTOR signaling are proteins involved in protein synthesis, including eukaryotic initiation factor 4E binding protein 1 (4E-BP1) and S6 kinase (S6K). In addition to activating mTOR, Akt activation also promotes protein synthesis through the inhibition of glycogen synthase kinase 3 (GSK3) activity (Cross et al.

1995; Hardt and Sadoshima 2002). GSK3 inhibits protein translation by phosphorylating eukaryotic initiation factor 2B (eIF2B) (Pap and Cooper 2002; Vary et al. 2002). GSK3 also phosphorylates the nuclear factor of activated T cells (NFAT), which excludes its entrance into the nucleus to induce muscle protein expression (Schulz and Yutzey 2004). In addition, insulin/IGF-1 signaling enhances extracellular signal-regulated kinase 1/2 (Erk1/2) signaling in skeletal muscle cells (Goodyear et al. 1996; Haddad and Adams 2004; Halevy and Cantley 2004), which contributes to insulin/IGF-1-induced muscle growth (Figure 10.1).

AMPK mediates the sensitivity of insulin signaling in skeletal muscle through at least three mechanisms:

1. AMPK phosphorylation of IRS-1 at Ser 789, which enhances the ability of IRS-1 to recruit downstream mediators, augmenting insulin signaling (Jakobsen et al. 2001) (Figure 10.1). However, this claim was later disputed (Qiao et al. 2002).
2. AMPK activation enhances fatty acid oxidation, which reduces intracellular fatty acids and their derivatives (Minokoshi et al. 2002). These fatty acid derivatives activate protein kinase C-θ (PKC-θ) signaling, which inhibits IRS-1-mediated signaling, leading to insulin resistance (Figure 10.1) (Kim et al. 2004a). The net result is that AMPK improves insulin sensitivity.
3. AMPK inhibits mTOR and S6K signaling (Aguilar et al. 2007). Because mTOR and S6K phosphorylate IRS-1 at Ser 636/639, which attenuates IRS-1 function and downstream insulin signaling (Khamzina et al. 2005; Um et al. 2004), the inhibition of mTOR/S6K by AMPK sensitizes insulin signaling (Figure 10.1).

In our previous study in pregnant sheep, maternal obesity and over-nutrition down-regulated insulin/IGF-1 signaling in fetal muscle, which was correlated with the down-regulation of AMPK in fetal muscle from obese mothers (Zhu et al. 2008). In rats, AMPK activity was reduced in the skeletal muscle of diet-induced obese rats and linked to insulin resistance (Sriwijitkamol et al. 2006). The role of AMPK in insulin signaling is also supported by recent reports that impaired insulin signaling accelerates muscle protein degradation (Wang et al. 2006). TNFα is known to induce insulin resistance and muscle atrophy (Ozes et al. 2001; Peraldi and Spiegelman 1998). Very recently, it was shown that TNFα induced insulin resistance through inhibition of AMPK (Steinberg et al. 2006b), thereby linking AMPK with insulin resistance and muscle atrophy.

mTOR signaling pathway regulates protein translation. AMPK negatively regulates protein translation through the inhibition of mTOR. Energy availability (AMP/ATP ratio in cells), working through AMPK, controls the activation of mTOR signaling and protein synthesis (Inoki et al. 2003) (Figure 10.1). The TSC complex is a target of AMPK. AMPK phosphorylates TSC2 at a site different from that phosphorylated by Akt. Unlike Akt, phosphorylation by AMPK enhances the stability of the TSC1/TSC2 complex and inhibits mTOR signaling (Corradetti et al. 2004; Inoki et al. 2003; Kimura et al. 2003). In addition, AMPK may also inhibit the activity of mTOR directly by phosphorylation at Thr 2446. The phosphorylation of mTOR at Thr 2446 and Ser 2448 is mutually exclusive and may act as a switch to integrate energy status with protein synthesis (Figure 10.1) (Cheng et al. 2004).

Furthermore, activated AMPK can directly phosphorylate eukaryotic elongation factor 2 (eEF2) kinase at several sites *in vitro*, including Ser 78, Ser 266, and Ser 398 (Browne et al. 2004; Browne and Proud 2004). Phosphorylation of eEF2 kinase at Ser 398 leads to activation of its kinase activity. Activated eEF2 kinase then phosphorylates and inhibits eEF2, which mediates translocation of the ribosome relative to the mRNA during mRNA translation (Browne and Proud 2002) and thus inhibits mRNA translation (Carlberg et al. 1990).

To summarize the role of AMPK in muscle protein synthesis, AMPK enhances protein synthesis through sensitization of insulin/IGF-1 signaling, but also negatively regulates protein translation through down-regulation of mTOR signaling. Sensitization of insulin/IGF-1 signaling by AMPK is associated with fatty acid oxidation and PKC-θ activation, which is a relatively slow response. However, AMPK inhibits mTOR and its associated signaling through direct phosphorylation, which is a fast, acute response. Therefore, the long-term effect of AMPK activation tends to promote muscle growth through sensitization of insulin/IGF-1 signaling, although the acute effect of AMPK activation is to inhibit protein synthesis.

10.4.2 AMPK and Muscle Protein Degradation

The insulin/IGF-1 signaling pathway also controls protein degradation (Figure 10.1). The first step in the degradation of myofibrillar proteins is their release from myofibrils. The aspirate-specific cysteinyl proteases (Caspase 3) and calpains play a major role in the release of myofibrillar proteins (Goll et al. 2003; McLoon et al. 2004; Menconi et al. 2004). Released myofibrillar proteins can either be re-assembled into myofibrils or degraded by ubiquitinization (Goll et al. 2003). The ubiquitin proteasome system degrades ubiquitinized myofibrillar proteins. Thus, ubiquitinization is a key control point in determining the fate of released myofibrillar proteins. Two ubiquitin ligases, Muscle Ring Finger 1 (MuRF 1) and Muscle Atrophy F-box (MAFbx), play essential roles in the ubiquitinization of myofibrillar proteins (Bodine et al. 2001). Their increased expression promotes apoptosis and muscle atrophy (Glass 2005). The Forkhead box-containing protein, O-subfamily (FOXO) transcription factors, are required for the up-regulation of MuRF1 and MAFbx expression. Transcription factors FOXO are downstream targets of Akt. Akt phosphorylates FOXO and leads to nuclear exclusion and blocks its function (Skurk et al. 2005), thus inhibiting muscle protein breakdown (Glass 2005).

Pharmacological activation of AMPK promotes the protein degradation through enhancing the expression of MuRF1 and MAFbx (Krawiec et al. 2007; Nakashima and Yakabe 2007). However, our data showed that IGF-1 in combination with a low degree of AMPK activation did not increase MuRF1 and MAFbx mRNA expression, and MuRF1 and MAFbx protein contents were not changed (unpublished data). Therefore, the effect of AMPK in protein degradation may be dependent on the level of AMPK activation, with a low degree of activation inhibiting protein degradation via the sensitization of insulin/IGF-1 signaling while a high degree of activation promotes protein degradation. The differential effect of low and high AMPK activation on downstream signaling was observed in a recent study in colon cancer cells (Park et al. 2006). Low levels of AMPK activation facilitates insulin/IGF-1 signaling inhibiting protein degradation, while high levels of AMPK activation promote protein degradation through the inhibition of mTOR. mTOR controls muscle protein degradation independent of FOXO (Latres et al. 2005).

10.5 AMPK IN ADIPOGENESIS AND FAT ACCUMULATION

10.5.1 AMPK in Lipid Metabolism

The key function of AMPK is to regulate the energy balance within the cell. Once activated, AMPK phosphorylates downstream substrates and the biological effects of AMPK are pleiotropic, switching off ATP-consuming processes such as biosynthetic pathways, and switching on catabolic processes to generate ATP, including glycolysis and fatty acid oxidation (Carling 2004; Hardie et al. 2006; Hardie et al. 2003). In addition to the acute regulation of energy metabolism, AMPK also alters both transcription (Foretz et al. 1998; Leclerc et al. 1998; Yang et al. 2001) and the translation of genes involved in energy metabolism (Fryer et al. 2002b; Winder et al. 2000a, b), exerting long-term effects (Carling 2004).

AMPK mediates lipid metabolism through regulating several key regulators of lipid metabolism. Acety-CoA carboxylase (ACC) and 3-hydroxyl-3-methylglutaryl (HMG)-CoA reductase were the first proteins identified as downstream targets of AMPK that are involved in lipogenesis (Beg et al. 1973; Carling et al. 1989; Carlson and Kim 1973; Henin et al. 1995; Park et al. 2002).

ACC is a rate-limiting enzyme for fatty acid synthesis (Carling et al. 1989; Carling et al. 1987). ACC catalyzes malonyl-CoA synthesis by the carboxylation of acetyl-CoA. Malonyl-CoA is the initial precursor for the *de novo* fatty acid synthesis, and also an important regulator of fatty acid oxidation via its allosteric inhibition of carnitine palmitoyl transferase 1 (CPT-1), which catalyzes the transport of fatty acids across the mitochondrial membrane, a rate-limiting step for fatty acid β-oxidation (McGarry and Brown 1997). The turnover of malonyl-CoA is controlled by the activity of both ACC and malonyl-CoA decarboxylase (MCD). ACC exists in two isoforms, ACC1 (ACCα) and ACC2 (ACCβ), which have different roles (Abuelheiga et al. 1995; Kim et al. 1998; Steinberg et al. 2006a). ACC1, present in the cytosol, is primarily responsible for fatty acid synthesis and is highly abundant in liver and adipose tissue (Iverson et al. 1990). ACC2, which is localized to the outer leaflet of mitochondria, plays a critical role in fatty acid β-oxidation and is the dominant isoform expressed in muscle (Abu-Elheiga et al. 2000).

ACC1 is one of the initial identified substrates of AMPK (Carling et al. 1987; Carlson and Kim 1973). AMPK phosphorylates ACC1 at multiple sites, corresponding to Ser 79, Ser 1200, and Ser 1215 (Davies et al. 1990; Munday et al. 1988). Phosphorylation of ACC1 at Ser 79, the only site exclusively phosphorylated by AMPK, correlates with inactivation of ACC1 (Ha et al. 1996). Incubation of isolated rat hepatocytes with AICAR inhibits fatty acid synthesis through phosphorylation and inhibition of ACC1 by activating AMPK (Corton et al. 1995; Henin et al. 1995). In adipocytes, AMPK inhibits ACC1, concomitant with a decreased lipogenic rate (Sullivan et al. 1994). Over-expression of a dominant negative form of AMPK in adipocytes precludes this phosphorylation of ACC1 at Ser 79 by AICAR (Daval et al. 2005). Endurance exercise activates AMPK, decreases malonyl-CoA concentration, and diminishes ACC activity in liver, adipose tissue, and muscle (Park et al. 2002). Exercise induces an increase in MCD activity, which further reduces malonyl-CoA concentration and decreases glycerol-3-phosphate acyl transferase (GPAT) activity, a key enzyme involved in triacylglycerol synthesis.

HMG-CoA reductase, an integral membrane protein of the smooth endoplasmic reticulum, catalyzes the committed and rate-limiting step in cholesterol biosynthesis, the conversion of HMG-CoA to mevalonate. HMG-CoA reductase, like ACC1, is phosphorylated and inactivated by AMPK when intracellular ATP falls (Beg et al. 1973; Ferrer et al. 1985). Activation of AMPK inhibits HMG-CoA activity and cholesterol biosynthesis in cultured rat hepatocytes (Corton et al. 1995; Henin et al. 1995).

AMP-activated protein kinase regulates triacylglycerol synthesis through GPAT. The first committed and rate-limiting step in phospholipid and triglyceride synthesis is catalyzed by GPAT. GPAT exists in two isoforms: microsomal GPAT and mitochondrial GPAT. Purified recombinant AMPK inhibits hepatic mitochondrial GPAT in a time- and ATP-dependent manner, but direct GPAT phosphorylation by AMPK was not observed, raising the possibility that an associated protein may be the substrate of AMPK responsible for the inhibition of GPAT (Muoio et al. 1999). Activation of AMPK by endurance exercise reduces mitochondrial GPAT activity without any effect on microsomal GPAT in liver and adipose tissue (Park et al. 2002). Over-expression of constitutively active AMPK by adenovirus infection reduces triglyceride esterification in primary human myotubes (Steinberg et al. 2006a). Activated AMPK in liver cells can also switch off the expression of enzymes involved in fatty acid biosynthesis, including fatty acid synthase and ACC1 (Woods et al. 2000). In short, AMPK can switch off fatty acid biosynthesis, both acutely (by phosphorylation of ACC1) and chronically (by reduced expression of fatty acid synthase and ACC1). In addition, AMPK directly regulates fatty acid oxidation. Therefore, cellular fatty acid level is very sensitive to AMPK regulation.

10.5.2 AMPK and Adipogenesis

As well as playing a very important role in fatty acid metabolism, AMPK has recently been implicated in the development of adipose tissue. As discussed in Chapter 3, transcriptional factors, peroxisome proliferators-activated receptor γ (PPARγ) and CCAAT/enhancer-binding protein α (C/EBPα), are crucial regulators of adipogenesis (Dagon et al. 2006a; Giri et al. 2006; Habinowski and Witters 2001). Considering that adipocyte differentiation and growth are energy-consuming processes that involve both lipid and protein synthesis, it is not surprising that AMPK activation inhibits adipocyte differentiation.

Skeletal muscle cells and adipocytes are both derived from mesenchymal pluripotent cells (Artaza et al. 2005; Poulos and Hausman 2006). In approximately mid-gestation, fetal skeletal muscle has a large number of pluripotent cells that can differentiate into either myogenic cells or adipogenic cells (Feve 2005; Gnanalingham et al. 2005; Muhlhausler et al. 2007). Enhancing adipogenesis in fetal muscle is expected to provide sites for fat accumulation in later life, increasing marbling (Tong et al. 2008). Previous studies in pigs and cattle indicate that fetal stage is important for the regulation of genes involved in intramuscular fat accumulation (Cagnazzo et al. 2006; Lehnert et al. 2007).

In fetal muscle through late gestation, there are no mature adipocytes (Casteilla et al. 1987; Lomax et al. 2007). However, the lack of mature adipocytes does not exclude the cells that are molecularly equipped to become adipocytes. Indeed, the initial events in adipogenesis start in mid-gestation (Feve 2005; Gnanalingham et al. 2005; Muhlhausler et al. 2007). The expression of PPARγ leads to adipogenic differentiation from pluripotent cells, and PPARγ is highly expressed in adipose tissue (Spiegelman et al. 2000). Preadipocyte factor-1 (Pref-1) is exclusively expressed in pre-adipocytes, not mature adipocytes (Fahrenkrug et al. 1999; Mei et al. 2002). Both PPARγ and Pref-1 are frequently used as markers for adipogenic studies.

AMPK activation inhibits the expression of PPARγ and C/EBPs in pre-adipocytes (Giri et al. 2006). Activation of AMPK in pre-adipocytes prevents adipocyte differentiation and blocks the expression of the late adipogenic markers, fatty acid synthase and ACC. We previously demonstrated that AMPK was negatively associated with marbling in beef cattle (Underwood et al. 2008). In 3T3-L1 cells, activation of AMPK induced a dose-dependent inhibition of PPARγ, eliciting that activation of AMPK inhibited adipogenesis in 3T3-L1 cells (Tong et al. 2008). These data in 3T3-L1 cells clearly demonstrate that AMPK has a crucial regulatory role in adipogenesis, which is in agreement with previous reports (Giri et al. 2006; Habinowski and Witters 2001). Furthermore, it has also been proposed that hypoxia, genistein, epigallocatechin gallate, or capsaicin inhibits adipocyte differentiation and proliferation via activating AMPK (Hwang et al. 2005; Kim et al. 2005).

It is well established that PPARγ expression induces adipogenic differentiation. This receptor functions as an obligate heterodimer with retinoid X receptor, and binds to DNA sequences called direct repeat (DR)-1 elements, which induces the expression of adipocyte-specific genes (Spiegelman et al. 2000). However, it remains unclear how AMPK regulates PPARγ expression and adipogenesis. Only a few reports have examined the possible mechanisms associated with AMPK in adipogenesis (Giri et al. 2006; Habinowski and Witters 2001). Activation of AMPK by AICAR induced down-regulation of key adipogenic genes, which suggested that this effect was mediated by phosphorylation of eukaryotic initiation factor 2α (eIF2α) (Dagon et al. 2006b). In summary, AMPK regulates adipogenesis but the underlying mechanisms remain vague (Dagon et al. 2006b).

10.6 AMPK AND WHOLE-BODY ENERGY METABOLISM

Much of the previous work on AMPK focused on its regulation of energy level within individual cells. Recently, it has become clear that AMPK also regulates whole-body energy metabolism in response to hormones and nutrient signals. Two adipocyte-derived hormones (leptin and adiponectin) have opposite effects on AMPK activity: activating AMPK in skeletal muscle but inhibiting AMPK in hypothalamus. These two hormones are secreted by adipocytes and thought to represent signals that body lipid stores are adequate. Leptin and adiponectin activates AMPK in skeletal muscle to increase fatty acid oxidation, whereas inhibition of AMPK in hypothalamus by these two hormones is responsible for their anorexigenic effects: reducing food intake. Different responses of

AMPK to leptin and adiponectin in the central nervous system and peripheral tissues are additive to the control of fat accumulation.

Increasing evidence demonstrates that hypothalamic AMPK activity regulates food intake and fat accumulation. Fasting increases AMPK activity in multiple hypothalamic regions, whereas feeding inhibits its activity (Minokoshi et al. 2004). Fasting and feeding regulation of AMPK in hypothalamus is mediated by the accompanied oscillations of hormones and nutrients, especially glucose. The anorexigenic hormones, leptin and insulin, are excreted after feeding, which represses food intake through the inhibition of AMPK, whereas AMPK in the hypothalamus is activated by ghrelin (Andersson et al. 2004; Minokoshi et al. 2004), a gut hormone of which the circulating level rises before a meal and falls following food ingestion. Ghrelin stimulates food intake. Over-expression of constitutively active AMPK by adenovirus injection in the medial basal hypothalamus increases expression of orexigenic neuropeptides, neuropeptide Y and agouti-related peptide, in the arcuate nucleus of hypothalamus, whereas over-expression of dominant negative AMPK suppresses the expression of these fasting-induced orexigenic neuropeptides in the hypothalamus (Minokoshi et al. 2004). Also, direct injection of AICAR into the hypothalamus increases hypothalamic AMPK activity and food intake (Andersson et al. 2004). Furthermore, the anorexigenic effect of ALA is prevented by intracerebroventricular injection of AICAR or by over-expressing a constitutively active AMPK in the hypothalamus (Kim et al. 2004b). This suggests that hypothalamic AMPK plays an important role in food intake regulation. The effects of these hormones on AMPK, coupled with its response to nutrient signal in both the central nervous system and periphery, provide evidence that this kinase is a crucial regulator of whole-body energy metabolism.

In addition to the acute effects on energy metabolism, AMPK regulates the expression of numerous genes. In liver cells, activation of AMPK decreases the expression of phosphoenolpyruvate carboxykinase and glucose-6-phosphatase, both of which are enzymes involved in gluconeogenesis (Lochhead et al. 2000; Yamauchi et al. 2002). In muscle cells, AMPK activation up-regulates the expression of glucose transporter-4 (GLUT4), hexokinase, and mitochondrial enzymes (Zheng et al. 2001). Similar changes in the expression of these genes are also seen in response to endurance training or to chronic ATP depletion in muscle by feeding rodents with a creatine analogue β-guanidinopropionic acid. Through up-regulating GLUT4, hexokinase and mitochondrial biogenesis via the AMPK pathway, the capacity of muscle to take up glucose and oxidize glucose and fatty acids is enhanced.

10.7 AMPK, GLYCOLYSIS, AND PSE MEAT

AMPK activation in response to physiological and pharmacological stimuli increases glucose metabolism in skeletal muscle by stimulating glucose uptake and glycolysis. The involvement of AMPK in glucose metabolism is first evidenced by a study showing that AICAR activation of AMPK in perfused rat hindlimb muscles increases both fatty acid oxidation and glucose uptake (Merrill et al. 1997). There is a strong correlation between AMPK activation and the increase in glucose uptake (Bergeron et al. 1999; Hayashi et al. 2000; Hayashi et al. 1998; Musi et al. 2001; Tian et al. 2001). This increased glucose uptake in response to AICAR stimulation, muscle contraction, or hypoxia is blocked by over-expression of a dominant-negative AMPK in skeletal muscle of AMPK kinase-inactive mice (Fryer et al. 2002b; Mu et al. 2001; Sakoda et al. 2002), indicating that AMPK is the mediator of increased glucose uptake. AMPK stimulates muscle glucose uptake in several ways: (1) AMPK activation increases the intrinsic activity of the glucose transporter GLUT1 by an unknown mechanism (Barnes et al. 2002); (2) AMPK enhances GLUT4 translocation from the intracellular storage pool to the plasma membrane (Kurth-Kraczek et al. 1999; Russell et al. 1999); and (3) AMPK activation up-regulates the transcription and/or translation of GLUTs (Fryer et al. 2002b; Holmes et al. 1999; Zheng et al. 2001).

AMPK also increases the intracellular glucose level by phosphorylating and activating phosphorylase kinase, which then phosphorylates and activates glycogen phosphorylase, leading to glycogenolysis (Fraser et al. 1999; Russell et al. 1999; Young et al. 1997).

Concomitant with increased glucose uptake and glycogenolysis, another mechanism by which activated AMPK increases glucose metabolism is the phosphorylation (Ser 466) and activation of phosphofructokinase-2 (PFK-2) (Marsin et al. 2000). PFK-2 catalyzes the production of fru-2, 6-P_2, which is a potent allosteric activator of PFK-1, the most important rate-controlling enzyme in glycolysis. *In vitro* phosphorylation of PFK-2 by purified AMPK further confirms the direct involvement of AMPK in glycolysis (Marsin et al. 2000). Based on the previous evidence, AMPK plays an important role at various stages of glucose metabolism to maintain intracellular energy levels.

As discussed in Chapter 9, PSE meat is a significant problem facing the meat industry. In our studies, we found that AMPK was activated earlier and more easily in muscle that became PSE meat compared to normal meat, showing that AMPK is associated with PSE (Shen et al. 2006d). To further assess the role of AMPK in postmortem glycolytic regulation, we used pre-slaughter forced swimming, intraperitoneal injection of AICAR, and a skeletal-muscle-specific AMPK knockout to alter the AMPK activity in postmortem mouse muscle (Shen and Du 2005; Shen et al. 2006b). AMPK activity is correlated with postmortem glycolysis (Shen et al. 2006c). Using a specific β-adrenoceptor blocker, propranolol, to block β-adrenoceptor, we demonstrated that stresses can accelerate postmortem glycolysis through β-adrenoceptor mediated AMPK activation (Du et al. 2005).

Pre-slaughter stress and the halothane gene are reported to increase the incidence of PSE meat. Studies in pigs showed that pre-slaughter transport and the halothane gene accelerated ATP depletion, leading to a lower energy status in muscle early postmortem, which was evidenced by a lower ATP concentration and a higher (AMP+IMP)/ATP ratio (Shen et al. 2007a). The lower energy status leads to earlier and fast AMPK activation, resulting in elevated postmortem glycolytic rates and increased PSE incidence. Higher fructo-2,6-biphosphate concentration in postmortem muscle with higher AMPK activity and glycolytic rate indicated that AMPK increased postmortem glycolysis by phosphorylating and activating PFK-2 (Shen et al. 2007a).

10.8 *RN⁻* PHENOTYPE IN HAMPSHIRE PIGS: A CLEAR DEMONSTRATION FOR THE ROLE OF AMPK IN MUSCLE GROWTH, FAT ACCUMULATION, AND MEAT QUALITY

Pigs have been selected for enhanced muscle growth and reduced fat deposition for generations. However, intensive genetic selection for pigs with high lean growth efficiency unintentionally leads to the spread of RN^- (*Rendement Napole*) gene. Based on the presence of RN^- gene, pigs can be separated into normal pigs (rn^+/rn^+), RN^- carriers (RN^-/rn^+), and RN^- positive (RN^-/RN^-). The RN^- gene is a dominant gene (Lebret et al. 1999b). The incidence of the RN^- gene in U.S. pigs, especially Hampshire pigs, is high (Miller et al. 2000b). Carcasses of RN^- gene carrier (RN^-/rn$^+$) pigs had decreased fat depth and a greater lean percentage than carcasses of normal (rn^+/rn^+) pigs, while no difference was observed between RN^-/rn$^+$ and RN^-/RN^- pigs due to the dominant effect of the RN^- gene (Carr et al. 2006; Enfalt et al. 1997; Moeller et al. 2003). In addition, pigs genotyped RN^-/rn$^+$ grew faster than rn^+/rn^+ pigs (Enfalt et al. 1997).

Recently, the RN^- gene was identified to result from a mutation in AMP-activated protein kinase (AMPK) γ3-subunit (Milan et al. 2000). Based on studies of a similar mutation of the AMPK γ-subunit in cardiac muscle and other related studies, this mutation leads to a constantly active AMPK (Daniel and Carling 2002; Sanders et al. 2007; Scott et al. 2004). In addition, this mutation was consistently shown to render mutated AMPK unresponsive to AMP by disrupting the AMP binding site (Arad et al. 2002; Daniel and Carling 2002; Sidhu et al. 2005). As the major function of AMPK *in vivo* is to sense the energy demand in cells through an AMP-dependent activation, the mutation in the γ-subunit renders it unresponsive to AMP and thus is a loss-of-function mutation (Daniel and Carling 2002; Sidhu et al. 2005).

The RN^- phenotype is characterized by a 70% increase in glycogen content in skeletal muscle (Estrade et al. 1993). Increased glycolytic potential is frequently used to identify RN^- phenotype

(Hamilton et al. 2003; Larzul et al. 1998; Miller et al. 2000a). The RN^- phenotype is associated with "acid meat" and RSE (red, soft, and exudative) pork from Hampshire pigs (Josell et al. 2003; Le Roy et al. 2000; Lebret et al. 1999a; Milan et al. 1996; Van Laack and Kauffman 1999). Meat from RN^- pigs has a low ultimate pH (24 hr postmortem), a reduced water-holding capacity, and reduced cooking yield (Enfalt et al. 1997; Estrade et al. 1993).

The mutation in the AMPK γ-subunit leads to hypertrophy of cardiac and skeletal muscle (Hardie and Hawley 2001; Milan et al. 2000; Murphy et al. 2005; Shen et al. 2006a; Sidhu et al. 2005). In Hampshire pigs, a point mutation, Arg^{200} to Gln in the AMPK γ3-subunit, results in the RN^- phenotype characterized by increased lean growth and inferior meat quality (Andersson 2003; Enfalt et al. 1997; Hardie and Hawley 2001; Milan et al. 2000). Interestingly, a mutation at the same location in the AMPK γ2-subunit, which is highly expressed in cardiac muscle, results in cardiac hypertrophy (Ashrafian et al. 2003; Murphy et al. 2005; Oliveira et al. 2003; Sidhu et al. 2005; Spindler et al. 1998). Hearts from mice with AMPK γ2-subunit mutation had an average weight of 297 mg while those from wildtype mice had an average weight of only 140 mg ($p < 0.001$) (Sidhu et al. 2005). The role of AMPK in muscle growth is further evidenced by the muscle fiber type specific hypertrophy of RN^- pigs. A larger cross-sectional area of Type IIa and IIb fibers was observed in longissimus dorsi muscle of RN^- pigs compared to normal (rn^+/rn^+) pigs, with no difference in the area of Type I fibers (Lebret et al. 1999b). This coincides with the fiber type specific expression of the AMPK γ3-subunit. The AMPK γ3-subunit has high expression levels in Type IIb and Type IIa fibers, whereas its expression is undetectable in Type I fibers (Yu et al. 2004).

The constantly active AMPK in the skeletal muscle of RN^- pigs increases fatty acid oxidation and decreases fatty acid synthesis, which reduces fatty acids available for adipose tissue and leads to the repartitioning of nutrients from fat accumulation to lean growth, enhancing growth efficiency in RN^- pigs. In short, the AMPK constantly active mutation enhances muscle growth, reduces fat deposition, and impairs the meat quality of RN^- pigs, providing a good summary for the role of AMPK in muscle growth, fat deposition, and meat quality.

10.9 SUMMARY

AMPK has key roles in regulating skeletal muscle growth, fat deposition, and meat quality. Therefore, AMPK provides a useful molecular target for improving animal performance. Recent studies show that AMPK inhibits adipogenesis in skeletal muscle, which is linked to marbling, a very important factor in determining meat quality. However, our understanding of AMPK in adipogenesis remains rudimentary. The role of AMPK in animal performance remains poorly understood; more studies are apparently warranted to understand how AMPK is involved in live performance.

REFERENCES

Abu-Elheiga, L., W.R. Brinkley, L. Zhong, S.S. Chirala, G.Woldegiorgis, and S.J. Wakil. 2000. The subcellular localization of acetyl-CoA carboxylase 2. *Proc. Natl. Acad. Sci. U.S.A.* 97:1444–1449.

Abuelheiga, L., A. Jayakumar, A. Baldini, S.S. Chirala, and S.J. Wakil. 1995. Human acetyl-CoA carboxylase — characterization, molecular-cloning, and evidence for 2 isoforms. *Proc. Natl. Acad. Sci. U.S.A.* 92:4011–4015.

Aguilar, V., S. Alliouachene, A. Sotiropoulos, A. Sobering, Y. Athea, F. Djouadi, S. Miraux, E. Thiaudiere, M. Foretz, B. Viollet, P. Diolez, J. Bastin, P. Benit, P. Rustin, D. Carling, M. Sandri, R. Ventura-Clapier, and M. Pende. 2007. S6 kinase deletion suppresses muscle growth adaptations to nutrient availability by activating AMP kinase. *Cell Metab.* 5:476–487.

Anderson, K.A., RL. Means, Q.H. Huang, B.E. Kemp, E.G. Goldstein, M.A. Selbert, A.M. Edelman, R.T. Fremeau, and A.R. Means. 1998. Components of a calmodulin-dependent protein kinase cascade. Molecular cloning, functional characterization and cellular localization of Ca2+/calmodulin-dependent protein kinase kinase beta. *J. Biol. Chem.* 273:31880–31889.

Andersson, L. 2003. Identification and characterization of AMPK gamma 3 mutations in the pig. *Biochem. Soc. Trans.* 31:232–235.

Andersson, U., K. Filipsson, C.R. Abbott, A. Woods, K. Smith, S.R. Bloom, D. Carling, and C. J. Small. 2004. AMP-activated protein kinase plays a role in the control of food intake. *J. Biol. Chem.* 279:12005–12008.

Arad, M., D.W. Benson, A.R. Perez-Atayde, W.J. McKenna, E.A. Sparks, R.J. Kanter, K. McGarry, J.G. Seidman, and C.E. Seidman. 2002. Constitutively active AMP kinase mutations cause glycogen storage disease mimicking hypertrophic cardiomyopathy. *J. Clin. Invest.* 109:357–362.

Artaza, J.N., S. Bhasin, T.R. Magee, S. Reisz-Porszasz, R. Shen, N.P. Groome, M.F. Meerasahib, and N.F. Gonzalez-Cadavid. 2005. Myostatin inhibits myogenesis and promotes adipogenesis in C3H 10T(1/2) mesenchymal multipotent cells. *Endocrinology* 146:3547–3557.

Ashrafian, H., C. Redwood, E. Blair, and H. Watkins. 2003. Hypertrophic cardiomyopathy: a paradigm for myocardial energy depletion. *Trends Genet.* 19:263–268.

Banerjee, R.R., S.M. Rangwala, J.S. Shapiro, A.S. Rich, B. Rhoades, Y. Qi, J. Wang, M.W. Rajala, A. Pocai, P.E. Scherer, C.M. Steppan, R.S. Ahima, S. Obici, L. Rossetti, and M. A. Lazar. 2004. Regulation of fasted blood glucose by resistin. *Science* 303:1195–1198.

Barnes, K., J.C. Ingram, O.H. Porras, L.F. Barros, E.R. Hudson, L.G.D. Fryer, F. Foufelle, D. Carling, D.G. Hardie, and S.A. Baldwin. 2002. Activation of GLUT1 by metabolic and osmotic stress: potential involvement of AMP-activated protein kinase (AMPK). *J. Cell Sci.* 115:2433–2442.

Beg, Z.H., D.W. Allmann, and D.M. Gibson. 1973. Modulation of 3-hydroxy-3-methylglutaryl Coenzyme A reductase-activity with cAMP and with protein fractions of rat-liver cytosol. *Biochem. Biophys. Res. Commun.* 54:1362–1369.

Bergeron, R., R.R. Russell, L.H. Young, J.M. Ren, M. Marcucci, A. Lee, and G.I. Shulman. 1999. Effect of AMPK activation on muscle glucose metabolism in conscious rats. *Am. J. Physiol. – Endocrinol. Metab.* 276:E938–E944.

Beri, R.K., A.E. Marley, C.G. See, W.F. Sopwith, K. Aguan, D. Carling, J. Scott, and F. Carey. 1994. Molecular-cloning, expression and chromosomal localization of human AMP-activated. *FEBS Lett.* 356:117–121.

Bodine, S.C., E. Latres, S. Baumhueter, V.K. Lai, L. Nunez, B.A. Clarke, W.T. Poueymirou, F.J. Panaro, E. Na, K. Dharmarajan, Z.Q. Pan, D.M. Valenzuela, T.M. DeChiara, T.N. Stitt, G.D. Yancopoulos, and D.J. Glass. 2001. Identification of ubiquitin ligases required for skeletal muscle atrophy. *Science* 294:1704–1708.

Browne, G.J., S.G. Finn, and C.G. Proud. 2004. Stimulation of the AMP-activated protein kinase leads to activation of eukaryotic elongation factor 2 kinase and to its phosphorylation at a novel site, Serine 398. *J. Biol. Chem.* 279:12220–12231.

Browne, G.J. and C.G. Proud. 2002. Regulation of peptide-chain elongation in mammalian cells. *Eur. J. Biochem.* 269:5360–5368.

Browne, G.J. and C.G. Proud. 2004. A novel mTOR-regulated phosphorylation site in elongation factor 2 kinase modulates the activity of the kinase and its binding to calmodulin. *Mol. Cell. Biol.* 24:2986–2997.

Bush, J.A., S.R. Kimball, P.M. O'Connor, A. Suryawan, R.A. Orellana, H.V. Nguyen, L.S. Jefferson, and T.A. Davis. 2003. Translational control of protein synthesis in muscle and liver of growth hormone-treated pigs. *Endocrinology* 144:1273–1283.

Cagnazzo, M., M.F. te Pas, J. Priem, A.A. de Wit, M.H. Pool, R. Davoli, and V. Russo. 2006. Comparison of prenatal muscle tissue expression profiles of two pig breeds differing in muscle characteristics. *J. Anim. Sci.* 84:1–10.

Carlberg, U., A. Nilsson, and O. Nygard. 1990. Functional properties of phosphorylated elongation factor-II. *Eur. J. Biochem.* 191:639–645.

Carling, D. 2004. The AMP-activated protein kinase cascade — a unifying system for energy control. *Trends Biochem. Sci.* 29:18–24.

Carling, D., P.R. Clarke, V.A. Zammit, and D.G. Hardie. 1989. Purification and characterization of the AMP-activated protein kinase — copurification of acetyl-CoA carboxylase kinase and 3-hydroxy-3-methylglutaryl-CoA reductase kinase activities. *Eur. J. Biochem.* 186:129–136.

Carling, D., V.A. Zammit, and D.G. Hardie. 1987. A common bicyclic protein-kinase cascade inactivates the regulatory enzymes of fatty-acid and cholesterol-biosynthesis. *FEBS Lett.* 223:217–222.

Carlson, C.A. and K. Kim. 1973. Regulation of hepatic acetyl coenzyme-A carboxylase by phosphorylation and dephosphorylation. *J. Biol. Chem.* 248:378–380.

Carr, C.C., J.B. Morgan, E.P. Berg, S.D. Carter, and F.K. Ray. 2006. Growth performance, carcass composition, quality, and enhancement treatment of fresh pork identified through deoxyribonucleic acid marker-assisted selection for the *Rendement Napole* gene. *J. Anim. Sci.* 84:910–917.

Casteilla, L., C. Forest, J. Robelin, D. Ricquier, A. Lombet, and G. Ailhaud. 1987. Characterization of mitochondrial-uncoupling protein in bovine fetus and newborn calf. *Am. J. Physiol.* 252:E627–636.

Cheng, S.W., L.G. Fryer, D. Carling, and P.R. Shepherd. 2004. Thr2446 is a novel mammalian target of rapamycin (mTOR) phosphorylation site regulated by nutrient status. *J. Biol. Chem.* 279:15719–15722.

Cheng, Z., T. Pang, M. Gu, A.H. Gao, C.M. Xie, J.Y. Li, F.J. Nan, and J. Li. 2006. Berberine-stimulated glucose uptake in L6 myotubes involves both AMPK and p38 MAPK. *Biochim. Biophys. Acta* 1760:1682–1689.

Cheung, P.C.F., I.P. Salt, S.P. Davies, D.G. Hardie, and D. Carling. 2000. Characterization of AMP-activated protein kinase gamma-subunit isoforms and their role in AMP binding. *Biochem. J.* 346:659–669.

Chiang, G.G. and R.T. Abraham. 2005. Phosphorylation of mammalian target of rapamycin (mTOR) at Ser-2448 is mediated by p70S6 kinase. *J. Biol. Chem.* 280:25485–25490.

Corradetti, M.N., K. Inoki, N. Bardeesy, R.A. DePinho, and K.L. Guan. 2004. Regulation of the TSC pathway by LKB1: evidence of a molecular link between tuberous sclerosis complex and Peutz-Jeghers syndrome. *Genes Dev.* 18:1533–1538.

Corton, J.M., J.G. Gillespie, S.A. Hawley, and D.G. Hardie. 1995. 5-Aminoimidazole-4-carboxamide ribonucleoside — a specific method for activating AMP-activated protein-kinase in intact-cells. *Eur. J. Biochem.* 229:558–565.

Cross, D.A., D.R. Alessi, P. Cohen, M. Andjelkovich, and B.A. Hemmings. 1995. Inhibition of glycogen synthase kinase-3 by insulin mediated by protein kinase B. *Nature* 378:785–789.

Crute, B.E., K. Seefeld, J. Gamble, B.E. Kemp, and L.A. Witters. 1998. Functional domains of the alpha1 catalytic subunit of the AMP-activated protein kinase. *J. Biol. Chem.* 273:35347–35354.

Dagon, Y., Y. Avraham, and E.M. Berry. 2006. AMPK activation regulates apoptosis, adipogenesis, and lipolysis by eIF2 alpha in adipocytes. *Biochem. Biophys. Res. Commun.* 340:43–47.

Daniel, T. and D. Carling. 2002. Functional analysis of mutations in the gamma 2 subunit of AMP-activated protein kinase associated with cardiac hypertrophy and Wolff-Parkinson-White syndrome. *J. Biol. Chem.* 277:51017–51024.

Daval, M., F. Diot-Dupuy, R. Bazin, I. Hainault, B. Viollet, S. Vaulont, E. Hajduch, P. Ferre, and F. Foufelle. 2005. Antilipolytic action of AMP-activated protein kinase in rodent adipocytes. *J. Biol. Chem.* 280:25250–25257.

Davies, S.P., N.R. Helps, P.T. Cohen, and D.G. Hardie. 1995. 5′-AMP inhibits dephosphorylation, as well as promoting phosphorylation, of the AMP-activated protein kinase. Studies using bacterially expressed human protein phosphatase-2C alpha and native bovine protein phosphatase-2AC. *FEBS Lett.* 377:421–425.

Davies, S.P., A.T.R. Sim, and D.G. Hardie. 1990. Location and function of 3 sites phosphorylated on rat acetyl-CoA carboxylase by the AMP-activated protein-kinase. *Eur. J. Biochem.* 187:183–190.

Du, M., Q.W. Shen, and M.J. Zhu. 2005. Role of alpha-adrenoceptor signaling and AMP-activated protein kinase in glycolysis of postmortem skeletal muscle. *J. Agric. Food Chem.* 53:3235–3239.

Enfalt, A.C., K.Lundstrom, I. Hansson, S. Johansen, and P.E. Nystrom. 1997c. Comparison of non-carriers and heterozygous carriers of the RN(−) allele for carcass composition, muscle distribution and technological meat quality in Hampshire-sired pigs. *Livestock Production Sci.* 47:221–229.

Estrade, M., X. Vignon, E. Rock, and G. Monin. 1993. Glycogen hyperaccumulation in white muscle-fibers of RN− carrier pigs — a biochemical and ultrastructural study. *Compar. Biochem. and Physiol. B – Biochem. Mol. Biol.* 104:321–326.

Fahrenkrug, S.C., B.A. Freking, and T.P. Smith. 1999. Genomic organization and genetic mapping of the bovine PREF-1 gene. *Biochem. Biophys. Res. Commun.* 264:662–667.

Ferrer, A., C. Caelles, N. Massot, and F.G. Hegardt. 1985. Activation of rat-liver cytosolic 3-hydroxy-3-methylglutaryl Coenzyme A reductase kinase by adenosine 5′-monophosphate. *Biochem. Biophysical. Res. Commun.* 132:497–504.

Feve, B. 2005. Adipogenesis: cellular and molecular aspects. *Best Pract. Res. Clin. Endocrinol. Metab.* 19:483–499.

Foretz, M., D. Carling, C. Guichard, P. Ferre, and F. Foufelle. 1998. AMP-activated protein kinase inhibits the glucose-activated expression of fatty acid synthase gene in rat hepatocytes. *J. Biol. Chem.* 273:14767–14771.

Fraser, H., G.D. Lopaschuk, and A.S. Clanachan. 1999. Alteration of glycogen and glucose metabolism in ischaemic and post-ischaemic working rat hearts by adenosine A(1) receptor stimulation. *Br. J. Pharmacol.* 128:197–205.

Fryer, L.G., A. Parbu-Patel, and D. Carling. 2002a. The anti-diabetic drugs rosiglitazone and metformin stimulate AMP-activated protein kinase through distinct signaling pathways. *J. Biol. Chem.* 277:25226–25232.

Fryer, L.G.D., F. Foufelle, K. Barnes, S.A. Baldwin, A. Woods, and D. Carling. 2002b. Characterization of the role of the AMP-activated protein kinase in the stimulation of glucose transport in skeletal muscle cells. *Biochem. J.* 363:167–174.

Giri, S., R. Rattan, E. Haq, M. Khan, R. Yasmin, J.S. Won, L. Key, A.K. Singh, and I. Singh. 2006a. AICAR inhibits adipocyte differentiation in 3T3L1 and restores metabolic alterations in diet-induced obesity mice model. *Nutr. Metab.* 3:31–51.

Glass, D.J. 2005. Skeletal muscle hypertrophy and atrophy signaling pathways. *Int. J. Biochem. Cell Biol.* 37:1974–1984.

Gnanalingham, M.G., A. Mostyn, M.E. Symonds, and T. Stephenson. 2005. Ontogeny and nutritional programming of adiposity in sheep: potential role of glucocorticoid action and uncoupling protein-2. *Am. J. Physiol. Regul. Integr. Comp. Physiol.* 289:R1407–1415.

Goll, D.E., V.F. Thompson, H. Li, W. Wei, and J. Cong. 2003. The calpain system. *Physiol. Rev.* 83:731–801.

Goodyear, L.J., P.Y. Chang, D.J. Sherwood, S.D. Dufresne, and D.E. Moller. 1996. Effects of exercise and insulin on mitogen-activated protein kinase signaling pathways in rat skeletal muscle. *Am. J. Physiol.*271:E403–E408.

Ha, J., J.K. Lee, K.S. Kim, L.A. Witters, and K.H. Kim. 1996. Cloning of human acetyl-CoA carboxylase-beta and its unique features. *Proc. Natl. Acad. Sci. U.S.A.* 93:11466–11470.

Habinowski, S.A. and L.A. Witters. 2001a. The effects of AICAR on adipocyte differentiation of 3T3-L1 cells. *Biochem. Biophys. Res. Commun.* 286:852–856.

Haddad, F. and G.R. Adams. 2004. Inhibition of MAP/ERK kinase prevents IGF-I-induced hypertrophy in rat muscles. *J. Appl. Physiol.* 96:203–210.

Halevy, O. and L.C. Cantley. 2004. Differential regulation of the phosphoinositide 3-kinase and MAP kinase pathways by hepatocyte growth factor vs. insulin-like growth factor-I in myogenic cells. *Exp. Cell Res.* 297:224–234.

Hamilton, D.N., K.D. Miller, M. Ellis, F.K. McKeith, and E.R. Wilson. 2003. Relationships between longissimus glycolytic potential and swine growth performance, carcass traits, and pork quality. *J. Anim. Sci.* 81:2206–2212.

Hardie, D.G. 2007. Biochemistry. Balancing cellular energy. *Science* 315:1671–1672.

Hardie, D.G., and D. Carling. 1997. The AMP-activated protein kinase — fuel gauge of the mammalian cell? *Eur. J. BioChem.* 246:259–273.

Hardie, D.G., D. Carling, and M. Carlson. 1998. The AMP-activated/SNF1 protein kinase subfamily: metabolic sensors of the eukaryotic cell? *Annu. Rev. Biochem.* 67:821–855.

Hardie, D.G., D. Carling, and A.T.R. Sim. 1989. The AMP-activated protein-kinase — a multisubstrate regulator of lipid metabolism. *Trends Biochem. Sci.* 14:20–23.

Hardie, D.G. and S.A. Hawley. 2001. AMP-activated protein kinase: the energy charge hypothesis revisited. *Bioessays* 23:1112–1119.

Hardie, D.G., S.A. Hawley, and J. Scott. 2006. AMP-activated protein kinase — development of the energy sensor concept. *J. Physiol. – London* 574:7–15.

Hardie, D.G., I.P. Salt, S.A. Hawley, and S.P. Davies. 1999. AMP-activated protein kinase: an ultrasensitive system for monitoring cellular energy charge. *Biochem. J.* 338: –717–722.

Hardie, D.G., J.W. Scott, D.A. Pan, and E.R. Hudson. 2003. Management of cellular energy by the AMP-activated protein kinase system. *FEBS Lett.* 546:113–120.

Hardt, S.E. and J. Sadoshima. 2002. Glycogen synthase kinase-3beta: a novel regulator of cardiac hypertrophy and development. *Circulation Res.* 90:1055–1063.

Hawley, S.A., J. Boudeau, J.L. Reid, K.J. Mustard, L. Udd, T.P. Makela, D.R. Alessi, and D.G. Hardie. 2003. Complexes between the LKB1 tumor suppressor, STRAD alpha/beta and MO25 alpha/beta are upstream kinases in the AMP-activated protein kinase cascade. *J. Biol.* 2:28.

Hawley, S.A., M. Davison, A. Woods, S.P. Davies, R.K. Beri, D. Carling, and D.G. Hardie. 1996. Characterization of the AMP-activated protein kinase kinase from rat liver and identification of threonine 172 as the major site at which it phosphorylates AMP-activated protein kinase. *J. Biol. Chem.* 271:27879–27887.

Hawley, S.A., A.E. Gadalla, G.S. Olsen, and D.G. Hardie. 2002. The antidiabetic drug metformin activates the AMP-activated protein kinase cascade via an adenine nucleotide-independent mechanism. *Diabetes* 51:2420–2425.

Hawley, S.A., D.A. Pan, K.J. Mustard, L. Ross, J. Bain, A.M. Edelman, B.G. Frenguelli, and D.G. Hardie. 2005. Calmodulin-dependent protein kinase kinase-beta is an alternative upstream kinase for AMP-activated protein kinase. *Cell Metab.* 2:9–19.

Hawley, S.A., M.A. Selbert, E.G. Goldstein, A.M. Edelman, D. Carling, and D.G. Hardie. 1995. 5′-AMP activates the AMP-activated protein kinase cascade and Ca2+/calmodulin activates the calmodulin-dependent protein kinase-I cascade, via 3 independent mechanisms. *J. Biol. Chem.* 270:27186–27191.

Hayashi, T., M.F. Hirshman, N. Fujii, S.A. Habinowski, L.A. Witters, and L.J. Goodyear. 2000. Metabolic stress and altered glucose transport — activation of AMP-activated protein kinase as a unifying coupling mechanism. *Diabetes* 49:527–531.

Hayashi, T., M.F. Hirshman, E.J. Kurth, W.W. Winder, and L.J. Goodyear. 1998. Evidence for 5′AMP-activated protein kinase mediation of the effect of muscle contraction on glucose transport. *Diabetes* 47:1369–1373.

Henin, N., M.F. Vincent, H.E. Gruber, and G. Vandenberghe. 1995. Inhibition of fatty-acid and cholesterol-synthesis by stimulation of AMP-activated protein-kinase. *FASEB J.* 9:541–546.

Holmes, B.F., E.J. Kurth-Kraczek, and W.W. Winder. 1999. Chronic activation of 5′-AMP-activated protein kinase increases GLUT-4, hexokinase, and glycogen in muscle. *J. Appl. Physiol.* 87:1990–1995.

Hurley, R.L., K.A. Anderson, J.M. Franzone, B.E. Kemp, A.R. Means, and L.A. Witters. 2005. The Ca2+/calmodulin-dependent protein kinase kinases are AMP-activated protein kinase kinases. *J. Biol. Chem.* 280:29060–29066.

Hwang, J.T., I.J. Park, J.I. Shin, Y.K. Lee, S.K. Lee, H.W. Baik, J. Ha, and O.J. Park. 2005. Genistein, EGCG, and capsaicin inhibit adipocyte differentiation process via activating AMP-activated protein kinase. *Biochem. Biophys. Res. Commun.* 338:694–699.

Inoki, K., T. Zhu, and K.L. Guan. 2003. TSC2 mediates cellular energy response to control cell growth and survival. *Cell* 115:577–590.

Iverson, A.J., A. Bianchi, A.C. Nordlund, and L.A. Witters. 1990. Immunological analysis of acetyl-CoA carboxylase mass, tissue distribution and subunit composition. *Biochem. J.* 269:365–371.

Jakobsen, S.N., D.G. Hardie, N. Morrice, and H.E. Tornqvist. 2001. 5′-AMP-activated protein kinase phosphorylates IRS-1 on Ser-789 in mouse C2C12 myotubes in response to 5-aminoimidazole-4-carboxamide riboside. *J. Biol. Chem.* 276:46912–46916.

Jenne, D.E., H. Reimann, J. Nezu, W. Friedel, S. Loff, R. Jeschke, D. Muller, W. Back, and M. Zimmer. 1998. Peutz-Jeghers syndrome is caused by mutations in a novel serine threonine kinase. *Nature Genetics* 18:38–44.

Jensen, T.E., A.J. Rose, Y. Hellsten, J.F. Wojtaszewski, and E.A. Richter. 2007. Caffeine-induced Ca(2+) release increases AMPK-dependent glucose uptake in rodent soleus muscle. *Am. J. Physiol. Endocrinol. Metab.* 293:E286–E292.

Johnson, L.N., M.E. Noble, and D.J. Owen. 1996. Active and inactive protein kinases: structural basis for regulation. *Cell* 85:149–158.

Josell, A., A.C. Enfalt, G. von Seth, G. Lindahl, I. Hedebro-Velander, L. Andersson, and K. Lundstrom. 2003. The influence of RN genotype, including the new V199I allele, on the eating quality of pork loin. *Meat Sci.* 65:1341–1351.

Khamzina, L., A. Veilleux, S. Bergeron, and A. Marette. 2005. Increased activation of the mammalian target of rapamycin pathway in liver and skeletal muscle of obese rats: possible involvement in obesity-linked insulin resistance. *Endocrinology* 146:1473–1481.

Kim, J.K., J.J. Fillmore, M.J. Sunshine, B. Albrecht, T. Higashimori, D.W. Kim, Z.X. Liu, T.J. Soos, G.W. Cline, W.R. O'Brien, D.R. Littman, and G.I. Shulman. 2004a. PKC-theta knockout mice are protected from fat-induced insulin resistance. *J. Clin. Invest.* 114:823–827.

Kim, J.M., M. Yoon, I. Kang, S.S. Kim, and J. Ha. 1998. Evidence that acetyl-CoA carboxylase isoforms play different biological roles in H9c2 cardiomyocyte. *Biochem. Biophys. Res. Commun.* 248:490–496.

Kim, K.H., M.J. Song, J. Chung, H. Park, and J.B. Kim. 2005. Hypoxia inhibits adipocyte differentiation in a HDAC-independent manner. *Biochem. Biophys. Res. Commun.* 333:1178–1184.

Kim, M.S., J.Y. Park, C. Namkoong, P.G. Jang, J.W. Ryu, H.S. Song, J.Y. Yun, I.S. Namgoong, J. Ha, I.S. Park, I.K. Lee, B. Viollet, J.H. Youn, H.K. Lee, and K.U. Lee. 2004b. Anti-obesity effects of alpha-lipoic acid mediated by suppression of hypothalamic AMP-activated protein kinase. *Nature Medicine* 10:727–733.

Kimura, N., C. Tokunaga, S. Dalal, C. Richardson, K. Yoshino, K. Hara, B.E. Kemp, L.A. Witters, O. Mimura, and K. Yonezawa. 2003. A possible linkage between AMP-activated protein kinase (AMPK) and mammalian target of rapamycin (mTOR) signalling pathway. *Genes Cells* 8:65–79.

Krawiec, B.J., G.J. Nystrom, R.A. Frost, L.S. Jefferson, and C.H. Lang. 2007. AMP-activated protein kinase agonists increase mRNA content of the muscle-specific ubiquitin ligases MAFbx and MuRF1 in C2C12 cells. *Am. J. Physiol. Endocrinol. Metab.* 292:E1555–E1567.

Kurth-Kraczek, E.J., M.F. Hirshman, L.J. Goodyear, and W.W. Winder. 1999. 5′ AMP-activated protein kinase activation causes GLUT4 translocation in skeletal muscle. *Diabetes* 48:1667–1671.

Larzul, C., P. Le Roy, G. Monin, and P. Sellier. 1998. Genetic variability of muscle glycolytic potential in pigs. *Productions Animales* 11:183–197.

Latres, E., A.R. Amini, A.A. Amini, J. Griffiths, F.J. Martin, Y. Wei, H. C. Lin, G.D. Yancopoulos, and D.J. Glass. 2005. Insulin-like growth factor-1 (IGF-1) inversely regulates atrophy-induced genes via the phosphatidylinositol 3-kinase/Akt/mammalian target of rapamycin (PI3K/Akt/mTOR) pathway. *J. Biol. Chem.* 280:2737–2744.

Le Roy, P., J.M. Elsen, J.C. Caritez, A. Talmant, H. Juin, P. Sellier, and G. Monin. 2000. Comparison between the three porcine RN genotypes for growth, carcass composition and meat quality traits. *Genet. Select. Evolut.* 32:165–186.

Lebret, B., L. Fefaucheur, and J. Mourot. 1999a. Pig meat quality. Influence of rearing factors on skeletal muscle traits. *Productions Animales* 12:11–28.

Lebret, B., P. Le Roy, G. Monin, L. Lefaucheur, J.C. Caritez, A. Talmant, J.M. Elsen, and P. Sellier. 1999b. Influence of the three RN genotypes on chemical composition, enzyme activities, and myofiber characteristics of porcine skeletal muscle. *J. Anim. Sci.* 77:1482–1489.

Leclerc, I., A. Kahn, and B. Doiron. 1998. The 5′-AMP-activated protein kinase inhibits the transcriptional stimulation by glucose in liver cells, acting through the glucose response complex. *FEBS Lett.* 431:180–184.

Lee, W.J., K.H. Song, E.H. Koh, J.C. Won, H.S. Kim, H.S. Park, M.S. Kim, S.W. Kim, K.U. Lee, and J.Y. Park. 2005. Alpha-lipoic acid increases insulin sensitivity by activating AMPK in skeletal muscle. *Biochem. Biophys. Res. Commun.* 332:885–891.

Lee, Y.S., W.S. Kim, K.H. Kim, M.J. Yoon, H.J. Cho, Y. Shen, J.M. Ye, C.H. Lee, W.K. Oh, C.T. Kim, C. Hohnen-Behrens, A. Gosby, E.W. Kraegen, D.E. James, and J.B. Kim. 2006. Berberine, a natural plant product, activates AMP-activated protein kinase with beneficial metabolic effects in diabetic and insulin-resistant states. *Diabetes* 55:2256–2264.

Lehnert, S.A., A. Reverter, K.A. Byrne, Y. Wang, G.S. Nattrass, N.J. Hudson, and P.L. Greenwood. 2007. Gene expression studies of developing bovine longissimus muscle from two different beef cattle breeds. *BMC Dev. Biol.* 7:95.

Lessard, S.J., Z.P. Chen, M.J. Watt, M. Hashem, J.J. Reid, M.A. Febbraio, B.E. Kemp, and J.A. Hawley. 2006. Chronic rosiglitazone treatment restores AMPK alpha 2 activity in insulin-resistant rat skeletal muscle. *Am. J. Physiol. – Endocrinol. Metab.* 290:E251–E257.

Lizcano, J.M., O. Goransson, R. Toth, M. Deak, N.A. Morrice, J. Boudeau, S.A. Hawley, L. Udd, T.P. Makela, D.G. Hardie, and D.R. Alessi. 2004. LKB1 is a master kinase that activates 13 kinases of the AMPK subfamily, including MARK/PAR-1. *EMBO J.* 23:833–843.

Lochhead, P.A., I.P. Salt, K.S. Walker, D.G. Hardie, and C. Sutherland. 2000. 5-Aminoimidazole-4-carboxamide riboside mimics the effects of insulin on the expression of the 2 key gluconeogenic genes PEPCK and glucose-6-phosphatase. *Diabetes* 49:896–903.

Lomax, M.A., F. Sadiq, G. Karamanlidis, A. Karamitri, P. Trayhurn, and D.G. Hazlerigg. 2007. Ontogenic loss of brown adipose tissue sensitivity to beta-adrenergic stimulation in the ovine. *Endocrinology* 148:461–468.

Mahlapuu, M., C. Johansson, K. Lindgren, G. Hjalm, B.R. Barnes, A. Krook, J.R. Zierath, L. Andersson, and S. Marklund. 2004. Expression profiling of the gamma-subunit isoforms of AMP-activated protein kinase suggests a major role for gamma 3 in white skeletal muscle. *Am. J. Physiol. – Endocrinol. Metab.* 286:E194–E200.

Marley, A.E., J.E. Sullivan, D. Carling, W.M. Abbott, G.J. Smith, I.W. Taylor, F. Carey, and R.K. Beri. 1996. Biochemical characterization and deletion analysis of recombinant human protein phosphatase 2C alpha. *Biochem. J.* 320(Pt. 3):801–806.

Marsin, A.S., L. Bertrand, M.H. Rider, J. Deprez, C. Beauloye, M.F. Vincent, G. Van den Berghe, D. Carling, and L. Hue. 2000. Phosphorylation and activation of heart PFK-2 by AMPK has a role in the stimulation of glycolysis during ischaemia. *Curr. Biol.* 10:1247–1255.

McGarry, J.D. and N.F. Brown. 1997. The mitochondrial carnitine palmitoyltransferase system — from concept to molecular analysis. *Eur. J. Biochem.* 244:1–14.

McLoon, L.K., J. Rowe, J. Wirtschafter, and K.M. McCormick. 2004. Continuous myofiber remodeling in uninjured extraocular myofibers: myonuclear turnover and evidence for apoptosis. *Muscle Nerve* 29:707–715.

McManus, E.J. and D.R. Alessi. 2002. TSC1-TSC2: a complex tale of PKB-mediated S6K regulation. *Nat. Cell Biol.* 4:E214–216.

Mei, B., L. Zhao, L. Chen, and H.S. Sul. 2002. Only the large soluble form of preadipocyte factor-1 (Pref-1), but not the small soluble and membrane forms, inhibits adipocyte differentiation: role of alternative splicing. *Biochem. J.* 364:137–144.

Menconi, M.J., W. Wei, H. Yang, C.J. Wray, and P.O. Hasselgren. 2004. Treatment of cultured myotubes with the calcium ionophore A23187 increases proteasome activity via a CaMK II-caspase-calpain-dependent mechanism. *Surgery* 136:135–142.

Merrill, G.F., E.J. Kurth, D.G. Hardie, and W.W. Winder. 1997. AICA riboside increases AMP-activated protein kinase, fatty acid oxidation, and glucose uptake in rat muscle. *Am. J. Physiol. - Endocrinol. Metab.* 36:E1107–E1112.

Milan, D., J.T. Jeon, C. Looft, V. Amarger, A. Robic, M. Thelander, C. Rogel-Gaillard, S. Paul, N. Iannuccelli, L. Rask, H. Ronne, K. Lundstrom, N. Reinsch, J. Gellin, E. Kalm, P.L. Roy, P. Chardon, and L. Andersson. 2000. A mutation in PRKAG3 associated with excess glycogen content in pig skeletal muscle. *Science* 288:1248–1251.

Milan, D., N. Woloszyn, M. Yerle, P. LeRoy, M. Bonnet, J. Riquet, Y. LahbibMansais, J.C. Caritez, A. Robic, P. Sellier, J.M. Elsen, and J. Gellin. 1996. Accurate mapping of the "acid meat" RN gene on genetic and physical maps of pig Chromosome 15. *Mammalian Genome* 7:47–51.

Miller, K.D., M. Ellis, F.K. McKeith, B.S. Bidner, and D.J. Meisinger. 2000. Frequency of the Rendement Napole RN^- allele in a population of American Hampshire pigs. *J. Anim. Sci.* 78:1811–1815.

Minokoshi, Y., T. Alquier, N. Furukawa, Y.B. Kim, A. Lee, B. Xue, J. Mu, F. Foufelle, P. Ferre, M.J. Birnbaum, B.J. Stuck, and B.B. Kahn. 2004. AMP-kinase regulates food intake by responding to hormonal and nutrient signals in the hypothalamus. *Nature* 428:569–574.

Minokoshi, Y., Y.B. Kim, O.D. Peroni, L.G. Fryer, C. Muller, D. Carling, and B.B. Kahn. 2002. Leptin stimulates fatty-acid oxidation by activating AMP-activated protein kinase. *Nature* 415:339–343.

Moeller, S.J., T.J. Baas, T.D. Leeds, R.S. Emnett, and K.M. Irvin. 2003. *Rendement Napole* gene effects and a comparison of glycolytic potential and DNA genotyping for classification of *Rendement Napole* status in Hampshire-sired pigs. *J. Anim. Sci.* 81:402–410.

Momcilovic, M., S.P. Hong, and M. Carlson. 2006. Mammalian TAK1 activates Snf1 protein kinase in yeast and phosphorylates AMP-activated protein kinase in vitro. *J. Biol. Chem.* 281:25336–25343.

Monin, G. and P. Sellier. 1985. Pork of low technological quality with a normal rate of muscle pH fall in the immediate post-mortem period: the case of the Hampshire breed. *Meat Sci.* 13:49–63.

Mu, J., J.T. Brozinick, O. Valladares, M. Bucan, and M.J. Birnbaum. 2001. A role for AMP-activated protein kinase in contraction- and hypoxia-regulated glucose transport in skeletal muscle. *Mol. Cell* 7:1085–1094.

Muhlhausler, B.S., J.A. Duffield, and I.C. McMillen. 2007. Increased maternal nutrition stimulates peroxisome proliferator activated receptor-γ (PPARγ), adiponectin and leptin mRNA expression in adipose tissue before birth. *Endocrinology* 148:878–885.

Munday, M.R., D.G. Campbell, D. Carling, and D.G. Hardie. 1988. Identification by amino acid sequencing of three major regulatory phosphorylation sites on rat acetyl-CoA carboxylase. *Eur. J. BioChem.* 175:331–338.

Muoio, D.M., K. Seefeld, L.A. Witters, and R.A. Coleman. 1999. AMP-activated kinase reciprocally regulates triacylglycerol synthesis and fatty acid oxidation in liver and muscle: evidence that sn-glycerol-3-phosphate acyltransferase is a novel target. *Biochem. J.* 338:783–791.

Murphy, R.T., J. Mogensen, K. McGarry, A. Bahl, A. Evans, E. Osman, P. Syrris, G. Gorman, M. Farrell, J.L. Holton, M.G. Hanna, S. Hughes, P.M. Elliott, C. A. Macrae, and W.J. McKenna. 2005. Adenosine monophosphate-activated protein kinase disease mimicks hypertrophic cardiomyopathy and Wolff-Parkinson-White syndrome: natural history. *J. Am. Coll. Cardiol.* 45:922–930.

Muse, E.D., S. Obici, S. Bhanot, B.P. Monia, R.A. McKay, M.W. Rajala, P.E. Scherer, and L. Rossetti. 2004. Role of resistin in diet-induced hepatic insulin resistance. *J. Clin. Invest.* 114:232–239.

Musi, N., T. Hayashi, N. Fujii, M.F. Hirshman, L.A. Witters, and L.J. Goodyear. 2001. AMP-activated protein kinase activity and glucose uptake in rat skeletal muscle. *Am. J. Physiol. – Endocrinol. Metab.* 280:E677–E684.

Nakashima, K. and Y. Yakabe. 2007. AMPK activation stimulates myofibrillar protein degradation and expression of atrophy-related ubiquitin ligases by increasing FOXO transcription factors in C2C12 myotubes. *Biosci. Biotechnol. Biochem.* 71:1650–1656.

Oliveira, S.M., J. Ehtisham, C.S. Redwood, I. Ostman-Smith, E.M. Blair, and H. Watkins. 2003. Mutation analysis of AMP-activated protein kinase subunits in inherited cardiomyopathies: implications for kinase function and disease pathogenesis. *J. Mol. Cell Cardiol.* 35:1251–1255.

Owen, M.R., E. Doran, and A.P. Halestrap. 2000. Evidence that metformin exerts its anti-diabetic effects through inhibition of complex 1 of the mitochondrial respiratory chain. *Biochem. J.* 348:607–614.

Ozes, O.N., H. Akca, L.D. Mayo, J.A. Gustin, T. Maehama, J.E. Dixon, and D.B. Donner. 2001. A phosphatidylinositol 3-kinase/Akt/mTOR pathway mediates and PTEN antagonizes tumor necrosis factor inhibition of insulin signaling through insulin receptor substrate-1. *Proc. Natl. Acad. Sci. U.S.A.* 98:4640–4645.

Packer, L., K. Kraemer, and G. Rimbach. 2001. Molecular aspects of lipoic acid in the prevention of diabetes complications. *Nutrition* 17:888–895.

Palanivel, R. and G. Sweeney. 2005. Regulation of fatty acid uptake and metabolism in L6 skeletal muscle cells by resistin. *FEBS Lett.* 579:5049–5054.

Pap, M. and G.M. Cooper. 2002. Role of translation initiation factor 2B in control of cell survival by the phosphatidylinositol 3-kinase/Akt/glycogen synthase kinase 3beta signaling pathway. *Mol. Cell Biol.* 22:578–586.

Park, H., V.K. Kaushik, S. Constant, M. Prentki, E. Przybytkowski, N.B. Ruderman, and A.K. Saha. 2002. Coordinate regulation of malonyl-CoA decarboxylase, sn-glycerol-3-phosphate acyltransferase, and acetyl-CoA carboxylase by AMP-activated protein kinase in rat tissues in response to exercise. *J. Biol. Chem.* 277:32571–32577.

Park, I.H., E. Erbay, P. Nuzzi, and J. Chen. 2005. Skeletal myocyte hypertrophy requires mTOR kinase activity and S6K1. *Exp. Cell Res.* 309:211–219.

Park, I.J., J.T. Hwang, Y.M. Kim, J. Ha, and O.J. Park. 2006. Differential modulation of AMPK signaling pathways by low or high levels of exogenous reactive oxygen species in colon cancer cells. *Ann. N.Y. Acad. Sci.* 1091:102–109.

Peraldi, P. and B. Spiegelman. 1998. TNF-alpha and insulin resistance: summary and future prospects. *Mol. Cell Biochem.* 182:169–175.

Poulos, S.P. and G.J. Hausman. 2006. A comparison of thiazolidinedione-induced adipogenesis and myogenesis in stromal-vascular cells from subcutaneous adipose tissue or semitendinosus muscle of postnatal pigs. *J. Anim. Sci.* 84:1076–1082.

Qiao, L.Y., R. Zhande, T.L. Jetton, G. Zhou, and X.J. Sun. 2002. *In vivo* phosphorylation of insulin receptor substrate 1 at serine 789 by a novel serine kinase in insulin-resistant rodents. *J. Biol. Chem.* 277:26530–26539.

Rose, A.J., B. Kiens, and E.A. Richter. 2006. Ca2+-calmodulin-dependent protein kinase expression and signalling in skeletal muscle during exercise. *J. Physiol. – London* 574:889–903.

Russell, R.R., R. Bergeron, G.I. Shulman, and L.H. Young. 1999. Translocation of myocardial GLUT-4 and increased glucose uptake through activation of AMPK by AICAR. *Am. J. Physiol. – Heart Circulat. Physiol.* 277:H643–H649.

Sabina, R.L., D. Patterson, and E.W. Holmes. 1985. 5-Amino-4-imidazolecarboxamide riboside (Z-riboside) metabolism in eukaryotic cells. *J. Biol. Chem.* 260:6107–6114.

Sakamoto, K., O. Goransson, D.G. Hardie, and D.R. Alessi. 2004. Activity of LKB1 and AMPK-related kinases in skeletal muscle: effects of contraction, phenformin, and AICAR. *Am. J. Physiol. – Endocrinol. Metab.* 287:E310–E317.

Sakamoto, K., A. McCarthy, D. Smith, K.A. Green, D.G. Hardie, A. Ashworth, and D.R. Alessi. 2005. Deficiency of LKB1 in skeletal muscle prevents AMPK activation and glucose uptake during contraction. *EMBO J.* 24:1810–1820.

Sakamoto, K., E. Zarrinpashneh, G.R. Budas, A.C. Pouleur, A. Dutta, A.R. Prescott, J.L. Vanoverschelde, A. Ashworth, A. Jovanovic, D.R. Alessi, and L. Bertrand. 2006. Deficiency of LKB1 in heart prevents ischemia-mediated activation of AMPK alpha 2 but not AMPK alpha 1. *Am. J. Physiol. – Endocrinol. Metab.* 290:E780–E788.

Sakoda, H., T. Ogihara, M. Anai, M. Fujishiro, H. Ono, Y. Onishi, H. Katagiri, M. Abe, Y. Fukushima, N. Shojima, K. Inukai, M. Kikuchi, Y. Oka, and T. Asano. 2002. Activation of AMPK is essential for AICAR-induced glucose uptake by skeletal muscle but not adipocytes. *Am. J. Physiol. – Endocrinol. Metab.* 282:E1239–E1244.

Sanders, M.J., P.O. Grondin, B.D. Hegarty, M.A. Snowden, and D. Carling. 2007. Investigating the mechanism for AMP activation of the AMP-activated protein kinase cascade. *Biochem. J.* 403:139–148.

Satoh, H., M.T.A. Nguyen, P.D.G. Miles, T. Imamura, I. Usui, and J.M. Olefsky. 2004. Adenovirus-mediated chronic "hyper-resistinemia" leads to *in vivo* insulin resistance in normal rats. *J. Clin. Invest.* 114:224–231.

Schulz, R.A. and K.E. Yutzey. 2004. Calcineurin signaling and NFAT activation in cardiovascular and skeletal muscle development. *Dev. Biol.* 266:1–16.

Scott, J.W., S.A. Hawley, K.A. Green, M. Anis, G. Stewart, G.A. Scullion, D.G. Norman, and D.G. Hardie. 2004. CBS domains form energy-sensing modules whose binding of adenosine ligands is disrupted by disease mutations. *J. Clin. Invest.* 113:274–284.

Shaw, R.J. and L.C. Cantley. 2006. Ras, PI(3)K and mTOR signalling controls tumour cell growth. *Nature* 441:424–430.

Shaw, R.J., M. Kosmatka, N. Bardeesy, R.L. Hurley, L.A. Witters, R.A. DePinho, and L.C. Cantley. 2004. The tumor suppressor LKB1 kinase directly activates AMP-activated kinase and regulates apoptosis in response to energy stress. *Proc. Natl. Acad. Sci. U.S.A.* 101:3329–3335.

Shaw, R.J., K.A. Lamia, D. Vasquez, S.H. Koo, N. Bardeesy, R.A. DePinho, M. Montminy, and L.C. Cantley. 2005. The kinase LKB1 mediates glucose homeostasis in liver and therapeutic effects of metformin. *Science* 310:1642–1646.

Shen, Q.W. and M. Du. 2005. Role of AMP-activated protein kinase in the glycolysis of postmortem muscle. *J. Sci. Food Agric.* 85:2401–2406.

Shen, Q.W., M. Du, and J. Ren. 2006a. Fitness and fatness: the debate continues for AMP activated protein kinase in heart function. *Curr. Cardiol. Rev.* 2:117–123.

Shen, Q.W., W.J. Means, S.A. Thompson, K.R. Underwood, M.J. Zhu, R.J. McCormick, S.P. Ford, and M. Du. 2006b. Pre-slaughter transport, AMP-activated protein kinase, glycolysis, and quality of pork loin. *Meat Sci.* 74:388–395.

Shen, Q.W., W.J. Means, K.R. Underwood, S.A. Thompson, M. J. Zhu, R. J. McCormick, S. P. Ford, M. Ellis, and M. Du. 2006c. Early post-mortem AMP-activated protein kinase (AMPK) activation leads to phosphofructokinase-2 and -1 (PFK-2 and PFK-1) phosphorylation and the development of pale, soft, and exudative (PSE) conditions in porcine longissimus muscle. *J. Agric. Food Chem.* 54:5583–5589.

Shen, Q.W., K.R. Underwood, W.J. Means, R.J. McCormick, and M. Du. 2007a. The halothane gene, energy metabolism, adenosine monophosphate-activated protein kinase, and glycolysis in postmortem pig longissimus dorsi muscle. *J. Anim. Sci.* 85:1054–1061.

Shen, Q.W., M.J. Zhu, J. Tong, J. Ren, and M. Du. 2007b. Ca2+/calmodulin-dependent protein kinase kinase is involved in AMP-activated protein kinase activation by alpha-lipoic acid in C2C12 myotubes. *Am. J. Physiol. Cell Physiol.* 293:C1395–1403.

Shen, Q.W., W.J. Means, K.R. Underwood, S.A. Thompson, M.J. Zhu, R.J. McCormick, S.P. Ford, M. Ellis, and M. Du. 2006d. Early post-mortem AMP-activated protein kinase (AMPK) activation leads to phosphofructokinase-2 and-1 (PFK-2 and PFK-1) phosphorylation and the development of pale, soft, and exudative (PSE) conditions in porcine longissimus muscle. *J. Agric. Food Chem.* 54:5583–5589.

Sidhu, J.S., Y.S. Rajawat, T.G. Rami, M.H. Gollob, Z. Wang, R. Yuan, A.J. Marian, F.J. DeMayo, D. Weilbacher, G.E. Taffet, J.K. Davies, D. Carling, D.S. Khoury, and R. Roberts. 2005. Transgenic mouse model of ventricular preexcitation and atrioventricular reentrant tachycardia induced by an AMP-activated protein kinase loss-of-function mutation responsible for Wolff-Parkinson-White syndrome. *Circulation* 111:21–29.

Skurk, C., Y. Izumiya, H. Maatz, P. Razeghi, I. Shiojima, M. Sandri, K. Sato, L. Zeng, S. Schiekofer, D. Pimentel, S. Lecker, H. Taegtmeyer, A.L. Goldberg, and K. Walsh. 2005. The FOXO3a transcription factor regulates cardiac myocyte size downstream of AKT signaling. *J. Biol. Chem.* 280:20814–20823.

Song, Y.H., M. Godard, Y. Li, S.R. Richmond, N. Rosenthal, and P. Delafontaine. 2005. Insulin-like growth factor I-mediated skeletal muscle hypertrophy is characterized by increased mTOR-p70S6K signaling without increased Akt phosphorylation. *J. Investig. Med.* 53:135–142.

Spiegelman, B.M., P. Puigserver, and Z. Wu. 2000. Regulation of adipogenesis and energy balance by PPARgamma and PGC-1. *Int. J. Obes. Relat. Metab. Disord.* 24:S8–S10.

Spindler, M., K.W. Saupe, M.E. Christe, H.L. Sweeney, C.E. Seidman, J.G. Seidman, and J.S. Ingwall. 1998. Diastolic dysfunction and altered energetics in the alphaMHC403/+ mouse model of familial hypertrophic cardiomyopathy. *J. Clin. Invest.* 101:1775–1783.

Sriwijitkamol, A., J.L. Ivy, C. Christ-Roberts, R.A. DeFronzo, L.J. Mandarino, and N. Musi. 2006. LKB1-AMPK signaling in muscle from obese insulin-resistant Zucker rats and effects of training. *Am. J. Physiol. Endocrinol. Metab.* 290:E925–932.

Stapleton, D., K.I. Mitchelhill, G. Gao, J. Widmer, B.J. Michell, T. Teh, C.M. House, C.S. Fernandez, T. Cox, L.A. Witters, and B.E. Kemp. 1996. Mammalian AMP-activated protein kinase subfamily. *J. Biol. Chem.* 271:611–614.

Stapleton, D., E. Woollatt, K.I. Mitchelhill, J.K. Nicholl, C.S. Fernandez, B.J. Michell, L.A. Witters, D.A. Power, G.R. Sutherland, and B.E. Kemp. 1997. AMP-activated protein kinase isoenzyme family: subunit structure and chromosomal location. *FEBS Lett.* 409:452–456.

Stein, S.C., A. Woods, N.A. Jones, M.D. Davison, and D. Carling. 2000. The regulation of AMP-activated protein kinase by phosphorylation. *Biochem. J.* 345:437–443.

Steinberg, G.R., S.L. Macaulay, M.A. Febbraio, and B.E. Kemp. 2006a. AMP-activated protein kinase — the fat controller of the energy railroad. *Can. J. Physiol. Pharmacol.* 84:655–665.

Steinberg, G.R., B.J. Michell, B.J. van Denderen, M.J. Watt, A.L. Carey, B.C. Fam, S. Andrikopoulos, J. Proietto, C.Z. Gorgun, D. Carling, G.S. Hotamisligil, M.A. Febbraio, T.W. Kay, and B.E. Kemp. 2006b. Tumor necrosis factor alpha-induced skeletal muscle insulin resistance involves suppression of AMP-kinase signaling. *Cell Metab.* 4:465–474.

Steinberg, G.R., J.W. E. Rush, and D.J. Dyck. 2003. AMPK expression and phosphorylation are increased in rodent muscle after chronic leptin treatment. *Am. J. Physiol. – Endocrinol. Metab.* 284:E648–E654.

Subramaniam, S., N. Shahani, J. Strelau, C. Laliberte, R. Brandt, D. Kaplan, and K. Unsicker. 2005. Insulin-like growth factor 1 inhibits extracellular signal-regulated kinase to promote neuronal survival via the phosphatidylinositol 3-kinase/protein kinase A/c-Raf pathway. *J. Neurosci.* 25:2838–2852.

Sullivan, J.E., K.J. Brocklehurst, A.E. Marley, F. Carey, D. Carling, and R.K. Beri. 1994. Inhibition of lipolysis and lipogenesis in isolated rat adipocytes with AICAR, a cell-permeable activator of AMP-activated protein kinase. *FEBS Lett.* 353:33–36.

Thornton, C., M.A. Snowden, and D. Carling. 1998. Identification of a novel AMP-activated protein kinase beta subunit isoform that is highly expressed in skeletal muscle. *J. Biol. Chem.* 273:12443–12450.

Tian, R., N. Musi, J. D'Agostino, M.F. Hirshman, and L.J. Goodyear. 2001. Increased adenosine monophosphate-activated protein kinase activity in rat hearts with pressure-overload hypertrophy. *Circulation* 104:1664–1669.

Tomas, E., T.S. Tsao, A.K. Saha, H.E. Murrey, C. Zhang , S.I. Itani, H.F. Lodish, and N.B. Ruderman. 2002. Enhanced muscle fat oxidation and glucose transport by ACRP30 globular domain: acetyl-CoA carboxylase inhibition and AMP-activated protein kinase activation. *Proc. Natl. Acad. Sci. U.S.A.* 99:16309–16313.

Tong, J., M.J. Zhu, K.R. Underwood, B.W. Hess, S.P. Ford, and M. Du. 2008. AMP-activated protein kinase and adipogenesis in sheep fetal skeletal muscle and 3T3-L1 cells. *J. Anim. Sci.* 86:1296–1305.

Um, S.H., F. Frigerio, M. Watanabe, F. Picard, M. Joaquin, M. Sticker, S. Fumagalli, P.R. Allegrini, S.C. Kozma, J. Auwerx, and G. Thomas. 2004. Absence of S6K1 protects against age- and diet-induced obesity while enhancing insulin sensitivity. *Nature* 431:200–205.

Underwood, K.R., W.J. Means, M.J. Zhu, S.P. Ford, B.W. Hess, and M. Du. 2008. AMP-activated protein kinase is negatively associated with intramuscular fat content in longissimus dorsi muscle of beef cattle. *Meat Sci.* 79:394–402.

van Laack, R.L. and R.G. Kauffman. 1999. Glycolytic potential of red, soft, exudative pork longissimus muscle. *J. Animal Sci.* 77:2971–2973.

Vary, T.C. 2006. IGF-I stimulates protein synthesis in skeletal muscle through multiple signaling pathways during sepsis. *Am. J. Physiol. Regul. Integr. Comp. Physiol.* 290:R313–321.

Vary, T.C., G. Deiter, and S.R. Kimball. 2002. Phosphorylation of eukaryotic initiation factor eIF2Bepsilon in skeletal muscle during sepsis. *Am. J. Physiol. Endocrinol. Metab.* 283:E1032–1039.

Vavvas, D., A. Apazidis, A.K. Saha, J. Gamble, A. Patel, B.E. Kemp, L.A. Witters, and N.B. Ruderman. 1997. Contraction-induced changes in acetyl-CoA carboxylase and 5′-AMP-activated kinase in skeletal muscle. *J. Biol. Chem.* 272:13255–13261.

Wang, X., Z. Hu, J. Hu, J. Du, and W.E. Mitch. 2006. Insulin resistance accelerates muscle protein degradation: activation of the ubiquitin-proteasome pathway by defects in muscle cell signaling. *Endocrinology* 147:4160–4168.

Weigert, C., M. Dufer, P. Simon, E. Debre, H. Runge, K. Brodbeck, H.U. Haring, and E.D. Schleicher. 2007. Upregulation of IL-6 mRNA by IL-6 in skeletal muscle cells: role of IL-6 mRNA stabilization and Ca^{2+}-dependent mechanisms. *Am. J. Physiol. Cell Physiol.* 293:C1139–1147.

Winder, W.W., and D.G. Hardie. 1999. AMP-activated protein kinase, a metabolic master switch: possible roles in type 2 diabetes. *Am. J. Physiol.* 277:E1–E10.

Winder, W.W., B.F. Holmes, D.S. Rubink, E.B. Jensen, M. Chen, and J.O. Holloszy. 2000a. Activation of AMP-activated protein kinase increases mitochondrial enzymes in skeletal muscle. *J. Appl. Physiol.* 88:2219–2226.

Winder, W.W., B.F. Holmes, D.S. Rubink, E.B. Jensen, M. Chen, and J.O. Holloszy. 2000b. Chronic chemical activation of AMP-activated protein kinase increases mitochondrial enzymes and GLUT4 in skeletal muscle of resting rats. *Diabetes* 49:A10.

Witczak, C.A., N. Fujii, J. Brandauer, M. Seifert, N. Nozaki, and L.J. Goodyear. 2006. CaMKK alpha activates AMPK signaling in mouse skeletal muscle *in vivo*. *FASEB J.* 20:A820–A820.

Woods, A., D. Azzout-Marniche, M. Foretz, S.C. Stein, P. Lemarchand, P. Ferre, F. Foufelle, and D. Carling. 2000. Characterization of the role of AMP-activated protein kinase in the regulation of glucose-activated gene expression using constitutively active and dominant negative forms of the kinase. *Mol. Cell. Biol.* 20:6704–6711.

Woods, A., K. Dickerson, R. Heath, S.P. Hong, M. Momcilovic, S.R. Johnstone, M. Carlson, and D. Carling. 2005. Ca^{2+}/calmodulin-dependent protein kinase kinase-beta acts upstream of AMP-activated protein kinase in mammalian cells. *Cell Metab.* 2:21–33.

Woods, A., S.R. Johnstone, K. Dickerson, F.C. Leiper, L.G.D. Fryer, D. Neumann, U. Schlattner, T. Wallimann, M. Carlson, and D. Carling. 2003a. LKB1 is the upstream kinase in the AMP-activated protein kinase cascade. *Curr. Biol.* 13:2004–2008.

Woods, A., I. Salt, J. Scott, D.G. Hardie, and D. Carling. 1996. The alpha1 and alpha2 isoforms of the AMP-activated protein kinase have similar activities in rat liver but exhibit differences in substrate specificity in vitro. *FEBS Lett.* 397:347–351.

Woods, A., D. Vertommen, D. Neumann, R. Turk, J. Bayliss, U. Schlattner, T. Wallimann, D. Carlling, and M.H. Rider. 2003b. Identification of phosphorylation sites in AMP-activated protein kinase (AMPK) for upstream AMPK kinases and study of their roles by site-directed mutagenesis. *J. Biol. Chem.* 278:28434–28442.

Wu, X.D., H. Motoshima, K. Mahadev, T.J. Stalker, R. Scalia, and B.J. Goldstein. 2003. Involvement of AMP-activated protein kinase in glucose uptake stimulated by the globular domain of adiponectin in primary rat adipocytes. *Diabetes* 52:1355–1363.

Wullschleger, S., R. Loewith, and M.N. Hall. 2006. TOR signaling in growth and metabolism. *Cell* 124:471–484.

Xie, M., D. Zhang, J.R. Dyck, Y. Li, H. Zhang, M. Morishima, D.L. Mann, G.E. Taffet, A. Baldini, D.S. Khoury, and M.D. Schneider. 2006. A pivotal role for endogenous TGF-beta-activated kinase-1 in the LKB1/AMP-activated protein kinase energy-sensor pathway. *Proc. Natl. Acad. Sci. U.S.A.* 103:17378–17383.

Yamauchi, T., J. Kamon, Y. Minokoshi, Y. Ito, H. Waki, S. Uchida, S. Yamashita, M. Noda, S. Kita, K. Ueki, K. Eto, Y. Akanuma, P. Froguel, F. Foufelle, P. Ferre, D. Carling, S. Kimura, R. Nagai, B.B. Kahn, and T. Kadowaki. 2002. Adiponectin stimulates glucose utilization and fatty-acid oxidation by activating AMP-activated protein kinase. *Nature Medicine* 8:1288–1295.

Yang, W.B., Y.H. Hong, X.Q. Shen, C. Frankowski, H.S. Camp, and T. Leff. 2001. Regulation of transcription by AMP-activated protein kinase — phosphorylation of p300 blocks its interaction with nuclear receptors. *J. Biol. Chem.* 276:38341–38344.

Yoon, M.J., G.Y. Lee, J.J. Chung, Y.H. Ahn, S.H. Hong, and J.B. Kim. 2006. Adiponectin increases fatty acid oxidation in skeletal muscle cells by sequential activation of AMP-activated protein kinase, p38 mitogen-activated protein kinase, and peroxisome proliferator-activated receptor alpha. *Diabetes* 55:2562–2570.

Young, L.H., Y. Renfu, R. Russell, X.Y. Hu, M. Caplan, J.M. Ren, G.I. Shulman, and A.J. Sinusas. 1997. Low-flow ischemia leads to translocation of canine heart GLUT-4 and GLUT-1 glucose transporters to the sarcolemma *in vivo*. *Circulation* 95:415–422.

Yu, H., N. Fujii, M.F. Hirshman, J.M. Pomerleau, and L.J. Goodyear. 2004. Cloning and characterization of mouse 5′-AMP-activated protein kinase gamma3 subunit. *Am. J. Physiol. Cell Physiol.* 286:C283–292.

Zheng, D.H., P.S. MacLean, S.C. Pohnert, J.B. Knight, A.L. Olson, W.W. Winder, and G.L. Dohm. 2001. Regulation of muscle GLUT-4 transcription by AMP-activated protein kinase. *J. Appl. Physiol.* 91:1073–1083.

Zhou, G.C., R. Myers, Y. Li, Y. L. Chen, X.L. Shen, J. Fenyk-Melody, M. Wu, J. Ventre, T. Doebber, N. Fujii, N. Musi, M.F. Hirshman, L.J. Goodyear, and D.E. Moller. 2001. Role of AMP-activated protein kinase in mechanism of metformin action. *J. Clin. Invest.* 108:1167–1174.

Zhu, M., B. Han, J. Tong, C. Ma, J.W. Kimzey, K.R. Underwood, Y. Xiao, B.W. Hess, S.P. Ford, P.W. Nathanielsz, and M. Du. 2008. AMP-activated protein kinase signalling pathways are down regulated and skeletal muscle development impaired in fetuses of obese, overnourished sheep. *J. Physiol.* 586:2651–2664.

Ziegler, D., M. Reljanovic, H. Mehnert, and F.A. Gries. 1999. alpha-Lipoic acid in the treatment of diabetic polyneuropathy in Germany: current evidence from clinical trials. *Exper. Clin. Endocrinol. Diabetes* 107:421–430.

Ziegler, D., H. Schatz, F. Conrad, F.A. Gries, H. Ulrich, and G. Reichel. 1997. Effects of treatment with the antioxidant alpha-lipoic acid on cardiac autonomic neuropathy in NIDDM patients — a 4-month randomized controlled multicenter trial (DEKAN study). *Diabetes Care* 20:369–373.

11 Meat Color

Richard A. Mancini

CONTENTS

11.1 INTRODUCTION

Meat color influences retail purchasing decisions more than any other quality factor because consumers use discoloration as an indicator of spoilage and shelf life. This results in annual revenue losses of $1 billion as nearly 15% of retail beef is discounted in price due to surface discoloration (Smith et al. 2000). Improvements in retail meat profit associated with increased product color life rely on several factors that determine meat color, particularly myoglobin redox chemistry.

11.2 MYOGLOBIN CHEMISTRY

11.2.1 Myoglobin

The pigment primarily responsible for meat color is myoglobin, a globular single-chain protein located in the sarcoplasm. Hemoglobin and cytochrome also play a small role in meat color; however, myoglobin is by far the major contributor and determinant. Located within the center of myoglobin is a non-protein prosthetic group often referred to as a porphyrin or heme structure. The redox state and ligand present on the iron centrally located within myoglobin's heme ring is responsible for visible color transformations that occur on the surface of meat products.

Myoglobin is approximately 17 kDa and contains 153 residues that are organized in eight α-helices linked by short nonhelical strands. Within the 153 amino acids that contribute to the proteins structure, two histidines (64 and 93) play a significant role in myoglobin function. In particular, iron's interaction with the proximal histidine (H93) is partially responsible for attaching the prosthetic heme group to the apoprotein. In addition, the distal histidine can interact with small molecules such as oxygen and aldehydes (lipid oxidation by-products), both of which influence muscle-food color stability.

The intrinsic chemistry of iron, in particular its ability to bind ligands and its valence state, plays a significant role in meat color. For example, iron located within the center of myoglobin's heme ring can reversibly as well as preferentially bind ligands, such as diatomic oxygen, carbon monoxide, or nitric oxide. Iron has the ability to transition between a reduced ferrous state (Fe^{++} or Fe^{2+}) and an oxidized ferric state (Fe^{+++} or Fe^{3+}). A combination of the ligand present and the redox state of iron determines fresh meat color via the formation of one of four chemical forms of myoglobin, known as deoxy-, oxy-, carboxy-, and metmyoglobin.

Further information regarding the role of myoglobin chemistry in meat color is available in: Giddings (1977), Seideman et al. (1984), Faustman and Cassens (1990), Cornforth (1994), Pegg and Shahidi (1997), Baron and Andersen (2002), and Mancini and Hunt (2005).

11.2.2 Deoxymyoglobin

Deoxymyoglobin results when no ligand is attached to myoglobin's ferrous heme iron (Fe^{2+} redox state). This myoglobin form is purplish-red (beef) or purplish-pink (pork) and as the "deoxy" prefix implies, it is associated with meat products that are not exposed to oxygen. Although this commonly occurs in vacuum packages, a thorough removal of oxygen is necessary as a very low oxygen partial pressure (<1.4 mmHg; Brooks 1935) is required to maintain myoglobin in a deoxygenated state. In addition, deoxymyoglobin is also found at depths within the interior of steaks and roasts where oxygen does not penetrate.

11.2.3 Oxygenation and Oxygen Consumption

Oxygenation occurs when deoxymyoglobin is exposed to oxygen, a process often referred to as bloom. The resulting color change from purple to red is often associated with opening vacuum packages, cutting steaks, and grinding meat, all of which allow myoglobin to bind oxygen. No change in iron's redox state occurs during oxygenation.

Competition for oxygen between myoglobin and mitochondria is possibly the most important factor influencing the formation of bright cherry-red meat. In particular, the use of oxygen by enzymes and proteins other than myoglobin results in a darker, deoxygenated meat color due to limited oxygen availability to myoglobin. Postmortem competition for oxygen between myoglobin and mitochondria is the primary factor responsible for dark-cutting beef. In general, long-term stress of an animal prior to slaughter depletes muscle glycogen, decreasing the amount of substrate available for anaerobic metabolism at the time of slaughter. This limits postmortem glycolysis and results in elevated ultimate muscle pH, a factor that promotes mitochondrial respiration and competition for oxygen in meat. Increased mitochondrial activity out-competes myoglobin for oxygen and therefore helps maintain pigments in a deoxygenated state. As a result, dark-cutting beef fails to bloom and does not form the characteristic bright cherry-red color associated with freshly oxygenated beef of normal pH (5.6) and typical oxygen consumption.

11.2.4 Oxymyoglobin

As the result of pigment oxygenation, diatomic oxygen is the ligand attached to iron within oxymyoglobin and meat is red in color. However, myoglobin remains in the ferrous redox state. In addition to oxygen binding to iron, the ligand is also linked to a helix within myoglobin via a hydrogen bond between the distal histidine and oxygen. This helps to stabilize O_2 and alters oxymyoglobin's structure compared with deoxymyoglobin. In particular, this interaction between histidine and bound oxygen brings the iron more in line with the protoporphyrin plane, making oxymyoglobin slightly more compact than deoxymyoglobin.

11.2.5 Metmyoglobin

Oxidation of myoglobin's heme iron results in metmyoglobin and the brown discoloration associated with consumer dissatisfaction. As atmospheric oxygen diffuses into meat products, it initially leaves a layer of bright-red oxymyoglobin on the surface where oxygen partial pressure is enough to maintain oxygenated myoglobin. As the depth beneath the meat surface increases, oxygen partial pressure decreases, resulting in a deoxygenated interior of a steak or roast. A thin layer of metmyoglobin forms beneath the surface where oxygen levels are not sufficient enough to form oxymyoglobin, yet above the deoxygenated interior where anaerobic oxygen levels maintain deoxymyoglobin. This layer of brown results from low oxygen partial pressure, which promotes myoglobin oxidation when levels are not sufficient for oxygenation and not low enough for deoxygenation. Metmyoglobin beneath the surface (located between superficial oxymyoglobin and interior deoxymyoglobin) will eventually thicken and migrate toward the surface, where it will be visible to consumers. Thus, discoloration (1) is due to both the amount of surface area covered by metmyoglobin as well as subsurface myoglobin, and (2) is not an "all-or-nothing" phenomenon, but rather ranges between no discoloration (0% metmyoglobin) and complete surface coverage (100% metmyoglobin). The rate of discoloration depends on several factors, including oxygen partial pressure, temperature, pH, postmortem age, reducing activity, and microbial growth.

11.2.6 Metmyoglobin Reduction

Maintenance of myoglobin in a ferrous state (Fe^{2+}) is essential to maximizing desirable meat color and shelf life. Because discoloration results from myoglobin oxidation (a loss of electrons), metmyoglobin reduction is a process that plays a vital role in meat color. Metmyoglobin reduction involves the transfer of an electron from a donating source, such as NADH, to the ferric heme of myoglobin. Electron transfer and mediation in postmortem muscle is supported by metmyoglobin reductase, which is a cytochrome b5 complex reductase enzyme system (Livingston et al. 1985). Thus, metmyoglobin reduction in meat is a process that is influenced by oxygen-scavenging enzymes, reducing enzyme systems, and the NADH pool, all of which are continually depleted with time postmortem (Faustman and Cassens 1990).

11.2.7 Deoxygenation and Subsequent Reoxygenation

Packaging bright-red (oxygenated) product in vacuum is a process of interest to the industry because the conversion of oxymyoglobin to deoxymyoglobin on the surface of fresh meat is a two-step process in which oxymyoglobin is not converted directly to deoxymyoglobin, but first is oxidized by low oxygen partial pressures. This is problematic because oxygen usage via endogenous oxygen consumption is required to promote oxygen partial pressures low enough to produce deoxymyoglobin. As a result of this two-step process, meat can regenerate deoxymyoglobin only if it is capable of both oxygen consumption and reduction of ferric to ferrous iron. From a meat-packaging standpoint, even if regeneration of deoxy- from oxymyoglobin is possible, subsequent rebloom is often troublesome. To counteract these problems, carbon monoxide can be added to anaerobic meat packages because it binds strongly to myoglobin and forms a bright cherry-red pigment known as carboxymyoglobin.

11.2.8 Carboxymyoglobin

Carboxymyoglobin is a redox state of myoglobin that is rarely included in "meat color triangles" as they traditionally have considered only interconversions between deoxy-, oxy-, and metmyoglobin. However, the use of carbon monoxide in modified atmosphere packaging makes carboxymyoglobin a significant part of meat color chemistry. At the level of carbon monoxide accepted for use in meat packaging (<0.4%), deoxymyoglobin is likely more readily converted to carboxymyoglobin than are oxymyoglobin or metmyoglobin from a fundamental molecular standpoint. However, both (1) the reaction of carbon monoxide with myoglobin and (2) the interconversion(s) between carboxy- and other myoglobin forms remain somewhat unclear when considering meat-packaging scenarios. In addition, although myoglobin's heme ring has a strong affinity for carbon monoxide, the stability of carboxymyoglobin in an oxygen-rich atmosphere is not straightforward. This idea of CO-ligand dissociation and subsequent replacement by an oxygen-ligand should be reviewed in order to better understand myoglobin's tendency to release carbon monoxide and subsequently bind oxygen once carboxymyoglobin is exposed to the atmosphere. In addition, the likelihood of replacing an oxygen ligand with carbon monoxide (conversion of oxy- to carboxymyoglobin) is particularly important when packaging previously bloomed steaks in carbon monoxide.

11.2.9 Cured Color and NO-Myoglobin

Nitrite is an ingredient added to cured meat products in order to (1) inhibit the outgrowth of *Clostridium botulinum* spores, (2) fix the characteristic pink color associated with cured meat, (3) impart flavor, and (4) act as an antioxidant. However, nitrite does not react with myoglobin. Rather, through a series of steps including reduction of nitrite, NO binds to myoglobin and forms ferrous NO-myoglobin. Heat-induced denaturation of NO-myoglobin exposes the centrally located porphyrin ring, which is pink due to its interaction with NO. The pink color of cured meat products is susceptible to light and oxygen, and will fade to gray if not vacuum packaged. Replacement of myoglobin's heme iron with zinc also is associated with the color of certain dry-cured hams.

11.2.10 Iridescence

Iridescence refers to a shiny, rainbow-like appearance most commonly associated with green, red, orange, and yellow colors on the surface of cooked meat products. This kaleidoscope-like appearance is a concern for the meat industry because it can be misinterpreted by consumers as an indication of microbial spoilage. The mechanism of iridescence is primarily associated with product surface microstructure and light diffraction, and therefore, it is likely that myoglobin has essentially no role in iridescence (Swatland 1984). In particular, structural uniformity on the surface of cooked meat will result in light diffraction patterns that are conducive to iridescence. Conversely, disruption of surface microstructure causes light to be reflected in a more irregular pattern and tends to minimize iridescence.

11.2.11 The Role of Histidine in Meat Color Chemistry

In addition to the basic roles of the distal and proximal histidine in myoglobin function, the interrelation between lipid and pigment oxidation is related to histidine chemistry. Lipid oxidation generates secondary products such as aldehydes that can accelerate heme protein oxidation. More specifically, 4-hydroxy-nonenal is a by-product of lipid oxidation that has the potential to cause covalent modification of myoglobin residues via adduction exclusively at six different nucleophilic histidine residues (HIS 24, 64, 93, 116, 119, and 152) via Michael addition (Faustman et al. 1999; Alderton et al. 2003; Lee et al. 2003; Suman et al. 2006; 2007).

11.3 FACTORS AFFECTING MEAT COLOR

11.3.1 Antemortem Factors

Diet and genetics can significantly influence meat color. In particular, changes in muscle lightness/darkness and color stability likely result from diet- and genetic-induced effects on muscle pH and/or myoglobin redox chemistry. For example, genetics play a significant role in pork color, primarily due to alterations in stress susceptibility and postmortem muscle metabolism. Genotypes commonly associated with pork color defects include halothane, ryanodine, and napole. Typically, genetics that predispose animals to stress influence postmortem muscle pH and result in a greater incidence of pale muscle (Channon et al. 2000; Fisher et al. 2000; Eggert et al. 2002; Fernandez et al. 2002; Moelich et al. 2003). Genetic susceptibility to stress-related color defects has been associated with mutations affecting alleles associated with ryanodine receptors (Bertram et al. 2000; Kuchenmeister et al. 2000; Piedrafita et al. 2001; Van Oeckel et al. 2001; Velarde et al. 2001; Moeller et al. 2003). Although genotypic effects on postmortem muscle temperature and pH often result in pale muscle, redness and degree of bloom can be enhanced by stress-induced increases in temperature, which can inactivate proteins involved in oxygen consumption. This results in greater surface oxygenation (greater degree of bloom) due to less competition for oxygen by enzymes (Rosenvold et al. 2003; Lindahl et al. 2004). However, stress-related elevation of postmortem muscle temperature also can inactivate metmyoglobin reductase and therefore decrease color stability during display.

The role of diet in meat color is primarily due to indirect effects on metabolism, glycogen storage and utilization, pH, chilling rate, and antioxidant accumulation. In particular, diets that promote oxidative metabolism, rather than anaerobic muscle metabolism and glycogen storage, can result in higher pH and darker muscle color (Vestergaard et al. 2000; Baublits et al. 2004). Increased muscle pH results in a darker color. In addition, myoglobin tends to be more stable against both oxidative and heat-induced changes at a higher than normal pH. This idea is particularly important in swine, where strategic finishing with low-carbohydrate diets can reduce muscle glycogen stores. In this case, less glycogen at the time of slaughter will minimize both postmortem glycolysis and pH decline, both of which should improve pork color and increase longissimus darkness in pigs susceptible to stress and pale muscle defects (Rosenvold et al. 2001a, b). Overall, muscle capacity for anaerobic metabolism and lactate production are directly related to pork lightness and longissimus quality (Meadus and MacInnis 2000; Hamilton et al. 2003). Because this is related to substrates found in muscle that can be converted to lactic acid, reducing postmortem glycolytic potential via manipulation of free glucose may improve pork muscle color (Juncher et al. 2001). Other husbandry practices such as housing environment can alter postmortem metabolism via changes in muscle fiber type. For example, housing that promotes physical activity can increase muscle pigmentation and darkness through changes in slow-contracting fiber content, vascularization, and oxidative metabolic capacity, all of which could (1) decrease lactate production, (2) increase pyruvate oxidation within mitochondria, (3) promote β-oxidation, and (4) lengthen time to muscle exhaustion (Vestergaard et al. 2000).

There is a relation between subcutaneous fat depth and both muscle lightness and pH because fat thickness influences postmortem chilling rate (Bruce et al. 2004). In general, increased fat causes a slower chilling rate, which is more conducive to anaerobic metabolism. Warmer muscle temperatures shortly after harvest promote glycolytic enzyme activity and pH decline, increasing protein denaturation and muscle lightness (Bruce et al. 2004). Thus, diet-induced effects on fat thickness can indirectly affect muscle lightness. Dietary effects on color stability are primarily attributed to antioxidant accumulation and the instability of polyunsaturated fatty acids, rather than the chilling rate. This concept is associated with the interrelationship between lipid and pigment oxidation, whereby a diet-induced decrease in lipid oxidation either through changes in fatty acid profile or changes in antioxidant accumulation should improve color stability (French et al. 2000; Muramoto et al. 2003; O'Sullivan et al. 2003).

11.3.2 Postmortem Factors

Proper cold chain management (1) is likely the most important factor in food preservation and (2) involves the maintenance of refrigerated temperatures during both storage and display in order to maximize meat product color and shelf life. A lack of cold chain management will promote pigment and lipid oxidation as well as microbial growth. In addition to the display case temperature, both lighting type and intensity will have an effect on surface discoloration.

Meat packaging plays a significant role in both meat color and consumer purchasing. As a result, several types of packaging are available, ranging from traditional polyvinyl chloride (PVC) over-wrap and Styrofoam trays to modified atmosphere technologies. Traditional PVC over-wrap is permeable to oxygen, and therefore allows for the development of bright cherry-red oxymyoglobin on the product surface. However, an increased demand for case-ready meat products has allowed processors to modify the atmosphere within a meat package. Typical modifications include the use of either high levels of oxygen or low levels of carbon monoxide. In addition, the oxygen can be removed from the package atmosphere, as in the case of vacuum packaging, which can improve color life compared with traditional PVC over-wrap.

High-oxygen packages typically contain 80% oxygen in order to promote both myoglobin oxygenation and depth of oxymyoglobin penetration, which prolongs the amount of time before metmyoglobin migrates to the muscle surface. Carbon dioxide is also used in high-oxygen packages to inhibit microbial growth. Although high-oxygen atmospheres improve product redness, they also are conducive to lipid oxidation and, as a result, rancidity often develops before surface discoloration (Jayasingh et al. 2002).

Removing oxygen from meat packages, either through the use of vacuum or ultra-low-oxygen atmospheres containing a blend of nitrogen and carbon dioxide, can minimize lipid oxidation and aerobic microorganism growth. However, there are several obstacles associated with anaerobic packaging, including (1) purplish deoxymyoglobin can negatively affect consumer purchasing, (2) levels of residual oxygen within anaerobic packages must remain less than 1% for pork and less than 0.05% for beef in order to limit metmyoglobin formation at low partial pressures, and (3) sufficient pigment-reducing capacity and oxygen consumption are necessary for both deoxygenation and post-storage surface oxygenation (rebloom). As a result, low levels of carbon monoxide can be added to packages because of its ability to form a bright cherry-red carboxymyoglobin and improve color life (Sørheim et al. 1999; Luno et al. 2000; Jayasingh et al. 2001; Hunt et al. 2004).

Cooking denatures myoglobin, resulting in the characteristic dull brown color associated with cooked meat products. Several intrinsic (myoglobin redox state, muscle source, and antioxidants) and extrinsic (packaging, storage, and cooking from a frozen state) factors influence the development of cooked color. In particular, cooked color is associated with packaging type and, as a result, it is possible that the effects of modified atmospheres on raw meat pigment chemistry can translate into a cooked color concern known as premature browning.

Although consumers generally assume that a brown color in the center of beef indicates that the product has reached a temperature sufficient to inactivate food-borne pathogens, cooked color is not necessarily a reliable indicator that beef has been pasteurized (Bigner-George and Berry 2000). This is of particular concern in ground beef, which can experience premature browning, a condition in which myoglobin denatures at a temperature lower than that necessary to inactivate pathogens. The practical implication of this is that premature browning, or the appearance of a cooked color on the interior of ground beef patties prior to pathogen destruction, can be misinterpreted as an indicator of thermal pasteurization. Meat should be cooked to an internal temperature of 71°C in order to ensure that *E. coli* O157:H7 is destroyed (USDA 1997). However, premature browning has been noted at 55°C, a temperature at which some pink color is normally expected (Lavelle et al. 1995).

The role of packaging in premature browning is primarily related to the redox state of myoglobin in raw beef prior to cooking (Hague et al. 1994). This is due to the effects of redox state on the temperature at which myoglobin denatures. Relative resistance of myoglobin to heat-induced denaturation is as follows: deoxy > oxy > metmyoglobin. Thus, packaging atmospheres that are conducive to metmyoglobin formation will likely result in more premature browning at a lower temperature than packaging atmospheres that maintain predominantly ferrous myoglobin (Warren et al. 1996). More specifically, maintenance of beef pigments in a deoxygenated state with vacuum packaging will decrease the occurrence of premature browning, whereas the tendency for beef to undergo premature browning is often greater when packaged in high-oxygen packaging (containing 80%

oxygen) (Lavelle et al. 1995; Seyfert et al. 2004a, b; Suman et al. 2005). In contrast, packaging containing 0.4% CO will minimize the occurrence of premature browning (John et al. 2004; 2005).

Although hemoglobin plays a minimal role in postmortem muscle color, the redox state of hemoglobin is responsible for bone marrow color changes (Mancini et al. 2004; 2005). For example, cutting bone-in steaks (1) disrupts red blood cells, (2) exposes deoxygenated hemoglobin located within marrow to oxygen, and (3) produces bright-red oxyhemoglobin, which can be further oxidized to grayish-black methemoglobin. Thus, maintenance of ferrous hemoglobin on the surface of cut bones is critical to improving beef and pork shelf life. Water-soluble reducing agents such as ascorbic acid and sodium erythorbate or modified atmosphere packaging containing 0.4% carbon monoxide can improve rib and vertebrae color stability, whereas high-oxygen packaging increases bone marrow discoloration (Mancini et al. 2005; Grobbel et al. 2006; Raines et al. 2006). Similarly, the redox state of residual hemoglobin (capillaries and/or hemorrhage) within subcutaneous fat determines color (Irie 2001). Discoloration of both bone marrow and fat can detrimentally affect beef and pork shelf life.

11.4 SUMMARY

Meat color is influenced by several factors that interact, making it difficult to troubleshoot meat color problems. Ante- and postmortem factors not only affect meat color, but also play a role in consumer satisfaction and shelf life. A solid understanding of myoglobin chemistry will maximize an individual's ability to utilize technology, such as injection-enhancement and modified atmosphere packaging, to their advantage in order to improve postmortem muscle quality.

REFERENCES

Alderton, A.L., C. Faustman, D.C. Liebler, and D.W. Hill. 2003. Induction of redox instability of bovine myoglobin by adduction with 4-hydroxy-2-nonenal. *Biochemistry* 42:4398–4405.

Baron, C.P. and H.J. Andersen. 2002. Myoglobin-induced lipid oxidation. A review. *J. Agric. Food Chem.* 50:3887–3897.

Baublits, R.T., A.H. Brown, Jr., F.W. Pohlman, Z.B. Johnson, D.O. Onks, and H.D. Loveday. 2004. Carcass and beef color characteristics of three biological types of cattle grazing cool-season forages supplemented with soyhulls. *Meat Sci.* 68:297–303.

Bertram, H.C., J.S. Petersen, and H.J. Andersen. 2000. Relationship between RN-genotype and drip loss in meat from Danish pigs. *Meat Sci.* 56:49–55.

Bigner-George, M.E. and B.W. Berry. 2000. Thawing prior to cooking affects sensory, shear force, and cooking properties of beef patties. *J. Food Sci.* 65:2–8.

Brooks, J. 1935. The oxidation of haemoglobin to methaemoglobin by oxygen. II. The relation between the rate of oxidation and the partial pressure of oxygen. *Proc. Roy. Soc., London, Ser. B* 118:560–577.

Bruce, H.L., J.L. Stark, and S.L. Beilken. 2004. The effects of finishing diet and postmortem ageing on the eating quality of the M. longissimus thoracis of electrically stimulated Brahman steer carcasses. *Meat Sci.* 67:261–268.

Channon, H.A., A.M. Payne, and R.D. Warner. 2000. Halothane genotype, pre-slaughter handling and stunning method all influence pork quality. *Meat Sci.* 56:291–299.

Cornforth, D. 1994. Color — its basis and importance. In A.M. Pearson and T.R. Dutson, (Eds.), *Advances in Meat Research,* Vol. 9, Blackie Academic and Professional, London (1994), pp. 34–78.

Faustman, C. and R.G. Cassens. 1990. The biochemical basis for meat discolouration in fresh meat: a review. *J. Muscle Foods* 1:217–243.

Faustman, C., D.C. Liebler, T.D. McClure, and Q. Sun. 1999. Alpha,beta-unsaturated aldehydes accelerate oxymyoglobin oxidation. *J. Agric. Food Chem.* 47:3140–3144.

Fernandez, X., E. Neyraud, T. Astruc, and V. Sante. 2002. Effects of halothane genotype and pre-slaughter treatment on pig meat quality. 1. Postmortem metabolism, meat quality indicators and sensory traits of m. *longissimus lumborum*. *Meat Sci.* 62:429–437.

Fisher, P., F.D. Mellett, and L.C. Hoffman. 2000. Halothane genotype and pork quality. 1. Carcass and meat quality characteristics of three halothane genotypes. *Meat Sci.* 54:97–105.

French, P., C. Stanton, F. Rawless, E.G. O'Riordan, F.J. Monahan, and P.J. Caffery. 2000. Fatty acid composition, including conjugated linoleic acid of intramuscular fat from steers offered grazed grass, grass silage, or concentrate-based diets. *J. Animal Sci.* 55:2849–2855.

Giddings, G.G. 1977. The basis of color in muscle foods. *CRC Crit. Rev. Food Sci. Nutr.* 9:81–114.

Gill, C.O. 1996. Extending the storage life of raw chilled meats. *Meat Sci.* 43:S99–S109.

Grobbel, J.P., M.E. Dikeman, J.S. Smith, D.H. Kropf, and G.A. Milliken. 2006. Effects of polyvinyl chloride overwrap film, high oxygen modified atmosphere packaging, or ultra-low modified atmosphere packaging on bone marrow discoloration in beef humerus, rib, thoracic vertebrae, and scapula. *J. Anim. Sci.* 84:694-701.

Hague, W.A., K.E. Warren, M.C. Hunt, D.H. Kropf, C.L. Kastner, S.L. Stroda, and D.E. Johnson. 1994. Endpoint-temperature, internal cooked color, and expressible juice color relationships in ground beef patties. *J. Food Sci.* 59:465-470

Hamilton, D.N., K.D. Miller, M. Ellis, F.K. McKeith, and E.R. Wilson. 2003. Relationships between longissimus glycolytic potential and swine growth performance, carcass traits, and pork quality. *J. Anim. Sci.* 81:2206–2212.

Hunt, M.C., R.A. Mancini, K.A. Hachmeister, D.H. Kropf, M. Merriman, and G. DelDuca. 2004. Carbon monoxide in modified atmosphere packaging affects color, shelf life, and microorganisms of beef steaks and ground beef. *J. Food Sci.* 69:C45–C52.

Irie, M. 2001. Optical evaluation of factors affecting appearance of bovine fat. *Meat Sci.* 57:19–22.

Jayasingh, P., D.P. Cornforth, C P. Brennand, C.E. Carpenter, and D.R. Whittier. 2002. Sensory evaluation of ground beef stored in high-oxygen modified atmosphere packaging. *J. Food Sci.* 67:3493–3496.

Jayasingh, P., D.P. Cornforth, C.E. Carpenter, and D. Whittier. 2001. Evaluation of carbon monoxide treatment in modified atmosphere packaging or vacuum packaging to increase color stability of fresh beef. *Meat Sci.* 59:317–324.

John, L., D.P. Cornforth, C.E. Carpenter, O. Sørheim, B.C. Pettee, and D.R. Whittier. 2004. Comparison of color and thiobarbituric acid values of cooked hamburger patties after storage of fresh beef chubs in modified atmospheres. *J. Food Sci.* 69:608–614.

John, L., D.P. Cornforth, C.E. Carpenter, O. Sørheim, B.C. Pettee, and D.R. Whittier. 2005. Color and thiobarbituric acid values of cooked top sirloin steaks packaged in modified atmospheres of 80% oxygen, or 0.4% carbon monoxide, or vacuum. *Meat Sci.* 69:441–449.

Juncher, D., B., Ronn, E.T. Mortensen, P. Henckel, A. Karlsson, and L.H. Skibsted. 2001. Effect of pre-slaughter physiological conditions on the oxidative stability of colour and lipid during chill storage of pork. *Meat Sci.* 58:347–357.

Krause, T.R., J.G. Sebranek, R.E. Rust, and M.S. Honeyman. 2003. Use of carbon monoxide packaging for improving the shelf life of pork. *J. Food Sci.* 68:2596–2603.

Kuchenmeister, U., G. Kuhn, and E. Ender. 2000. Seasonal effects on Ca+2 transport of sarcoplasmic reticulum and on meat quality of pigs with different malignant hyperthermia status. *Meat Sci.* 55:239–245.

Lavelle C.L., M.C. Hunt, and D.H. Kropf. 1995. Display life and internal cooked color of ground beef from vitamin E-supplemented steers. *J. Food Sci.* 60:1175–1178, 1196.

Lee, S., A.L. Phillips, D.C. Liebler, and C. Faustman. 2003. Porcine oxymyoglobin and lipid oxidation *in vitro*. *Meat Sci.* 63:241–247.

Lindahl, G., A. Enfalt, G. von Seth, A. Joseli, I. Hedebro-Velander, and H.J. Andersen. 2004. A second mutant allele (V199I) and the PRKAG3 (RN) locus II. Effect of colour characteristics of pork loin. *Meat Sci.* 66:621–627.

Livingston, D.J., S J. McLachlan, G.N. LaMar, and W.D. Brown. 1985. Myoglobin: cytochrome b5 interactions and the kinetic mechanism of metmyoglobin reductase. *J. Biol. Chem.* 260:15699–15707.

Luno, M., P. Roncales, D. Djenane, and J.A. Beltran. 2000. Beef shelf life in low O_2 and high CO_2 atmospheres containing different low CO concentrations. *Meat Sci.* 55:413–419.

Mancini, R.A. and M.C. Hunt. 2005. Current research in meat color. *Meat Sci.* 71:100–121.

Mancini, R.A., M.C. Hunt, K.A. Hachmeister, D.H. Kropf, and D.E. Johnson. 2004. Ascorbic acid minimizes lumbar vertebrae discoloration. *Meat Sci.* 68:339–345.

Mancini, R.A., M.C. Hunt, K.A. Hachmeister, D.H. Kropf, and D.E. Johnson. 2005. Exclusion of oxygen from modified atmosphere packages limits beef rib and lumbar vertebrae marrow discoloration during display and storage. *Meat Sci.* 69:493–500.

Meadus, W.J. and R. MacInnis. 2000. Testing for the RN- gene in retail pork chops. *Meat Sci.* 54:231–237.

Moelich, E., L.C. Hoffman, and P.J. Conradie. 2003. Sensory and functional meat quality characteristics of pork derived from three halothane genotypes. *Meat Sci.* 63:333–338.

Moeller, S.J., T.J. Baas, T.D. Leeds, R.S. Emnett, and K.M. Irvin. 2003. *Rendement Napole* gene effects and a comparison of glycolytic potential and DNA genotyping for classification of *Rendement Napole* status in Hampshire-sired pigs. *J. Anim. Sci.* 81:402–410.

Muramoto, T., N. Nakanishi, M. Shibata, and A. Aikawa. 2003. Effect of dietary β-carotene supplementation on beef color stability during display of two muscles from Japanese Black steers. *Meat Sci.* 63:39–42.

O'Sullivan, A., K. Galvin, A.P. Moloney, D.J. Troy, K. O'Sullivan, and J.P. Kerry. 2003. Effect of pre-slaughter rations of forage and/or concentrates on the composition and quality of retail packaged beef. *Meat Sci.* 63:279–286.

Pegg, R.B. and F. Shahidi. 1997. Unraveling the chemical identity of meat pigments. *CRC Crit. Rev. Food Sci. Nutr.* 37:561–589.

Piedrafita, J., L.L. Christian, and S.M. Lonergan. 2001. Fatty acid profiles in three stress genotypes of swine and relationships with performance, carcass and meat quality traits. *Meat Sci.* 57:71–77.

Raines, C.R., M.E. Dikeman, J.P. Grobbel, and E.J. Yancey. 2006. Effects of ascorbic acid and Origanox™ in different packaging systems to prevent pork lumbar vertebrae discoloration. *Meat Sci.* 74:267–271.

Rosenvold, K. and H.J. Andersen. 2003. The significance of preslaughter stress and diet on colour and colour stability of pork. *Meat Sci.* 63:199–209.

Rosenvold, K., H.N. Laerke, S.K. Jensen, A.H. Karlsson, K. Lundstrom, and H.J. Andersen. 2001a. Strategic finishing feeding as a tool in the control of pork quality. *Meat Sci.* 59:397–406.

Rosenvold, K., J.S. Peterson, H.N. Laerke, S.K. Jensen, M. Therkildsen, and A.H. Karlsson. 2001b. Muscle glycogen stores and meat quality as affected by strategic finishing feeding of slaughter pigs. *J. Animal Sci.* 79:382–391.

Seideman, S.C., H.R. Cross, G.C. Smith, and P.R. Durland. 1984. Factors associated with fresh meat color: a review. *J. Food Quality*, 6:211–237.

Seyfert, M., M.C. Hunt, R.A. Mancini, D.H. Kropf, and S.L. Stroda. 2004a. Internal premature browning in cooked steaks from enhanced beef round muscles packaged in high-oxygen and ultralow oxygen modified atmospheres. *J. Food Sci.* 69:142–146.

Seyfert, M., R.A. Mancini, and M.C. Hunt. 2004b. Internal premature browning in cooked ground beef patties from high-oxygen modified-atmosphere packaging. *J. Food Sci.* 69:721–725.

Smith, G.C., K.E. Belk, J.N. Sofos, J.D., Tatum, and S.N. Williams. 2000. Economic implications of improved color stability in beef. In E.A. Decker, C. Faustman, and C.J. Lopez-Bote (Eds.), *Antioxidants in Muscle Foods: Nutritional Strategies to Improve Quality* New York: Wiley Interscience, p. 397–426.

Sørheim, O., H. Nissen, and T. Nesbakken. 1999. The storage life of beef and pork packaged in an atmosphere with low carbon monoxide and high carbon dioxide. *Meat Sci.* 52:157–164.

Suman, S.P., C. Faustman, S. Lee, J. Tang, H.A. Sepe, P. Vasudevan, T. Annamalai, M. Manojkumar, P. Marek and K.S. Venkitanarayanan. 2005. Effect of erythorbate, storage and high-oxygen packaging on premature browning in ground beef. *Meat Sci.* 69:363–369.

Suman S.P., C. Faustman, S.L. Stamer, and D.C. Liebler. 2006. Redox instability induced by 4- hydroxy-2-nonenal in porcine and bovine myoglobins at pH 5.6 and 4°C. *J. Agric. Food Chem.* 54:3402–3408.

Suman, S.P., C. Faustman, S.L. Stamer, and D.C. Liebler. 2007. Proteomics of lipid oxidation-induced oxidation of porcine and bovine oxymyoglobins. *Proteomics* 7:628–640.

Swatland, H.J. 1984. Optical characteristics of natural iridescence in meat. *J. Food Sci.* 49:685–686.

Van Oeckel, M.J., N. Warrants, C.V. Boucque, P. Delputte, and J. Depuydt. 2001. The preference of the consumer for pork from homozygous or heterozygous halothane negative animals. *Meat Sci.* 58:247–251.

Velarde, A., M. Gispert, L. Faucitano, P., Alonso, X., Manteca, and A. Diestre. 2001. Effects of the stunning procedure and the halothane genotype on meat quality and incidence of haemorrhages in pigs. *Meat Sci.* 58:313–319.

Vestergaard, M., N. Oksberg, and P. Henckel. 2000. Influence of feeding intensity, grazing and finishing feeding on muscle fibre characteristics and meat colour of semitendinosus, longissimus dorsi and supraspinatus muscles of young bulls. *Meat Sci.* 54:177–185.

Warren, K.E., M.C. Hunt, and D.H. Kropf. 1996. Myoglobin oxidative state affects internal cooked color development in ground beef patties. *J. Food Sci.* 61:513–515, 519.

12 Lipid Oxidation and Flavor

D.U. Ahn, K.C. Nam, and E.J. Lee

CONTENTS

12.1 INTRODUCTION

Lipid oxidation produces various primary and secondary by-products with food quality, nutritional, and health implications (Pearson et al. 1983). Lipid oxidation can occur in any foods that contain lipids but meat is among the most susceptible foods to lipid oxidation. The rate of lipid oxidation varies, depending on various internal conditions of meat such as fat content, fatty acid composition, antioxidants, heme pigments, and iron content, as well as the external conditions of meat. Lipid oxidation in meat produces warmed-over flavor, deteriorates meat color, causes loss of functional properties and nutritional values, and forms possible toxic compounds. Lipid oxidation can occur by autoxidation, photoxidation, and enzymatic oxidation processes. In enzymatic lipid oxidation, cyclooxygenase and lipoxygenase catalyze the reactions between oxygen and polyunsaturated fatty acids. The main function of these enzymes is to generate superoxide, which serves as a reducing agent for ferric iron complexes in microsomes. Enzymatic lipid oxidation in skeletal muscle microsomes is dependent upon NADH or NADPH, and requires ADP and Fe (II) or Fe (III) for maximum rate, and Fe (II) and its complexes stimulate membrane peroxidation more than Fe (III) does. In the presence of sensitizers such as chlorophyll, porphyrins, myoglobin, riboflavin, bilirubin, erythrosine, rose bengal, or methylene blue, singlet oxygen initiates photoxidation. Photoxidation plays an important role in the quality of oils and milk powder but its role in meat oxidation is not well documented. Photoxidation can be inhibited by carotenoids through interference with the formation of singlet oxygen from the oxygen molecule. Tocopherols can also inhibit photoxidation by quenching the previously formed singlet oxygen. Autoxidation is the major oxidation mechanism in meat and is initiated by reactive oxygen species such as hydroxyl radicals that abstract hydrogen atoms from the fatty acyl group of polyunsaturated fatty acids and form lipid radicals. Because the lipid radicals formed from the initial reactions have a strong enough reactivity, they can also transfer free radicals to adjacent lipids and thus the reaction can perpetuate unless the free radicals are quenched or reactions are interrupted by chain-breaking antioxidants. Various primary and secondary byproducts of lipid oxidation can be formed from meat, depending upon the type of meat, pro- and antioxidants, processing, cooking, packaging, and storage conditions. Some of the lipid oxidation products are known to be toxic or carcinogenic, and others alter meat flavor. Therefore, prevention of lipid oxidation is very important not only for the meat industry, but also for consumers.

12.2 MEAT FLAVOR

The measurable flavor perception from eating food products is determined by three components: volatile odor and nonvolatile taste compounds present, and the availability of these components (Overbosch 1991). Nonvolatile nitrogen and sulfur compounds play important roles in the development of characteristic flavors of red meats (Mabrouk 1977). Hsieh (1981) reported that certain precursor compounds, including simple sugars, 5′-nucleotides, glycoprotein, and S-containing amino acids, were necessary for the development of basic meat flavor. Schutte (1997) distinguished three types of meat flavor: (1) taste mixtures composed of nonvolatiles like monosodium glutamate, ribonucleotides, and organic acids; (2) mixtures of volatile compounds ("top-notes") such as furanones, thiophenes, and aldehydes, which give a specific meaty odor; and (3) so-called "reaction flavors," which comprise precursors that react upon heating to yield mixtures with a basic meat flavor that balances the rather harsh top-notes. The precursors of meat flavor generally consist of sulfur compounds like cysteine, methionine, thiamine, or H_2S releasers and carbonyl compounds like monosaccharides, aldehydes, or furanones. Each of the three flavor types makes its specific contribution to the formation of meat flavor. A well-balanced combination of the three flavor types leads to a total meat flavor.

Taste compounds are water-soluble substances with sensation properties on the tongue. Basic tastes include salty, sweet, sour, and bitter. In addition to those basic tastes, "umami," which is used to describe a characteristic pleasant taste for meats by the Japanese, is very important for meat flavor. Glutamate is the main amino acid corresponding to this "umami" taste. Inosine monophosphate and guanosine monophosphate work synergistically with glutamate. Maga et al. (1994) described that a 1:1 mixture of glutamate and the above nucleotides provides a flavor that is 30 times greater in intensity then that of monoglutamate. A salty taste is caused by sodium chloride and has a significant enhancing effect on the "umami" taste of meat.

Hundreds of volatile and aroma compounds can be produced from meat during processing and storage. The amount and kinds of volatiles produced from meat can be influenced by animal species, muscles within an animal, diets, age of animal, lipid content, and pH of muscle (Calkins et al. 2007). The characteristic flavor of cooked meat derives from thermally induced reactions that occur during heating, principally the Maillard reaction and the degradation of lipids. During heating, meat flavor precursors, including free amino acids, peptides, sugars, nucleotides, organic acids, and fatty acids, react with each other or degrade to produce large quantities of volatiles and form the typical aroma of cooked meat. Both sulfur and carbonyl compounds in volatiles are considered important in cooked meaty, savory, roast, and boiled flavors. Lipid degradation by heat produces compounds that give fatty aromas specific to the animal species (Ramaswamy and Richards 1982; Mottram 1998). Ang and Liu (1996a) indicated that more than 50% of volatiles increased linearly or quadratically as cooking end-point temperature increased. Most of those volatile increases are related to increased lipid oxidation as the temperature increased. Bailey and Um (1992) and Imafidon and Spanier (1994) reported that at temperatures above 77°C, lipid oxidation is inhibited due to the antioxidant effect of Maillard reactions. The Maillard reaction and other reactions during heating contribute significantly to the formation of volatiles that comprise the cooked meat aroma. Several hundred volatiles were detected by analyzing the headspace of cooked meat (Ramarathnam et al. 1993). But most of them are with high thresholds, and thus might not contribute significantly to the aroma of meat. Among the lipid oxidation-related volatiles, aldehydes contributed the most to the warmed-over flavor of meat; heptanal has the lowest threshold, followed by 2,4-decadienal, pentanal, and hexanal (Brewer and Vega 1995). Dietze et al. (2007) reported that hexanal, 1-octene-3-ol, and hexanoic acid play an important role in warmed-over flavor development in turkey meat. The odor of raw meat is quite weak compared with cooked meat, and sulfur compounds may determine the normal aroma of raw meat. By examining the volatiles in the package headspace of modified atmosphere packaged raw chicken legs, Eilamo et al. (1998) found that butene, ethanol, acetone, pentane, dimethyl sulfide, carbon disulfide, and dimethyl disulfide were the main volatiles. However, the final perception of meat flavor is the result of complex interactions among the various volatiles present in the meat.

Warmed-over flavor (WOF) is most frequently used to describe the unpleasant flavor of reheated cooked meat associated with oxidation, often characterized as "cardboard" and having a "paint"-like odor (Love and Pearson 1971; Asghar et al. 1988). Warmed-over flavor is closely related to the increase in lipid oxidation-derived off-flavor and odor, such as "rancid' and "sulfur/rubber" sensory notes, with a concurrent decrease in meaty flavor (Byrne et al. 2002; Campo et al. 2006). Byrne et al. (2001) reported that the reduction in "meatiness" during the early stage of WOF development was attributed to the degradation of unstable sulfur-containing amino acids and sulfur-containing meaty aroma compounds; the later stage of WOF development was caused by an increase in lipid oxidation by-products. Byrne et al. (2003) also reported that increasing the cooking temperature produced meats with a more "roasted," "toasted," and "bitter" taste, and the changes in sulfur compounds and volatiles that produced the roasted flavor were related to WOF development in cooked meat during chill-storage. The sensory notes of those compounds were sulfury/rubbery and associated with WOF. Campo et al. (2006) reported that sensory and analytical attributes such as TBARS values had strong correlations and indicated that a beef with a TBARS value greater than 2 was unacceptable. The speed of oxidation depends on the unsaturation of fatty acid and the presence of prooxidants and antioxidants. Effects to increase the n-3 polyunsaturated fatty acids in poultry in order to improve nutritional value exacerbate the problem of oxidation. N-3 PUFA in poultry meat can be increased by feeding fish oil, fish meal, algae, or by feeding vegetable oils high in linolenic acid (Pinchasov and Nir 1992). In meats, the concentration of unsaturated fatty acids in phospholipids is much higher than in triglycerides. Thus, phospholipids played a main role in lipid oxidation (Igene and Pearson 1979; Igene et al. 1980). Oxidation is much quicker in cooked meat due to phospholipid structure damage during cooking (Ahn et al. 1992). Wu and Sheldon (1988) suggested that WOF in cooked meat is related to lipid oxidation, and that phospholipids are the main source of this fatty acid oxidation. Additionally, any degree of lipid oxidation in raw meat accelerates the development of oxidized off-flavors in cooked meat, due to the free-radical chain reaction nature of lipid oxidation (Rhee 1989).

Malonaldehyde is a secondary oxidation product of polyunsaturated fatty acids. TBARS measures the malonaldehyde content in meat. TBARS analysis is the most frequently used test to determine lipid oxidation. A potential problem with the TBARS test is that some other components in meat may interfere with measurement. However, the TBARS value has been shown to be positively related to WOF (Igene et al. 1985). Warmed-over and rancid flavor and aroma scores correlated highly and consistently with TBARS values (Nolan et al. 1989). Hexanal, an oxidative product of linoleic acid, has also been used as an indicator of lipid oxidation (Shahidi 1989). Shu et al. (1995) suggested that hexanal and TBARS are highly correlated with the WOF of meat, but Daze et al. (2005) reported that hexanal was a better indicator for lipid oxidation than TBARS in Iberian pig muscle. Significant correlations were found between TBARS values and pentanal, hexanal, heptanal, and total volatiles (Su et al. 1991). Lai et al. (1995) also showed that TBARS was related to hexanal and off-flavor development, as judged by sensory panelists. Ang and Lyon (1990) illustrated that the intensities of cardboard, warmed-over, rancid/painty, and overall off-flavor characteristics increased and were associated with TBARS and headspace GC aldehyde content. The propanal, pentanal, hexanal, and total volatiles were highly correlated with the TBARS values of cooked meat (Ahn et al. 1998a).

Irradiation is approved in meat to control food-borne pathogens (Farkas 1998). However, ionizing radiation generates free radicals that may induce lipid oxidation and other chemical changes. Patterson and Stevenson (1995) isolated those volatiles from irradiated raw chicken by gas chromatography in conjunction with olfactory assessment of the effluent carrier gas to locate compounds with strong smells, and found that dimethyl trisulfide was the most potent and obnoxious compound, followed by *cis*-3*trans*-6-nonenals, 1-octenone, and bis(methylthio-) methane. Ahn et al. (2000b) reported that dimethyl disulfide was found in volatiles induced by irradiation and might be related to irradiation odor. Hashim et al. (1995) showed that raw irradiated chicken meat had higher "fresh chickeny aroma" than nonirradiated samples, and cooked meat after irradiation had more chicken flavor. The reason may be due to the presence of increased sulfur compounds after irradiation. It has already been shown that sulfur compounds are important for the characteristic chicken flavor, and increased sulfur compounds after irradiation strengthen the chicken flavor. Ahn et al. (1998b) observed that irradiation accelerated lipid oxidation in turkey meat but that dietary α-tocopheryl acetate minimized the oxidation and off-odor generation, suggesting that lipid oxidation induced by irradiation might also contribute to irradiation odor.

In addition to lipid oxidation and irradiation, other off-odors may be caused by bacteria, diet, sex, or contamination of environmental odors during storage. For example, intestinal microflora may influence the flavor of poultry meat. Harris et al. (1968) showed that broilers raised under germ-free conditions had a significantly different odor compared with chickens raised under conventional conditions. Mead et al. (1983) reported that birds had higher counts of *E. coli* and fecal *streptococci,* with the meat described as "richer", "sweeter," and a stronger overall flavor. Fish oils in feed tend to cause a fishy taint in the resulting poultry meat (Poste 1991; Leskanich and Noble 1997).

12.3 ANALYTIC METHODS FOR AROMA AND OFF-ODOR OF MEAT

12.3.1 Instrumental Analysis

Dynamic headspace coupled with gas chromatography-mass spectrometry (GC-MS) is the most powerful method for analyzing volatiles in meat samples. One difficulty with instrumental analysis of volatiles was volatile collection. Prior to the 1990s, most published articles on volatile analysis employed a volatile extraction method using either distillation, solvent extraction, or direct isolation (Tanchotikul and Hsieh 1991; Ajuyah et al. 1993a; Specht and Baltes 1994). However, there are many artifacts during the volatile extraction and concentration steps that put the results in doubt as to whether they truly represent the actual volatiles of sample. Fortunately, in recent years, purge-and-trap and solid-phase microextraction techniques for collecting headspace volatiles became available (Brunton et al. 2000). Because many volatiles are unstable, they are easily oxidized or subject to other changes; thus, headspace volatiles may still change due to oxidation or other chemical reactions during holding for analysis. Ahn et al. (1999a, b) performed a detailed study of volatile stability at different sample holding times, temperatures, oxygen levels, and purge temperatures. Hexanal was used as an indicator for volatile changes in those studies. Figure 12.1 shows the effect of headspace oxygen and holding time on the volatiles detected in both raw and cooked turkey breast meat. Helium flush and oxygen absorber were used to reduce the oxygen present in the headspace. From Figure 12.1a, without helium flush and oxygen absorber, the hexanal content for raw meat remained almost constant for 320 min of holding before analysis. The increase in hexanal content in volatiles means the oxidation of meat sample and volatiles, and thus 320 min seemed to be the maximum holding time when oxygen was available. But after helium flush, the content of hexanal decreased significantly, indicating that the stability of volatiles during sample holding for analysis improved significantly. After adding the oxygen absorber, there was virtually no increase in hexanal content during the sample holding period. However, volatiles from the oxygen absorber itself were a

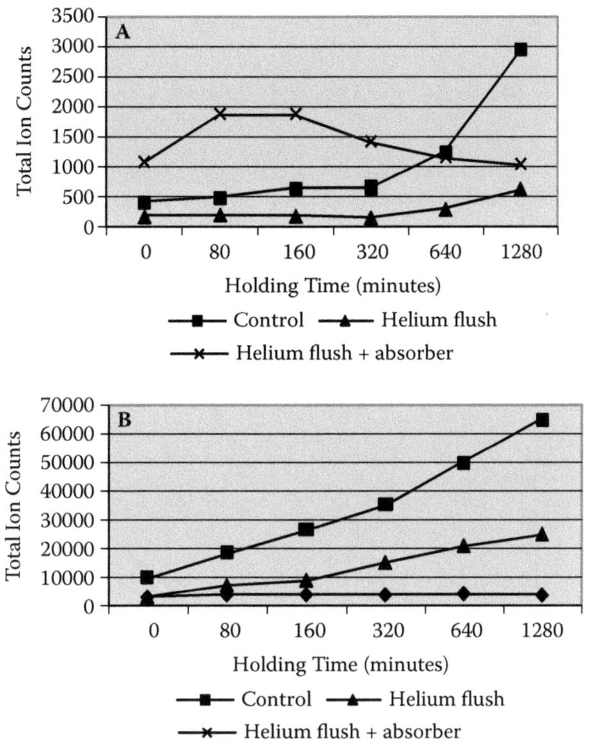

FIGURE 12.1 The effect of headspace oxygen on hexanal content in volatiles from turkey thigh meat. (A) Hexanal content in raw turkey meat; and (B) hexanal content in cooked turkey meat. Control means no additional treatment for the headspace oxygen during sample preparation; helium flush means helium (99.999%) was used to flush the headspace to remove oxygen; helium flush + absorber means that in addition to helium flush, an oxygen absorber was also used to remove oxygen. The hexanal content for raw meat with helium flush + absorber was mixed with 2,6-dimethyl heptane. Total ion counts: ion counts × 10^4.

problem. 2,6-Dimethyl heptane resulted from the oxygen absorber mixed with hexane and interfered with the hexanal analysis. Because the volatiles of raw meat are relatively stable and the total volatile content of raw meat is quite low, low amounts of volatiles from the oxygen absorber interferes with volatile analysis, and thus adding oxygen absorber is not appropriate for raw meat. However, cooked meat can be easily oxidized when oxygen is available (Ahn et al. 1992; Nam et al. 2004). As shown in Figure 12.1b, without any treatment to remove the oxygen in the headspace of sample, the hexanal content in the volatiles of cooked meat increased quickly as the holding time increased. After helium flush, the production of hexanal significantly decreased. With the addition of oxygen absorber, however, the hexanal content in the sample was almost constant. For cooked meat, therefore, both helium flush and oxygen absorber are necessary in order to remove headspace oxygen.

Figure 12.2 shows the increased aldehyde content with increasing purge temperature. As the purge temperature increases, the content of aldehydes in volatiles tended to increase, especially for cooked meat. Therefore, in order to minimize the heat-induced production of volatiles during purging, a purging temperature of 40 to 50°C is recommended. The column chosen for volatile analysis influences the separation of volatiles. Ahn et al. (2000a, b) showed that combinations of two or three columns with different polarities could improve the separation of volatiles significantly. In addition, adding a cryofocusing unit before injection also improved volatile separation (Ahn et al. 2000b). Olsen et al. (2005) reported that fluorescence and dynamic headspace GC-MS methods detected oxidative changes in pork fat and mechanically recovered meat earlier than the sensory and electronic nose, and the sensory results showed high correlations with the instrumental results.

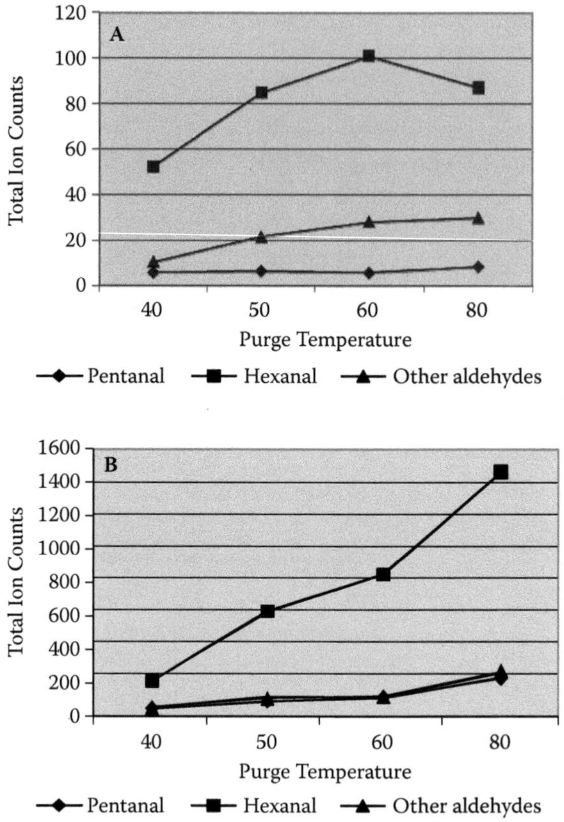

FIGURE 12.2 The effect of purge temperature on the production of aldehydes in turkey thigh meat. (A) Volatile changes in raw turkey meat; and (B) volatile changes in cooked turkey meat. Total ion counts: ion counts × 10^4.

12.3.2 Sensory Analysis

For the sensory analysis of off-odor, quantitative descriptive analysis (QDA) is most commonly used. The basic procedure for QDA is first to characterize the off-odors present in samples. To do this, 10 to 12 panelists are needed to work under a sensory professional to explore and describe those off-odors in meat. A sensory terminology was developed to describe those off-odors and also establish evaluation procedures. Each off-odor characterized by the panelists may be evaluated for the intensity. Panelists are trained to familiarize themselves with those off-odors and evaluation procedures before evaluating samples. However, WOF and irradiation odor have already been well-defined, and thus the procedure to develop terminology for that may not be necessary. Instead, sensory panelists can be trained directly with the characteristic oxidation or irradiation odor and let them familiarize themselves with that off-odor, and then judge the strength of off-odor by line scaling, category scaling, or magnitude estimation. Line scaling is most commonly used. Panelists are asked to judge the intensity of characterized off-odor on a 15-cm line scale extending beyond fixed verbal endpoints. If the sensitivity of evaluation is important, the number of sensory panelists should be increased. In addition to evaluating off-odor analytically by trained panelists, hedonic scaling or preference tests can be used for judging consumer acceptance of off-odor (Lawless and Heymann 1998).

12.3.3 Mechanisms of Off-Odor Formation

The mechanism for lipid oxidation is well defined (Gray and Pearson 1994). Lipid oxidation is initiated when a hydrogen atom is abstracted from a methylene group by reactive oxygen species. The presence of a double bond on the fatty acid weakens the C-H bonds on the carbon atom adjacent to the double bond and thus makes hydrogen atom removal easier. The abstraction of the hydrogen atom leaves behind an unpaired electron on the carbon, forming a carbon radical. This free radical initiates the formation of other radicals by abstracting a hydrogen atom from the double bond site of fatty acids and forms more radicals, which is called proliferation of radicals. Oxygen can react with radicals to form peroxyl radicals or hydroperoxides and greatly accelerates oxidation (Halliwell and Gutteridge 1989; Esterbauer et al. 1992). As a result of lipid oxidation, complex mixtures of aldehydes, ketones, hydrocarbons, esters, lactones, and alcohols are produced and form the oxidative off-odor. Apparently, the proportion of unsaturated fatty acids in fat significantly influences lipid oxidation.

The Maillard reaction is a reaction between carbonyls and amines, and is very important for the flavor formation of cooked meat. Farmer and Mottram (1990) reported that the Maillard reaction formed numerous volatiles. Sulfur compounds formed through the reaction of sulfur-containing amino acids with reducing sugars apparently plays an important role in cooked chicken aroma formation (Farmer et al. 1989). Similar mechanisms for the formation of sulfur compounds during irradiation may exist. However, the overall importance of the Maillard reaction in off-odor formation is questionable.

Degradation of thiamine might also play a role in the off-odor formation. Buttery et al. (1984) showed that thermal degradation of thiamine formed sulfur-containing volatiles, of which 2-methyl-3-furanthiol was associated with the meaty aroma of chicken. Enzymes, both exogenous and endogenous, may catalyze reactions associated with flavor and off-odor formation. Exogenous enzymes, mainly from contaminating microorganisms, cause the degradation of amino acids, fatty acids, and nucleotides, and thus form an unpleasant smell.

Bacteria influence the flavor of meat. The total microbial counts and the slime off-odor of cooked meat and raw chicken meat were significantly correlated, but weak (Chen et al. 1992). Sheldon and Essary (1983) indicated that by receiving penicillin and combination antibiotics, birds exhibited several germ-free-like characteristics, which changed the intestinal microflora and its metabolites, and thus changed to broiler meat broth flavor. Urlings et al. (1993) showed that the volatiles of poultry viscera and breast meat from amino acid degradation were not only related to metabolites of amino acid degradation by bacteria, but also as a result of initial enzymatic activity. Dickens et al. (1994) reported that acetic acid dip of chicken carcasses did not affect the odor of cooked chicken meat while reducing the bacteria counts in the surface of carcasses. Light may accelerate oxidation, especially for short wavelength light, and thus related to off-odor formation. Dietary energy, age, and sex may influence the palatability of chicken (Sonaiya et al. 1990).

12.4 FACTORS AFFECTING LIPID OXIDATION AND OFF-ODOR FORMATION IN MEAT

12.4.1 Dietary Fatty Acids

The adverse effect of fatty acids on the flavor of meat is mainly due to lipid oxidation. The higher the unsaturated fatty acid content in meats, the more the meats tend to be oxidized. Most reports on dietary fatty acids and poultry meat quality have focused on fish oil and linseed oil (Hargis and Van Elswyck 1993). This is apparently due to economic reasons, since feeding fish oil and linseed oil can increase the EPA and DHA content in poultry meat or eggs. While EPA and DHA are good for health, they also cause problems related to lipid oxidation in meats. Poste (1991) showed that increasing fishmeal in diet reduced the stability of cooked chicken meat. Although there was no difference between treatments immediately after cooking in sensory characteristics, following overnight storage, chicken flavor decreased and fish-off flavor increased as the fishmeal content in diet increased. Leskanich and Noble (1997) suggested that fish meal and fish oil in broiler diets should not exceed 12% and 1% (respectively) of feed weight in order to avoid fishy taints. In another study, broilers were assigned to diets containing 8% of three different fat sources (tallow, lard, or sunflower oil) and it was found that a dietary fat source rich in linoleic acid produces a marked effect on fat consistency, but only a moderately higher susceptibility to lipid oxidation of meat compared to tallow or lard (Sanz et al. 1999). Ajuyah et al. (1993b) observed that after feeding 150 g/kg of linseed oil, a significant rancid off-odor presented in chicken meat, as judged by sensory panel, and increased further after 5 days of storage at 4°C. Sarraga and Garcia-Regueiro (1999) reported that broiler thigh meats gave higher levels of oxidation than breast meat after feeding sunflower oil. Fat content also influenced the oxidative stability of poultry meat. Ang and Young (1992) showed that cooked chicken thigh meat patties with high fat contents had reduced oxidative stability, as indicated by a higher TBARS value after storage.

Several reports have been published on the effect of conjugated linoleic acid on fatty acid oxidation and off-odor formation. Conjugated linoleic acid (CLA) reduced arachidonic acid, linoleic acid, and oleic acid content in fat, and shifted the whole fatty acid composition to the saturated side. Because the CLA itself is quite stable in meat and does not participate in oxidation, the net effect of CLA on flavor is improving the oxidative stability of meat (Du et al. 2000; 2001b). As the dietary CLA level increased, the TBARS values and hexanal content of meat patties after storage at aerobic condition decreased, showing that dietary CLA improved the oxidative stability of chicken meat (Du et al. 2002c).

12.4.2 Irradiation

Ionizing radiation generates hydroxyl radicals, the most reactive oxygen species in nature, by splitting water molecules. Meat contains 75% or more water, and thus irradiation is expected to initiate lipid oxidation in meat. The amounts of free radicals formed in meat by irradiation and the degree of oxidative chemical changes by irradiation are dose dependent (Ahn et al. 1999c). Diehl (1995) indicated that irradiation of aqueous systems produced hydrogen peroxide in the presence of oxygen, which can produce hydroxyl radical in the presence of metal catalysts via the Fenton reaction. Merritt et al. (1978) postulated that irradiation produces carbonyls through the reactions of hydrocarbon radicals with molecular oxygen, which follows the same pathway as normal lipid oxidation. Nawar (1986) reported that irradiation of unsaturated triacylglycerols at 60 kGy under vacuum conditions generated a series of dienes, trienes, and tetraenes, and that the hydrocarbons produced are related to the fatty acid composition of the fat (Champaign and Nawar 1969). Among the lipid oxidation by-products, aldehydes contributed most to the oxidation flavor and rancidity in cooked meat and hexanal was the major volatile aldehyde (Shahidi and Pegg 1994).

The presence of oxygen was very important in the development of oxidation and odor production not only in nonirradiated, but also in irradiated meat. Ahn et al. (2000e) and Du et al. (2001a) reported that irradiation increased TBARS in raw and cooked meat only under aerobic packaging conditions; its effects on lipid oxidation of raw and cooked meat even under aerobic conditions were minor (Du et al. 2002b). Storage of irradiated cooked meat under aerobic conditions accelerated lipid oxidation (Nam et al. 2003). Under aerobic conditions, TBARS values highly correlated with the amount of aldehydes, total volatiles, and ketones in aerobically packaged irradiated meat. Without oxygen, lipid oxidation in raw and cooked meat did not progress, and TBARS values and volatiles of vacuum-packaged irradiated raw and cooked meat did not correlate well (Ahn et al. 1998a, b). Ahn et al. (1998d) reported that preventing oxygen exposure after cooking was more important for cooked meat quality than packaging, irradiation, or storage conditions of raw meat before and after irradiation (Ahn et al. 1999c; 2000a).

Jo and Ahn (2000) reported that irradiation of oil emulsions increased the production of aldehydes such as hexanal, heptanal, octanal, and nonanal. However, Lee and Ahn (2003) reported that irradiation had a minimal effect on aldehydes and TBARS values of oil emulsions immediately after irradiation. After 10 days of storage, irradiated samples developed higher TBARS values than non-irradiated emulsions — especially arachidonic acid, linolenic acid, and fish oil, which have a high proportion of multi-double-bonded fatty acids, had accelerated lipid oxidation after irradiation.

Irradiation increased lipid oxidation of pork patties and turkey breast under frozen conditions, but the rate was slower than that under refrigerated conditions (Nam et al. 2002b, c). Luchsinger et al. (1997) also showed that the TBARS values of frozen boneless pork chops were stable, regardless of display day, dose, and irradiation sources. These results indicated that the radiation chemistry of refrigerated and frozen meat could be different. Under frozen conditions, the reactive intermediates of water radiolysis were trapped in deep-frozen materials and kept from reacting with each other or with the substrates (Taub et al. 1979; Tarte et al. 1996). During the warming process, however, they tend to react with each other rather than with the substrates (Diehl 1995).

Irradiated meat produced characteristic irradiation odor regardless of the degree of lipid oxidation (Ahn et al. 1997, 1999). Huber et al. (1953) characterized the irradiation odor as "metallic," "sulfide," "wet dog," "wet grain," or "burnt," and assumed that the off-odor was the result of free-radical-dependent oxidation, which was initiated by the irradiation process. Sensory results, however, clearly indicated that irradiation odor was totally different from warmed-over flavor, and the amount of volatiles related to lipid oxidation had no direct relationship with irradiation. In recent years, other researchers described the irradiated meat odor as a "bloody and sweet" (Hashim et al. 1995), "hot fat," "burned oil" or "burned feathers" (Heath et al. 1990), and "barbecued corn-like" flavor (Ahn et al. 2000b).

Batzer and Doty (1955) reported that methyl mercaptan and hydrogen sulfide produced from the water-soluble sulfur compounds in meat were important for the irradiation odor. More recent studies, however, showed that irradiation not only increased some hydrocarbons, ketones, and aldehydes, but also produced many new volatile compounds such as 2-methyl butanal, 3-methyl butanal, 1-hexene, 1-heptene, 1-octene, 1-nonene, *cis*-3- and *trans*-6-nonenals, oct-1-en-3-one, hydrogen sulfide, sulfur dioxide, mercaptomethane, dimethyl sulfide, bis(methylthio)methane, methyl thioacetate, dimethyl disulfide, and trimethyl sulfide from meat (Patterson and Stevens 1995; Jo et al. 1999; Jo and Ahn 2000; Fan et al. 2002; Nam et al. 2002a). However, most of those volatiles were produced via the radiolytic degradation of amino acids or lipids rather than by lipid oxidation (Ahn 2002; Ahn and Lee 2002).

The odor intensity of sulfur-compounds was much stronger and stringent than that of other compounds because sulfur compounds have a very low odor threshold (Angelini et al. 1975). Volatiles from lipids accounted for only a small part of the off-odor in irradiated meat (Lee and Ahn 2003). This indicated that sulfur compounds would be the major volatile components responsible for the characteristic off-odor in irradiated meat, and supported the concept that the changes that occur following irradiation were distinctly different from those of warmed-over flavor in oxidized meat (Ahn et al. 2000a, b).

12.4.3 Antioxidants

Antioxidants can both quench and react with peroxyl or alkoxyl radicals and terminate the chain reaction of peroxidation by scavenging chain-propagating radicals and thus preventing lipids from oxidation (Morel 1989). Vitamin E can react with lipid peroxyl and alkoxyl radicals by donating labile hydrogen and protect oxidation. The tocopheryl radical formed is insufficiently reactive to abstract a hydrogen atom from fatty acids because the unpaired electron on the oxygen atom can be delocalized into the aromatic ring structure, thus increasing its stability (Fukuzawa 1983). However, when vitamin E is totally consumed, it may have a prooxidant effect and act as a chain-transfer agent (Kontush et al. 1996). Supplementation of tocopherol in the diet to prevent oxidative off-odor, such as fishy taints and WOF, has been well established. Beef from cattle raised on pasture had higher linolenic acid, polyunsaturated fatty acids and vitamin E content, and produced lower amounts of TBARS and major off-odor aldehydes such as hexanal, pentanal, heptanal, octanal, and 3-methyl butanal than that from grain-fed (Descalzo et al. 2005). Adding vitamin E to the diet inhibited the off-odor development of frozen turkey drumsticks and beef during storage (Bartov et al. 1983; Arnold et al. 1993). Vitamin E supplementation improved oxidative stability and the sensory quality of chicken and turkey meats, and developed significantly less WOF and TBARS during refrigerated storage; the effect of vitamin E in reducing off-odor volatiles was more distinct in previously frozen than in fresh meat (Winne and Dirinck 1996; Sheldon et al. 1997; Higgins et al. 1999; Nam et al. 2003; Carreras et al. 2004). Descriptive flavor profiling analysis indicated that the off-odor notes lowered as the dietary vitamin E level increased. The stability of meat from birds fed linseed or linseed oil improved after dietary supplementation with vitamin E (Ajuyah et al. 1992; 1993b; Ahn et al. 1995a).

Supplementation of diet with 100 mg vitamin E per kilogram increased tissue vitamin E content in pigs and reduced lipid oxidation in fresh, cooked, and uncured and cured pork products (Buckley 1995; Cannon et al. 1996; Channon et al. 2006). Morrissey et al. (1997) reported that a supplemental diet of 200 mg α-tocopheryl acetate resulted in a saturation level of α-tocopherol in plasma after 1 week of feeding and in tissues within 3 to 4 weeks of feeding. The results show that feeding 200 mg α-tocopheryl acetate/kg feed to chicks for at least 4 weeks prior to slaughter is necessary to optimize muscle content and stability against lipid peroxidation (Morrissey et al. 1997). Ahn et al. (1998d) reported that dietary tocopherol stabilized raw turkey meat, but that the antioxidant strength of tocopherol in muscles was not enough to control lipid oxidation and off-odor production in cooked meat during storage under aerobic conditions. Galvin et al. (1998) pointed out that irradiation had little effect on lipid stability in α-tocopherol-supplemented meat following cooking and storage. Dietary supplementation of selenium in combination with tocopherol was more effective in reducing oxidative changes in broiler and turkey meats during storage (Ryu et al. 2005; Yan et al. 2006). Dietary tea catechins supplementation at levels above 100 mg/kg feed also inhibited lipid oxidation of chicken breast (Tang et al. 2000). They further showed that tea catechins at 300 mg/kg in diet were superior to vitamin E supplementation at a level of 200 mg/kg in preventing lipid oxidation (Tang et al., 2000; 2002). Du et al. (2002a) observed that dietary sorghum with high tannin content can improve the oxidation stability of chicken meat. The high contents of tannin and phenolic compounds in sorghum should contribute to the improved oxidative stability.

In addition to vitamin E, numerous other antioxidants are reported to be effective in preventing oxidative off-odor. Ascorbate can regenerate the chain-breaking antioxidant α-tocopherol in biological systems. Vitamins C and E have cooperative effects *in vivo*. However, ascorbate itself seems ineffective in preventing lipid oxidation in meats. Craig et al. (1997) reported that 0.1% sodium ascorbate could slightly reduce the oxidative changes of ground turkey. Vitamin C supplementation appears to have little, if any, beneficial effects on meat stability (King et al. 1995; Morrissey et al. 1998). Lipid oxidation in cooked meat was significantly influenced by the degree of lipid oxidation in raw meat, and the effectiveness of added antioxidants varied with the presence of prooxidants and packaging conditions. Free radical terminators were more efficient antioxidants than iron chelators in cooked meat, but combinations of hot packaging or vacuum-packaging with antioxidants were better than using either treatment alone (St. Angelo et al. 1990; Ahn et al. 1993a).

Carpenter et al. (2007) reported that grapeseed and bearberry extracts had strong antioxidant effects in raw and cooked pork. Carnosine inhibited lipid oxidation in salted pork and chicken (Decker and Crum 1991; O'Neill et al. 1998), and inclusion of 0.25 to 10 mM carnosine inhibited copper- and iron-catalyzed oxidation of liposome model systems (Decker et al. 1992; Kansci et al. 1997). Therefore, post-slaughter carnosine addition may be an effective means of improving lipid stability in processed meats (Morrissey et al. 1998). Incorporation of plum extract was effective in controlling lipid oxidation and aldehydes production in turkey rolls (Lee and Ahn 2005). Clove and Maillard reaction products exhibited very good antioxidant effects in preventing WOF in chicken and were very effective in arresting the buildup of secondary oxidation products during refrigerator storage (Jayathilakan et al. 1997).

Phenols and aromatic amines are also used as antioxidants. Many plant extracts from rosemary, tea, oregano, onion, coffee, and grapeseed have high levels of polyphenols, and the addition of plant extracts showed strong antioxidant effects in raw and cooked meat products (Nissen et al. 2004; Mitsumoto et al. 2005; Nam et al. 2007; Rojas et al. 2007; Tang and Cronion 2007). Dietze et al. (2007) reported that the WOF reducing efficiency of various natural antioxidants were rosemary extract > quercetine dehydrate > tartaric acid > onion extract. Rababah et al. (2006) reported that infusion with tea extract, grapeseed extract, or their combination decreased the TBARS values as well as hexanal and pentanal content in irradiated cooked chicken breast meat. Addition of plant extracts or in combination with other antioxidants such as vitamin C, vitamin E, or sodium tripolyphosphate (STP), however, produced higher sensory scores and lower TBARS values for frozen, pre-cooked beef, and beef steaks than using them alone (Sanchez-Escalante et al. 2001; Djeane et al. 2003; Reverte et al. 2003; Thongtan et al. 2005; Hua and Ockerman 2007). The combination of rosemary oleoresin with sodium tripolyphosphate (STP) or rosemary oleoresin extract alone was as effective as the TBHQ/STP or BHA/BHT combination in retarding lipid oxidation in chicken nuggets or beef during frozen storage (Lai et al. 1991; McBride et al. 2007). Sesamol, gallate, and trolox were used as antioxidants to prevent lipid oxidation and off-odor in pork patties and homogenates, and found that they distinctively reduced lipid oxidation and volatiles of irradiated pork patties and homogenates (Nam and Ahn 2003a, b). Nam and Ahn (2003c) reported that a combination of sesamol and tocopherol, or gallate and tocopherol, was excellent in preventing lipid oxidation in irradiated and nonirradiated raw and cooked turkey during storage in aerobic packaging. Craig et al. (1996) reported that ground turkey and ground beef added with inorganic phosphate and ascorbate salts had more meaty flavor and less stale and rancid flavor and aroma, and contained less hexanal than controls after 3 days of storage.

Tumbling beef with antioxidants such as phosphate, ascorbate, and tocopherol significantly decreased oxidized, cardboardy, and oxidation aroma during storage (Cheng et al. 2007). Nitrite shows a strong antioxidant effect in meat (Shahidi 1989, 1992) and can prevent the development of WOF in cooked meat (Gray et al. 1981). Ramarathan et al. (1991) reported that nitrite significantly reduced the production of hexanal and other carbonyl compounds in pork. Fan et al. (2004) reported that addition of rosemary extract or sodium nitrite inhibited lipid oxidation of irradiated ready-to-eat turkey bologna.

12.4.4 Prooxidants

Iron can accelerate lipid peroxidation (Ahn and Kim 1998b). Iron or copper salts cause fission of relatively stable lipid peroxides to form alkoxyl radicals, or peroxyl radicals, which further degrade to secondary products of lipid oxidation in the presence of catalysts and oxygen. In the presence of Fe^{3+}, the addition of low concentrations of ascorbate may accelerate lipid oxidation (Kansci et al. 2000). Fe^{2+} and its complexes stimulate lipid oxidation more than Fe^{3+} does (Ahn et al. 1993a). In oil emulsion and cooked meat homogenates, ferrous iron and hemoglobin had strong prooxidant effects, but ferritin became prooxidant only when ascorbate was present. The prooxidant effect of ferrous iron in oil emulsion and cooked meat homogenate disappears in the presence of superoxide, H_2O_2, or xanthine oxidase systems. Hemoglobin and ferritin had no prooxidant effect in raw meat homogenates (Ahn et al. 1998a). Free iron is more effective in inducing oxidation than bound iron (Ahn et al. 1993b). Clove and Maillard reaction products were found to affect the extent of release of non-heme iron during meat cooking and this should contribute in part to their antioxidative effects on lipid oxidation (Jayathilakan et al. 1997).

Metal ions and free radical generating systems are the main catalysts of lipid oxidation in meats (St. Angelo 1993). Both ferrous and ferric iron have strong prooxidant effects but structural iron such as the iron in heme pigments and bound or chelated iron had little prooxidant effect in raw meat (Kanner 1992; Ahn et al. 1993a). Ferric iron could not catalyze lipid oxidation in oil emulsion and cooked meat unless reducing agents were present. Ferrous and hemoglobin showed strong catalytic effects on the oxidation of oil emulsion and cooked meat homogenates (Ahn and Kim 1998b). Yancy et al. (2006) reported that livery flavor increased and beefy flavor decreased when total iron and myoglobin content in meat increased. Adding chelators with strong or weak chelation ability, such as diethylenetriaminepentaacetic acid, citrate, or phosphate, were effective in preventing the formation of TBARS in meat and meat homogenates (Ahn et al. 1993a,b; McGee et al. 2003). Bartov and Kanner (1996) showed that while dietary supplements with high levels of Fe did not affect the oxidative stability of thigh meat of turkey, stability was reduced by injecting ferrous sulfate. Ruiz et al. (2000) showed that the removal of iron and copper from the diet reduced oxidation values in cooked broiler leg meat, as measured by the TBARS value. The prooxidant effect of iron (Fe) was enhanced when ascorbate was added together in the chicken muscle system (Sista et al. 2000). Ahn et al. (1995) reported that dietary α-linolenic acid increased the degree of polyunsaturation in lipids and the susceptibility of the meat to lipid oxidation but that "hot vacuum packaging" protected the meat from sensory and quality deterioration due to lipid oxidation.

12.4.5 Cooking, Storage, and Reheating

The sensory profile changes in broiler tissues due to cooking, storage, and reheating. Storage significantly increased the off-odor of chicken thigh (Lyon 1994). However, during frozen storage, little change was observed in lipid oxidation, color, and volatiles in vacuum-packaged irradiated raw turkey (Nam et al. 2002b). Su et al. (1991) compared cooking chicken breast meat patties in water bath or oven to an internal temperature of 83°C and then stored at 3°C for 3 days, reheated, and evaluated by headspace GC and TBA analysis. Results showed that more severe changes occurred in oven-cooked patties and suggested that the reason may be due to the lower moisture content of oven-cooked patties. Lyon and Ang (1990) examined the effect of reheating method on the off-odor of refrigerated chicken patties and showed that the chickeny and brothy notes decreased with storage time but were more intense in microwave-reheated patties. Cardboard and rancid/painty character notes increased with storage time. The overall off-flavor intensity was significantly higher in conventional oven reheating samples, while there were no differences in GC-headspace components between microwave and oven reheating methods. The more rapid heating rate decreased sulfur-containing compounds but did not influence other volatile concentrations. Storage resulted in a decline in fresh roasted or meaty flavor notes, an increase in WOF notes, and quantities of alcohols, aldehydes and ketones, hydrocarbons, and total headspace volatiles (Turner and Larick 1996). The end-temperature of cooking influenced the volatiles significantly: when the central temperature of chicken increased from 60 to 80°C, volatiles from lipid oxidation increased greatly. Those volatiles included propanal, pentanal, hexanal, heptanal, and octanal (Ang and Liu 1996a, b). The amount of lipid oxidation products, especially those of n-alkanal, 2-alkenals, 1-alkanols, and allylfuran, increased during cooking as the amount of polyunsaturated fatty acid and fat content in the meat increased (Elmore et al. 1999; Jo et al. 1999).

12.4.6 pH of Meat

The pH values of meat also influenced the yield of sulfur compounds produced during cooking. Rao et al. (1977) prepared blended broiler breast and thigh meat, and cooked it at pH values of 2.4, 4.0, 7.0, 9.0, and 10.0. Volatiles were analyzed and indicated that the production of sulfur compounds was least at neutral pH and increased greatly as the pH increased. But for carbonyls, the yield was highest at neutral pH, and decreased as meat pH increased. Nam et al. (2002c) showed that dark-firm-dry (DFD) meat, with a higher pH than normal, was very stable and resistant to oxidative changes compared to normal and pale-soft-exudative (PSE) meat. The total volatiles produced in normal and PSE meat were higher than in DFD meat. Furthermore, irradiating vacuumed-packaged meats with different pH values, normal meats produced highest sulfur-containing compounds (Nam et al. 2002c). Although beef has more saturated fatty acids than pork and poultry, it is more susceptible to lipid oxidation than poultry because of its low antioxidant power and high heme pigment content (Kim et al. 2002; Min et al. 2007).

12.4.7 High-Pressure Treatment

High-pressure treatment induces lipid oxidation (Cheah and Ledward 1997; Orlien et al. 2000). Orlien et al. (2000) indicated that chicken breast meat exposed to pressures at or below 500 MPa showed no indication of lipid oxidation, but that pressure treatment at 800 MPa induced lipid oxidation to the same extent as the heat treatment. Pressure treatment at 600 MPa and 700 MPa resulted in less oxidation and thus suggested that 500 MPa is the critical pressure for the pressure treatment of chicken breast meat in order to prevent oxidation (Orlien et al. 2000). Cheah and Ledward (1997) reported that the addition of 0.02% citric acid inhibited the increased rate of lipid oxidation found in rendered pork fat after high-pressure treatment.

12.4.8 Packaging

The importance of packaging on lipid oxidation was studied extensively over the years. Poultry meat patties packaged and stored under vacuum conditions had significantly lower lipid oxidation than those aerobically packaged during storage (Ahn et al. 1998a, c; Ahn et al. 2000; Du et al. 2000; Nam and Ahn 2003a). The availability of oxygen to cooked meat was more critical to the stability of meat than raw meat during storage. Ahn et al. (1992) reported that minimizing oxygen contact by packaging the cooked meat immediately after cooking while it was hot ("hot-packaging") prevented oxidative changes during storage. Ahn et al. (1993c, d) showed that elimination of oxygen from meat was the most critical factor in preventing lipid oxidation of cooked meat. They showed that "hot-packaging" virtually stopped lipid oxidation in precooked meat added with various prooxidants such as ionic iron, hemoglobin, and NaCl and/or fat with highly unsaturated fatty acids during refrigerated storage. Preventing oxygen contact with meat after cooking was more important for preventing lipid oxidation and off-odor volatiles such as hexanal, pentanal, propanal, and 1-propanol than packaging, irradiation, and storage conditions of raw meat (Ahn et al. 1998a). Vacuum-packaging of pre-cooked pork and turkey, and dry-cured hams was better than CO_2 and N_2-modified atmosphere packaging or aerobic packaging in maintaining meaty flavor and minimizing WOF during storage (Nolan et al. 1989; Cilla et al. 2006), and WOF, rancid flavor, and aroma scores correlated well with TBARS values (Nolan et al. 1989). Dawson et al. (1995) examined the effects of the packaging film oxygen transmission rate on the odor, color, aerobic plate count, and cooked volatile compounds in ground chicken after storage, and found that packaging film with the lowest oxygen transmission rate was the lowest in off-odor. Meat samples stored under CO_2 and N_2 modified packaging were less oxidized, as indicated by sensory scores and TBARS values, than those stored in air (Patsias et al. 2006) but were more oxidized than vacuum-packaged meat (Pettersen et al. 2004; Seyfert et al. 2005).

12.5 SUMMARY

Lipid oxidation is a major cause of flavor and color deterioration in meat. Lipid oxidation in meat occurs mainly via the autoxidation process, and the rate varies depending upon various internal and external conditions of the meat. Autoxidation is the major oxidation mechanism in meat and is initiated by reactive oxygen species such as hydroxyl radicals, which abstract hydrogen atoms from the fatty acyl group of polyunsaturated fatty acids and form lipid radicals. Because the lipid radicals formed from the initial reactions have a strong enough reactivity, they can also transfer free radicals to adjacent lipids, and thus the reaction can perpetuate unless the free radicals are quenched or reactions are interrupted by chain-breaking antioxidants. Various primary and secondary by-products of lipid oxidation can be formed from meat, depending upon the type of meat, pro- and antioxidants, processing, cooking, packaging, and storage conditions. Some lipid oxidation products are known to be toxic or carcinogenic, and others alter meat flavor. Therefore, prevention of lipid oxidation is very important not only for meat industry, but also for consumers.

REFERENCES

Ahn, D.U. 2002. Production of volatiles from amino acid homopolymers by irradiation. *J. Food Sci.* 67:2565–2570.

Ahn, D.U., A. Ajuyah, F.H. Wolfe, and J.S. Sim. 1993d. Oxygen availability affects in prooxidant catalyzed lipid oxidation of cooked turkey patties. *J. Food Sci.* 58:278–282.

Ahn, D.U. and S.M. Kim. 1998a. Effect of superoxide and superoxide-generating systems on the prooxidant effect of iron in oil emulsion and raw turkey homogenates. *Poult. Sci.* 77:1428–1435.

Ahn, D.U. and S.M. Kim. 1998b. Prooxidant effects of ferrous iron, hemoglobin, and ferritin in oil emulsion and cooked-meat homogenates are different from those in raw-meat homogenates. *Poult. Sci.* 77348–77355.

Ahn, D.U., C. Jo, and D.G. Olson. 1999a. Headspace oxygen in sample vials affects volatiles production of meat during the automated purge-and-trap/GC analyses. *J. Agric. Food Chem.* 47:2776–2781.

Ahn, D.U., C. Jo, and D.G. Olson, 1999b. Volatile profiles of raw and cooked turkey thigh as affected by purge temperature and holding time before purge. *J. Food Sci.* 64:230–233.

Ahn, D.U., C., Jo, M. Du, D.G. Olson, and K.C. Nam. 2000a. Quality characteristics of pork patties irradiated and stored in different packaging and storage conditions. *Meat Sci.* 56:203–209.

Ahn, D.U., C. Jo, and D.G. Olson. 2000b. Analysis of volatile components and the sensory characteristics of irradiated raw pork. *Meat Sci.* 54:209–215.

Ahn, D.U. and E.J. Lee. 2002. Production of off-odor volatiles from liposome-containing amino acid homopolymers by irradiation. *J. Food Sci.* 67:2659–2665.

Ahn, D.U., D.G. Olson, C. Jo, X. Chen, C. Wu, and J.I. Lee. 1998c. Effect of muscle type, packaging, and irradiation on lipid oxidation, volatile production, and color in raw pork patties. *Meat Sci.* 49:27–39.

Ahn, D.U., D.G. Olson, C. Jo, J. Love, and S.K. Jin. 1999c. Volatile production and lipid oxidation in irradiated cooked sausage as related to packaging and storage. *J. Food Sci.* 64:226–229.

Ahn, D.U., D.G. Olson, J.I. Lee, C. Jo, C., C. Wu, and X. Chen. 1998a. Packaging and irradiation effects on lipid oxidation and volatiles in pork patties. *J. Food Sci.* 63:15–19.

Ahn, D.U., J.L. Sell, X. Chen, C. Wu, and J.L. Lee. 1998b. Effects of dietary vitamin E supplementation on lipid oxidation and volatiles content of irradiated, cooked turkey meat patties with different packaging. *Poult. Sci.* 77:912–920.

Ahn, D.U., J.L. Sell, M. Jeffery, C. Jo, X. Chen, C. Wu, and J.I. Lee. 1997. Dietary vitamin E affects lipid oxidation and total volatiles of irradiated raw turkey meat. *J. Food Sci.* 62:954–959.

Ahn, D.U., F.H. Wolfe, and J.S. Sim. 1993a. The effect of metal chelators, hydroxyl radical scavengers, and enzyme systems on the lipid peroxidation of raw turkey meat. *Poult. Sci.* 72:1972–1980.

Ahn, D.U., F.H. Wolfe, and J.S. Sim. 1993b. The effect of free and bound iron on lipid peroxidation in turkey meat. *Poult. Sci.* 72:209–215.

Ahn, D.U., F.H. Wolfe, and J.S. Sim. 1993c. Prevention of lipid oxidation in pre-cooked turkey meat patties with hot packaging and antioxidant combinations. *J. Food Sci.* 58:283–287.

Ahn, D.U., F.H. Wolfe, and J.S. Sim. 1995. Dietary α-linolenic acid and mixed tocopherols and packaging influences on lipid stability in broiler chicken breast and leg muscle. *J. Food Sci.* 60:1013–1018.

Ahn, D.U., F.H. Wolfe, J.S. Sim, and D.H. Kim. 1992. Packaging cooked turkey meat patties while hot reduced lipid oxidation. *J. Food Sci.* 57:1075–1077, 1115.

Ajuyah, A.O., T.W. Fenton, R.T. Hardin, and J.S. Sim. 1993a. Measuring lipid oxidation volatiles in meats. *J. Food Sci.* 58:270–273.

Ajuyah, A.O., R.T. Hardin, and J.S. Sim. 1993b. Effect of dietary full fat flaxseed with and without antioxidant on the fatty acid composition of major lipid classes of chicken meats. *Poult. Sci.* 72:125–136.

Ajuyah, A.O., R.T. Hardin, K. Cheung, and J.S. Sim. 1992. Yield, lipid, cholesterol and fatty acid composition of spent hens fed full-fat seeds and fish meal diets. *J. Food Sci.* 57:338–341.

Ang, C.Y.W. and F. Liu. 1996a. Static headspace capillary GC method for determining changes of volatiles from heated chicken breast meat. *J. Muscle Foods* 7:131–138.

Ang, C.Y.W. and F. Liu. 1996b. Influence of heating end-point temperature on volatiles from poultry meat — a review. *J. Muscle Foods* 7:291–32.

Ang, C.Y.W. and B.G. Lyon. 1990. Evaluations of warmed-over flavor during chill storage of cooked broiler breast, thigh and skin by chemical, instrumental and sensory methods. *J. Food Sci.* 55:644–648.

Ang, C.Y.W. and L.L. Young. 1992. Influence of fat content on oxidative stability of cooked thigh patties during frozen storage. *Poult. Sci.* 71:1794–1796.

Angelini, P., C. Merritt, Jr., J.M. Mendelshon, and F.J. King. 1975. Effect of irradiation on volatile constituents of stored haddock flesh. *J. Food Sci.* 40:197–199.

Arnold, R.N., K.K. Scheller, S.C. Arp, S.N. Williams, and D.M. Schaefer. 1993. Dietary alpha-tocopheryl acetate enhances beef quality in Holstein and beef breed steers. *J. Food Sci.* 58:28–33.

Asghar, A., J.I. Gray, D.J. Buckley, A.M. Pearson, and A.M. Booren. 1988. Perspectives on warmed-over flavor. *Food Technol.* 42:102–109.

Bailey, M.E. and K.W. Um. 1992. Maillard reaction products and lipid oxidation. *Am. Oil Chem. Soc. Symp. Ser., American Chemists' Society* 500:122–139.

Bartov, I., D. Basker, and S. Angel. 1983. Effect of dietary vitamin E on the stability and sensory quality of turkey meat. *Poult. Sci.* 62:1224–1230.

Bartov, I. and J. Kanner. 1996. Effect of high levels of dietary iron, iron injection, and dietary vitamin E on the oxidative stability of turkey meat during storage. *Poult. Sci.* 75:1039–1046.

Batzer, O.F. and D.M. Doty. 1955. Nature of undesirable odors formed by gamma irradiation of beef. *J. Agric. Food Chem.* 3:64.

Brewer, M.S. and J.D. Vega. 1995. Detectable odor thresholds of selected lipid oxidation compounds in a meat model system. *J. Food Sci.* 60:592–595.

Brunton, N.P. D.A. Cronin, F.J. Monahan, and R.A. Durcan. 2000. Comparison of solid-phase microextraction (SPME) fibres for measurement of hexanal and pentanal in cooked turkey. *Food Chem.* 68: 339–345.

Buckley, D.J., P.A. Morrissey, and J.I. Gray. 1995. Influence of dietary vitamin E on the oxidative stability and quality of pig meat. *J Anim. Sci.* 73:3122–3130.

Buttery, R.G., W.F. Haddon, R.M. Seifert, and J.G. Turnbaugh. 1984. Thiamin odor and bis (2-methyl-3-furyl) disulphate. *J. Agric. Food Chem.* 32:674–676.

Byrne, D.V., W.L.P. Bredie, L.S. Bak, G. Bertesen, H. Martens, and M. Martens. 2001. Sensory and chemical analysis of cooked porcine meat patties in relation to warmed-over flavour and pre-slaughter stress. *Meat Sci.* 59:229–249.

Byrne, D.V., W.J.P. Bredie, D.S. Mottram, and M. Martens. 2002. Sensory and chemical investigations on the effect of oven cooking on warmed-over flavour development in chicken meat. *Meat Sci.* 61:127–139.

Byrne, D.V., M.G. O'Sullivan, W.L.P. Bredie, H.J. Andersen, and M. Martens. 2003. Descriptive sensory profiling and physical/chemical analyses of warmed-over flavor in pork patties from carriers and non-carriers of the RN^- allele. *Meat Sci.* 63:211–224.

Calkins, C.R. and J.M. Hodgen. 2007. A fresh look at meat flavor. *Meat Sci.* 77:63–80.

Campo, M.M., G.R. Nute, S.I. Hughes, M. Enser, J.D. Wood, and R.I. Richardson. 2006. Flavour perception of oxidation in beef. *Meat Sci.* 72:303–311.

Cannon, J.E., J.B. Morgan, G.R. Schmidt, J.D. Tatum, J.N. Sofos, G.C. Smith, R.J. Delmore, and S.N. Williams. 1996. Growth and fresh meat quality characteristics of pigs supplemented with vitamin E. *J. Anim. Sci.* 74:98–105.

Carpenter, R., M.N. O'Grady, Y.C. O'Callaghan, N.M. O'Brien, and J.P. Kerry. 2007. Evaluation of the antioxidant potential of grape seed and bearberry extracts in raw and cooked pork. *Meat Sci.* 76:604–610.

Carreras, I., L. Guerrero, M.D. Guardia, E. Esteve-Garcia, J.A. Garcia-Regueiro, and C. Sarraga. 2004. Vitamin E levels, thiobarbituric acid test and sensory evaluation of breast muscles from broilers fed α-tocopheryl acetate- and β-carotene-supplemented diets. *J. Sci. Food Agric.* 84:313–317.

Champaign, J.R. and W.W. Nawar. 1969. The volatile components of irradiated beef and pork fats. *J. Food Sci.* 34:335–340.

Channon, H.A. and G.R. Trout. 2002. Effect of tocopherol concentration on rancidity development during frozen storage of a cured and an uncured processed pork product. *Meat Sci.* 62:9–17.

Cheah, P.B. and D.A. Ledward. 1997. Catalytic mechanism of lipid oxidation following high pressure treatment in pork fat and meat. *J. Food Sci.* 62:1135–1138, 1141.

Chen, Y.H., M.J. Woodburn, and M.W. Kelsey. 1992. Microbial and sensory quality of refrigerated market fryers. *J. Food Protection* 54:704–710.

Cheng, J.H. and H.W. Ockerman. 2007. Effects of phosphate, ascorbic acid, α-tocopherol and salt with nonvacuum tumbling on lipid oxidation and warmed-over flavor or roast beef. *J. Muscle Foods* 18:313–329.

Cilla, I., L. Martinez, J.A. Beltran, and P. Roncales. 2006. Dry-cured ham quality and acceptability as affected by the preservation system used for retail sale. *Meat Sci.* 73:581589.

Craig, J.A., J.A. Bowers, X.Y. Wang, and P.A. Seib. 1997. Inhibition of lipid oxidation in meats by inorganic phosphate and ascorbate salts. *J. Food Sci.* 61:1062–1067.

Dawson, P.L., I.Y. Han, L.M. Voller, C.B. Clardy, R.M. Martinez, and J.C. Acton. 1995. Film oxygen transmission rate effects on ground chicken meat quality. *Poult. Sci.* 74:1381–1387.

Daza, A., A.I. Rey, J. Ruiz, and C.J. Lopez-Bote. 2005. Effects of feeding in free-range conditions or in confinement with different dietary MUFA. *Meat Sci.* 69:151–163.

Decker, E.A. and A.D. Crum. 1991. Inhibition of oxidative rancidity in salted ground pork by carnosine. *Food Sci.* 56:1179–1181.

Decker, E.A., A.D. Crum, and J.T. Calvert. 1992. Differences in the antioxidant mechanism of carnosine in the presence of copper and iron. *J. Agric. Food Chem.* 40:756–759.

Descalzo, A.M., E.M. Insani, A. Biolatto, A.M. Sancho, P.T. Garcia, N.A. Pensel, and J.A. Josifovich. 2005. Influence of pasture or grain-based diets supplemented with vitamin E on antioxidant. *Meat Sci.* 70:35–44.

Dickens, J.A., B.G. Lyon, A.D. Whittemore, and C.E. Lyon. 1994. The effect of an acetic acid dip on carcass appearance, microbiological quality, and cooked breast meat texture and flavor. *Poult. Sci.* 73:576–581.

Diehl, J.F. 1995. In *Safety of Irradiated Foods, 2nd ed.* Marcel Dekker, New York.

Dietze, K., M. Lammers, and W. Ternes. 2007. Effects of natural additives on warmed-over flavor of dark turkey meat analyzed with dynamic headspace GC/MS. *Fleischwirtschaft Int.* 5:67–70.

Du, M., D.U. Ahn, K.C. Nam, and J.L. Sell. 2000. Influence of dietary conjugated linoleic acid on volatile profiles, color and lipid oxidation of irradiated raw chicken meat. *Meat Sci.* 56:387–395.

Du, M., D.U. Ahn, K.C. Nam, and J.L. Sell. 2001b. Volatile profiles and lipid oxidation of irradiated cooked chicken meat from laying hens fed diets containing conjugated linoleic acid. *Poult. Sci.* 80:235–241.

Du, M., G. Cherian, and D.U. Ahn. 2002a. Effect of dietary sorghum cultivars on the storage stability of broiler breast and thigh meats. *Poult. Sci.* 81:1385–1391.

Du, M., S.J. Hur, and D.U. Ahn. 2002b. Raw-meat packaging and storage affect of color and odor of irradiated broiler breast fillets after cooking. *Meat Sci.* 61:49–54.

Du, M., S.J. Hur, K.C. Nam, H. Ismail, and D.U. Ahn. 2001a. Volatiles, color, and lipid oxidation of broiler breast fillets irradiated before and after cooking. *Poult. Sci.* 80:1748–1753.

Du, M., K.C. Nam, S.J. Hur, H. Ismail, and D.U. Ahn. 2002c. Effect of dietary conjugated linoleic acid, irradiation, and packaging conditions on the quality characteristics of raw broiler breast fillets. *Meat Sci.* 60:9–15.

Eilamo, M., A. Kinnunen, K. Latva-Kala, and R. Ahnvenainen. 1998. Effects of packaging and storage conditions on volatile compounds in gas-packed poultry meat. *Food Additives Contaminants* 15:217–228.

Elmore, J.S., D.S. Mottram, M. Enser, and J.D. Wood. 1999. Effect of the polyunsaturated fatty acid composition of beef muscle on the profile of aroma volatiles. *J. Agric. Food Chem.* 47:1619–1625.

Esterbauer, H., J. Gebicki, H. Puhl, and G. Juergens. 1992. The role of lipid peroxidation and antioxidants in oxidative modification of LDL. *Free Rad. Biol. Med.* 13:341–390.

Fan, X., C.H. Sommers, and K.J.B. Sokorai. 2004. Ionizing radiation and antioxidants affect volatile sulfur compounds, lipid oxidation, and color of ready-to-eat turkey bologna. *J. Agric. Food Chem.* 52:3509–3515.

Fan, X., C.H. Sommers, D.W. Thayer, and S.J. Lehotay. 2002. Volatile sulfur compounds in irradiated precooked turkey breast analyzed with pulsed flame photometric detection. *J. Agric. Food Chem.* 50:4257–4261.

Farkas, J. 1998. Irradiation as a method for decontaminating food. *Int. J. Food Microbiol.* 44:189–204.

Farmer, L.J. and D.S. Mottram. 1990. Interaction of lipid in the Maillard reaction between cysteine and ribose: effect of a triglyceride and three phospholipids on the volatile products. *J. Sci. Food Agric.* 53:505–525.

Farmer, L.J., D.S. Mottram, and F.B. Whitfield. 1989. Volatile compounds produced in Maillard reactions involving cysteine, ribose and phospholipid. *J. Sci. Food Agric.* 49:347–368.

Fukuzawa, K. and J.M. Gebicki. 1983. Oxidation of α-tocopherol in micelles and liposomes by the hydroxyl, perhydroxyl and superoxyl free radicals. *Arch. Biochem. Biophys.* 226:242–251.

Galvin, K., P.A. Morrissey, and D.J. Buckley. 1998. Effect of dietary α-tocopherol supplementation and gamma-irradiation on α-tocopherol retention and lipid oxidation in cooked minced chicken. *Food Chem.* 62:185–190.

Gray, J.I., B. MacDonald, A.M. Pearson, and I.D. Morton. 1981. Role of nitrite in cured meat flavor: a review. *J. Food Protection* 44:302–312, 319.

Gray, J.I. and A.M. Pearson. 1994. Lipid-derived off-flavors in meat — formation and inhibition. In *Flavor of Meat and Meat Products*. Shahidi, F. (Ed.). Chapman & Hall, Glasgow, p. 116–143.

Halliwell, B. and J.M.C. Gutteridge. 1989. Lipid peroxidation: a radical chain reaction. In *Free Radical in Biology and Medicine, 2nd ed.* Clarendon Press, London, p. 189.

Hargis, P.S. and M.E. van Elswyck. 1993. Manipulation of the fatty acid composition of poultry meat and eggs for the health conscious consumer. *World Poult. Sci. J.* 49:251–264.

Harris, N.D., D.H. Strong, and M.L. Sunde. 1968. Intestinal flora and chicken flavor. *J. Food Sci.* 33:543–547.

Hashim, I.B., A.V.A. Resurreccion, and K.H. McWatters. 1995. Descriptive sensory analysis of irradiated frozen or refrigerated chicken. *J. Food Sci.* 60:664–666.

Heath, J.L., S.L. Owens, S. Tesch, and K.W. Hannah. 1990. Effect of high-energy electron irradiation of chicken on thiobarbituric acid values, shear values, odor, and cook yield. *Poult. Sci.* 69:313–319.

Higgins, F.M., J.P. Kerry, D.J. Buckley, and P.A. Morrissey. 1999. Effects of α-tocopheryl acetate supplementation and salt addition on the oxidative stability (TBARS) and warmed-over flavor (WOF) of cooked turkey meat. *Br. Poult. Sci.* 40:59–64.

Hsieh, Y.P.C. 1981. Development and characterization of a synthetic meat flavor mixture. *Diss. Abstr. Int.* 40:661.

Hua, C.J. and H.W. Ockerman. 2007. Effects of phosphate, ascorbic acid, α-tocopherol and salt with non-vacuum tumbling on lipid oxidation and warmed-over flavor of roast beef. *J. Muscle Foods* 18:313–329.

Igene, J.O. and A.M. Pearson. 1979. Role of phospholipids and triglycerides in warmed-over flavour development in meat model systems. *J. Food Sci.* 44:1285–1290.

Igene, J.O., A.M. Pearson, L.R. Dugan, and J.F. Price. 1980. Role of triglycerides and phospholipids on development of rancidity in model meat systems during frozen storage. *Food Chem.* 5:263–276.

Igene, J.O., K. Yamauchi, A.M. Pearson, J.I. Gray, and S.D. Aust. 1985. Evaluation of 2-thiobarbituric acid reactive substance (TBARS) in relation to warmed-over flavour development in cooked meat. *J. Agric. Food Chem.* 33:364–367.

Imafidon, G.I. and A.M. Spanier. 1994. Unraveling the secret of meat flavor. *Trends in Food Sci. Technol.* 5:315–321.

Jayathilakan, K., T.S. Vasundhara, and K.V. Kumudavally. 1997. Effect of spices and Maillard reaction products on rancidity development in precooked refrigerated meat. *J. Food Sci. Technol.* 34:128–131.

Jo, C. and D.U. Ahn. 2000. Production of volatile compounds from irradiated oil emulsion containing amino acids or proteins. *J. Food Sci.* 65:612–616.

Jo, C., J.I. Lee, and D.U. Ahn. 1999. Lipid oxidation, color changes and volatiles production in irradiated pork sausage with different fat content and packaging during storage. *Meat Sci.* 51:355–361.

Kanner, J. 1992. Mechanism of nonenzymic lipid peroxidation in muscle foods. *Am. Oil Chem. Soc. Symp. Ser., American Chemists' Society* 500:55–73.

Kansci, G., C. Genot, A. Meynier, and G. Gandemer. 1997. The antioxidant activity of carnosine and its consequences on the volatile profiles of liposomes during iron/ascorbate induced phospholipid oxidation. *Food Chem.* 60:165–175.

Kim, Y.H., K.C. Nam, and D.U. Ahn. 2002. Volatile profiles, lipid oxidation and sensory characteristics of irradiated meat from different animal species. *Meat Sci.* 61:257—265.

King, A.J., T.G. Uijttenboogaart, and A.W. de Vries. 1995. α-Tocopherol, β-carotene and ascorbic acid as antioxidants in stored poultry muscle. *J. Food Sci.* 60:1009–1012.

Kontush, A., B. Finckh, B. Karten, and A. Kohlschuetter. 1996. Antioxidant and prooxidant activity of alpha-tocopherol in human plasma and low density lipoprotein. *J. Lipid Res.* 37:1436–1448.

Lai, S.M., J.L. Gray, A.M. Booren, R.L. Crackel, and J.L. Gill. 1995. Assessment of off-flavour development on restructured chicken nuggets using hexanal and TBARS measurements and sensory evaluation. *J. Sci. Food Agric.* 67:447–452.

Lai, S.M., J.I. Gray, D.M. Smith, A.M. Booren, R.L. Crackel, and D.J. Buckley. 1991. Effects of oleoresin rosemary, tertiary butylhydroquinone, and sodium tripolyphosphate on the development of oxidative rancidity in restructured chicken nuggets. *J. Food Sci.* 56:616–620.

Lawless, H.T. and H. Heymann. 1998. *Sensory Evaluation of Food.* Chapman & Hall, New York.

Lee, E.J. and D.U. Ahn. 2003. Production of off-odor volatiles from fatty acids and oils by irradiation. *J. Food Sci.* 68:70–75.

Lee, E.J. and D.U. Ahn. 2005. Quality characteristics of irradiated turkey breast rolls formulated with plum extract. *Meat Sci.* 71:300–305.

Leskanich, C.O. and R.C. Noble. 1997. Manipulation of the n-3 polyunsaturated fatty acid composition of avian eggs and meat. *World's Poult. Sci. J.* 53:155–183.

Love, J.D. and A.M. Pearson. 1971. Lipid oxidation in meat and meat products — a review. *J. Am. Oil Chem. Soc.* 48:547–549.

Luchsinger, S.E., D.H. Kropf, C. Garcia-Zepeda, M.C. Hunt, S.L. Stroda, J.L. Marsden, and C.L. Kastner 1997. Color and oxidative properties of irradiated ground beef patties. *J. Muscle Foods* 8:445–464.

Lyon, B.G. 1994. Sensory profile changes in broiler tissues due to cooking, storage, and reheating. *Poult. Sci.* 72:1981–1988.

Lyon, B.G. and C.Y.W. Ang. 1990. Effects of reheating method on off-flavor development in precooked, stored chicken patties. *Poult. Sci.* 69:320–328.

Mabrouk, A.F. 1977. Non-volatile nitrogen and sulfur compounds in red meats and their relation to flavor and taste. *American Chemical Society* 170:AGFD 23.

Maga, J.A. 1998. Umami flavor in meat. In Shahidi, F., *Flavor of Meat Products and Seafoods. 2nd ed.* Blackie Academic & Professional, London, p. 197–215.

McBride, N.T.M., S.A. Hogan, and J.P. Kerry. 2007. Comparative addition of rosemary extract and additives on sensory and antioxidant properties of retail packaged beef. *Int. J. Food Sci. Technol.* 42:1201–1207.

McGee, M.R., K.L. Henry, J.C. Brooks, F.K. Ray, and J.B. Morgan. 2003. Injection of sodium chloride, sodium tripolyphosphate, and sodium lactate improves Warner-Bratzler shear and sensory characteristics of pre-cooked inside round roasts. *Meat Sci.* 64:273–277.

Mead, G.C., N.M. Griffiths, C.S. Impey, and J.C. Coplestone. 1983. Influence of diet on the intestinal microflora and meat flavour of intensively-reared broiler chickens. *Br. Poult. Sci.* 24:261–272.

Merritt, C. Jr., P. Angelini, and R.A. Graham. 1978. Effect of radiation parameters on the formation of radiolysis products in meat and meat substances. *J. Agric. Food Chem.* 26:29–36.

Mitsumoto, M., M.N. O'Grady, J.P. Kerry, and D.J. Buckley. 2005. Addition of tea catechins and vitamin C on sensory evaluation, colour and lipid stability during chilled storage in cooked or raw beef and chicken patties. *Meat Sci.* 69:773–779.

Morel, D.W. and G.M. Chisolm. 1989. Antioxidant treatment of diabetic rats inhibits lipoprotein oxidation and cytotoxicity. *J. Lipid Res.* 30:1827–1834.

Morrissey, P.A., S. Brabdon, D.J. Buckley, P.J.A. Sheehy, and M. Frigg. 1997. Tissue content of α-tocopherol and oxidative stability of broilers receiving dietary α-tocopherol acetate supplement for various periods pre-slaughter. *Br. Poult. Sci.* 38:84–88.

Morrissey, P.A., P.J.A. Sheehy, K. Galvin, J.P. Kerry, and D.J. Buckley. 1998. Lipid stability in meat and meat products. *Meat Sci.* 49:S73–S86.

Mottram, D.S. 1998. Flavor formation in meat and meat products: a review. *Food Chem.* 62:415–424.

Nam, K.C. and D.U. Ahn. 2003a. Combination of aerobic and vacuum packaging to control lipid oxidation and off-odor volatiles of irradiated raw turkey breast. *Meat Sci.* 63:389–395.

Nam, K. C. and D.U. Ahn. 2003b. Use of antioxidants to reduce lipid oxidation and off-odor volatiles of irradiated pork homogenates and patties. *Meat Sci.* 63:1–8.

Nam, K.C. and D.U. Ahn. 2003c. Use of double packaging and antioxidant combinations to improve color, lipid oxidation, and volatiles of irradiated raw and cooked turkey breast patties. *Poult. Sci.* 82:850–857.

Nam, K.C., D.U. Ahn, M. Du, and C. Jo. 2001. Lipid oxidation, color, volatiles, and sensory characteristics of aerobically packaged and irradiated pork with different ultimate pH. *Food Sci.* 66:1225–1229.

Nam, K.C., J. Cordray, and D.U. Ahn. 2004. Automated dynamic headspace. *J. Agric. Food Chem.* 52:1735–1741.

Nam, K.C., M. Du, C. Jo, and D.U. Ahn. 2002a. Quality characteristics of vacuum-packaged, irradiated normal, PSE, and DFD pork. *Innovative Food Sci. Emerging Technol.* 3:73–79.

Nam, K.C., S.J. Hur, H. Ismail, and D.U. Ahn. 2002b. Lipid oxidation, volatiles, and color changes in irradiated raw turkey breast during frozen storage. *J. Food Sci.* 67:2061–2066.

Nam, K.C., Y. H. Kim, M. Du, and D.U. Ahn. 2002c. Off-odor volatiles and pink color development in precooked, irradiated turkey breast during frozen storage. *Poult. Sci.* 81:269–275.

Nam, K.C., K.Y. Ko, B.R. Min, H. Ismail, E.J. Lee, J. Cordray, and D.U. Ahn. 2007. Effects of oleoresin-tocopherol combinations on lipid oxidation, off-odor, and color of irradiated raw and cooked pork patties. *Meat Sci.* 75:61–70.

Nam, K.C., B.R. Min, S.C. Lee, J. Cordray, and D.U. Ahn. 2004. Prevention of pinking, off-odor, and lipid oxidation in irradiated pork loin using double packaging. *J. Food Sci.* 69:FCT214–FCT219.

Nam, K.C., B.R. Min, H. Yan, E.J. Lee, A. Mendonca, I. Wesley, and D.U. Ahn. 2003. Effect of dietary vitamin E and irradiation on lipid oxidation, color, and volatiles of fresh and previously frozen turkey breast patties. *Meat Sci.* 65:513–521.

Nawar, W.W. 1986. Volatiles from food irradiation. *Food Rev. Int.* 2:45–78.

Nissen, L.R., D.V. Byrne, G. Bertelsen, and L.H. Skibsted. 2004. The antioxidative activity of plant extracts in cooked pork patties as evaluated by descriptive sensory profiling and chemical analysis. *Meat Sci.* 68:485–495.

Nolan, N.L., J.A. Bowers, and D.H. Kropf. 1989. Lipid oxidation and sensory analysis of cooked pork and turkey stored under modified atmospheres. *J. Food Sci.* 54:846–849.

O'Neill, L.M., K. Galvin, P.A. Morrissey, and D.J. Buckley. 1998a. Inhibition of lipid oxidation in chicken by carnosine and dietary alpha-tocopherol supplementation and its determination by spectrophotometry. *Meat Sci.* 50:479–488.

O'Neill, L.M., K. Galvin, P.A. Morrissey, and D.J. Buckley. 1998b. Comparison of effects of dietary olive oil, tallow and vitamin E on the quality of broiler meat and meat products. *Br. Poult. Sci.* 39:365–371.

Olsen, E., G. Vogt, D. Ekeberg, M. Sandbakk, J. Pettersen, and A. Nilsson. 2005. Analysis of the early stages of lipid oxidation in freeze-stored pork back fat and mechanically recovered poultry meat. *J. Agric. Food Chem.* 53:338–348.

Overbosch, P., W.G.M. Afterof, and P.G.M. Haring. 1991. Flavor release in the mouth. *Food Rev. Int.* 7:137–184.

Patsias, A., I. Chouliara, A. Badeka, I.N. Savvaidis, and M.G. Kontominas. 2006. Shelf-life of a chilled precooked chicken product stored in air and under modified atmospheres: microbiological, chemical, sensory attributes. *Food Microbiol.* 23:423–429.

Patterson, R.L.S. and M.H. Stevenson. 1995. Irradiation induced off-odour in chicken and its possible control. *Br. Poult. Sci.* 36:425–441.

Pearson, A.M., J.I. Gray, A.M. Wolzak, and N.A. Horenstein. 1983. Institute of Food Technologists IFT Muscle Foods Division Annual Meeting (42nd). Safety implications of oxidized lipids in muscle foods. *Food Technol.* 37:121–129.

Pettersen, M.K., M.B. Mielnik, T. Eie, G. Skrede, and A. Nilsson. 2004. Lipid oxidation in frozen, mechanically deboned turkey meat as affected by packaging parameters and storage conditions. *Poult. Sci.* 83:1240–1248.

Pinchasov, Y. and I. Nir. 1992. Effect of dietary polyunsaturated fatty acid concentration on performance, fat deposition and carcass fatty acid composition in broiler chickens. *Poult. Sci.* 71:1504–1512.

Poste, L.M. 1991. A sensory perspective of effect of feeds on flavor in meats: poultry meats. *J. Anim. Sci.* 68:4414–4420.

Rababah, T., N.S. Hettiarachchy, R. Horax, M.J. Cho, B. Davis, and J. Dickson. 2006. Thiobarbituric acid reactive substances and volatile compounds in chicken breast meat infused with plant extracts and subjected to electron beam irradiation. *Poult. Sci.* 85:1107–1113.

Ramarathnam, N., L.J. Rubin, and L.L. Diosady. 1993. Studies on meat flavor. 4. Fractionation, characterization, and quantitation of volatiles form uncured and cured beef and chicken. *J. Agric. Food Chem.* 41:939–945.

Ramaswamy, H.S. and J.F. Richards. 1982. Flavor of poultry meat — a review. *Can. Inst. Food Sci. Technol. J.* 15:7–18.

Rao, C.S., E.J. Day, and T.C. Chen. 1977. Effects of pH on the flavor volatiles of poultry meat during cooking. *Poult. Sci.* 56:1034–1035.

Reverte, D., Y.L. Xiong, and W.G. Moody. 2003. Properties of restructured beef steaks from forage- and grain-fed cattle as affected by antioxidant and flavoring agents. *Meat Sci.* 65:539–546.

Rhee, K.S. 1989. Chemistry of meat flavor. In *Flavor Chemistry of Lipid Foods*, Min. D.B. and T.H. Smouse (Eds.). American Oil Chemists' Society, Champaign, IL, p. 166–189.

Rojas, M.C. and M.S. Brewer. 2007. Effect of natural antioxidants on oxidative stability of cooked, refrigerated beef and pork. *J. Food Sci.* 72:S282–S288.

Ruiz, J.A., A.M. Perez-Vendrell, and E. Esteve-Garcia. 2000. Effect of dietary iron and copper on performance and oxidative stability in broiler leg meat. *Br. Poult. Sci.* 41:163–167

Ryu, Y.C., M.S. Rhee, K.M. Lee, and B.C. Kim. 2005. Effects of different levels of dietary supplemental selenium on performance, lipid oxidation, and color stability of broiler chicks. *Poult. Sci.* 84:809–815.

Sanchez-Escalante, A., D. Djenane, G. Torrescano, J.A. Beltran, and P. Roncales. 2001. The effects of ascorbic acid, taurine, carnosine and rosemary powder on colour and lipid stability of beef patties packaged in modified atmosphere. *Meat Sci.* 58:421–429.

Sanz, M., A. Flores, and C.J. Lopez-Bote. 1999. Effect of fatty acid saturation in broiler diets on abdominal fat and breast muscle fatty acid composition and susceptibility to lipid oxidation. *Poult. Sci.* 78:378–382.

Sarraga, C. and J.A. Garcia-Regueiro. 1999. Membrane lipid oxidation and proteolytic activity in thigh muscles from broilers fed different diets. *Meat Sci.* 52:213–219.

Schutte, L. 1977. Flavor precursors in food systems. *2nd Joint Conf., The Chemical Institute of Canada/American Chemical Society,* Montreal, May 29–June 2, 1977. Abstracts-of-Papers 170.

Seyfert, M., M.C. Hunt, R.A. Mancini, K.A. Hachmeister, D.H. Kropf, J.A. Unruh, and T.M. Loughin. 2005. Beef quadriceps hot boning and modified-atmosphere packaging influence properties of injection-enhanced beef round muscles. *J. Anim. Sci.* 83:686–693.

Shahidi, F. 1989. Flavor of cooked meats. *Am. Oil Chem. Soc. Symp. Ser., American Chemists' Society* 388:188–201.

Shahidi, F. 1992. Prevention of lipid oxidation in muscle foods by nitrite and nitrite-free compositions. *Am. Oil Chem. Soc. Symp. Ser., American Chemists' Society* 500:161–182.

Shahidi, F. and R.B. Pegg. 1994. In *Lipids in Food Flavors.* American Oil Chemists' Society Symposium Series, American Chemists' Society, Washington, D.C., p. 256–279.

Sheldon, B.M., P.A. Curtis, P.L. Dawson, and P.R. Ferket. 1997. Effect of dietary vitamin E on the oxidative stability, flavor, color, and volatile profiles of refrigerated and frozen turkey breast meat. *Poult. Sci.* 76:634–641.

Sheldon, B.W. and E.O. Essary. 1983. Effect of antibiotics and intestinal microflora and flavor of broiler meat. *Poult. Sci.* 61:280–287.

Shu, M.L., J.L. Gray, A.M. Booren, R L. Cracket, and J.L. Gill. 1995. Assessment of off-flavor development in restructured chicken nuggets using hexanal and TBARS measurements and sensory evaluation. *J. Sci. Food Agric.* 67:447–452.

Sista, R.V., M.C. Erickson, and R.L. Shewfelt. 2000. Lipid oxidation in a chicken muscle model system: oxidative response of lipid classes to iron ascorbate or methemoglobin catalysis. *J. Agric. Food Chem.* 48:1421–1426.

Sonaiya, E.B., M. Ristic, and F.W. Klein. 1990. Effect of environmental temperature, dietary energy, age and sex on broiler carcass portions and palatability. *Br. Poult. Sci.* 31:121–128.

Spacht, K. and W. Baltes. 1994. Identification of volatile flavor compounds with high aroma values from shallow-fried beef. *J. Agric. Food Chem.* 42:2246–2253.

St. Angelo, A.J. 1993. Natural and synthetic compounds for flavor quality maintenance. *Am. Oil Chem. Soc. Symp. Ser., American Chemists' Society* 528:56–77.

St. Angelo, A.J., K.L. Crippen, H.P. Dupuy, and C. James, Jr. 1990. Chemical and sensory studies of antioxidant-treated beef. *J Food Sci.* 55:1501–1505.

Su, Y., C.Y.W. Ang, and D.A. Lillard. 1991. Precooking method affects warmed-over flavor of broiler breast patties. *J. Food Sci.* 56:881–883.

Tanchotikul, U. and T.C.Y. Hsieh. 1991. Analysis of volatile flavor components in steamed rangia clam by dynamic headspace sampling and simultaneous distillation and extraction. *J. Food Sci.* 56:327–331.

Tang, S.Z., J.P. Kerry, D. Sheehan, and D.J. Buckley. 2002. Antioxidative mechanisms of tea catechins in chicken meat systems. *Food Chem.* 76:45–51.

Tang, S.Z., J.P. Kerry, D. Sheehan, D.J. Buckley, and P.A. Morrissey, P.A. 2000. Dietary tea catechins and iron-induced lipid oxidation in chicken meat, liver and heart. *Meat Sci.* 56:285–290.

Tang, X. and D.A. Cronin. 2007. The effects of brined onion extracts on lipid oxidation and sensory quality in refrigerated cooked turkey breast rolls during storage. *Food Chem.* 100:712–718.

Tarte, R. 1996. Sensitivity of *Listeria* to Irradiation in Raw Ground Meat, as Affected by Type of Radiation, Product Temperature, Packaging Atmosphere, and Recovery Medium. Ph.D. dissertation, Iowa State University, Ames, IA.

Taub, I.A., R.A. Karielian, J.W. Halliday, J.E. Walker, P. Angeline, and C. Merritt. 1979. Factors affecting radiolytic effects of food. *Radiat. Phys. Chem.* 14:639–653.

Thongtan, K., R.B. Toma, W. Reiboldt, and A.Z. Daoud. 2005. Effect of rosemary extract on lipid oxidation and sensory evaluation of frozen, precooked beef patties. *Foodservice Res. Int.* 16:93–104.

Turner, B.E. and D.K. Larick. 1996. Palatability of *sous vide* processed chicken breast. *Poult. Sci.* 75:1056–1063.

Urlings, H.A.P., N.G. Fransen, P.G.H. Bijker, and J.G. Logtestijn. 1993. Proteolysis and amino acid breakdown of heated and irradiated poultry byproducts and muscle tissue. *J. Anim. Sci.* 71:2432–2438.

Winne, A. and P. Dirinck. 1996. Studies on vitamin E and meat quality. 2. Effect of feeding high vitamin E levels on chicken meat quality. *J. Agric. Food Chem.* 44:1691–1696.

Wu, T.C. and B.W. Sheldon. 1988. Flavor components and factors associated with the development of off-flavors in cooked turkey rolls. *J. Food Sci.* 53:49–54.

Yan, H.J., E.J. Lee, K.C. Nam, B.R. Min, and D.U. Ahn. 2006. Dietary functional ingredients: performance of animals and quality and storage stability of irradiated raw turkey breast. *Poult. Sci.* 85:1829–1837.

Yancey, E.J., J.P. Grobbel, M.E. Dikeman, J.S. Smith, K.A. Hachmeister, I.V. Chambers, P. Gadgil, G.A. Milliken, and E.A. Dressler. 2006. Effects of total iron, myoglobin, hemoglobin, and lipid oxidation of uncooked muscles on livery flavor development and volatiles of cooked beef steaks. *Meat Sci.* 73:680–686.

13 Lipids in Muscle
Structure, Composition, and Metabolism

Daniel C. Rule

CONTENTS

13.1 INTRODUCTION

The purpose of this chapter is to provide a review of the types of lipids found in muscle and adipose tissue, along with structural features and their locations. Muscle becomes food once harvested from the animal as meat, and contains numerous nutrients that originated as complex chemical components that functioned in supporting the contractile nature of the organ. Intra- and extracellular water occupies most of the mass characteristics of fresh meat, followed by protein. Contractile protein, along with lesser concentrations of bioactive proteins and peptides, provide for the protein of meat. However, connective tissue of several varieties must not be ignored. Collagen, for example, exists in numerous forms designed to function in support processes, including the matrix that houses adipocytes, which within a particular depot constitutes adipose tissue. Within muscle of meat animals, adipose tissue can exist and is commonly referred to as marbling fat (composed of interfascicular adipocytes). Intramuscular lipid would then consist of lipid of muscle cell membranes, any intramuscular triacylglycerols, and the lipid of marbling fat. Thus, muscle can include varying concentrations of lipid based on the level of marbling of that particular muscle. Meat from muscle essentially devoid of marbling will contain a small proportion of total mass as lipid, usually about 1.0%; whereas meat that contains visible marbling will contain up to about 3% to 4% lipid, with ranges of 2.6% to 15.7% reported in early work by Link and co-workers (1970). The lipid contained within muscle without marbling adipose tissue functions primarily as muscle cell membranes and constitutes most of the polar or phospholipids (glycerophospholipids); concentrations of these lipids are nearly constant during growth and deposition of marbling fat. Thus, lipids associated with the muscle cell provide for structure of the organ by supporting membrane morphology and integrity. Lipids contained within the adipocytes of marbling and the other fat depots primarily function in fatty acid storage, in addition to the membrane needs of the adipocyte. Fat cells function beyond fatty acid storage in that an increasingly well-defined role as an endocrine gland has emerged in the past decade. For example, adipocyte-derived peptides, such as adiponectin and leptin, influence energy metabolism, diabetes, and food intake by mechanisms involving other physiological systems, including neurological systems. Thus, meat stems from muscle and associated adipose tissue that forms an extremely complex system that supports the life of the host organism. Numerous reviews, book chapters, as well as entire books have been published on the topic of lipids. The focus of the present chapter is devoted to the structure and function of lipids in muscle, and how lipids can be altered by pre-harvest management of meat animals. Much current interest emphasizes altering fatty acid profiles to meet the nutritional needs of consumers for omega-3 (ω-3) fatty acids, as well as to increase conjugated linoleic acid intake. Influencing proportions of fatty acids in beef, lamb, and pork has been a significant research goal of many laboratories for many years, with the greatest challenge being with cattle and sheep because of their gastrointestinal tract, which, because of the rumen and associated bacteria, readily modifies dietary unsaturated fatty acids by converting them to their saturated counterparts. This chapter provides a review of the types of lipids found in muscle and adipose tissue, along with structural features and their locations. A review of the function of various lipids shows how they interact metabolically during different physiological states, namely feeding, fasting, and activity. Finally, pre-harvest management strategies, in particular grass-fed beef, is discussed to update readers on current systems to alter fatty acid profiles of muscle to improve the nutritional quality of meat for consumers.

13.2 TYPES AND STRUCTURAL CHARACTERISTICS OF MUSCLE-ASSOCIATED LIPIDS

Most biology, chemistry, biochemistry, and nutrition textbooks contain the structures of the various types of lipids of interest to meat science students and researchers. The principal lipids of muscle include triacylglycerols (TAGs), various glycerophospholipids, and the fatty acids contained within these lipid types. Diacylglycerol (DAG) and cholesterol occur at lower concentrations in muscle, and cholesterol esters also have been reported in muscle of cattle, sheep, and pigs. Cholesterol esters are more prevalent in liver and blood lipids; thus, contamination of muscle samples with blood could influence the analysis. Relevant examples are given in subsequent figures to provide readers with reference structures pertinent to subsequent discussion on their metabolism.

13.2.1 Triacylglycerol (TAG) and Diacylglycerol (DAG)

In (depot) fat-free muscle, TAGs would likely occur as lipid droplets occurring within muscle cells; however, these TAGs do not occur at sufficient levels to affect lipid composition of fat-free muscle. Virtually all of the TAGs in muscle tissue would occur from seam fat, or within a particular muscle from marbling fat. In adipose tissue of cattle and sheep, TAG and DAG occur from 81% to 92% and 1% to 6%, respectively, as illustrated in a comprehensive review by Christie (1981). Diacylglycerol occurs in depot fat to a small extent, is the intermediate in the synthesis and hydrolysis of TAG, and also serves as an intermediate in specific cell signaling pathways. During synthesis of adipose tissue-associated lipids in porcine adipose tissue, incorporation of radio-labeled fatty acid into TAG was 85% to 92% of total radioactivity, with about 8% to 10% as DAG (Rule et al. 1988). The structures of TAG and DAG are illustrated in Figure 13.1.

In the generic TAG and DAG structures in the example, the R groups represent fatty acids, which would vary according to the dietary composition of the animal, location of the tissue, and rate of synthesis or hydrolysis of fatty acids. Note the bond that attaches the fatty acid to the three-carbon backbone of both TAG and DAG; this bond is the carboxyl ester that occurs in a reaction between a carboxylic acid (the fatty acid) and an alcohol. The alcohol occurs as one of the hydroxyl groups of glycerol, and the glycerol backbone originates from glycerol-3-phosphate, a downstream product

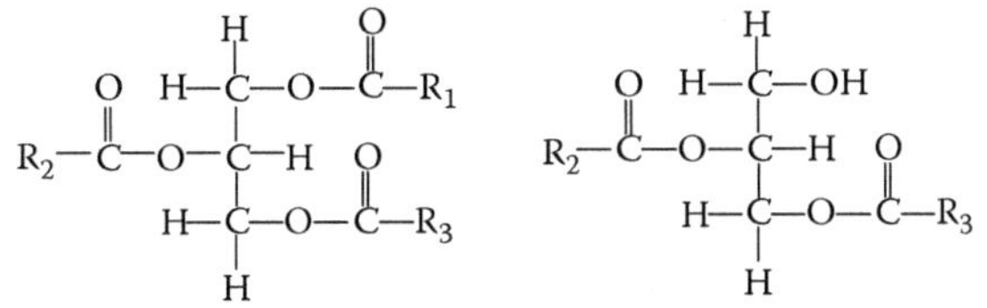

Triacylglycerol Diacylglycerol

FIGURE 13.1 Examples of triacylglycerols (TAG) and diacylglycerols (DAG). The "R" groups represent fatty acids. For DAG, fatty acids could be located at either sn-1,2, sn-2,3, or sn-1,3 positions.

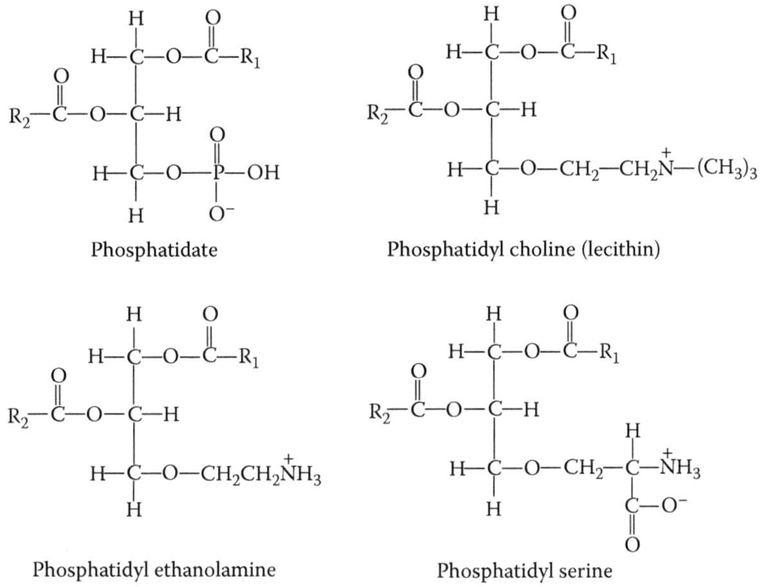

FIGURE 13.2 Glycerophospholipids.

of dihydroxyacetone phosphate, which is derived from glucose and produced during glycolysis. A description of fatty acid structure will be presented after the complex lipids have been illustrated and discussed.

13.2.2 Glycerophospholipids

The glycerophospholipids are the major components of all cell membranes, as well as the membranes of intracellular organelles such as mitochondria. Specific types of glycerophospholipids occur, and the vast majority of them are represented by phosphatidylcholine and phosphatidylethanolamine. Phosphatidylserine, phosphatidylinositol, cardiolipin, and sphingomyelin occur in lesser but significant proportions of muscle membrane glycerophospholipids. Structures of phosphotidate, the building block of some glycerophospholipids and intermediate in TAG and DAG synthesis in adipose tissue and liver, as well as phosphatidyl-choline, -serine, and -ethanolamine, are shown in Figure 13.2. The fatty acids making up the "R" groups in Figure 13.2 vary but generally will be more highly unsaturated than fatty acids of depot fat. As will be discussed later in this chapter on pre-harvest management, fatty acids of membrane glycerophospholipids are the target lipid for affecting the long-chain polyunsaturated fatty acid content of meat animals. Depot fat fatty acids tend to not harbor certain fatty acids unless the animal is a non-ruminant, or feeding of particular fatty acids that can bypass ruminal degradation/saturation.

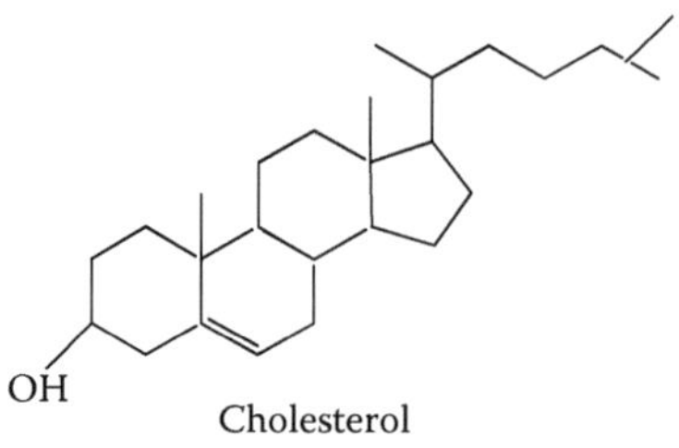

FIGURE 13.3 The structure of cholesterol.

The structure of the glycerophospholipids allows for amphipathic characteristics necessary for inclusion into a membrane. The lipid bilayer requires that hydrophobic fatty acid hydrocarbons exist in the middle; whereas, the outer parts of the membrane, both facing the outside or inside of the cell, contain charged groups that are miscible with the aqueous intra- and extracellular environment. The hydrophobic nature of the inner membrane area allows it to act as a gasket, preventing free movement of water-borne materials in and out of the cell. Imbedded within the cell membrane are the various transport proteins, transmembrane receptors, other proteins and channels, and cholesterol, the fluidizing regulator of the cell.

Cholesterol appears to affect the fluid characteristics and stability of the cell membrane. Cholesterol cannot develop into membranes but does increase the stability of the membrane with critical proportions of glycerophospholipids. The structure of cholesterol is given in Figure 13.3. Cholesterol is hydrophobic but contains the hydroxyl group at carbon-3 of the molecule, giving it limited hydrophilic properties. Cholesterol elutes near DAG in thin-layer chromatography using a predominately neutral solvent on silica gel G.

Glycerophospholipids make up the majority of the membrane's structure. In Figure 13.4 an example of this is illustrated showing the general proximity of the fatty acid "tails" that make up the hydrophobic aspect, and the phospho- "group" aligned with one of the several possible polar "head" groups (choline, ethanolamine, etc.). The heavily charged head group is miscible with the aqueous environment adjacent to the immediate interior or exterior of the cell, while the fatty acids provide the hydrophobic interior. The dichotomy of polarity illustrated here clearly shows the amphipathic nature of the glycerophospholipids.

The cell membrane is a metabolically active part of the cell. With regard to lipid metabolism by the cell, the membrane is a barrier separating the inner and outer parts of the cell, and regulatory factors are in place to control the movement of materials into and out of the cell, including lipids. For adipose tissue and muscle, fatty acids are the primary lipids for which movement across the membrane is needed, either as an uptake or storage mechanism in adipocytes, or uptake and oxidation in muscle cells. For a more in-depth understanding of membrane structure, function, and metabolism, the reader is referred to reviews by Cullis and Hope (1985), Thompson (1992), and Yeagle (1992).

13.2.3 Fatty Acids

Fatty acids represent a class of hydrocarbon molecules of varying numbers of carbon atoms arranged with primarily single carbon-carbon covalent bonds, as well as with varying numbers of carbon-carbon double bonds. The common feature of all fatty acids is the characteristic functional group — the carboxylic acid — as shown in Figure 13.5.

The carboxylic acid consists of the carboxyl-carbon, the acidic "OH" group, and the carboxyl oxygen. The polar negative nature of the oxygen groups imparts a positive polarity to the carboxyl-carbon, which increases the reactivity of this electrophilic carbon in the metabolism of fatty acids. The acidic nature of these compounds occurs because the carboxyl group hydrogen is donated to the

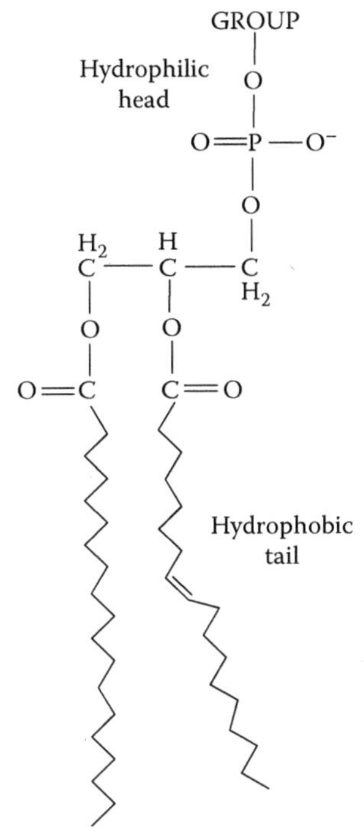

FIGURE 13.4 Illustration of the structure of the glycerophospholipid arrangement within the cell membrane. The hydrophilic head group would consist of highly charged groups characteristic of choline, ethanolamine, serine, and others. The fatty acids make up the hydrophobic tails that would comprise the inner part of the membrane.

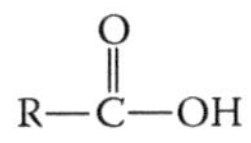

FIGURE 13.5 The common structure of carboxylic acid.

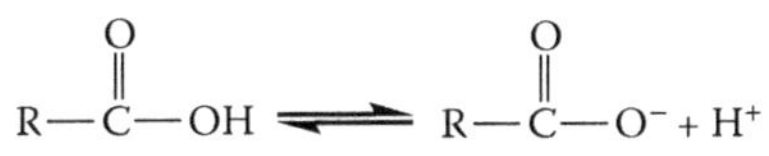

FIGURE 13.6 Fatty acids are weak acids at cellular conditions due to dissociation of hydrogen.

solution, and thus these compounds fit the definition of a weak acid. As a weak acid, the hydrogen dissociates in equilibrium according to the reaction shown in Figure 13.6. Because dissociation is affected by the pH of the medium, and the pH of the cell is slightly alkaline, most fatty acids are in the anionic form (Figure 13.6).

The “R” group in this example represents the variable number of $–CH_2–$ groups (methylene groups) in the hydrocarbon chain of the fatty acid. Based on the number of carbon atoms, “R” can be 0 to over 30 with from 0 to 6 or more carbon-carbon double bonds. The smallest fatty acid is formic acid, where “R” is hydrogen. In meat and associated fat, fatty acids typically range from 12 to 22 carbon atoms, including the carboxyl-carbon; fatty acids with 16 and 18 carbon atoms make up most of the fatty acids in cellular membranes and depot fat.

Within the variety of fatty acids there exist those that are either saturated or unsaturated. This terminology refers to the number of carbon-carbon double bonds. Figure 13.7 shows a 16-C fatty acid without any carbon-carbon double bonds.

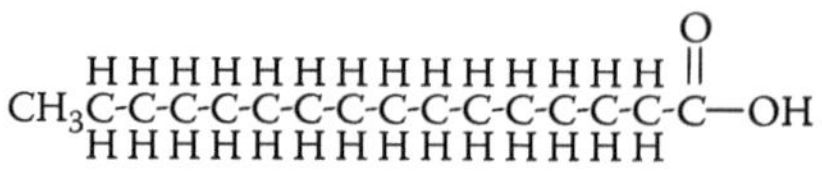

FIGURE 13.7 The structure of a 16-C saturated fatty acid (palmitic acid).

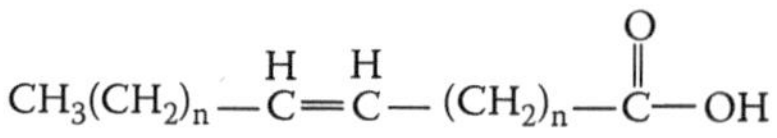

FIGURE 13.8 A fatty acid with one carbon-carbon double bond in the middle.

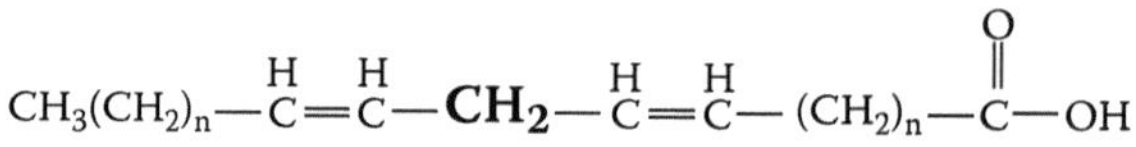

FIGURE 13.9 Two carbon-carbon double bonds of a fatty acid are separated by a single methylene group.

FIGURE 13.10 Conjugated carbon-carbon double bonds.

Except for the carboxyl-carbon, each carbon atom is bonded to two hydrogen atoms. Within the hydrocarbon chain, two hydrogen atoms is the maximum possible because each carbon atom must also bond to two other carbon atoms, the terminal carbon atom notwithstanding. A saturated fatty acid, therefore, is one in which all of the carbon atoms are saturated with hydrogen atoms. In contrast, an unsaturated fatty acid is one in which at least two adjacent carbon atoms are each bonded to a single hydrogen atom. Of course, because of the sp^3 hybrid orbital nature of the carbon atom, two pairs of electrons will need to be shared between these two carbon atoms, resulting in a carbon-carbon double bond, as shown in Figure 13.8.

Fatty acids with one carbon-carbon double bond are referred to as monounsaturated fatty acids. Fatty acids with two carbon-carbon double bonds are di-unsaturated, and those with three or more carbon-carbon double bonds are referred to as polyunsaturated fatty acids. Fatty acids with two or more carbon-carbon double bonds and are of plant or animal origin contain the double bonds oriented such that each carbon-carbon double bond is separated from the other by a single methylene (CH_2) group, as shown in Figure 13.9.

With the intervening methylene group the fatty acid contains a "methylene interrupted" sequence of carbon-carbon double bonds. Di- and polyunsaturated fatty acids of bacterial origin, or of plant or animal origin subsequently modified by bacteria, can contain multiple carbon-carbon double bonds without the methylene interruption. If we consider the carbon-carbon double bond as a unit, then without the methylene interruption, two of these "units" would be connected — or in chemical terminology, they would be "conjugated" — as shown in Figure 13.10. There is a class of fatty acids that occurs as a consequence of bacterial metabolism and is referred to as "conjugated linoleic acid" and numerous isomers therein; these are discussed later in this chapter.

The nomenclature of fatty acids is based on two fundamental approaches. First, the structure of the fatty acid follows standard IUPAC guidelines, in which the carboxyl-carbon is carbon-1 (the only functional group), and the location of any carbon-carbon double bonds or other substitutions are numerated from the carboxyl end. This also has been given the "Δ" designation. To abbreviate fatty acids with the Δ designation, indicate the number of carbon atoms in the fatty acid and then the number of carbon-carbon double bonds next to this number separated by the grammatical colon (:). For example, a six-carbon saturated fatty acid would be referred to as 6:0. A six-carbon fatty acid with a carbon-carbon double bond at carbon-3 from the carboxyl end would be referred to as 6:1Δ-3.

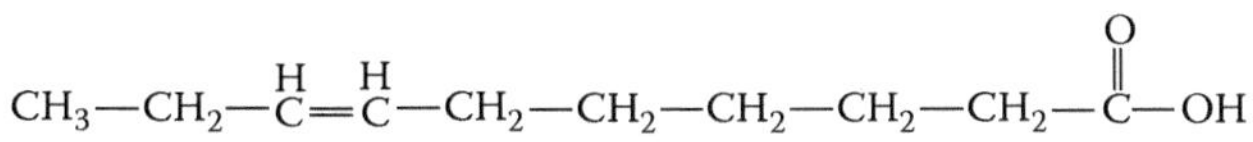

10:1, Δ-7, an ω-3 fatty acid

FIGURE 13.11 An example of both Δ and the ω in fatty acid nomenclature.

cis-double bond *trans*-double bond

FIGURE 13.12 *Cis*- and *trans*-double bond structures.

Second, a fatty acid that has one or more carbon-carbon double bonds could be one of several fatty acids based on the number of carbon atoms in which the carbon-carbon double bond occurs counting in from the methyl end (i.e., the end opposite the carboxyl-carbon) of the fatty acid. Fatty acids classified as such are named using the "ω" designation. A simplified example of both the Δ and ω in fatty acid nomenclature is given in Figure 13.11.

In addition to methylene interruption or conjugation of carbon-carbon double bonds, the geometry of the hydrogen atoms bonded to the respective carbon atoms is specific. Fatty acids of animal and plant origin contain carbon-carbon double bonds in which the hydrogen atoms are in the *cis* configuration. Fatty acids of bacterial origin can contain carbon-carbon double bonds in which the hydrogen atoms are in the *trans* configuration. The significance of the geometric configuration of unsaturated fatty acids is that the physical characteristics of the fatty acid change, thereby affecting the nutritional impact of this class of nutrient. Consumption of *trans*-fatty acids is now known to be associated with cardiovascular disease, and efforts to reduce *trans*-fats in processed foods have become major political and societal issues. From a meat science perspective, it is important to understand that ruminant animals naturally produce small amounts of *trans*-fatty acids through modification of dietary fat of plant origin, which either through direct deposition or modification in fat tissue before deposition, result in small concentrations of mono- and di-unsaturated *trans*-fatty acids containing conjugated carbon-carbon double bonds. These fatty acids, however, do not have a negative impact on the consumer. In contrast to industrial *trans*-fats, those of ruminant animal origin, namely *trans*-vaccenic acid and conjugated linoleic acid, have a number of beneficial effects in experimental animals, and would be expected to have some beneficial effect in consumers as well. The *cis* and *trans* configurations are illustrated in Figure 13.12.

Fatty acids of plant and animal origin share many similarities. For example, a number of saturated fatty acids are known to be synthesized *de novo* from both plants and animals, and oleic acid (an 18-carbon monounsaturated fatty acid) also is synthesized *de novo* by most plants and adipocytes of meat animals. However, the di- and polyunsaturated fatty acids that are required by animals and consumers must be consumed intact, or the precursor must be consumed from plant sources or from animals that obtained them from plants or algae. These fatty acids are the dietary essential fatty acids and include linoleic (18:2*cis* Δ-9, *cis* Δ-12) acid and linolenic acid (18:3*cis* Δ-9, *cis* Δ-12, *cis* Δ-15). Structures of these two fatty acids, as well as others of importance to consumers, are shown in Figure 13.13. The reason for the dietarily essential fatty acids is that animals do not produce the correct enzymes to place the carbon-carbon double bonds beyond the ninth carbon atom in an 18-carbon fatty acid; only plants can do this. Enzymes that catalyze the introduction of carbon-carbon double bonds are called "desaturases" and several distinct desaturase enzymes have been identified. Linoleic acid, for example, contains double bonds at carbon-9 and carbon-12. Only plants can desaturate at carbon-12, yet all animals require arachidonic acid, which contains carbon-carbon double bonds that would originate at carbon-12 in the 18-carbon precursor (linoleic acid). In contrast to plants, animals produce Δ-5 and Δ-6 desaturases, which will insert carbon-carbon double bonds appropriately to

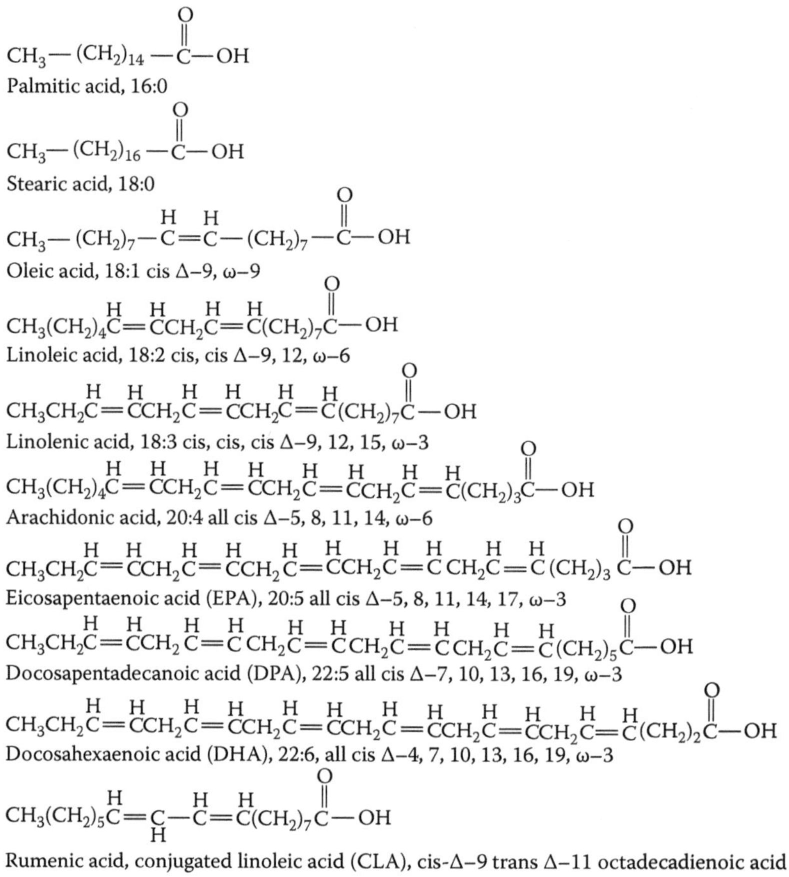

FIGURE 13.13 Common fatty acids relevant to meat and adipose tissue. Each fatty acid is represented by the structure and nomenclature above.

accommodate the requirements for arachidonic acid, as well as modification of key ω-3 fatty acids. Reviews by Cook (1985) and Whelan and Rust (2006) discuss elongation and desaturation of unsaturated fatty acids.

13.3 LIPID METABOLISM IN PRE-HARVEST MUSCLE

13.3.1 Lipid Metabolism

Lipids in pre-harvest muscle are considered nutrients when the muscle becomes meat. In the living muscle, lipids have functions that support the cell and its overall contribution to the tissue, organ, and ultimate success of the animal. In defining muscle lipids in the previous section, we saw what and where various lipids occur, but not how their metabolic involvement occurs. We did see where glycerophospholipids occur in membranes and discussed to a limited extent the other lipids. Some lipids interact in cellular metabolism in signal transduction. For example, a form of phosphatidylinositol, phosphatidyl-4,5-bisphosphate, is converted to DAG and a phosphoinositol that will influence cellular metabolism, which is discussed below in more detail. These lipids interact with cellular proteins in pathways affecting cell stimulants, for example cell surface receptor responses to a number of metabolic hormones. Membrane glycerophospholipids harbor bioactive fatty acids (for example, ω-3 and ω-6 fatty acids) known to be involved in eicosanoid metabolism. Elongation and desaturation of linoleic acid produces arachidonic acid, which serves as a substrate

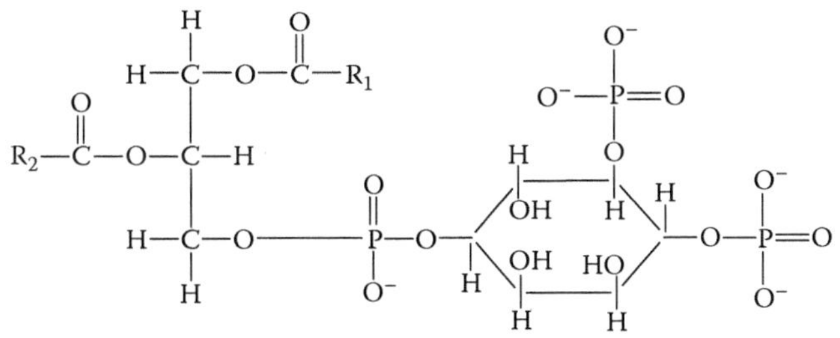

Phosphatidylinositol-4,5-bisphosphate

FIGURE 13.14 Structure of phosphatidylinositol-4,5-bisphosphate.

for prostaglandins via cyclooxygenase enzymes. Several prostaglandins are pro-inflammatory and an overabundance can be linked to inflammatory disorders. Consumption of ω-3 fatty acids has been shown to compete with arachidonic acid, resulting in development of several different types of eicosanoids. The ω-3 fatty-acid-related eicosanoids tend to have anti-inflammatory properties, and are desired in red meat to contribute to a daily intake. Fish muscle contains lipids truly rich in long-chain ω-3 polyunsaturated fatty acids, namely EPA and DHA, and a diet that includes fish, or at least a supplement containing these fatty acids, has been deemed beneficial. Current efforts to increase ω-3 fatty acids in beef and lamb have become a major focus in some aspects of the livestock industry and will be the subject of discussion in the final section of this chapter. One of the challenges of increasing long-chain polyunsaturated fatty acids in muscle stems from the strict regulation of membrane fatty acid composition to maintain function, and that long-chain polyunsaturated fatty acids tend to be a membrane glycerophospholipid component. In a review by De Smet (2004) this was pointed out by comparing percentages of polyunsaturated fatty acids in glycerophospholipids with that in TAG in both beef and pork longissimus muscle. In beef, the percentage of polyunsaturated fatty acids in glycerophospholipids ranged from 21.5% to 48.8%; whereas, in beef longissimus TAG, they ranged from 0.11% to 0.98%. The same comparison in pork longissimus glycerophospholipids was 35.3% to 48.2%, but in pork longissimus TAG, it was 7.3% to 14.5%. The difference in distribution between the two species' longissimus lipids is rooted in their gastrointestinal tracts. Cattle are ruminants, and dietary polyunsaturated fatty acids are lost through biohydrogenation induced by ruminal microorganisms; however, pigs are monogastric animals and consumed fatty acids are largely absorbed intact, thus allowing for greater deposition in muscle-associated TAG. The reader is referred to Jenkins et al. (2008) for a review of the factors affecting biohydrogenation of dietary fatty acids in ruminant animals.

Metabolically, muscle cell membranes contain certain glycerophospholipids and polyunsaturated fatty acids that occur as a reservoir for bioactive lipids and fatty acids involved in second messenger mechanisms, as well as precursors for hormones and gene regulators. Second messenger systems are complex and beyond the scope of this chapter. However, involvement of lipids has been well characterized for the mechanism involving G protein activation in response to several hormones. Phosphatidylinositol-4,5-bisphosphate is shown in Figure 13.14. This glycerophospholipid is located in the cell membrane. Upon hormone binding, G protein activation of phospholipase C in the plasma membrane results in hydrolysis of the glycerophosphate ester bond of phosphatidylinositol-4,5-bisphosphate to yield DAG and inositol-1,4,5-trisphosphate (IP3). Calcium ion release from the endoplasmic reticulum is stimulated by IP3. At the plasma membrane, the protein kinase-C is activated by DAG and can then become involved in downstream phosphorylation reactions in response to hormone binding occurring at the cell surface. Most current biochemistry textbooks (e.g., Lehninger, Nelson, and Cox 1993) present this and other examples in much more depth, as well as current review articles on cell signaling and second messenger systems, which illustrate the vast discoveries in this area during the past decade.

Long-chain polyunsaturated fatty acids, mainly EPA and DHA, along with conjugated linoleic acid (CLA) affect metabolism through anti-inflammatory mechanisms. Increased deposition of these classes of fatty acids in muscle cell membranes would most likely play a localized role unless their release allows for use elsewhere in the body. From a meat nutrient perspective, increasing these fatty acids in meat would be advantageous to consumers and producers alike. The major feature of EPA and DHA are their effects on arachidonic acid metabolism and eicosanoid synthesis. Prostaglandins are common products of virtually all cells of the body and act as lipid mediators synthesized from arachidonic acid. Adipose tissue typically contains very little arachidonic acid, whereas muscle tissue contains 1% to 3% arachidonic acid. This fatty acid is retained by the glycerophospholipids until needed for *de novo* prostaglandin synthesis, and is hydrolyzed most commonly from the sn-2 position by cytosolic phospholipase A_2. The arachidonic acid is substrate for the enzyme prostaglandin H synthase, which is more commonly referred to as cyclooxygenase (COX). Two isoforms of COX occur: COX1, which catalyzes the basal, or non-stimulated production of prostaglandin; and COX2, which increases prostaglandin synthesis in response to pro-inflammatory stimuli. Thus, COX1 is considered the constitutive form and COX2 is the inducible form. The products of COX enzymes are largely prostaglandin H_2 and thromboxanes. Another eicosanoid product of arachidonic acid metabolism includes the leukotrienes, which occur mainly in macrophages, mast cells, and polymorphonuclear leukocytes, all examples of inflammatory cells. By consuming EPA and DHA, some arachidonic acid is displaced and COX enzymes can use the former fatty acids as substrate. Moreover, both EPA and DHA are inhibitors of COX (Ringbom et al. 2001). Regardless, consumption of EPA and DHA will lead to a shift in proportions of products formed. There are a host of leukotrienes that are produced, with some less pro-inflammatory than others. For example, leukotriene B_4 is favored as a product when EPA substitutes for arachidonic acid, and this leukotriene is far less pro-inflammatory than others. Thus, EPA would be "anti-inflammatory." Generally speaking, the eicosanoids are considered lipid mediators that represent a complex set of cell signaling molecules that regulate many cellular processes. A number of dietary sources of long-chain ω-3 fatty acids, including red meat, as well as the pathway of elongation and desaturation, were summarized by Whelan and Rust (2006).

In addition to effects on eicosanoid synthesis, long-chain polyunsaturated fatty acids also affect gene transcription of materials important to cardiovascular health, as well as overall lipid synthesis. Wahle et al. (2003) have reviewed this subject; however, little direct evidence for effects within skeletal muscle is provided. Nevertheless, the current state of understanding would imply that cells engaged in lipid synthesis will be affected by the presence of long-chain polyunsaturated fatty acids, as well as CLA. Initial observations were made in rat hepatocytes where *de novo* lipogenesis as well as lipogenic protein expression were down-regulated in animals fed high-carbohydrate diets that also contained linoleic and linolenic acids. Subsequent work showed conclusively that certain polyunsaturated fatty acids did in fact influence gene transcription in hepatocytes and adipocytes.

Polyunsaturated fatty acids influence fatty acid uptake as well as the synthesis of fatty acids by adipocytes through influence on specific gene expression. As discussed by Wahle et al. (2003), polyunsaturated fatty acids down-regulate gene expression of fatty acid synthase and stearoyl-CoA desaturase 1 (SCD1). Accretion of lipid (TAG) in adipose tissue of meat animals occurs via the uptake of preformed fatty acids from the circulation and by *de novo* synthesis in the adipocyte. Fatty acids synthesized *de novo* utilize the multi-enzyme complex fatty acid synthase to produce a growing hydrocarbon chain in which a two-carbon unit is added sequentially until palmitate occurs. The palmitate is released and elongated to stearate, which can then be desaturated in a reaction catalyzed by SCD1, which currently is recognized as an essential marker of fatty acid biosynthesis. The observation that long-chain fatty acids down-regulate gene transcription for these three pivotal enzymes presents clear evidence for a molecular mechanism for down-regulation of lipid synthesis and deposition in adipocytes.

It is also important to understand that some enzymes and cellular proteins can be up-regulated or induced by long-chain polyunsaturated fatty acids in adipose tissue. Enzymes and proteins that support oxidation, storage, and metabolism of fatty acids tend to be up-regulated by polyunsaturated fatty acids. In addition, enzymes and proteins supporting fat cell differentiation are up-regulated by polyunsaturated fatty acids. For example, the enzyme lipoprotein lipase (LPL), is synthesized by the adipocyte and then secreted into the capillary endothelium, where it becomes anchored adjacent to the cell of origin. This enzyme catalyzes hydrolysis of fatty acids from circulating lipoprotein TAGs. These TAGs originate from the diet as chylomicrons or from the liver as very-low-density lipoproteins (VLDLs). Hydrolysis of the fatty acid is necessary for transport into the adipocyte. Thus, LPL plays a critical role in lipid accretion (storage) in adipocytes of meat animals. Fatty acid binding protein, acyl-CoA synthetase, and fatty acyl-CoA transferase are additional examples of enzymes and proteins for which gene transcription is up-regulated in adipocytes in response to polyunsaturated fatty acids. Fatty acid binding protein is a family of relatively small proteins that transport fatty acids within the intracellular environment to locations of further metabolism. A greater influx of fatty acids into a cell would necessitate a greater capacity to transport them for storage or metabolism. Metabolism or storage requires that fatty acids be "activated," which involves developing the thioester bond between the carboxyl group of the fatty acid and the thiol (-SH group) of co-enzyme A; this reaction is catalyzed by acyl-CoA synthetase. The thioester is thought to increase reactivity of the ester carbon atom such that esterification reactions occur more readily. Furthermore, the downstream reactions would require energy to occur; fatty acid activation utilizes energy of ATP: ATP $\rightarrow$ AMP + 2PPi. Thus, the energy to drive the reaction would come from the hydrolysis of the thioester during the reaction creating the new carboxyl ester bond.

How do fatty acids affect gene transcription? This area of research is ongoing, and a clearer picture of the particular mechanisms continues to emerge. Wahle et al. (2003) have summarized literature current to date of publication on this topic. Polyunsaturated fatty acids and associated metabolites have been shown to activate receptors and various kinases to cause "signal cascades." As a result, transcription factors are activated; nuclear translocations can occur, even involving polyunsaturated fatty acid response elements, with the overall effect of up- or down-regulation of mRNA transcription. Additionally, the fatty acids themselves can bind to transcription factors affecting recognition at promoter regions of target genes. Fatty acids could affect the stability of transcription factor mRNA as well as DNA binding capacity. The reader is referred to the review by Wahle et al. (2003) for details of these possible mechanisms, as well as greater elaboration of specific points.

13.3.2 Conjugated Linoleic Acid

Conjugated linoleic acid (CLA) represents a large group of isomers of C18:2 fatty acids in which the conjugated double bonds occur at different locations, as well as with different *cis* and *trans* orientations about the carbon-carbon double bonds. CLA intake by laboratory animals has demonstrated largely positive health effects that, if similarly positive in humans at moderate intakes, would represent significant benefit to consumers. In the past two decades, research on CLA has skyrocketed because of the initial observations of the potential health benefits of isolates obtained from grilled beef. Michael Pariza is credited with the initial discoveries that eventually included elucidation of the causative agent's identity as a fatty acid, which subsequently was determined to be CLA. Because CLA is a natural product of ruminant body fat and milk fat, and because it has been shown to have remarkable health benefits, a great deal of research and hundreds of papers have been published to date on numerous factors that could increase CLA in foods, especially in meat and milk, as well as investigations into other effects and mechanisms of action of CLA. About 20 years of research has been summarized in two volumes of *Advances in Conjugated Linoleic Acid Research* (Volume 1, Yurawecz et al., 1999; AOCS Press; Volume 2, Sebedio et al., 2003, AOCS Press). Of particular interest to meat scientists and dairy scientists is the fact that CLA is a natural product of ruminant metabolism of dietary fatty acids, starting in the rumen and ending in either adipose tissue or the lactating mammary gland. However, trace amounts of CLA can be found in the meat of other species (for example, pork, chicken, turkey, and fish) (Watkins and Li 2003).

The most abundant isomer of CLA in milk and adipose tissue of cattle is 18:2 *cis* Δ-9, *trans* Δ-11, which has the common name of rumenic acid (Figure 13.13). When a ruminant animal consumes dietary linoleic acid, the fatty acid is typically esterified to a glycerolipid, for example, TAG, glycerophospholipids, or galactolipids of forages. Ruminal microbes are numerous and hundreds of strains exist in the rumen. Bacteria produce enzymes that catalyze hydrolysis of these esters; upon availability of the free carboxyl group of the free fatty acid, microbial-induced saturation of the unsaturated fatty acids begins. Although saturation is a quite complex process, it essentially occurs by first isomerizing linoleic acid to rumenic acid, which is short lived because the *cis*-9 double bond is rapidly saturated producing *trans*-vaccenic acid (TVA; 18:1 *trans* Δ-11). Saturation of TVA appears to be less rapid than CLA because more TVA is usually observed in the duodenal contents of sheep fed supplemental linoleic acid than CLA (Kucuk et al. 2001; Atkinson et al. 2006). The gastrointestinal tract of all animals is a dynamic organ such that digesta flow is continuous. The hydrogenation of unsaturated fatty acids that occurs in the rumen, termed biohydrogenation because the process is initiated by living organisms, occurs through multiple steps, which infers the presence of intermediates. The rate at which intermediates are modified to produce downstream intermediates and finally end products will affect the rate at which the intermediates and final products occur in the intestine where they will be available for absorption by the small intestine. Thus, some level of biohydrogenation intermediates will occur in the rumen. The occurrence of CLA is quite low relative to TVA, which are both far lower than the final end-product, stearic acid (Kucuk et al. 2001; Atkinson et al. 2006). The duodenal flow of CLA and TVA can be influenced by the dietary intake of high linoleic acid oils (Kucuk et al. 2001; Atkinson et al. 2006), yet duodenal CLA remains quite low. So, where does the CLA in milk fat and adipose tissue come from? The TVA arises from saturation of CLA in the rumen. In adipose tissue and lactating mammary gland cells, the enzyme SCD occurs and plays a very significant role in lipogenesis in both tissues. For fatty acids that are synthesized *de novo*, stearic acid would be desaturated to form oleic acid, the most abundant fatty acid occurring in adipocyte TAG. However, if TVA is the substrate for SCD, the desaturation end-product would be rumenic acid because the carbon-carbon double bond will be re-inserted at the Δ9-*cis* position. As mentioned earlier in this chapter, carbon-carbon double bonds in animals are always *cis* when occurring during fatty acid biosynthesis; for the most part, only bacteria produce *trans* double bonds. Thus, the more TVA available, the more CLA can occur in adipose tissue or in milk fat.

Rumenic acid represents one of many isomers of CLA. Another important isomer is 18:2 Δ10-*trans* Δ12-*cis*. This isomer has been positively identified as a potent fatty acid synthesis inhibitor and contributes substantially to down-regulation of milk fat synthesis. Milk fat depression in dairy cows has been the subject of intense investigation for many years, and up until the discovery that this isomer of CLA was involved, the mechanism remained elusive. Bauman and co-workers (2008) have recently reviewed the regulation of fat synthesis by CLA. Additional review has been provided in *Advances in Conjugated Linoleic Acid Research*. The mechanism of down-regulation of fatty acid synthesis has now been shown to involve mRNA expression for important enzymes involved in the complex process of *de novo* fatty acid synthesis. The Δ10-*trans* Δ12-*cis* CLA isomer also down-regulates fatty acid synthesis in adipose tissue; however, about 40 times more of this isomer is required to accomplish this compared with the concentration needed in the lactating bovine mammary gland (Bauman et al. 2008). This observation implies that a dietary supplement could serve an anti-obesity function in humans and animals, but substantial amounts of the fatty acid would be needed. For most red meat consumed today, there is little Δ10-*trans* Δ12-*cis* CLA observed and, depending on the animal's diet, much more rumenic acid will occur. Nevertheless, understanding that both isomers, and very likely others yet to be discovered, play a very significant role in animal food production is imperative for the successful meat scientist. For advances in meat nutritional quality, there have been numerous attempts to increase CLA. However, by feeding CLA to ruminants, the fatty acid would be saturated and stearic acid would be the absorbed product. Protecting the CLA from further saturation would allow for more of it to be absorbed and deposited within the fat of the meat but this would be quite costly to achieve. By feeding CLA to non-ruminants, for example, pigs and poultry, more CLA can be realized because no pre-duodenal saturation would occur. Watkins and Li (2003) have summarized efforts to accomplish CLA enrichment in non-ruminant meat products. In ruminants, efforts to increase CLA in meat have included feeding supplemental vegetable oils rich in linoleic acid. When growing lambs were fed a diet based on beet pulp pellets and supplemented with safflower oil, meat contained increasing concentrations of both TVA and CLA as the dietary level of oil was increased (Bolte et al. 2002). In cattle, most efforts seldom yield CLA greater than 1% of the total fatty acids, indicating that species differences exist in the ability to raise the concentration of this fatty acid. Other post-harvest strategies to augment CLA, as well as polyunsaturated fatty acids, in beef are presented later in this chapter.

13.3.3 Metabolism of Fatty Acids in Skeletal Muscle

Fatty acids utilized by muscle can arise from several possible sources: uptake of circulating free fatty acids and lipoprotein TAG, and mobilization of fatty acids within lipid droplets associated with the muscle cell. If the animal is not fasting, muscle energy substrates will arise from the diet. In cattle and other ruminants, the energy substrates will consist primarily of acetate; whereas, in non-ruminants, such as hogs, glucose will constitute the bulk of energy substrates. If primary energy substrates are adequate, dietary fat in most species will be stored in adipocytes. Excess energy substrates will be converted to fatty acids and stored in adipocytes as a survival tactic to attempt to ensure ample nutrients for ATP production during times of inadequate energy intake. Consider the fat depth over the rump in wild ungulates after a spring, summer, and fall in non-drought conditions where lush forages are available. Fat depths of 5 cm or more are not uncommon. Following a typical winter, especially in the inter-mountain West, this animal will have very little of this fat left by the onset of the subsequent spring. The reduction of the fat depth was a direct result of mobilization of fatty acids from adipocytes, which were transported to the various organs, including muscle, for uptake and eventual oxidation to provide metabolic intermediates needed for ATP synthesis within the mitochondria. Thus, fat tissue serves as the primary repository of energy substrate reserves and can mobilize the fatty acids within seconds of the appropriate metabolic stimulus.

Fatty acids stored in fat cells arise from deposition or by *de novo* synthesis. In ruminants, absorbed acetate is the major starting material, whereas in non-ruminants, glucose produces the acetate after glycolysis and pyruvate dehydrogenase activities. Fatty acid synthesis in adipocytes yields mainly three fatty acids: palmitic, stearic, and oleic acids (see Figure 13.13 for structures). The biochemical process of fatty acid biosynthesis is well documented in most biochemistry and nutrition textbooks. The process involves converting acetate to longer-chain fatty acids by sequentially adding two-carbon units originating from acetyl-CoA until a 16-carbon product, palmitic acid, is formed. The initial reaction converts acetyl-CoA to malonyl-CoA in a carboxylation reaction catalyzed by acetyl-CoA carboxylase, which in many species is the rate-limiting step in this process. An acetyl-CoA and malonyl-CoA are converted to acetyl-ACP and malonyl-ACP (acyl-carrier protein) before interacting with the multi-enzyme complex called fatty acid synthase, which is composed of seven enzymes. The incoming two-carbon units occur as malonyl-ACP. The initial condensation of two-carbon units consists of the first bond, followed by modifications to yield a four-carbon fatty acid product that remains attached to fatty acid synthase and is ready to condense with another two-carbon unit entering again as malonyl-ACP as shown in Figure 13.15. After enough cycles to yield a 16-carbon fatty acid, the palmitic acid is released, activated, and then either esterified to the glycerol backbone of a developing TAG or elongated to stearoyl-CoA; the latter also can be either esterified to the developing TAG backbone or desaturated to oleoyl-CoA. Elongation of palmitic acid typically results in desaturation of the stearic acid but the oleic acid cannot be elongated further or desaturated further out toward the methyl end of the fatty acid because of the lack of appropriate enzymes. As a result, oleic acid accumulates; however, the intermediates can also be esterified, which sequesters the fatty acids. Consequently, the fatty acids of bovine subcutaneous adipose tissue contain the

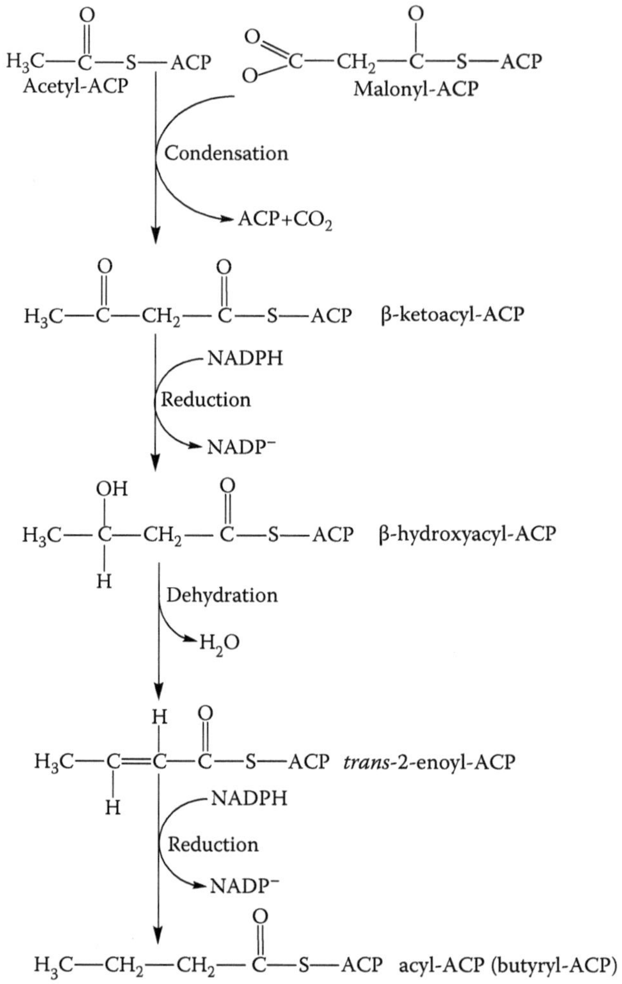

FIGURE 13.15 One "cycle" of the fatty acid biosynthesis pathway. Acetyl-CoA is converted to acetyl-ACP (acyl-carrier protein), which receives the next incoming 2-C unit from malonyl-ACP.

following profile: palmitic, 30%; stearic, 15%; oleic 48%; with the remainder consisting of shorter-chain saturated intermediates of 12 and 14 carbons, as well as some shorter chain mono-unsaturated fatty acids such as 14:1 and 16:1. Dietary and bacterial fatty acids, along with biohydrogenation intermediates, also contribute to the fatty acid profile in ruminants. In non-ruminants, a diet rich in fat would depress *de novo* synthesis, resulting in an adipose tissue fatty acid profile similar to the dietary source. This is not possible to the same extent in ruminants because too much fat in the diet depresses feed intake; and because of biohydrogenation, most of the unsaturated fatty acids are converted to their saturated counterparts and then deposited. Moreover, typical ruminant diets are quite low in fat. The profile of fatty acids in muscle and adipose tissue lipids of meat animals is influenced by diet and other factors, which has been previously reviewed (Christie 1981; Rule et al. 1995). Current efforts to modify the fatty acid composition of beef adipose tissue and muscle lipids are discussed in the final section of this chapter.

The location of adipose tissue depots varies with species, diet, and genetics, especially in ungulates. For example, beef cattle have been selected for a number of production traits, including marbling. Marbling adipose tissue can be readily observed as visible fat deposits within a given muscle, which the M. longissimus dorsi typically constitutes the industry standard for grading purposes.

Variation in the amount of marbling within cattle, per se, continues to be an issue of economic importance to the beef industry. Nutritionally induced marbling is possible if enough dietary energy excess is provided for a long enough period of time. However, dietary influences that affect the availability of glucose for absorption by the small intestine could impact the development of marbling. Texas A&M University research (S.B. Smith, personal communication) has suggested that corn processing that increases ruminal fermentation could reduce the amount of marbling because less starch would be available for small intestinal digestion and glucose absorption compared with feeding cracked corn. Glucose is a preferred substrate for fatty acid synthesis in marbling adipocytes, whereas acetate is the preferred substrate in the other depots. Thus, greater glucose availability could induce greater marbling, given that the adipocytes are present within this depot. Genetics also likely plays a role in the development of marbling. Currently, grass-fed beef has been gaining in popularity, and many grass-fed beef producers are directing their herd genetics toward smaller framed cattle that could develop marbling during grass fattening. Changes in herd genetics seem at present to be a retrograde type of selection for frame sizes popular during the mid-twentieth century. At present, documentation in the meats literature on carcass composition of these grass-fed cattle is lacking, leaving conclusions more appropriate for later reviews. In other ungulate species (e.g., deer, elk, and bison), marbling does not occur, indicating that marbling fat is not necessary for the success of the muscle or the animal. Delivery of nutrients to muscle for continued contractile activity would provide the muscle with its needs during stress, exertion, and fasting.

Under conditions of stress, exertion, and bouts of fasting, muscle will rely on the oxidation of stored fatty acids for ultimate generation of ATP. Fatty acids mobilized from adipocytes are delivered to the muscle cell bound to albumin. Muscle-associated LPL also contributes to the fatty acid pool available to muscle cells during these conditions. Fatty acids are taken up and transported through the muscle cell by several mechanisms to be metabolized first by activation and second by transport into mitochondria for oxidation and conversion to acetyl-CoA. The Krebs cycle completes the oxidation process, and the reduced proton carriers, NADH and $FADH_2$, interact with the electron transport system where protons are "pumped" from the mitochondrial matrix to the intermembrane space as part of the complex oxidation and reduction pathway. As the protons flow back into the matrix through specialized gates associated with ATP synthetase, the energy of the H^+ (proton, or hydrogen ion) flow powers phosphorylation of ADP to produce the ATP.

Important aspects of fatty acid metabolism in ruminant skeletal muscle were reviewed by Drackley (2005). Fatty acid oxidation requires mitochondria; thus, red muscle fibers (high mitochondrial concentrations) utilize fatty acids as opposed to white muscle fibers, which are considered glycolytic (low mitochondrial concentrations). Within the red muscle fiber, more fatty acid oxidation occurs in the intermyofibrillar mitochondria than in the subsarcolemmal mitochondria. Because fatty acid oxidation occurs in the mitochondria, the fatty acids must traverse both the sarcolemmal and mitochondrial membranes. An overview of fatty acid transport into the muscle cell and mitochondria along with points of regulation are shown in Figure 13.16.

The first step in fatty acid metabolism in muscle cells is the uptake of the fatty acids and transport through the cell membrane. This step has long been thought to occur by diffusion of the fatty acids; however, a large body of evidence supports the presence of fatty acid transport proteins that facilitate their uptake, especially under the influence of several stimuli. This topic was recently reviewed by Koonen et al. (2005). Regulation of long-chain fatty acid transport through acute stimuli, such as muscle contraction or insulin, involves at least one such transport protein, fatty acid translocase (FAT)/CD36. This protein resides within the cytosol, and under the appropriate stimulus it is translocated to the plasma membrane where it facilitates fatty acid transport. There are several other fatty acid transport proteins that have been discovered besides FAT/CD36, and include plasmalemmal fatty acid binding protein (FABPpm) and fatty acid transport protein (FATP). Interestingly, FABPpm is identical to mitochondrial aspartate aminotransferase, and FATP1 has acyl-CoA synthetase activity, making it a bifunctional protein. The relationship between the two functions is not known; however, that fatty acid acyl-CoA cannot diffuse back across the plasma membrane suggests

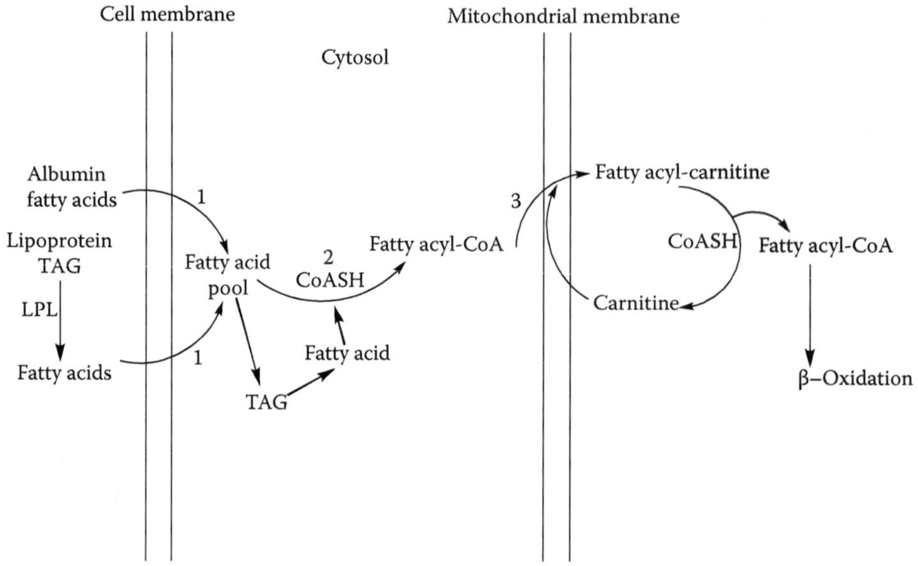

FIGURE 13.16 Overview of fatty acid uptake into the muscle cell and mitochondria. Major regulatory control points occur at: 1) plasma membrane fatty acid transport; 2) cytosolic fatty acid transport; 3) mitochondrial fatty acid transport through acyl-carnitine synthesis. Excess fatty acid would likely be diverted to TAG synthesis, which would be hydrolyzed to fatty acids under appropriate conditions.

intriguing possibilities. FAT/CD36 is also homologous to other proteins, for example, human glycoprotein IIIb and IV. Moreover, FAT/CD36 may influence mitochondrial acyl-CoA uptake.

Regarding fatty acid uptake by muscle cells, it is plausible to suggest that fatty acids will diffuse through the lipid bilayer readily because of the hydrophobic nature of both structures. However, facilitation of fatty acid transport is now generally accepted, and several proteins have been identified in this role as indicated already. An alternative role of FAT/CD36 and FABPpm, as indicated by Koonen et al. (2005), is to accelerate fatty acid dissociation from albumin as well as stimulate integration of long-chain fatty acids into the outer membrane layer. These factors would serve to concentrate long-chain fatty acids in the outer transmembrane layer creating a diffusion gradient. Coupled with the acyl-CoA synthetase activity of FATP1, a clearer picture emerges of plasma membrane fatty acid transport of long-chain fatty acids. Another important element in the acylation of fatty acids upon transport is that the fatty acid becomes "trapped" because this form does not diffuse out back across the plasma membrane.

An interesting model of regulation of long-chain fatty acid uptake in muscle involves translocation of FAT/CD36 from cytosolic compartments to the plasma membrane, as pointed out by Koonen et al. (2005). Bonen et al. (2000) showed that long-chain fatty acid uptake was regulated by the amount of FAT/CD36 occurring at the plasma membrane surface. In skeletal muscle, contraction causes a shift in location of FAT/CD36 from intracellular to plasma membrane regions.

Muscle contraction is associated with numerous intracellular events affecting fatty acid metabolism, which are influenced by intracellular signaling pathways involving several kinases. Protein kinase A (PKA), adenosine monophosphate activator protein kinase (AMPK), protein kinase C (PKC), and mitogen-activated protein kinase (MAPK) have been studied in relation to muscle metabolism. PKA and PKC have been shown to affect Ca^{2+}, and PKC (a family consisting of numerous isoforms) itself is regulated by upstream DAG and/or Ca^{2+}, which serve to activate the kinase. The downstream effect of activation of these kinases involves phosphorylation of acetyl-CoA carboxylase (ACC), which decreases its activity (discussed further below). Uptake of long-chain fatty acids mediated by FAT/CD36 could also be affected by MAPK because physical exertion stimulates this signaling pathway.

Contraction-induced as well as hormonal stimulation of AMPK increases long-chain fatty acid uptake and oxidation in muscle cells. Translocation of FAT/CD36 from the intracellular region to the plasma membrane occurs in response to AMPK activation. This concept was recently illustrated by Habets et al. (2007) using cardiac myocytes of mice that were either CD36 +/+ or CD36 –/–. Upon contraction, mimetic stimulation with oligomycin, a 2.5-fold increase in long-chain fatty acid uptake occurred in the CD36 +/+ mice myocytes, whereas long-chain fatty acid uptake stimulation was lost in the CD36 –/– mice myocytes, even with intact AMPK signaling.

Cytosolic transport of fatty acid and fatty-acyl-CoA is accomplished with intracellular fatty acid binding protein (FABP). For review of FABP the reader is referred to Storch and Thumser (2000) and Storch and Corsico (2008). Numerous isoforms of FABP have been discovered during the past three decades. Isoforms specific to some tissues have been identified, including skeletal muscle. Structurally, the isoforms of FABP belong to a 14- to 15-kDa FABP gene family. In skeletal muscle, heart (H)-FABP occurs, as well as in cardiac muscle, brain, mammary gland, kidney, adrenals, ovaries, and testes. The function of FABPs has been developed with a large body of indirect and direct evidence associating it with cellular metabolism. Binding affinity studies have shown that different fatty acids bind with more or less affinity, depending on fatty acid size and degree of saturation. Acyl-CoA also binds to FABP with a net effect of long-chain fatty acids and fatty acyl-CoA being "shuttled" to points of further metabolism. For muscle cells, this would be to mitochondria for oxidation, as well as to the endoplasmic reticulum for TAG biosynthesis. The latter occurs through the glycerolipid biosynthesis pathway, in which fatty acyl-CoA is esterified to glycerol-3-phosphate, sequentially followed by hydrolysis of the sn-3 phosphate, and finally esterification of a fatty acid at the sn-3 position to yield TAG. This pathway occurs in response to insulin in contrast to the effect of contraction or catecholamines. During fasting, for example, hormone-sensitive lipase (HSL) would be stimulated, which then would catalyze hydrolysis of the carboxyl esters of the intracellular TAG (Jeukendrup 2002).

Functionally, FABP not only would bind to fatty acids for "shuttle" purposes, but would also encourage the concentration gradient needed for optimal uptake of long-chain fatty acids by the muscle cell. Thus, as fatty acids diffuse into the cell, along with fatty acids transported by membrane fatty acid transport proteins that also catalyze acyl-CoA formation, FABP will bind and sequester the fatty acids.

The primary need for fatty acids by muscle cells is for a source of energy substrates. Once transported across the plasma membrane through the cytosol, and to activation to the acyl-CoA, the fatty acid must be transported into the mitochondria for oxidation (termed β-oxidation because of the carbon position relative to the carboxyl group). Transport into the mitochondria represents the most prevalent point of regulation of fatty acid metabolism in the muscle cell. This topic was discussed in detail for liver by Odle (2005) and other tissues in numerous reviews and textbooks. However, by summarizing much of the information on transport of fatty acids, an interesting picture develops about fatty acid transport/metabolism in muscle cells.

The key enzyme involved in mitochondrial uptake is carnitine-palmitoyl-CoA transferase 1 (CPT-I). This enzyme is regulated through inhibition by malonyl-CoA, the product of carboxylation of acetyl-CoA and catalyzed by acetyl-CoA carboxylase-2 (ACC_2; ACC_1 is the adipocyte isoform). As introduced in Figure 13.17, CPT-I catalyzes transfer of carnitine for CoASH of the fatty acyl-CoA. The acyl-carnitine then translocates into the mitochondria where CPTII catalyzes transfer of CoASH for carnitine. Because CPT-I is regulated by malonyl-CoA, factors influencing ACC_2 will be important to understand in order to appreciate fatty acid metabolism in muscle cells.

Hormonal effects (for example, when insulin secretion occurs in the fed state) will impact ACC in both adipose tissue and muscle. In both tissues, malonyl-CoA will be produced; however, the purpose for malonyl-CoA is different. In adipose tissue, it serves as a substrate for fatty acid synthase during elongation of the growing fatty acid chain. During feeding when energy substrates are in excess, *de novo* fatty acid synthesis would be expected to occur. Under fed conditions, dietary energy substrates (for example, acetate in ruminants or glucose in non-ruminants) will be used by muscle cells for energy needs. At rest, when the muscle cellular energy charge is high (high ratio of ATP:ADP and NADH:NAD^+), dietary energy substrates will be diverted further to adipose

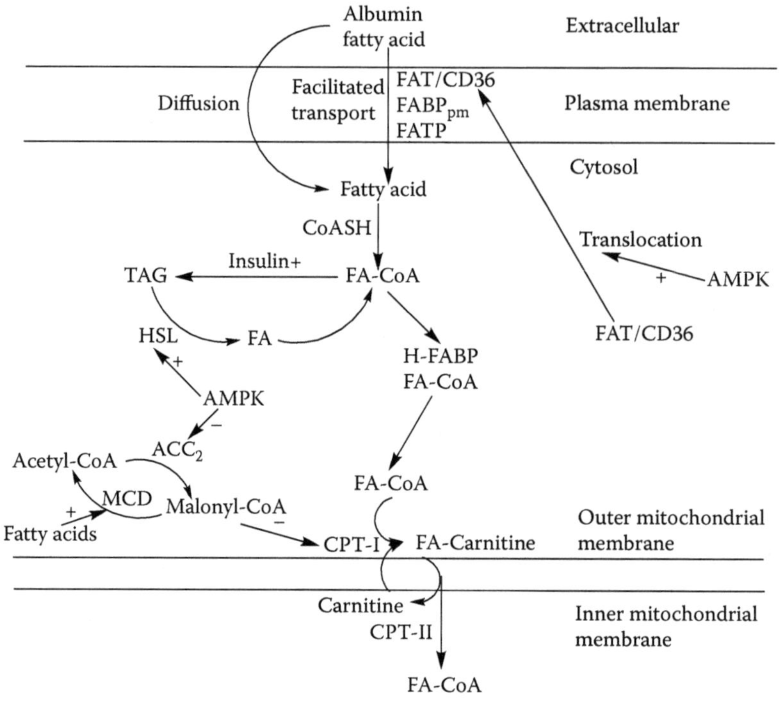

FIGURE 13.17 Model of contraction stimulation of fatty acid metabolism in muscle. During contraction, AMPK is stimulated → FAT/CD36 translocates to the cell membrane → with FABPpm and FATP fatty acid removal from albumin increases and fatty acid uptake increases → acyl-CoA to shuttle fatty acids and maintain concentration gradient. AMPK → phosphorylates HSL (stimulates) and ACC2 (inhibits) → TAG hydrolysis increases and malonyl-CoA decreases near CPT-1 binding site → CPT-1 activity increases → increased mitochondrial uptake of fatty acids. Increased fatty acid uptake causes up-regulation of MCD. Insulin decreases the phosphorylation cascade → reduces plasma membrane fatty acid transport, decreases HSL activity, increases ACC2 activity → increased malonyl-CoA → decreases CPT-1 activity → decreased mitochondrial fatty acid uptake.

tissue for fatty acid synthesis and subsequent storage. Under this condition, it is unnecessary for adipose tissue to mobilize fatty acids and for muscle cells to be oxidizing them. In the muscle cell, malonyl-CoA produced from highly active ACC_2 will inhibit CPT-I such that little, if any, fatty acid transport into mitochondria occurs. Furthermore, under these conditions, FAT/CD36 will be greater in its intracellular compartment than the plasma membrane, further reducing fatty acid uptake. Generally, the environment established discourages fatty acid uptake and oxidation.

During fasting (when insulin is low) or stress, sensitivity to catecholamines is greatest. Phosphorylation cascades occur in response to elevated cAMP and numerous kinases become activated. Under these conditions, especially with fasting, storage of nutrients decreases and the need to mobilize fatty acids from adipose tissue increases. Acute regulation of ACC occurs through inhibition by phosphorylation. In adipose tissue, phosphorylation of ACC decreases its activity; phosphorylation of HSL increases its activity, resulting in low fatty acid synthesis and increased fatty acid mobilization. In addition to HSL, a recently discovered lipase called desnutrin — an adipose tissue triglyceride lipase — contributes significantly to fatty acid mobilization in adipocytes and may play a role in myocyte intracellular lipolysis along with HSL (See Duncan et al., 2007 for review). In muscle, ACC activity is also decreased by phosphorylation, causing malonyl-CoA production to decline substantially, which removes inhibition of CPT-I activity. In addition to the down-regulation of ACC by phosphorylation, increased intracellular fatty acids stimulate expression and activity of malonyl-CoA decarboxylase, which catalyzes conversion of malonyl-CoA back to acetyl-CoA, thereby providing additional down-regulation of malonyl-CoA (Young et al. 2001). Phosphorylation of muscle cell HSL would likely occur, resulting in hydrolysis of fatty acid esters from an intracellular TAG. The overall effect would be increased mitochondrial transport of fatty acid through greater production of acyl-carnitine. Moreover, the phosphorylation cascade would result in greater translocation of FAT/CD36 to the plasma membrane, where it interacts with FABPpm to facilitate uptake of fatty acids.

Muscle contraction also stimulates fatty acid uptake and oxidation. Rasmussen and Wolf (1999) discuss a plausible model. In this model, muscle contraction stimulates AMPK activity, which stimulates phosphorylation of ACC, resulting in lower malonyl-CoA concentrations. Moreover, AMPK signaling increases translocation of FAT/CD36 to the plasma membrane to facilitate fatty acid transport into the cell. A potential pitfall in this model is related to the level of malonyl-CoA occurring after fasting or contraction, which remains high enough to completely inhibit CPT-I. Kim (1997) reviewed regulation of ACC and proposed that malonyl-CoA can be produced near the CPT-I binding site for malonyl-CoA. This would suggest that an increase in malonyl-CoA can be regulated near CPT-I to provide tight control of fatty acid uptake. Figure 13.17 summarizes the concepts outlined above.

13.4 PRE-HARVEST STRATEGIES TO INCREASE ω-3 FATTY ACIDS AND CLA IN MEAT

Red meat is a very good source of valuable nutrients, especially protein. Lean meat offers consumers the opportunity for a high-protein food source that also contains important vitamins and minerals. The fat associated with red meat, especially beef, also provides consumers with valuable unsaturated fatty acids; however, as depot fat increases in the meat, so do the amounts and proportions of less-desirable saturated fatty acids. Fatty acids of particular interest currently include the ω-3 fatty acids and CLA, which occur to a modest extent in beef and can be influenced by the diet the animal has consumed. The ω-3 fatty acids must come from the diet because the animal lacks enzymes needed to synthesize them *de novo*. Similarly, CLA — in particular, rumenic acid — requires precursor fatty acids that originate in the diet, as discussed previously in this chapter. The only viable strategy to increase the concentrations of these fatty acids in meat, therefore, is through dietary manipulation. Conventional feedlot-type diets contain modest levels of fat. For example, a feedlot finishing diet based mostly on corn could contain 3% fat, with most coming from the oil associated with the corn. This would provide these animals with nearly 60% of their fat as linoleic acid (18:2 Δ9 *cis*, Δ12 *cis*). However, because of biohydrogenation, most of the linoleic acid will be converted to stearic acid (18:0). Nevertheless, enough linoleic acid will bypass rumen microbial biohydrogenation to provide the animal with adequate amounts of this essential fatty acid. Moreover, a portion of the biohydrogenation intermediate, *trans*-vaccenic acid, will be converted to rumenic acid in adipose tissue, allowing for the appearance of this fatty acid in the intramuscular fat. By increasing the dietary intake of linoleic acid, greater *trans*-vaccenic acid and rumenic acid will be deposited in the intramuscular fat; however, not all studies show marked increases in CLA.

Increasing the concentration of ω-3 fatty acids poses similar challenges because of biohydrogenation; however, like linoleic acid, linolenic acid can also be increased by dietary means. Although linolenic acid provides important anti-cholesterolemic protection, this effect is not considered much greater than that provided by linoleic acid. The greatest benefit of linolenic acid is as a precursor for the long-chain polyunsaturated fatty acids, EPA and DHA. Unfortunately, the conversion of linolenic acid to EPA or DHA is not very great, as discussed further below. Supplementation of cattle diets with high-linolenic acid sources (for example, safflower or linseed oil) has merit; however, the current trend in cattle nutrition to increase ω-3 fatty acids is production of grass-fed beef. Scollan et al. (2005) provided a review that discusses grass-fed beef and the impact of grass feeding on fatty acids, CLA, and effects on the eating quality of the meat. Of significant interest to meat scientists is the observation that meat with elevated ω-3 fatty acids often tastes fishy, likely because of oxidation of the polyunsaturated fatty acids that occurs with cooking. Indeed, this area of endeavor by significant numbers livestock producers will provide interesting challenges to meats researchers. Another purpose for production of grass-fed beef, in addition to the purported nutritional benefits, is that it is appealing to the ever-increasing population of organic food consumers because of the more natural appearance of the production style, as well as the impression of environmental compatibility. In the interest of the topic of this book and the focus of this chapter, discussion will be limited to grass-fed beef and the impact on muscle and meat composition. The following discussion provides the reader with a general view based on the scientific literature. However, the reader is also encouraged to read the reviews of Scollan et al. (2005, 2006) as well.

Beef represents one of just a few sources of long-chain ω-3 fatty acids (seafood and eggs are also good sources). However, lean meat from grass-fed beef can provide appreciable amounts of ω-3 fatty acids (Scollan et al. 2006). A study by Rule et al. (2002) indicated that the proportion of total fatty acids represented by long-chain ω-3 fatty acids was substantially higher in the rib-eye steak of grass-fed beef compared with feedlot-fed beef (2.90% vs. 0.64%) and bison (5.35% vs. 1.51%), but the concentration of total fatty acids in the meat was substantially lower for the grass-fed animals. Moreover, recent reports on fatty acids in meat of grass-fed beef clearly illustrate that ω-3 fatty acids will increase by dietary inputs. Compared with concentrate-finished cattle, grass-fed beef had substantially greater proportions of all ω-3 fatty acids in the meat, with differences as high as eightfold reported (Lorenz et al. 2002; Nuernberg et al. 2002; Dannenberger et al. 2004; Nuernberg et al. 2005).

Differences in fatty acids between grass-fed beef and feedlot-fed beef notwithstanding, levels of ω-3 fatty acids in grass-fed beef are subject to variation depending on if the forage was grazed or fed after harvested, as well as variation in forage fatty acids due to plant genetics and season of the year (Dewhurst et al. 2003). In a review, Dewhurst et al. (2003) revealed that levels of 18:3ω-3, the primary fatty acid in forage, decreased in the forage when harvested, and animals that grazed the forage deposited more 18:3ω-3 in the meat and milk than when they consumed harvested forage. For example, the rib-eye steaks of grass-fed cattle that grazed had higher proportions of 18:3ω-3 (1.13%) compared with cattle fed the silage (0.71%; French et al. 2000). Elgersma et al. (2003) reported that when perennial ryegrass is ensiled, most of the fatty acids are hydrolyzed from the plant's lipids, forming free fatty acids, which are oxidized. Elgersma et al. (2003) concluded that 18:1ω-9 and 18:3ω-3 were the fatty acids primarily lost after ensiling. Moreover, as forages mature through the growing season, levels of 18:3ω-3 change. Dewhurst et al. (2001) reported that in ryegrass, levels of 18:3ω-3 decreased from 12.9 g/kg of dry matter in April to 4.4 g/kg in June, but then increased to 10.4 g/kg by November. These authors attributed the change in fatty acid composition to conversion to vegetative stages in which flowering caused reduced fatty acid levels. The magnitude by which 18:3ω-3 decreased during the forage growing season was illustrated by Weston (2007), where up to twofold decreases in concentration of 18:3ω-3 in Western Wheat Grass, Needle and Thread, Needle Leaf Sedge, Blue Grama, Dalmation Toad Flax, and Scarlet Globemallow from May through October were observed.

Clearly, the proportion of ω-3 fatty acids in the meat lipids of grass-fed beef increases because of the 18:3ω-3 consumed with the forage. Nutritionally for humans, ω-6 fatty acid intake levels are adequate, if not excessive, and deficiency is not a problem. However, consumption of ω-3 fatty acids generally should be increased (Scollan et al. 2001a; Wijendran and Hayes 2004). Consumption of 18:3ω-3 is equally effective as ω-6 fatty acids for cardiovascular health (Sanderson et al. 2002); however, consumption of ω-3 fatty acids as EPA and DHA should be increased. Fish oil is rich in these fatty acids, and their health benefits have been well documented. Also, these fatty acids are commercially produced using marine algae, which does not have the negative attributes of fish oil (odor, fishy flavor, etc.).

When an animal consumes and deposits 18:3ω-3 in tissues, EPA and DHA can be detected without consumption of the latter two fatty acids because of elongation and desaturation of some of the 18:3ω-3 (Kinsella et al. 1990). However, the efficiency of conversion of 18:3ω-3 to EPA and DHA is about 20% at best (Kinsella et al. 1990). Nonetheless, recent reports showing increased 18:3ω-3 in grass-fed beef also show increased EPA + DHA (Dannenberger et al. 2004; Nuernberg et al. 2005). In view of the low conversion of 18:3ω-3 to the longer-chain ω-3 fatty acids, enhancing EPA and DHA concentrations in beef muscle would occur most efficiently by feeding EPA and DHA directly. However, feeding unsaturated fatty acids to ruminants results in the loss of many of these fatty acids because of biohydrogenation. To restate this phenomenon, the rumen is the part of the stomach that first encounters consumed feeds, and harbors the bacteria and protozoa that produce the enzymes needed to catalyze the breakdown of the nutrients consumed by the animal. The microbes also produce enzymes that catalyze the saturation of unsaturated fatty acids, or biohydrogenation. The reason for this is not clear but it is known that unsaturated fatty acids can be toxic to ruminal microorganisms. Biohydrogenation has also been thought to provide a means to sequester excess hydrogen ions in the anaerobic rumen environment. Biohydrogenation is a multistep process that converts 18:2ω-6 into 18:0 if the process goes to completion. Biohydrogenation of 18:3ω-3 also occurs and results in the loss of much of this fatty acid when consumed by ruminants because it is converted to 18:0. When sheep consumed high-forage diets, the flow of 18:3ω-3 through the rumen and into the small intestine was quite low (Kucuk et al. 2001). Work with beef heifers grazing bromegrass pasture showed that intake of 30 g/d of 18:3ω-3 was associated with a duodenal flow of 5.5 g/d (Lake et al. 2002) in one study and an intake of 17 g/d was associated with a duodenal flow of 2.4 g/d in another study (Carter et al. 2002). Wachira et al. (2000) reported values for biohydrogenation of 18:3ω-3 of 80.2% in sheep fed a basal diet of only pelleted dried grass, but when the pelleted grass was supplemented with linseed to provide 3% of dry matter intake as 18:3ω-3, biohydrogenation was 92.4%. It can be concluded that feeding 18:3ω-3 from intact forage allows for a greater percentage of the fatty acid to escape ruminal biohydrogenation. On the other hand, supplementing linseeds provided about 10% more 18:3ω-3 flowing to the small intestine than without the supplement, indicating that not all of the 18:3ω-3 will be lost, and some will be absorbed and deposited in the muscle lipids of the animal. Wachira et al. (2000) suggested that 18:3ω-3 in plant lipids occurs as glycolipids, which are less susceptible to ruminal biohydrogenation than when fed as an oil (triacylglycerols). In addition to the chemical form of the ω-3 fatty acid, other aspects of plant biochemistry have been shown to affect the stability of unsaturated fatty acids. For example, red clover contains an enzyme, polyphenol oxidase, that is thought to provide some protection from biohydrogenation and was proposed to contribute to a doubling of 18:3ω-3 in grass-fed beef M. longissimus dorsi compared with cattle fed grass silage (Scollan et al. 2006). Also discussed in the review by Scollan et al. (2006), sheep raised on alpine pasture in Norway produced milk with 25% greater 18:3ω-3 compared with sheep raised on low-land pastures; thus an "alpine factor" that decreases ruminal biohydrogenation has been proposed.

The question of current relevance is whether biohydrogenation of ω-3 fatty acids, in particular EPA and DHA, is influenced by consumption of grass at different times during the forage growing season. In a report by Ashes et al. (1992), biohydrogenation of EPA and DHA was very low when assessed *in vitro*. Later, Gulati et al. (1999) reported that biohydrogenation of EPA and DHA was not substantial, implying that supplementation would result in significant deposition of these fatty acids in muscle lipids of cattle. Additional evidence that EPA and DHA are relatively resistant to ruminal biohydrogenation was provided by Kitessa et al. (2001), who observed substantial amounts of EPA and DHA in abomasal contents of sheep fed supplemental tuna oil. Moreover, Wachira et al. (2002) observed increased proportions of EPA and DHA in muscle of sheep fed 3% fish oil. Scollan et al. (2001b) supplemented the diets of beef steers with 3% of diet dry matter as linseed oil or fish oil and found 18:3ω-3, EPA, and DHA to be increased in muscle by up to twofold. However, results of Scollan et al. (2001b) indicated that biohydrogenation of EPA and DHA was over 90% in cattle supplemented with fish oil. On the other hand, Wachira et al. (2000) reported biohydrogenation of EPA at 76% and of DHA at 72% when 3% of the dietary dry matter as fish oil was fed to sheep consuming a diet of pelleted dried grass. Both Wachira et al. (2000) and Scollan et al. (2001b) attributed the differences between their results and the results of previously published data to the greater dietary fiber fed by Wachira et al. (2000) and Gulati et al. (1999). Apparently, variation exists in the biohydrogenation of these long-chain ω-3 fatty acids, and it may be related to the basal diet consumed by the animal. Scollan et al. (2001b) suggested that the higher fiber diets supported less biohydrogenation of these fatty acids, which is interesting because on a high-fiber (high-forage) diet, the ruminal microorganisms responsible for producing the biohydrogenation enzymes are highest (Latham et al. 1972). However, no mention was made of possible associative effects with the forage at particular phenological stages of forage maturity and the potential impact on biohydrogenation of unsaturated fatty acids. Of the grass-fed beef literature currently available, only the study of Noci et al. (2005) has evaluated length of time grazing on intramuscular fatty acids in cattle with particular emphasis on ω-3 fatty acids. In this study, grazing for 158 days resulted in increased ω-3 fatty acids from 0.66% to 0.96% of intramuscular lipids. Cattle were grazed for 0, 59, 118, and 158 days but the animals grazed for 59 and 118 days were started on pasture later in the season, and all tissues were obtained when the cattle were harvested. Research is needed to determine if changes in ω-3 fatty acids can be monitored throughout grazing. If resistance to biohydrogenation is possible during the early grazing season, then increased ω-3 in general, and further increases in EPA and DHA in the supplemented cattle could be expected.

Substantially increasing ω-3 fatty acids in beef could be negatively impacted if the rate of transfer of dietary ω-3 fatty acids into muscle cell membranes is very tightly regulated. These fatty acids would be found in greatest abundance in the polar lipids of muscle cell membranes (Scollan et al. 2001a). Dietary ω-3 fatty acids would have to replace existing fatty acids in the cell membranes or become part of newly developing membrane lipids. Generally, 18:2ω-6 in meat is found in muscle cell membranes (Marmer et al. 1984), and transfer of ω-3 fatty acids tends to replace the 18:2ω-6 but they will replace other fatty acids as well (Scollan et al. 2001a). However, the cell membrane must regulate a characteristic known as membrane fluidity, which must be maintained at specific levels to ensure proper cell function. Thus, transfer rates probably will not reflect total intake of ω-3 fatty acids by cattle.

The above discussion suggests the potential for grass-feeding along with appropriate supplements fed at critical times to augment ω-3 fatty acids, as well as CLA in beef. This information should be reviewed by meat scientists to remain current with new trends (or old systems revisited) in livestock production. Of particular importance to understanding lipids in meat is for meat scientists to evaluate composition on a true concentration basis. As illustrated for much of the literature on fatty acids in meats, especially in grass-fed beef literature, the proportion of a particular fatty acid could be high, while at the same time the concentration of this fatty acid in the meat product to be consumed remains quite low. This is illustrated in Figure 13.18.

In this example, the beef samples represent a steak from a steer that graded USDA Choice and contained 3% fat within the fat-trimmed core muscle. The sample on the right represents a steak from a grass-fed steer with minimal fat and very slight visible marbling, and would be typical of a piece of meat containing 1.5% fat. Each steak sample represents a 226-g sample (8 oz.), resulting in a fat pool of 6.8 and 3.4 g of fat, respectively. Based on data published from the author's laboratory (Rule et al. 2002), values typical for total saturated, ω-3, ω-6, and CLA fatty acids are shown as percentages (weight percentages, or grams of each type of fatty acid per 100 g of total fatty acids in the analysis), as well as amounts to indicate the contrast in results, which can create confusion with interpretation of the data. Percentages of total saturated fatty acids was 2.3 percentage units lower for the grass-fed beef sample, whereas ω-3 fatty acid percentages were 4.5-fold greater and ω-6 fatty acids were 1.5-fold greater for the grass-fed than for conventionally fed beef samples, respectively. Percentages of CLA also were greater for the grass-fed beef example and ranged from 1.5- to 2.7-fold higher than for the conventional beef example. When the data are viewed on a concentration basis, the differences change quite significantly. The amount of total saturated fatty acids is about double for the conventional beef compared with the grass-fed beef example. The amount of ω-3 fatty acids is now twofold greater in the grass-fed sample but the ω-6 fatty acids were 1.4-fold greater in the conventional beef sample than in the grass-fed sample. Amounts of CLA were similar for both types of meat. Greater total saturated fatty acids would be expected in the higher fat cut simply because this type of fat was of similar proportion for each type compared. The ω-3 fatty

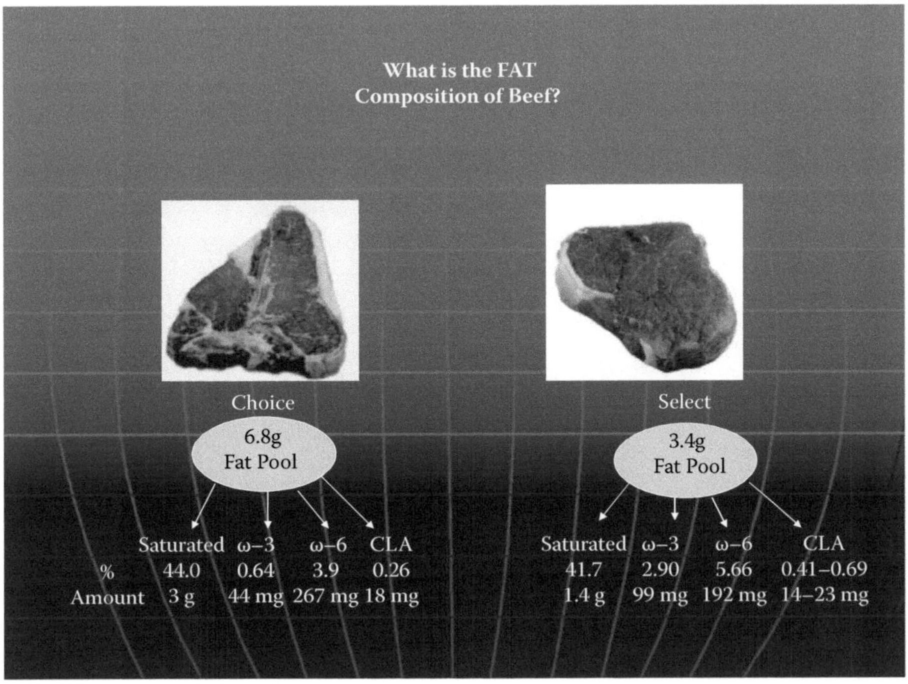

FIGURE 13.18 (See color insert following page 148.) Fatty acid composition of choice and select beef.

acids are a membrane fatty acid and tend to not deposit as much into the adipose tissue. Therefore, greater intake of ω-3 fatty acids by grass-fed beef would be expected to increase these fatty acids into the muscle cell membranes. However, because the amount of fat in the grass-fed sample was much lower than the conventional meat sample, representing this type of fat as a percentage inflates the true difference between the two types of meat. The same principle holds for ω-6 fatty acid amounts; however, in this example, the amounts were greater for the conventionally fed beef. This would be expected because on corn-based finishing diets, intake of linoleic acid would be quite high within the dietary fat, and this fatty acid will deposit in the adipose tissue lipids. The amount of CLA was not different for the two cuts of meat in this example. Because CLA tends to be associated with adipose tissue (neutral lipids such as TAG), the leaner cuts would be expected to contain less concentration, but when expressed as a percentage, the value for this fatty acid also can be inflated. Scollan et al. (2005) discussed a similar comparison between concentrate- and pasture-fed German Holstein and German Simmental bulls from data published by Nuernberg et al. (2004). These data showed that percentages of CLA (rumenic acid) were about 1.2% higher for the pasture-fed bulls on average, but the concentration within the muscle was slightly lower for the samples of pasture-fed bulls compared to concentrate-fed bulls.

Another current trend in sales of grass-fed beef is to market ground meat with quite high proportions of fat, with 25% to 30% fat product common. Grass-fed beef producers suggest, anecdotally, that the ω-3 and CLA contents of ground meat of grass-fed beef are high and the elevated level of total fat provides a nutritional advantage for consumers. This author heard one producer admit that he indicates to his customers that the nutrition in grass-fed beef "is not in the lean but its all in the fat." If a similar evaluation of the composition of ground beef is done, there are increases in ω-3 fatty acids and CLA because of the extra fat that is added to the lean, as shown in Figure 13.19.

In this example, a ¼ pound sample of 30% ground beef, purchased from a mail-order and Web based grass-fed beef supplier, contained a total fat pool of 33 g raw and 22 g after cooking. About one third of the fat was lost during cooking, not uncommon for high-fat ground beef. As a percentage, ω-3 fatty acids were not high, but not uncharacteristically because this would be a muscle cell

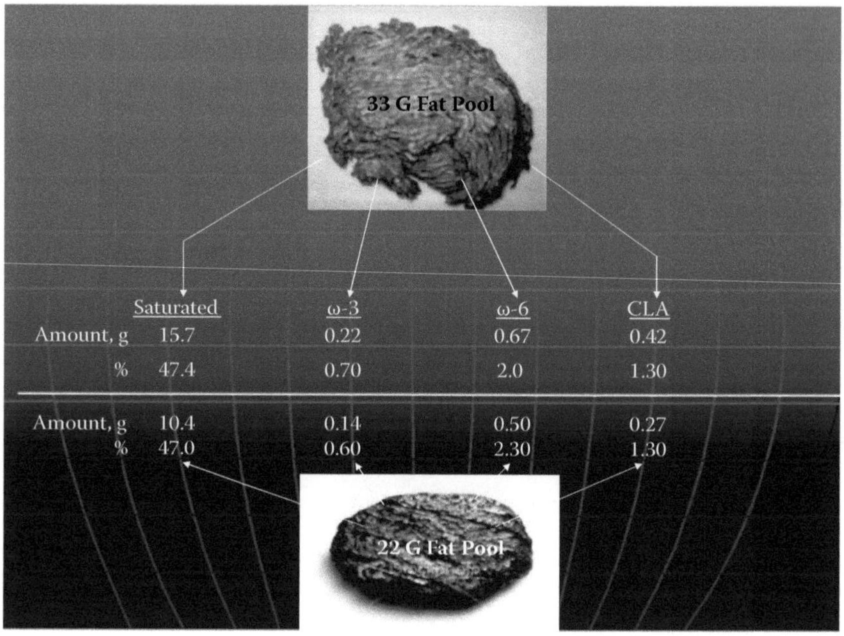

FIGURE 13.19 Cooking increases ω-3 fatty acids and CLA in ground beef.

membrane associated fatty acid, and would have been somewhat diluted by the excess fat. Notice how the percentage of CLA for cooked and raw ground meat did not change. Because CLA would be associated mostly with the fat, the lack of change in CLA percentage would not be surprising. Regarding the suggestion by grass-fed beef producers that the increased ω-3 fatty acids and CLA should support much greater intake of this beef fat, these claims have not been tested to date in controlled biomedical studies with human or animal subjects. Until tested by nutrition researchers, these claims should not be given much credence. The primary reason for caution here is that compared with total saturated fatty acids, intakes of the ω-3 fatty acids and CLA are still quite low. The ratios of total saturated fat to ω-3, ω-6, and CLA in this example are: 74, 21, and 39, respectively. If we evaluate the actual intake of the ω-3 and ω-6 fatty acids compared with current recommendations, the grass-fed beef will not be sufficient but it will contribute to the dietary intake of them. Current recommendations are based on modest fat intakes; and if fat intake increases, especially saturated fatty acids, then the intake of ω-3 and ω-6 fatty acids will have to increase further. Wajendram and Hays (2004) provided some guidelines for this in their review. For the ω-6 fatty acids, intake should be 5 to 8 en% (percentage of daily dietary metabolizable energy) or 5 to 8 g per 2000 Cal. For ω-3 fatty acids, intake should be about 1 en% or 2.2 g per 2000 Cal of energy intake with EPA + DHA intake at 0.5 to 1 g. However, if saturated fatty acid intake is high, then more of the unsaturated fatty acids should be consumed.

13.5 SUMMARY

This chapter was designed to provide students of meat science with a direction toward understanding the lipids associated with muscle and how they impact the animal, pre-harvest, and how they might impact consumers of the meat we all work so hard to provide for them. Understanding the biological phenomena in which lipids contribute to the functions of meat allows us to understand the limits by which we could manipulate meat lipid composition. Examples of current interest in grass-fed beef should be reviewed when this type of beef becomes an issue for you, the meat scientist. Can we "improve" on the inherent nutritional quality of beef by increasing the ω-3 fatty acid concentration? Can we even increase the ω-3 content to a point where the meat's lipid profile would contribute significantly to the consumer's desired endpoint for this fatty acid? What are the mechanisms controlling the biological limits to this goal? Furthermore, the meat scientist must also be a student of nutrition in order to understand the research needed to ascertain the impact of changing the composition of meat. Will the subtle changes — or even the large changes for that matter — that we can create in meat lipid composition be advantageous to consumers? Or, will they be harmful? What kinds of nutritional studies should be conducted to test hypotheses regarding new developments in the composition of meat? From my experience, I would predict that most assumptions are wrong, or at least extremely weak; therefore, the need for research to test the nutritional ramifications of altered meat must be integrated into meat science research. As consumers become more educated, it will become more important to provide advice that is based on research data rather than on assumptions from tenuous extrapolations.

REFERENCES

Ashes, J.R., B.D. Siebert, S.K. Gulati, A.Z. Cuthbertson, and T.W. Scott. 1992. Incorporation of ω-3 fatty acids of fish oil into tissue and serum lipids of ruminants. *Lipids* 27:629–631.

Atkinson, R.L., E.J. Scholljegerdes, S.L. Lake, V. Nayigihugu, B.W. Hess, and D.C. Rule. 2006. Site and extent of digestion and duodenal and ileal flow of total and esterified fatty acids in sheep fed a high-concentrate diet supplemented with high-linoleate safflower oil. *J. Anim. Sci.* 84:387–396.

Bauman, D.E., J.W. Perfield, K.J. Harvatine, and L.H. Baumgard. 2008. Regulation of fat synthesis by conjugated linoleic acid: lactation and the ruminant model. *J. Nutr.* 138:403–409.

Bolte, M R., B.W. Hess, W.J. Means, G.E. Moss, and D.C. Rule. 2002. Feeding lambs high-oleate or high-linoleate safflower seeds differentially influences carcass fatty acid composition. *J. Anim. Sci.* 80: 609–616.

Bonen, A., J.J. Luiken, Y. Arumugam, J.F. Glatz, and N.N. Tandon. 2000. Acute regulation of fatty acid uptake involves the cellular redistribution of fatty acid translocase. *J. Biol. Chem.* 275:14501–14508.

Carter, D.E., B.W. Hess, E.J. Scholljegerdes, and D.C. Rule. 2002. Effect of supplemental soybean oil on duodenal flow of long chain fatty acids in beef heifers grazing summer pasture. *Proc. West. Sect. Am. Soc. Anim. Sci.* 53:646–650.

Christie, W.W. 1981. The composition, structure and function of lipids in the tissues of ruminant animals. In *Lipid Metabolism in Ruminant Animals*. W.W. Christie, Ed., Pergamon Press, New York, p. 95–191.

Cook, H.W. 1985. Fatty acid desaturation and chain elongation in eukaryotes. In *Biochemistry of Lipids and Membranes*. D.E. Vance and J.E. Vance, Eds., The Benjamin/Cummings Publishing Co., Menlo Park, CA, p. 181–212.

Cullis, P.R. and M.J. Hope. 1985. Physical properties and functional roles of lipids in membranes. In *Biochemistry of Lipids and Membranes*. D.E. Vance and J.E. Vance, Eds., The Benjamin/Cummings Publishing Co., Menlo Park, CA, p. 25–72.

Dannenberger, D., G. Nuernberg, N. Scollan, W. Schabbel, H. Steinhart, K. Ender, and K. Nuernberg. 2004. Effect of diet on the deposition of ω-3 fatty acids, conjugated linoleic and C18:*trans* fatty acid isomers in muscle lipids of German Holstein bulls. *J. Agric. Food Chem.* 52:6607–6615.

De Smet, S., K Raes, and D. Demeyer. 2004. Meat fatty acid composition as affected by fatness and genetic factors: a review. *Anim. Res.* 53:81–98.

Dewhurst, R.J., N.D. Scollan, M.R.F. Lee, J.J. Ougham, and M.O. Humphreys. 2003. Forage breeding and management to increase the beneficial fatty acid content of ruminant products. *Proc. Nutr. Soc.* 62:329–336.

Dewhurst, R.J., N.D. Scollan, S.J. Youell, J.K. S. Tweed, and M.O. Humphreys. 2001. Influence of species, cutting date and cutting interval on the fatty acid composition of grasses. *Grass Forage Sci.* 56:68–74.

Drackley, J.K. 2005. Interorgan lipid and fatty acid metabolism in growing ruminants. In *Biology of Metabolism in Growing Animals*. D.G. Burrin and H.J. Mersman, Eds., Elsevier, New York, p. 323–350.

Duncan, R.E., M. Ahmadian, K. Jaworski, E. Sarkadi-Nagy, and H.S. Sul. 2007. Regulation of lipolysis in adipocytes. *Annu. Rev. Nutri.* 27:79–101.

Elgersma, A.G. Ellen, H. Van der Horst, B.G. Muuse, H. Doer, and S. Tamminga. 2003. Comparison of the fatty acid composition of fresh and ensiled perennial ryegrass (*Lolium perenne L.*), affected by cultivar and regrowth interval. *Anim. Feed Sci. Technol.* 108:191–205.

French, P., C. Stanton, F. Lawless, E.G. O'Riordan, F.J. Monahan, P.J. Caffrey, and A.P. Moloney. 2000. Fatty acid composition, including conjugated linoleic acid, of intramuscular fat from steers offered grazed grass, grass silage, or concentrate-based diets. *J. Anim. Sci.* 78:2849–2855.

Gulati, S.K., J.R. Ashes, and T.W. Scott. 1999. Hydrogenation of eicosapentaenoic and docosahexaenoic acids and their incorporation into milk fat. *Anim. Feed Sci. Technol.* 79:57–64.

Habets, D.D.J., W.A. Coumans, P.J. Voxhol, M.A.M. den Boer, M. Febbraio, A. Bonen, J.F.C. Glatz, and J.J.F.P. Luiken. 2007. AMPK-mediated increase in myocardial long-chain fatty acid uptake critically depends on sarcolemmal CD36. *Biochem. Biophys. Res. Commun.* 355:204–210.

Jenkins, T.C., R.J. Wallace, P.J. Moate, and E.E. Mosley. 2008. Recent advances in biohydrogenation of unsaturated fatty acids within the rumen microbial ecosystem. *J. Anim. Sci.* 86:397–412.

Jeukendrup, A.E. 2002. Regulation of fat metabolism in skeletal muscle. *Ann. N.Y. Acad. Sci.* 967:217–235.

Kim, K.H. 1997. Regulation of mammalian acetyl-coenzyme A carboxylase. *Annu. Rev. Nutr.* 17:77–99.

Kinsella, J.E., K.S. Broughton, and J.W. Whelan. 1990. Dietary unsaturated fatty acids: interactions and possible needs in relation to eicosanoid synthesis. *J. Nutr. Biochem.* 1:123–141.

Koonen, D.P.Y., J.F. C. Glatz, A. Bonen, and J.J.F.P. Luiken. 2005. Long-chain fatty acid uptake and FAT/CD36 translocation in heart and skeletal muscle. *Biochim. Biophys. Acta* 1736:163–180.

Kucuk, O., B.W. Hess, P.A. Ludden, and D.C. Rule. 2001. Effect of forage to concentrate ratio on ruminal digestion and duodenal flow of fatty acids in ewes. *J. Anim. Sci.* 79:2233–2240.

Lake, S.L., B.W. Hess, E.J. Scholljegerdes, L. Brokaw, and D.C. Rule. 2002. Effects of supplemental corn and high-oil corn on duodenal fatty acid flows for beef heifers grazing summer pasture. *Proc. West. Sect. Am. Soc. Anim. Sci.* 53:628–630.

Latham, M.J., J.E. Story, and M.E. Sharp. 1972. Effect of low-roughage diets on the microflora and lipid metabolism in the rumen. *Appl. Microbiol.* 24:871–877.

Lehninger, A.L., D.L. Nelson, and M.M. Cox. 1993. Lipids. In A.L. Lehninger, D.L. Nelson, and M.M. Cox, Eds., *Principles of Biochemistry*. Worth Publishers, New York, p. 240–267.

Link, B.A., R.W. Bray, R.G. Cassens and R.G. Kauffman. 1970. Lipid deposition in bovine skeletal muscle during growth. *J. Anim. Sci.* 30:6–9

Lorenz, S., A. Buettner, K. Ender, G. Nurnberg, H.J. Papstein, P. Schieberle, and K. Nurnberg. 2002. Influence of keeping system on the fatty acid composition in the longissimus muscle of bulls and odorants formed after pressure-cooking. *Eur. Food Res. Technol.* 214:112–118.

Marmer, W.N., R.J. Maxwell, and J.E. Williams. 1984. Effects of dietary regimen and tissue site on bovine fatty acid profiles. *J. Anim. Sci.* 59:109–121.

Noci, F., F.J. Monahan, P. French, and A.P. Moloney. 2005. The fatty acid composition of muscle fat and subcutaneous adipose tissue of pasture-fed beef heifers: influence of the duration of grazing. *J. Anim. Sci.* 83:1167–1178.

Nuernberg, K., D. Dannenberger, G. Nuernberg, K. Ender, J. Voigt, N.D. Scollan, J.D. Wood, G.R. Nute, and R.I. Richardson. 2005. Effect of a grass-based and a concentrate feeding system on meat quality characteristics and fatty acid composition of longissimus muscle in different cattle breeds. *Livestock Prod. Sci.* 94:137–147.

Nuernberg, K., G. Nuernberg, K. Ender, S. Lorenz, K. Winkler, R. Rickert, and H. Steinhart. 2002. ω-3 Fatty acids and conjugated linoleic acids of longissimus muscle in beef cattle. *Eur. J. Lipid Sci. Technol.* 104:463–471.

Odle, J., P. Lyvers-Peffer, and X. Lin. 2005. Hepatic fatty acid oxidation and ketogenesis in young pigs. In *Biology of Metabolism in Growing Animals*. D.G. Burrin and H.J. Mersmann, Eds., Elsevier, New York, p. 221–234.

Rasmussen, B.B. and R.R. Wolfe. 1999. Regulation of fatty acid oxidation in skeletal muscle. *Annu. Rev. Nutr.* 19:463–484.

Ringbom, T., U. Huss, L Stenholm, S. Flock, L. Skattebol, P. Perera, and L. Bohlin. 2001. COX-2 inhibitory effects of naturally occurring and modified fatty acids. *J. Nat. Prod.* 64:745–749.

Rule, D.C., K.S. Broughton, S.M. Shellito, and G. Maiorano. 2002. Comparison of muscle fatty acid profiles and cholesterol concentrations of bison, beef cattle, elk, and chicken. *J. Anim. Sci.* 80:1202–1211.

Rule, D.C., S.B. Smith, and H.J. Mersmann. 1988. Glycerolipid biosynthesis in swine adipose tissue *in vitro*. II. Synthesis by various types of cellular preparations. *J. Anim. Sci.* 66:1665–1675.

Rule, D.C., S.B. Smith, and J.R. Romans. 1995. Fatty acid composition of muscle and adipose tissue of meat animals. In *The Biology of Fat in Meat Animals, Current Advances*. S.B. Smith and D.R. Smith, Eds., American Society of Animal Science, Champaign, IL, p. 144–165.

Sanderson, P.Y.E. Finnegan, C.M. Williams, P.C. Calder, G.C. Burdge, S.A. Wootton, B.A. Griffin, D.J. Millward, N.C. Pegge, and W.J.E. Bemelmans. 2002. UK food standards agency alpha-linolenic acid workshop report. *Br. J. Nutr.* 88:573–579.

Scollan, N.D., N.J. Choi, E. Kurt, A.V. Fisher, M. Enser, and J.D. Wood. 2001a. Manipulating the fatty acid composition of muscle and adipose tissue in beef cattle. *Br. J. Nutr.* 85:115–124.

Scollan, N.D., M.S. Dhanoa, J.J. Choi, W.J. Maeng, M. Enser, and J.D. Wood. 2001b. Biohydrogenation and digestion of long chain fatty acids in steers fed on different sources of lipid. *J. Agric. Sci.* 136:345–355.

Scollan, N., J.F. Hocquette, K. Nuernberg, D. Dannenberger, I. Richardson, and A. Moloney. 2006. Innovations in beef production systems that enhance the nutritional and health value of beef lipids and their relationship with meat quality. *Meat Sci.* 74:17–33.

Scollan, N.J.F., I. Richardson, S. De Smet, A.P. Moloney, M. Doreau, D. Bauchart, and K. Nuernberg. 2005. Enhancing the content of beneficial fatty acids in beef and consequences for meat quality. In *Indicators of Milk and Beef Quality*. J.F. Hocquette and S. Gigli, Eds., Wageningen Academic Publishers, Wageningen, The Netherlands, p. 151–162.

Storch, J. and B. Corsico. 2008. The emerging functions and mechanisms of mammalian fatty acid-binding proteins. *Annu. Rev. Nutri.* 28:73–96.

Storch, J. and A.E.A. Thumser. 2000. The fatty acid transport function of fatty acid-binding proteins. *Biochim. Biophys. Acta* 1486:28–44.

Thompson, G.A. 1992. Rationale governing the regulation of lipid metabolism. In G.A. Thompson, Ed., *The Regulation of Membrane Lipid Metabolism*. CRC Press, Boca Raton, FL, p. 1–20.

Wachira, A.M., L.A. Sinclair, R.G. Wilkinson, M4. Enser, J.D. Wood, and A.V. Fisher. 2002. Effects of dietary fat source and breed on the carcass composition, ω-3 polyunsaturated fatty acid and conjugated linoleic acid content of sheep meat and adipose tissue. *Br. J. Nutr.* 88:697–709.

Wachira, A.M., L.A. Sinclair, R.G. Wilkinson, K. Hallett, M. Enser, and J.D. Wood. 2000. Rumen biohydrogenation of ω-3 polyunsaturated fatty acids and their effects on microbial efficiency and nutrient digestibility in sheep. *J. Agric. Sci.* 135:419–428.

Wahle, K.W.J., D. Rotondo, and S.D. Heys. 2003. Polyunsaturated fatty acids and gene expression in mammalian systems. *Proc. Nutr. Soc.* 62:349–360.

Watkins, B.A. and Y. Li. 2003. CLA in functional food: enrichment of animal products. In *Advances in Conjugated Linoleic Acid Research,* Vol. 2. J.L. Sebedio, W.W. Christie, and R. Adlof, Eds., AOCS Press, Champaign, IL, p. 174–188.

Whelan, J. and C. Rust. 2006. Innovative dietary sources of ω-3 fatty acids. *Annu. Rev. Nutri.* 26:75–103.

Wijendran, V. and K.C. Hayes. 2004. Dietary ω-6 and ω-3 fatty acid balance and cardiovascular health. *Annu. Rev. Nutr.* 24:597–615.

Weston, T.R. 2007. Nutrient Composition of Rangeland Forage Harvested in Southeast Wyoming. M.S. thesis. University of Wyoming, Department of Animal Science.

Yeagle, P.L. 1991. The dynamics of membrane lipids. In *The Structure of Biological Membranes*. P. Yeagle, Ed., CRC Press, Boca Raton, FL, p. 157–174.

Young, M.E., G.W. Goodwin, J. Ying, P. Githrie, C.R. Wilson, F.A. Laws, and H. Taegtmeyer. 2001. Regulation of cardiac and skeletal muscle malonyl-CoA decarboxylase by fatty acids. *Am. J. Physiol.* 280:E471–E479.

14 Fish Muscle

Peng Zhou and Joe M. Regenstein

CONTENTS

14.1 INTRODUCTION

In fish there are two major muscle fiber types, which are normally classified by their color: (1) white muscle fibers and (2) red muscle fibers. Unlike other vertebrates, the fiber types in fish are present in different positions within the myotome and are easy to distinguish. The white fibers are employed for high-speed, short-time swimming using anaerobic metabolic pathways to provide energy; while the red muscle fibers are employed for routine activity such as migration using aerobic metabolic pathways to provide sustained energy (Johnson 1999).

In addition to water, fish muscles are composed mostly of proteins and lipids. The content and composition of these components are very critical for the quality and stability of fish flesh. However, the composition of fishes varies, depending on species, seasons, life cycle, and many other factors. Thus, this chapter provides a brief review of the literature for the following main topics: the composition of fish muscle and its importance to the quality of fish meat, together with factors causing the variation in fish muscle composition; and effects of various storage and processing conditions on the quality of fish flesh. This chapter does not discuss surimi technology as several good reviews of this field have already been published (Lanier and Lee 1992; Park 2000).

14.2 THE COMPOSITION OF FISH MUSCLES

14.2.1 Fish Muscle Proteins

Fish muscles mainly contain three classes of proteins: (1) myofibrillar proteins, (2) sarcoplasmic proteins, and (3) stroma proteins.

14.2.1.1 Myofibrillar Proteins

Myofibrillar proteins constitute more than 60% of the total muscle proteins and include mainly myosin and actin. Together with collagen, they control the structure and the specific rheological properties of fish muscle (Dunajski 1979). Myosin comprises approximately 55% to 60% of the myofibrillar protein, and is the major component responsible for the texture properties of fish flesh (Lanier et al. 2005; Park et al. 2006). It contains two identical heavy-chain subunits of approximately 200 kDa and two pairs of light-chain subunits ranging from 17 to 22 kDa (McLachlan and Karn 1982; Craig and Woodhead 2006). An excellent and more detailed review of myosin filaments was prepared by Craig and Woodhead (2006).

There are various myosin isoforms in muscle tissues from different origins (Schiaffino and Reggiani 1996; Carpene et al. 1998). Myosins from fish species closely resemble those from white and red mammalian muscle (Rowlerson et al. 1985; Martinez et al. 1990). However, compared to myosin from mammals, fish myosin is less thermostable.

Although there are about 20,000 different species of fish living at a variety of environmental temperatures (Kawabata et al. 2000), based on their living environments, they can be subdivided into several groups: hot-water fish such as tilapia, warm-water fish such as freshwater mullet, cold-water fish such as cod and haddock, and ice-water fish such as Arctic char (Regenstein and Zhou 2007). Previous studies suggest that the thermostability of fish myosin correlates well with the habitat temperature of fish (Rodgers et al. 1987; Matsuura et al. 1988; Hamai and Konno 1990; Kawabata et al. 2000; Iwami et al. 2002), and these fish myosin isoforms differ mainly in the content and position of specific amino acid residues such as Pro and Gly (McLachlan and Karn 1982; Iwami et al. 2002). This adaptation of fish myosins to their living conditions allows the muscle to function efficiently without denaturing under the fish's normal environmental temperatures (Kawabata et al. 2000).

Another important myofibrillar protein is actin, which is a globular protein of about 40 kDa. It is the monomeric subunit and major component of thin filaments. Unlike myosin, actin is one of the most highly conserved proteins in the animal kingdom.

14.2.1.2 Sarcoplasmic Proteins

Sarcoplasmic proteins are a group of muscle proteins that are soluble in water or salt solutions of low ionic strength. Many metabolic enzymes and myoglobin belong to this class. These sarcoplasmic proteins have significant effects on the texture, flavor, and color of fish flesh (Chen et al. 1997; Haard et al. 1994; Chaijan et al. 2004; Chaijan et al. 2007).

The fresh quality of fish flesh is related to the biochemical changes taking place during the postmortem period, with endogenous enzymes playing a very critical role in many of these early changes. Unlike mammalian muscle, postmortem tenderization of fish muscle by proteases is one of the most unfavorable quality changes, as fish generally do not need to be more tender (Haard et al. 1994; Bigelow and Lee 2007). Although there is still debate about the particular role of each specific protease in postmortem changes, it is generally agreed that the breakdown of fish muscle proteins is synergistically caused by the involvement of several categories of proteases. Two major types of proteases have been widely studied: namely, the calpains and the cathepsins (Ladrat et al. 2003). Other proteases, such as a myofibril-bound serine protease, also participate in postmortem fish muscle changes (Cao et al. 1999). Calpains are involved in postmortem tenderization through a cytosolic calcium-dependent degradative pathway (Ladrat et al. 2003). These enzymes initiate the disintegration of myofibrils and make them more susceptible to other proteases (Goll et al. 1992; Papa et al. 1996; Lamare et al. 2002; Delbarre-Ladrat et al. 2004; Michalczyk and Surówka 2007). Cathepsins are a group of proteases located in the lysosomes. Cathepsins B, D, L, and H are the major cathepsins found within the fish lysosomes (Yamashita and Konagaya 1990; Aoki and Ueno 1997; Ogata et al. 1998; Hopkins and Thompson 2002; Ladrat et al. 2003), and previous studies have suggested that they can cause the degradation of myosin, α-actinin, desmin, actin, troponin-T, troponin-I, and tropomyosin as part of the muscle softening process (Jiang et al. 1992; Ladrat et al. 2003; Delbarre-Ladrat et al. 2004).

Transglutaminase is a Ca^{2+}-dependent enzyme that catalyzes an acyl transfer reaction between the γ-carboxy amide groups of glutamyl residues as the acyl donor and primary amines as the acyl acceptor (Folk 1980), which if this reaction occurs between two peptide chains results in the covalent crosslinking of muscle proteins (Hemung and Yongsawatdigul 2005). Transglutaminase has been found and characterized in fish muscle, and was found to mediate the cross-linking of myosin heavy chains (Kamath et al. 1992; Araki and Seki 1993; Benjakula et al. 2003). It contributes to the improved gel quality during surimi manufacture (Benjakula et al. 2003), and possibly, the polymerization of fish muscles during storage.

In addition, collagenases, nucleotide-degrading enzymes, and phospholipases are also important in the quality and stability of fish flesh. Collagenases contribute to the process responsible for the gaping phenomenon in which the muscle fibers are gradually disconnected from the myocommata (the unique collagen structure between myofibers of fish) during chilled storage resulting from collagen fibers breaking down. The latter two enzymes were found to participate in nucleotides breakdown and lipid hydrolysis/oxidation, respectively, in fish muscle, causing the loss of fresh fish flavor and the production of off-flavor (Sato et al. 1991; 1997, Yata et al. 2001), although initially the breakdown of nucleotides leads to a desirable flavor, particularly with the accumulation of IMP, which is a positive flavor potentiator associated with the fifth flavor, unami.

Myoglobin is a 17 kDa globular protein made up almost exclusively of α-helices with a heme prosthetic group. As a pigment protein in the sarcoplasmic fraction (Hashimoto et al. 1979; Chaijan et al. 2007), total myoglobin content has a significant impact on muscle color. The relative proportions of oxymyoglobin/oxyhemoglobin (bright red), myoglobin/hemoglobin (dark red), and metmyoglobin/methemoglobin (gray-brown) produce various colors in fish flesh. It was suggested that denaturation of these proteins would significantly affect fish muscle color (Yagiz et al. 2007). Fish myoglobins are more readily oxidized than mammalian myoglobins, resulting in the more rapid discoloration of fish muscle during frozen storage. In fresh fish, myoglobin can be partly removed from muscle by various washing processes, especially when comminuted, resulting in an increased whiteness of the final minced products (e.g., surimi). However, the denaturation of myoglobin and the interaction of myoglobin with myofibrillar proteins during fish storage make myoglobin less soluble and harder to remove.

14.2.1.3 Stroma Proteins

Collagen is the most important stroma protein. Compared with the muscles from mammals, the collagen content in fish muscles is relatively low; however, the content and composition of collagen still contributes greatly to the texture of fish muscle. Its content in muscle is well correlated with the firmness of the raw fish flesh (Hatae et al. 1986; Sato et al. 1986); and after heat processing, the texture of cooked fish meat with a high collagen content was more elastic than those of species with relatively low collagen content (Sato et al. 1986; Mizuta et al. 1999). Two genetic types of collagen, Type I and V, have been identified in the intramuscular connective tissue of fish muscles. The proportion of these two types of collagens is important for the thickness and stability of collagen fibril in fish muscle. The higher proportion of Type V collagen in fish muscle results in the formation of thinner collagen fibrils (Adachi and Hayashi 1986; Birk et al. 1990; Ando et al. 1999). The diameter of collagen fibrils in the pelagic fish was smaller than those in the demersal species, which was believed to result from more Type V collagen in pelagic fish pericellular collagen fibrils (Adachi and

Hayashi 1986; Birk et al. 1990; Ando et al. 1999). In addition, compared with Type I collagen, Type V is more labile to enzymatic degradation during cold storage and causes postmortem softening in fishes such as rainbow trout and sardine (Sato et al. 1991; 1997; Yata et al. 2001). On the other hand, collagen may undergo aggregation during long-term frozen storage, and can then contribute to the hardening of fish flesh (Borderías and Montero 1985; Montero et al. 2003).

14.2.2 Lipids in Fish Muscle

Compared with mammals, fish have a high concentration of polyunsaturated fatty acids (PUFAs). Fatty fish species like mackerel contain more PUFAs than lean fish species such as pollock (Chopin et al. 2007), and red fish muscle has a higher amount of unsaturated lipids and is more susceptible to lipid oxidation than white muscle (Yagiz et al. 2007). Lipid oxidation is one of the main processes contributing to the deterioration of fish flesh during storage and processing. It is related to negative changes in flavor, texture, nutritional value, and protein functionality (Erickson 1997; Tironi et al. 2002). Lipid oxidation products can react with proteins and polypeptides, and form large, covalent crosslinked structures (Aubourg 1993; Aubourg et al. 1998; Tironi et al. 2002; Chopin et al. 2007) that have an impact on the properties of proteins, including their solubility and state of aggregation. These processes result in changes in muscle structure and texture (Saeed and Howell 2002; Chopin et al. 2007). This deteriorative effect is more important in fish muscle than in mammalian muscle, mostly due to the high PUFA and low collagen contents in fish muscle (Chopin et al. 2007).

Malonaldehyde (MDA) is a major lipid oxidation product, and its effects on the chemical, physical, and textural changes of muscles have been investigated. Study on trout myosin found that MDA reacted preferentially with basic amino acids on proteins and the reaction rate was faster at room and freezing temperatures rather than at 0°C (Tironi et al. 2003). Using a model system containing myofibrillar proteins of sea salmon and MDA, the formation of protein aggregates occurred mainly through non-disulfide covalent crosslinks involving the myosin heavy chain. The solubility of collagen decreased and a different muscle network with the appearance of big filaments was observed (Tironi et al. 2002; 2003). In addition, formaldehyde (FA) and dimethylamine (DMA) can also be formed from trimethylamine oxide during frozen storage of fish, and the FA causes denaturation and aggregation of the myofibrillar proteins, resulting in changes in the structure and texture of fish muscle (Careche and Li-chan 1997; Barroso et al. 1998).

14.2.3 Factors Causing the Variation in Fish Muscle Composition

The compositions of fish muscles differ from one another, due to the variation in fish species, the specific muscle type, the seasonal changes, the physiological state, and many others factors (Guillerm-Regost et al. 2006).

14.2.3.1 Fish Species and Muscle Types

There is a difference between fish species living at different temperatures. Fish myosins from fish species living in warm water and hot water contain higher proportions of Gly but less Pro than cold-water fishes in their primary amino acid sequence. In addition, myosins in hot-water fishes and warm-water fishes have a high probability of forming coiled-coil structures around the COOH-terminal end of the rod, which enable them to adapt to a living environment with relatively higher temperatures than cold-water fishes (McLachlan and Karn 1982; Kawabata et al. 2000; Iwami et al. 2002). In the same fish, the red muscle fibers contain more PUFAs, more myoglobin, as well as more trimethylamine oxide, which make them less stable compared with the white muscle (Martinez et al. 2007). The red muscle is firmer than the white muscle in the raw salmon flesh but this difference is not significant in heated samples (Kong et al. 2007).

14.2.3.2 Seasonal Changes

Seasonal change is another important factor for the composition and properties of fish muscle. Studies on silver carp have suggested that there are at least two types of myosin in the muscle that are seasonally dependent, and the summer-type myosin is much more stable than the winter-type. With an increase in the environment temperature, the proportion of summer-type isomer increases and dominates in the summer season from July to September. On the contrary, the expression of winter-type myosin increases from summer to winter, and is the dominant type of myosin during the winter season (Yuan et al. 2006). This expression of myosin isomers with different thermal stabilities as a result of temperature acclimation/seasonal changes was also observed with common carp (Hwang et al. 1990; Guo and Watabe 1993; Imai et al. 1997; Watabe et al. 1998; Yuan et al. 2006).

14.2.3.3 Physiological State

The biochemical, physicochemical, and functional properties of fish muscles were also affected by the physiological state of fish. The isoforms of the myofibrillar proteins, particularly myosin, are usually expressed in a developmental-stage-specific manner. During studies on the developmental changes in fish myofibrillar protein composition of the African catfish, it was observed that "the synthesis of certain myofibrillar protein isoforms at the given developmental stage reflected the functional requirements of swimming muscles in the course of fish development" (Huriaux et al. 1999). Both the total lipid content and the proportion of nonpolar lipids in muscle were higher in the pre-spawned hake than those from post-spawned hake (Busalmen et al. 1995). In addition, actomyosin from pre-spawned hake had a lower relative percent of myosin and a decrease in the affinity between myosin and actin, which resulted in the loss of the filamentous structure of actomyosin in pre-spawned hake (Beas et al. 1988; Montecchia et al. 1990; Roura et al. 1990; Roura et al. 1992; Roura and Crupkin 1995; Montecchia et al. 1997).

14.2.3.4 Wild, Farm-Reared, and Ocean-Ranched Fishes

With the demand for more fish products and proteins, the proportion of fish from aquaculture is increasing (Sigurgisladottir et al. 1997; Gómez-Guillén et al. 2000). Wild fish usually contains less fat than fish from aquaculture. Wild cod also has a different protein expression pattern than farmed cod, which could cause differences in postmortem muscle tenderization (Martinez et al. 2007). Ocean-ranched salmon (a branch of aquaculture involving the cultivation of marine organisms in the open ocean or in an enclosed section of the ocean) contains higher levels of salt-soluble protein and has significantly higher water-holding capacity than farmed fish. The muscle is firmer in farmed salmon than in ocean-ranched salmon, due to the fact that myofibrillar protein aggregation and collagen content are relatively higher in farmed fish (Borderías and Montero 1985; Montero et al. 2003). However, a study by Sigurgisladottir et al. (2000) showed no significant differences in the texture between ocean-ranched and farmed salmon.

14.2.3.5 Pre-slaughter Starvation and Handling Stress

Pre-slaughter starvation and handling stress are observed to have significant effects on the chemical composition of fish muscle and its protein functionalities. Starvation causes the increase of intermolecular crosslinks in collagens, resulting in an increase in the insoluble fraction of collagen (Gómez-Guillén et al. 2000). This increase in collagen crosslinks, together with the change in muscle composition (increase in the relative ratio of collagen compared with myofibrillar and sarcoplasmic protein content), cause Atlantic salmon and brown trout muscle hardening, with shear strength significantly higher in starved specimens than in the equivalent non-starved lots (Gómez-Guillén et al. 2000; Bugeon et al. 2004). Pre-slaughter handling stress causes the denaturation of myofibrillar proteins, which has negative effects on the texture properties of muscle and results in a more rapid softening of flesh during postmortem tenderization (Gómez-Guillén et al. 2000).

14.3 EFFECTS OF VARIOUS STORAGE AND PROCESSING CONDITIONS ON THE PROPERTIES AND QUALITY OF FISH MUSCLES

14.3.1 Cold Storage in Ice

Following the death of fish, the muscles pass through several stages: rigor mortis, dissolution of rigor mortis, autolysis, and bacterial spoilage. Compared with mammalian muscles, the muscles of fish are more susceptible to microbial contaminants and autolysis by endogenous proteases in muscles (Lambert et al. 1991; Ashie and Simpson 1996; Simpson 1998; Yongswawatdigul and Park 2002; Yagiz et al. 2007). To slow down this undesirable process, fish are usually kept in ice to sustain the best quality during a relatively short-term fresh storage period.

Fish muscle has a particular structure, different from mammals. The muscular sheets, called "myotomes," are separated, and thereby defined by the myocomatta containing collagen. During post-harvest storage in ice, the pressure introduced by the contraction of fish muscle myotomes may cause the separation of these myotomes from the myocomatta, which is usually referred as "gaping" (Bremner and Hallett 1985; Hallett and Bremner 1988; Lavety et al. 1988; Bremner 1992; Chéret et al. 2005). Another important deterioration during ice storage of fish is the enzymatic degradation of myofibrils and collagen. Myofibrils of fish are liable to the action of endogenous proteases, primarily calpains and cathepsins (Montero 1997; Chéret et al. 2005), while the disintegration of the extracellular matrix structure mainly results from the function of collagenases (Bremner 1992; Ando 1997; Montero 1997; Chéret et al. 2005). The biochemical changes caused negative changes in fish muscle structure and texture, and may also facilitate the growth of microorganisms.

In addition to the breakdown in primary structure, changes in actomyosin conformation in fish muscles were also observed during ice storage (Yongswawatdigul and Park 2002), which resulted in an increased surface hydrophobicity and decreased Ca^{2+}–ATPase activity. Moreover, during this cold storage, discoloration of the muscle may also occur due to the oxidation and denaturation of myoglobin (Ladrat et al. 2003).

14.3.2 Frozen Storage

Freezing is an important method for maintaining the quality of fish for long-term storage. During the freezing and frozen storage of fish, the growth of microorganisms is inhibited, the chemical and biochemical reactions are significantly slowed down;many enzymatic rate constants approach zero, depending on the temperature of frozen storage. However, there is still some progressive deterioration in the textural and sensory properties of fish muscles during frozen storage due to changes in lipids and proteins, and the increased concentration of solutes in the unfrozen water (Reynolds et al. 2002).

Fish muscle proteins are less stable during frozen storage than muscle proteins from other vertebrates such as poultry and mammals (Ramírez et al. 2000). The aggregation of proteins during frozen storage causes the increased toughness in muscle structure and the loss of water-holding capacity (Montecchia et al. 1997). The formation of insoluble aggregates follow two steps: (1) the formation of soluble aggregates due to protein denaturation, and (2) the further crosslinking of these soluble aggregates to give insoluble aggregates (Montecchia et al. 1997). Protein denaturation during long-term frozen storage may be attributed to several possible causes: cell disruption by ice crystal formation, increased solute concentration, and change in pH (Jittinandana et al. 2005). For fatty fish, the products of lipid oxidation accelerate the formation of high-molecular-weight polymers through covalent crosslinks during frozen storage (Ang and Hultin 1989; Careche and Li-chan 1997). Transmission electron microscope results show that the protein aggregates grow in both number and size during frozen storage (Tejada et al. 1996; Careche and Li-chan 1997).

Intermolecular crosslinks, including hydrophobic interaction, hydrogen bonds, disulfide bond, and/or non-disulfide covalent bonds, are involved in the aggregation of muscle proteins during the frozen storage of fish (Huidobro et al. 1998; Bigelow and Lee 2007). A study of the structural changes of hake muscle during frozen storage also illustrated the increase in the relative content of β-sheets at the expense of α-helices in the secondary structure of the proteins (García et al. 1999; Careche et al. 2002).

14.3.3 High-Pressure Processing

Among the nonthermal processing techniques, high-pressure processing (HPP) is a promising seafood preservation method that has the advantages of destroying microorganisms with limited degradation of heat-labile compounds. High-pressure treatment induces a reversible pH fall (Hayert et al. 1999) and changes in protein structure through modification of hydrogen and hydrophobic bonds. The modifications of water bonds, protease activity, aggregation, and/or gelation of myosin and the sarcoplasmic proteins further affect the texture and other properties of seafood (Heremans 1982; Anguspanish and Ledward 1998; Chéret et al. 2005). Pressure treatment above 300 MPa improved the textural quality of chilled, stored fish fillets and resulted in higher fish hardness during refrigerated storage than in the untreated samples (Chéret et al. 2005). A cooked appearance of fillets is very common in the HPP of muscle systems, and the denatured sarcoplasmic and myofibrillar proteins in muscle may serve as catalysts (pro-oxidants) for lipid oxidation (Angsupanich and Ledward 1998).

The impact of HPP on seafood processing is dependent on the species and processing parameters. HPP at 150 MPa resulted in a completely cooked appearance in cold-water species such as rainbow trout but not in warm-water species. It seems that the proteins in warm-water species are more tolerant of HPP than proteins in cold-water species (Hastings et al. 1985; Yagiz et al. 2007). The conditions for HPP affected the discoloration mechanism for fish muscle. It was suggested that in the range of 200 to 350 MPa for HPP, the discoloration of muscle was due to the denaturation of myoglobin proteins and heme displacement/release; whereas above 400 MPa, oxidation of ferrous myoglobin to ferric metmyoglobin was the major cause of the discoloration (Carlez et al. 1995; Yagiz et al. 2007). HPP in the range of 50 to 150 MPa resulted in the unfolding of tilapia myosin and formation of inseparable network structures. At HPP of 200 MPa, the myosin formed irregular aggregates (Hsu and Ko 2001).

14.3.4 Heat Treatment

About 40% of the salmon harvested in the United States is canned or pouched using conventional retort processes. During retorting, the packaged salmon meat is heated around 120°C for various time periods ranging from 20 to 90 min, depending on the package size and shape (Kong et al. 2007). During cooking (<100°C), the shear force increased as temperature increased, and a higher number of fibers with smaller diameters were found in the cooked muscle compared to the raw muscle (Skrede and Storebakken 1986). Increases in the visual lightness and reduction in both the redness and the yellowness of muscle were observed during cooking (Bhattacharya et al. 1993; Kong et al. 2007). Juice expulsion typically occurs during heating of fish muscle as the heat-denatured actomyosin and collagen lose their ability to hold water in muscle (Bircan and Barringer 2002). Muscle from the central dorsal region had a lower cook loss and less shrinkage than samples from either the anterior or posterior region following heating (Sigurgisladottir et al. 1999; Kong et al. 2007), possibly because the overall heat treatment is lower.

14.4 SUMMARY

Compared to mammalian muscles, fish muscle has its own characteristics, including (1) a lower stability of the muscle myofibrillar proteins, particularly myosin; (2) a lower content of connective proteins, especially collagen; and (3) a higher proportion of polyunsaturated fatty acids. Moreover, the composition of fish muscle varies, depending on fish species, muscle type, seasonal changes, physiological state, and many others factors. Therefore, all of the above factors need to be taken into consideration during the storage and processing of fish and related products.

REFERENCES

Adachi, E. and T. Hayashi. 1986. *In vitro* formation of hybrid fibrils of type V collagen and type I collagen. *Connect Tissue Res.* 14:257–266.

Ando, M. 1997. The softening mechanisms of fish meat during refrigeration after death. *Rec. Res. Dev. Agric. Food Chem.* 1:49–59.

Ando, M., A. Nishiyabu, Y. Tsukamasa, Y. Makinodan. 1999. Post-mortem softening of fish muscle during chilled storage as affected by bleeding. *J. Food Sci.* 64(3):423–428

Ang, J.F. and H.O. Hultin. 1989. Denaturation of cod myosin during freezing after modification with formaldehyde. *J. Food Sci.* 54:814–818.

Angsupanich, K. and D.A. Ledward. 1998. High-pressure treatment effects of cod (*Gadus morhua*) muscle. *Food Chem.* 63:39–50.

Aoki, T. and R. Ueno. 1997. Involvement of cathepsins B and L in the post-mortem autolysis of mackerel muscle. *Food Res. Int.* 30:585–591.

Araki, H. and N. Seki. 1993. Comparison of reactivity of transglutamianse to various fish actomyosin. *Nippon Suisan Gakkaishi* 59:711–716.

Ashie, I.N.A. and B.K. Simpson. 1996. Application of high hydrostatic pressure to control enzyme related fresh seafood texture deterioration. *Food Res. Int.* 29:569–575.

Aubourg, S.P. 1993. Review: interaction of malondialdehyde with biological molecules — new trends about reactivity and significance. *Int. J. Food Sci. & Technol.* 28:323–335.

Aubourg, S.P., C.G. Sotelo, and R. Pérez-Martin. 1998. Assessment of quality changes in frozen sardine (*Sardina pilchardus*) by fluorescence detection. *J. Am. Oil Chem. Soc.* 75:578–580.

Barroso, M., M. Careche, L. Barrios, and A.J. Borderias. 1998. Frozen hake fillets quality as related to texture and viscosity by mechanical methods. *J. Food Sci.* 63:1–4.

Beas, V.E., M. Crupkin, and R.E. Trucco. 1988. Gelling properties of actomyosin from pre- and post-spawning hake (*Merluccius hubbsi*). *J. Food Sci.* 53:1322–1326.

Benjakula, S., C. Chantarasuwan, and W. Visessanguan. 2003. Effect of medium temperature setting on gelling characteristics of surimi from some tropical fish. *Food Chem.* 82:567–574.

Bhattacharya, S., G.S. Choudhury, and S. Studebaker. 1993. Hydrothermal processing of Pacific chum salmon: effects on texture and in vitro digestibility. *J. Food Qual.* 16:243–251.

Bigelow, W. and C.M. Lee. 2007. Evaluation of various infused cryoprotective ingredients for their freeze–thaw stabilizing and texture improving properties in frozen red hake muscle. *J. Food Sci.* 72:56–64.

Bircan, C. and S.A. Barringer. 2002. Determination of protein denaturation of muscle foods using the dielectric properties. *J. Food Sci.* 67:202–205.

Birk, D.E., M.J. Fitch, P.J. Babiarz, and F.T. Linsenmayer. 1990. Fibrogenesis *in vitro*: interaction of type I and V collagen regulates fibril diameter. *J. Cell Sci.* 95:649–657.

Borderías, A.J. and P. Montero. 1985. Changes in fish muscle collagen during frozen storage. In: *Storage Lives of Chilled and Frozen Fish Products*. Paris: International Institute of Refrigeration, p. 65–71.

Bremner, A.H. and C.I. Hallett. 1985. Muscle fiber connective tissue junctions in the fish blue grenadier (*Macruronus novaezelandae*): a scanning electron microscope study. *J. Food Sci.* 50:975–980.

Bremner, H.A. 1992. Fish flesh structure and the role of collagen: its post mortem aspects and implications for fish processing. In: Huss HH, Jakobsen M, Liston J, Editors. *Quality Assurance in the Fish Industry*. London: Elsevier Sci Pub, p. 39–62.

Bugeon, J., F., Lefèvre, and B. Fauconneau. 2004. Correlated changes in skeletal muscle connective tissue and flesh texture during starvation and re-feeding in brown trout (*Salmo trutta*) reared in seawater. *J. Sci. Food Agric.* 84:1433–1441.

Busalmen, J.P., S.I. Roura, H. Roldan, and M. Crupkin. 1995. Changes in lipids and biochemical properties of actomyosin from pre- and post-spawned hake (*Merluccius hubbsi Marini*). *Comp. Biochem. Physiol.* 112B:743–748.

Cao, M.J., Hara, K., Osatomi, K., Tachibana, K., Izumi, T., and T. Ishihara. 1999. Myofibril-bound serine proteinase (MBP) and its degradation of myofibrillar proteins. *J. Food Sci.* 64:644–647.

Careche, M., M.L. García, A. Herrero, M.T. Solas, and P. Carmona. 2002. Structural properties of aggregates from frozen stored hake muscle proteins. *J. Food Sci.* 67:2827–2832.

Careche, M. and E.C.Y. Li-chan. 1997. Structural changes in cod myosin after modification with formaldehyde or frozen storage. *J. Food Sci.* 62:717–723.
Carlez, A., T. Vecianariogues, and J.C. Cheftel. 1995. Changes in color and myoglobin of minced beef meat due to high-pressure processing. *Food Sci. Technol.—Lebens-Wissens & Technol.* 28:528–538.
Carpene, E., B. Martin, and L.D. Libera. 1998. Biochemical differences in lateral muscle of wild and farmed gilthead sea bream (*Sparus aurata L.*). *Fish Physiol. Biochem.* 19:229–238.
Chaijan, M., S. Benjakul, W. Visessanguan, and C. Faustman. 2004. Characteristics and gel properties of muscles from sardine (*Sardinella gibbosa*) and mackerel (*Rastrelliger kanagurta*) caught in Thailand. *Food Res. Int.* 37:1021–1030.
Chaijan, M., S. Benjakul, W. Visessanguan, and C. Faustman. 2007. Interaction between fish myoglobin and myosin *in vitro*. *Food Chem.* 103:1168–1175.
Chen, H.H., E.M. Chiu, and J.R. Huang. 1997. Color and gel-forming properties of horse mackerel (*Trachurus japonucus*) as related to washing conditions. *J. Food Sci.* 62:985–991
Chéret, R., N. Chapleau, C. Delbarre-Ladrat, V. Verrez-Bagnis, and M.D. Lamballerie. 2005. Effects of high pressure on texture and microstructure of sea bass (*Dicentrarchus labrax L.*) fillets. *J. Food Sci.* 70:477–483.
Craig, R. and J.L. Woodhead. 2006. Structure and function of myosin filaments. *Curr. Opin. Struct. Biol.* 16:204–212.
Delbarre-Ladrat, C., V. Verrez-Bagnis, J. Noel, and J. Fleurence. 2004. Relative contribution of calpain and cathepsins to protein degradation in muscle of sea bass (*Dicentrarchus labrax L.*). *Food Chem.* 88:389–395.
Dunajski, E. 1979. Texture of fish muscle. *J. Texture Stud.* 10:301–318.
Erickson, M.C. 1997. Lipid oxidation: flavor and nutritional quality deterioration in frozen foods. In: *Quality in Frozen Foods*. New York: Chapman & Hall, p. 141–173.
Folk, J.E. 1980. Transglutaminase. *Annu. Rev. Biochem.* 49:517–531.
Goll, D.E, V.F. Thompson, R.G. Taylor, and J.A. Christiansen. 1992. Role of the calpain system in muscle growth. *Biochimie* 74:225–237.
Gómez-Guillén, M.C., P. Montero, O. Hurtado, and A.J. Borderías. 2000. Biological characteristics affect the quality of farmed Atlantic salmon and smoked muscle. *J. Food Sci.* 65:53–60.
Guillerm-Regost, C., T. Haugen, R. Nortvedt, M. Carlehög, B.T. Lunestad, A. Kiessling, and A.M.B. Rørå. 2006. Quality characterization of farmed Atlantic halibut during ice storage. *J. Food Sci.* 71:83–90.
Guo, X.F. and S. Watabe. 1993. ATPase activity and thermostability of actomyosins from thermally acclimated carp. *Nippon Suisan Gakkaishi* 59:363–369.
Haard, N.F., B.K. Simpson, and B.S. Pan. 1994. Sarcoplasmic proteins and other nitrogenous compounds. In Sikorski ZE, Pan BS, Shahidi F, Editors. *Seafood Proteins*. New York: Chapman and Hall, p. 3–39.
Hallett, I.C. and H.A. Bremner. 1988. Fine structure of the myocommata-muscle fiber junction in hoki (*Macruronus novaezelandiae*). *J. Sci. Food Agric.* 44:245–261.
Hamai, M. and K. Konno. 1990. Structural stability of fish myosin subfragment-1. *Comp. Biochem. Physiol.* 95B:255–259.
Hashimoto, K., S. Watabe, M. Kono, and K. Shiro. 1979. Muscle protein composition of sardine and mackerel. *Bull. Jap. Soc. Sci. Fish* 45:1435–1441.
Hastings, R.J., G.W. Rodger, R. Park, A.D. Matthews, and E.M. Anderson. 1985. Differential scanning calorimetry of fish muscle: the effect of processing and species variation. *J. Food Sci.* 50:503–506.
Hatae, K., A. Tobimatsu, M. Takeyama, and J.J. Matsumoto. 1986. Contribution of the connective tissues on the textural difference of various fish species. *Bull. Jap. Soc. Sci. Fish* 52:2001–2007.
Hayert, M., J.M. Perrier-Cornet, and P. Gervais. 1999. A simple method for measuring the pH of acid solution under high-pressure. *J. Phys. Chem. A.* 103:1785–1789.
Hemung, B. and J. Yongsawatdigul. 2005. Ca^{2+} affects physicochemical and conformational changes of threadfin bream myosin and actin in a setting model. *J. Food Sci.* 70(8):455–460.
Heremans, K. 1982. High pressure effects on the proteins and other biomolecules. *Annu. Rev. Biophys. Bioeng.* 11:1–21.
Hopkins, D. and J. Thompson. 2002. Factors contributing to proteolysis and disruption of myofibrillar proteins and the impact on tenderization in beef and sheep meat. *Austr. J. Agric. Res.* 2:149–166.
Hsu, K.C. and W.C. Ko. 2001. Effect of hydrostatic pressure on aggregation and viscoelastic properties of tilapia (*Orechromis niloticus*) myosin. *J. Food Sci.* 66:1158–1162.
Huidobro, A., C. Alvarez, and M. Tejada. 1998. Hake muscle altered by frozen storage as affected by added ingredients. *J. Food Sci.* 63:638–643.

Huriaux, F., P. Vandewalle, E. Baras, M. Legendre, and B. Focant. 1999. Myofibrillar proteins in white muscle of the developing African catfish *Heterobranchus longifilis (Siluriforms, Clariidae). Fish Physiol. Biochem.* 21:287–301.

Hwang, G.C., S. Watabe, and K. Hashimoto. 1990. Changes in carp myosin ATPase induced by temperature acclimation. *J. Comp. Physiol. B.* 160:233–239.

Imai, J., Y. Hrayama, K. Kikuchi, and M. Kakinuma, Watabe S. 1997. cDNA cloning of myosin heavy chain isoforms from carp fast skeletal muscle and their gene expression associated with temperature acclimation. *J. Exp. Biol.* 200:27–34.

Iwami, Y., T. Ojima, A. Inoue, and K. Nishita. 2002. Primary structure of myosin heavy chain from fast skeletal muscle of chum salmon *Oncorhynchus keta. Comp. Biochem. Physiol. B.* 133:257–267.

Jiang, S.T., Y.T. Wang, and C.S. Chen. 1992. Lysosomal enzyme effects on the post mortem changes in tilapia muscle myofibrils. *J. Food Sci.* 57:277–279.

Jittinandana, S., P.B. Kenney, and S.D. Slider. 2005. Cryoprotectants affect physical properties of restructured trout during frozen storage. *J. Food Sci.* 70:35–42.

Johnson, I.A. 1999. Muscle development and growth: potential implications for flesh quality in fish. *Aquaculture* 177:99–115.

Kamath, G.G., T.C. Lanier, E.A. Foegeding, and D.D. Hamann. 1992. Nondisulfide covalent cross-linking of myosin heavy chain in setting of Alaska pollock and Atlantic croaker surimi. *J. Food Biochem.* 16:151–172.

Kawabata, R., N. Kanzawa, M. Ogawa, and T. Tsuchiya. 2000. Determination of primary structure of amberjack myosin heavy chain and its relationship with structural stability of various fish myosin rods. *Fish Physiol. Biochem.* 23:283–294.

Kong, F., J. Tang, B. Rasco, C. Crapo, and S. Smiley. 2007. Quality changes of salmon (*Oncorhynchus gorbuscha*) muscle during thermal processing. *J. Food Sci.* 72:103–111.

Ladrat, C., V. Verrez-Bagnis, J. Noe, and J. Fleurence. 2003. *In vitro* proteolysis of myofibrillar and sarcoplasmic proteins of white muscle of sea bass (*Dicentrarchus labrax L.*): effects of cathepsins B, D and L. *Food Chem.* 81:517–525.

Lamare, M., R.G. Taylor, L. Farout, Y. Briand, and M. Briand. 2002. Changes in proteasome activity during postmortem aging of bovine muscle. *Meat Sci.* 61:199–204.

Lambert, A.D., J.P. Smith, and K.L. Dodds. 1991. Shelf-life extension and microbiological safety of freshmeat — a review. *Food Microbiol.* 8:267–297.

Lanier, T.C., P. Carvajal, and J. Yongsawatdigul. 2005. Surimi gelation chemistry. In: Park JW, Editor. *Surimi and Surimi Seafood.* Boca Raton, FL: Taylor & Francis, p. 435–489.

Lanier, T.C. and C.M. Lee. 1992. *Surimi Technology.* New York: Marcel Dekker.

Lavety, J., O.A. Afolabi, and R.M. Love. 1988. The connective tissues of fish. IX. Gaping in farmed species. *Int. J. Food Technol.* 23:23–30.

Martinez, I., R. Ofstad, and R.L. Olsen. 1990. Myosin isoforms in red and white muscles of some marine teleost fishes. *J. Muscle Res. Cell Motil.* 11:489–495.

Martinez, I., R., Šližytė, E. Daukšas. 2007. High resolution two-dimensional electrophoresis as a tool to differentiate wild from farmed cod (*Gadus morhua*) and to assess the protein composition of klipfish. *Food Chem.* 102:504–510.

Matsuura, M., K. Konno, and K. Arai. 1988. Thick filaments of fish myosin and its actin-activated Mg-ATPase activity. *Comp. Biochem. Physiol.* 90B:803–808.

McLachlan, A.D. and J. Karn. 1982. Periodic charge distributions in the myosin rod amino acid sequence match cross-bridge spacings in muscle. *Nature* 299:226–231.

Michalczyk, M. and K. Surówka. 2007. Changes in protein fractions of rainbow trout (*Oncorhynchus mykiss*) gravads during production and storage. *Food Chem.* 104:1006–1013.

Mizuta, S., Y. Yamada, T. Miyagi, and R. Yoshinaka. 1999. Histological changes in collagen related to textural development of prawn meat during heat processing. *J. Food Sci.* 64:991–995.

Montecchia, C.L., M. Crupkin, and R.E. Trucco. 1990. Seasonal variations in biochemical and physicochemical properties of actomyosin and energy content of the liver, gonads and muscle of mature Argentinian hake, *Merluccius hubbsi Marini. J. Fish Biol.* 37:837–843.

Montecchia, C.L., S.I. Roura, H. Roldan, O. Perez-Borla, and M. Crupkin. 1997. Biochemical and physicochemical properties of actomyosin from frozen pre- and post-spawned hake. *J. Food Sci.* 62:491–495.

Montero, P. 1997. Involvement of collagen and connective tissue in postmortem changes in seafoods. *Eval. Fish Freshness* 12–14:190–197.

Montero, P., M.C. Gómez-Guillén, and A.J. Borderias. 2003. Influence of salmon provenance and smoking process on muscle functional characteristics. *J. Food Sci.* 68:1155–1160.

Ogata, H., F. Aranishi, K. Hara, K. Osatomi, and T. Ishihara. 1998. Proteolytic degradation of myofibrillar components by carp cathepsin L. *J. Sci. Food Agric.* 76:499–504.

Papa, I., C. Alvarez, V. Verrez-Bagnis, J. Fleurence, and Y. Benyamin. 1996. Postmortem release of fish white muscle a-actinin as a marker of disorganisation. *J. Sci. Food Agric.* 72:63–70.

Park, J.D., J. Yongsawatdigul, Y.J. Choi, and J.W. Park. 2006. Biochemical and conformational changes of myosin purified from Pacific sardine at various pHs. *J. Food Sci.* 73(3):191–197.

Park, J.W. 2000. *Surimi and Surimi Seafood*. New York: Marcel Dekker.

Ramírez, G.A., M.O. Martín-Polo, and E. Bandman. 2000. Fish myosin aggregation as affected by freezing and initial physical state. *J. Food Sci.* 65:556–560.

Regenstein, J.M. and P. Zhou. 2007. Collagen and gelatin from marine by-product. In Shahidi F, Editor. *Maximising the Value of Marine By-products*. New York: CRC Press, p. 279–303.

Reynolds, J., J.W. Park, and Y.J. Choi. 2002. Physicochemical properties of Pacific whiting surimi as affected by various freezing and storage conditions. *J. Food Sci.* 67(6):2072–2078.

Rodgers, M.E., T. Karr, K. Biedermann, H. Ueno, and W.F. Harrington. 1987. Thermal stability of myosin rod from various species. *Biochemistry* 26:8703–8708.

Roura, S., C.L. Montecchia, A.L. Goldemberg, R.E. Trucco, and M. Crupkin. 1990. Biochemical and physicochemical properties of actomyosin from pre- and post-spawned hake (*Merluccius hubbsi*) stored on ice. *J. Food Sci.* 55:688–692.

Roura, S.I. and M. Crupkin. 1995. Biochemical and functional properties of myofibrils from pre- and post-spawned hake (*Merluccius hubbsi Marini*) stored on ice. *J. Food Sci.* 60:269–272.

Roura, S.I., J. Pecci Saavedra, R.E. Trucco, and M. Crupkin. 1992. Electron microscopical and biochemical studies of actomyosin from pre- and post-spawned hake. *Comp. Biochem. Physiol.* 101B:361–365.

Rowlerson, A., P.A. Scapolo, F. Mascarello, E. Carpene, and A. Veggetti. 1985. Comparative study of myosins present in the lateral muscle of some fish: species variations in myosin isoforms and their distribution in red, pink and white muscle. *J. Muscle Res. Cell Motil.* 6:601–640.

Saeed, S. and N.K. Howell. 2002. 1,2-Lipoxygenase activity in the muscle tissue of Atlantic mackerel (*Scomber scombrus*) and its prevention by antioxidants. *J. Sci. Food Agric.* 81:745–750.

Sato, K., M. Ando, S. Kubota, K. Origasa, H. Kawase, H. Toyohara, M. Sakaguchi, T. Nakagawa, Y. Makinoda, K. Ohtsuki, and M. Kawabata. 1997. Involvement of type V collagen in softening of fish muscle during short-term chilled storage. *J. Agric. Food Chem.* 45:343–348.

Sato, K., C. Ohashi, K. Ohtsuki, and M. Kawabata. 1991. Type V collagen in trout (*Salmo gaidneri*) muscle and its solubility change during chilled storage of muscle. *J. Agric. Food Chem.* 39:1222–1225.

Sato, K., R. Yoshinaka, M. Sato, and Y. Shimizu. 1986. Collagen content in the muscle of fishes in association with their swimming movement and meat texture. *Bull. Jap. Soc. Sci. Fish* 52:1595–1600

Schiaffino, S. and C. Reggiani. 1996. Molecular diversity of myofibrillar proteins: gene regulation and functional significance. *Physiol. Rev.* 76:371–423

Sigurgisladottir, S., H. Hafsteinsson, A. Jonsson, Ø Lie, R. Nortvedt, M. Thomassen, and O. Torrissen. 1999. Textural properties of raw salmon fillets as related to sampling method. *J. Food Sci.* 64:93–98.

Sigurgisladottir, S, H. Ingvarsdottir, O.J. Torrissen, M. Cardinal, and H. Hafsteinsson. 2000. Effects of freezing/thawing on the microstructure and the texture of smoked Atlantic salmon (*Salmo salar*). *Food Res. Int.* 33:857–65.

Sigurgisladottir, S., O. Torrisen, O. Lie, M. Thomassen, and H. Hafsteinsson. 1997. Salmon quality: methods to determine the quality parameters. *Rev. Fish Sci.* 5:223–252.

Simpson, B.K. 1998. High pressure processing of fresh seafoods. *Process-induced Chem Changes Food* 434:67–80.

Skrede, G. and T. Storebakken. 1986. Characteristics of color in raw, baked and smoked wild and pen-reared Atlantic salmon. *J. Food Sci.* 51:804–808.

Tejada, M., M. Careche, P. Torrejon, M.L. Del Mazo, M.T. Solas, M.L. Garcia, and C. Barba. 1996. Protein extracts and aggregates forming in minced cod (*Gadus morhua*) during frozen storage. *J. Agric. Food Chem.* 44:3308–3314.

Tironi, V.A., M.C. Tomás, and M.C. Añón. 2002. Structural and functional changes in myofibrillar proteins of sea salmon (*Pseudopercis semifasciata*) by interaction with malonaldehyde (RI). *J. Food Sci.* 67:930–935.

Tironi, V.A., M.C. Tomás, and M.C. Añón. 2003. Effect of malonaldehyde on the gelation properties of myofibrillar proteins of sea salmon. *J. Food Sci.* 68:42–47.

Watabe, S., Y. Hirayama, M. Nakaya, M. Kakinuma, K. Kikuchi, X. Guo, S. Kanoh, S. Chaen, and T. Ooi. 1998. Carp expresses fast skeletal myosin isoforms with altered motor functions and structural stabilities to compensate for changes in environmental temperatures. *J. Therm. Biol.* 22:375–390.

Yagiz, Y., H.G. Kristinsson, M.O. Balaban, and M.R. Marshall. 2007. Effect of high pressure treatment on the quality of rainbow trout (*Oncorhynchus mykiss*) and mahi mahi (*Coryphaena hippurus*). *J. Food Sci.* 72:509–515.

Yamashita, M. and S. Konagaya. 1990. Participation of cathepsin L into extensive softening of the muscle of chum salmon caught during spawning migration. *Nippon Suisan Gakkaishi* 56:1271–1277.

Yata, M., C. Yoshida, S. Fujisawa, S. Mizuta, and R. Yoshinaka. 2001. Identification and characterization of molecular species of collagen in fish skin. *J. Food Sci.* 66:247–251.

Yongswawatdigul, J. and J.W. Park. 2002. Biochemical and conformation changes of actomyosin from threadfin bream stored in ice. *J. Food Sci.* 67:985–990.

Yuan, C., M. Kaneniwa, X. Wang, S. Chen, Y. Cheng, Y. Qu, Y. Fukuda, and K. Konno. 2006. Seasonal expression of 2 types of myosin with different thermostability in silver carp muscle (*Hypophthalmichthys molitrix*). *J. Food Sci.* 71:39–43.

15 Molecular Mapping and Marker-Assisted Breeding for Muscle Growth and Meat Quality

Penny K. Riggs and Clare A. Gill

CONTENTS

15.1 INTRODUCTION

Cattle coat color traits were selected for by the Romans, and ancient Egyptians were known to have polled cattle (Castle 1920). Selection for meat quantity and quality has probably occurred to varying degrees in domesticated meat animal species during the past 10,000 years, although identification of animals to be selected was a more difficult task than a simple choice of coat color. The effects on meat quality after selection of double-muscled animals to increase hindquarter size was discussed in the literature as early as the beginning of the nineteenth century ((Cully 1807; Youatt 1834) as reviewed in (Oliver and Cartwright 1969)). Bradford (1967) suggested that selection methods would be driven more by the availability of measurement techniques than by the size of the heritability coefficient, and that prediction seems to have held true. Cover et al. (1956) tested the belief that the tenderness and juiciness of steaks from yearling steers resulted from the fatness of the animal. In this particular study, estimated carcass fat was not associated with tenderness, although ether extract of fat from the ribeye was correlated. The authors determined, however, that fatness alone was not responsible for all of the variability observed in tenderness and juiciness. A subsequent investigation (Cover et al. 1957) demonstrated the influence of heredity on tenderness, but the authors noted that "the evaluation of breeding animals is a problem since no suitable, quick method of live animal evaluation for tenderness is now available." Now, more than half a century later, a relatively small but still growing number of DNA marker tests can be utilized for selection of breeding stock for traits including meat quality. The molecular tools and gene maps that can be used for marker-assisted selection and breeding strategies to improve meat production have only developed over the past 25 years, especially during the past decade (reviewed in Womack 2005). In cattle, for example, initial efforts began with construction of synteny maps to identify groups of genes located on the same chromosome (e.g., Womack and Moll 1986). These early maps were later combined with physical maps generated by *in situ* hybridization (ISH) or other techniques (e.g., Fries et al. 1986; Gaye et al. 1986), and linkage maps that were constructed as polymorphic markers and hybridization probes became available (e.g., Hallerman et al. 1987).

15.2 PRINCIPLES OF GENE MAPPING

Although Hutt initially published a chicken gene map in 1936, few additional genes were mapped in meat animal species until somatic cell hybridization techniques were exploited for mapping some 50 years later (reviewed in Womack 1996). With the advent of somatic cell mapping, gene maps were initially built with enzyme markers. Subsequently, additional genes were placed on maps as then new recombinant DNA technology resulted in the availability of cDNA, cosmids, or bacterial artificial chromosome (BAC) probes for those genes. In other cases, genes were chosen to be mapped because of an investigator's specific interest in a disease state, phenotypic trait, or physiological investigation.

15.2.1 Somatic Cell and Radiation Hybrid Panels

Synteny mapping, assignment of genes to the same chromosome, can be accomplished by somatic cell hybrid panel mapping and, more recently, radiation hybrid panel mapping. Both methods have been used for gene mapping in numerous mammalian species. For construction of the *Bos taurus* somatic cell hybrid panel, bovine leukocytes were fused in the presence of polyethylene glycol with mutant rodent cell lines deficient for a necessary enzyme (Womack and Moll 1986; Womack 1988). In general, cells are grown in a selective culture medium in which the rodent cells require complementation by the appropriate chromosome from the species of interest for survival. With continued proliferation, some of the bovine (or other species of interest) chromosomes are lost from the hybrid cells. In this example, cloning of individual hybrid cells results in a panel of clones that segregate different bovine chromosomes. Each cell line can be karyotyped to identify which bovine chromosomes or chromosome fragments are retained. DNA or protein homogenates can be extracted from the cell lines and screened for the presence or absence of individual genes or gene products. Pairs or sets of genes can be used to screen each cell line clone for "presence" or "absence" and analyzed for concordance of retention to identify unordered groups of genes that are present on the same chromosome.

Radiation hybrid (RH) mapping panels are similar to somatic cell panels but the cells of interest are exposed to x-ray or gamma-irradiation prior to fusion (Cox et al. 1990). This technique results in the fragmentation of chromosomes of the species of interest so that smaller fragments are retained in each hybrid clone, allowing ordering of gene maps at greater resolution. RH panels are currently popular tools for genome mapping, and are often the first step in the development of a genome map. In some species whose genomes have been sequenced, RH maps have provided the reference framework necessary for assembling genome sequence data. Whole-genome RH mapping panels have been produced for livestock animals, including pig (Hawken et al. 1999), chicken (Morisson et al. 2002), horse (Chowdhary et al. 2002), cattle (Womack et al. 1997), sheep (Wu et al. 2007), and river buffalo (Amaral et al. 2007). The basic method of RH mapping (also applicable to somatic cell hybrid mapping) is illustrated in Figure 15.1.

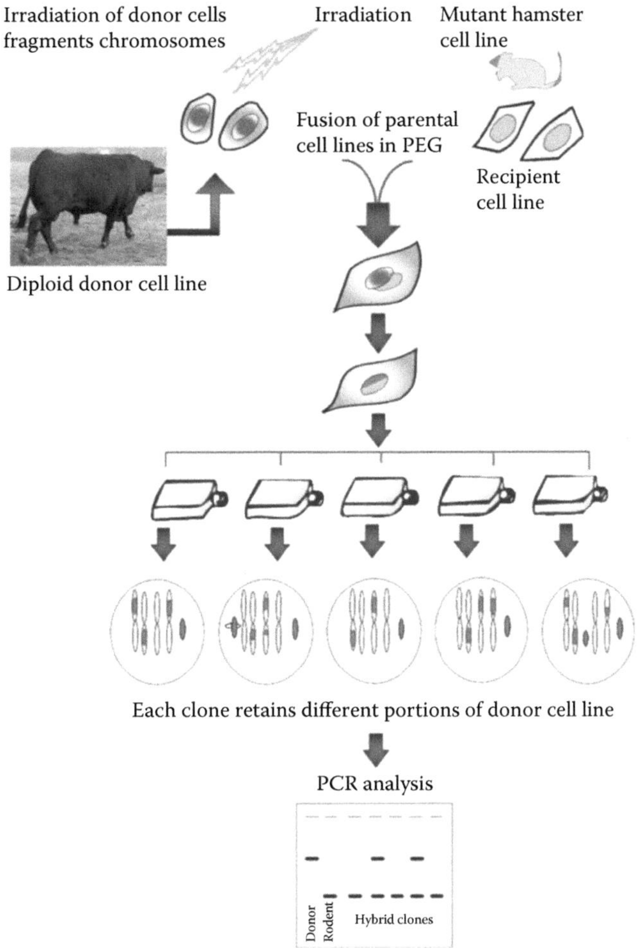

FIGURE 15.1 Radiation hybrid mapping schematic drawing. In this example, donor cells from a bull are irradiated then fused with a mutant hamster cell line. Individual hamster cells retain different portions of the bovine genome. This method results in a panel of cell line clones, each representing a proportional fraction of the bovine genome. After subsequent screening of DNA from individual clones, genes can be mapped by concordance analysis. (*Source:* Image modified from figures kindly provided by Dr. James E. Womack and Dr. M. Elisabete J. Amaral.)

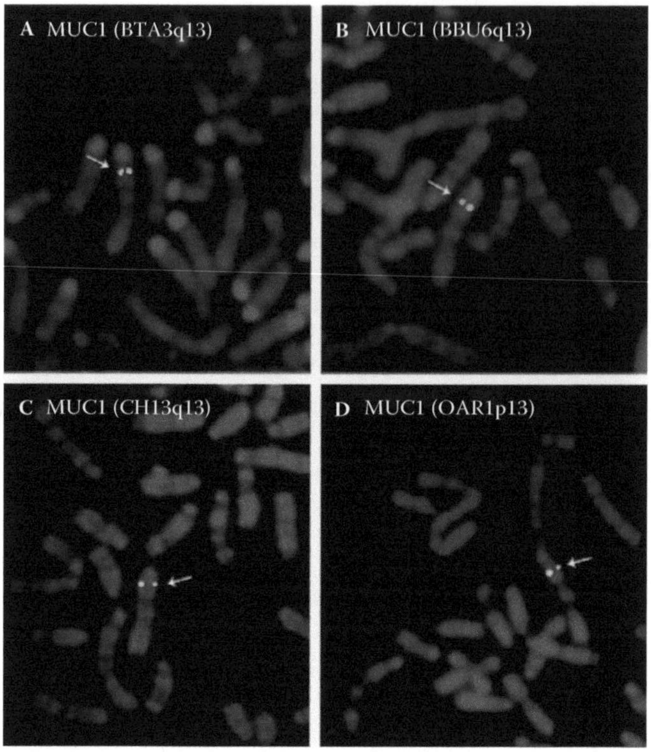

FIGURE 15.2 Fluorescence *in situ* hybridization. Details of cattle (a), river buffalo (b), goat (c), and sheep (d) metaphase plates showing the simultaneous visualization of R-banding and FISH mapping of *MUC1* to cattle, river buffalo, sheep, and goat chromosomes (BTA3q13, BBU6q13, OAR1p13, and CHI3q13, respectively). Arrows denote hybridization signal. (*Source:* Image obtained from Perucatti et al., *Cytogenet. Genome Res.* 2006. 112:103–105. With permission of S. Karger AG, Basel.)

15.2.2 *In Situ* Hybridization

Fluorescence *in situ* hybridization (FISH) is a method of choice for mapping genes cytogenetically. A DNA probe for the gene or genomic region of interest is labeled with a modified nucleotide such as digoxigenin-d-UTP and hybridized to denatured metaphase chromosomes on a glass slide. Following hybridization, the hybridized signal is detected by subsequent binding of a fluor-labeled anti-digoxigenin antibody (Riggs and Hahn 1998). Mapping by FISH allows direct microscopic visualization of hybridization to a specific chromosomal location. This technique is also useful for integrating genetic and physical maps when markers present on a linkage map are also mapped by FISH. An example of comparative FISH mapping (Perucatti et al. 2006), illustrating mapping of the same gene in several species, is shown in Figure 15.2.

15.2.3 Linkage Analysis

One disadvantage of early physical gene maps was that many of the genes were placed on the map, but not ordered. Genetic mapping by linkage analysis is a well-established technique accomplished in part by observation of meiotic recombination events between genes or markers of interest (Sturtevant 1913). This method of analysis requires that genetic variation in the genes of interest exists between the parents of families of animals studied. Prior to the availability of sequence data, polymorphisms or sequence variations could be revealed in genes based on restriction fragment length polymorphism (RFLP) analysis (Botstein et al. 1980). For example, the restriction enzyme

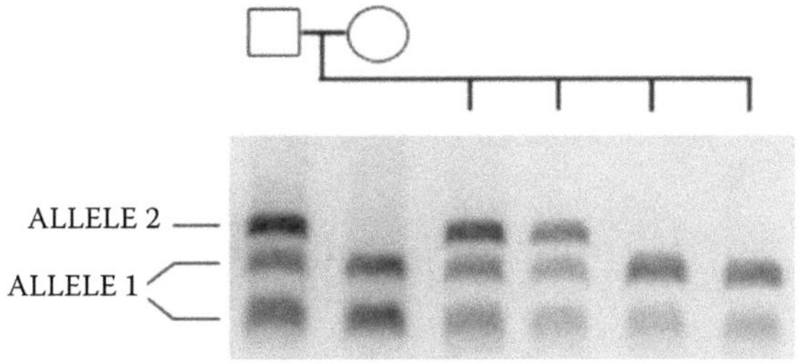

FIGURE 15.3 AFLP. The α-actin gene (*ACTA1*) was digested with restriction enzyme *HinF*1. The dam's *ACTA1* allele is digested by the enzyme, resulting in two DNA fragments. In contrast, the sire has two alleles: one is digested by the enzyme and the second, larger fragment is not digested by the enzyme, indicating that a sequence polymorphism has eliminated the enzyme recognition site. Two offspring inherited the second allele from the father, while the remaining offspring are homozygous for the allele containing the restriction site. (*Source:* Image provided by P.K. Riggs.)

*BamH*I cuts DNA anywhere the specific sequence pattern GGATCC is found. In a pedigreed family of animals, one could extract genomic DNA from a blood sample of each individual and digest the DNA with restriction enzymes. By gel electrophoresis, the fragments can be separated by size and charge. If a gene has two alleles, or variations of the same gene, that differ at the *BamH*I recognition site, different fragment sizes would be observed between the allele that was cut by the restriction enzyme and the allele that was not cut. The electrophoresed products can be transferred from the gel to a nylon membrane. The RFLP can be visualized by autoradiographic detection following hydridization with a gene-specific probe. Similarly, as shown in Figure 15.3, PCR products can be digested to reveal amplified fragment length polymorphism (AFLP). Patterns of inheritance of the alleles from different genes can be analyzed, and the frequency of recombination, or rearrangement of the inheritance pattern due to non-random segregation of markers at meiosis, is proportional to the distance between genes. By convention, map distances are reported in centiMorgans.

Following the advent of the polymerase chain reaction (PCR; Saiki et al. 1985; Mullis et al. 1986), it became common to use highly polymorphic genetic markers such as simple tandem repeats, or microsatellites (Tautz 1989), that are distributed throughout the genome for linkage analysis to increase representation of the genome on linkage maps. Most recently, single nucleotide polymorphisms (SNPs) identified through genome sequencing initiatives have become a popular marker because they occur about once every 100 to 300 nucleotides in mammals and genotyping of them is easily automated. In the past 10 years, the density of markers on the linkage maps of meat animal species has increased markedly. There are now more than 7400 markers on the linkage map of cattle (Ihara et al. 2004), more than 3500 in swine (reviewed by Rohrer 2004), over 1880 in chicken (Groenen et al. 2000), and more than 1200 in sheep (reviewed by Cockett 2003). Assays of tens of thousands of SNP markers are now commercially available for cattle, chicken, sheep, and swine, so the density of linkage maps will continue to dramatically increase.

15.2.4 QTL Analysis

Many measurable phenotypic traits reflect the cumulative influence of the inheritance of combinations of multiple genes or regions of DNA known as quantitative trait loci (QTL). Some traits, such as the meat phenotype observed in porcine stress syndrome, are affected by a single, major gene. Other traits, such as meat tenderness, for example, are affected by the DNA sequence composition at multiple locations across the genome. For QTL mapping of traits, one needs a population of animals that expresses variability for a quantitative trait and a linkage map for the population, with markers dispersed across the genome. A common procedure to identify QTL is to scan the entire genome at centiMorgan intervals using multiple linear regression (Haley and Knott 1992) and test for statistical significance of the trait at each position conditional on the genotypes of the flanking

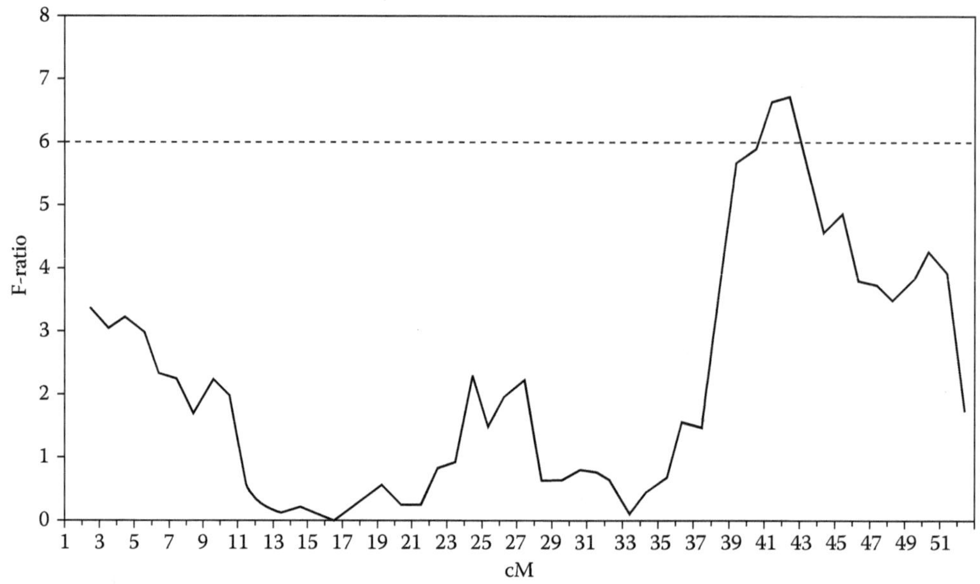

FIGURE 15.4 QTL analysis. Example of a QTL scan for Warner-Bratzler shear force tenderness on bovine chromosome 29. F-ratio profile was generated using GridQTL (Seaton et al. 2002). Horizontal dashed line indicates $P < 0.05$ significance threshold. (Image provided by C.A. Gill.)

markers (as represented graphically in Figure 15.4). This approach results in a good approximation to maximum likelihood estimates of the QTL location but is less computationally intensive. Empirical significance thresholds can be established by permutation (Churchill and Doerge 1994), with confidence intervals determined by non-parametric bootstrapping (Lebreton and Visscher 1998). To be able to detect QTL, the population must be segregating for the genetic markers and the genes responsible for the observed variation in the phenotype. Consequently, in the meat animal species, crosses between breeds that are widely divergent for traits of interest have often been used in QTL mapping. This approach identifies QTL alleles that differ in frequency between the two breeds, rather than within the breeds. Although numerous QTL studies have been reported in the literature, the confidence intervals associated with QTL locations are typically very broad, spanning many centiMorgans or even entire chromosomes. This has limited progress toward identifying the underlying genes. Causative mutations, or quantitative trait nucleotides (QTNs), have been identified in only a few cases (reviewed by Ron and Weller 2007). Recently, approaches that combine linkage and linkage disequilibrium (association) methods in a single analysis (Meuwissen and Goddard 2002; Uleberg and Meuwissen 2007) have become popular because they facilitate the use of non-structured, outbred populations for QTL mapping and typically result in narrower confidence intervals (e.g., Olsen et al. 2005).

15.2.5 Marker-Assisted Breeding and Selection

Historically, selection decisions have been made on observable phenotypes. For traits that are not easily observable in an individual prior to the time that selection decisions are made, like meat quality, this means that little progress has been made to improve the trait. Genetic markers that are predictive of an animal's breeding value for a trait of interest have utility in selective breeding programs because DNA can be collected and tested early in life (reviewed in Dekkers and Hospital 2002). Despite the considerable progress that has been made in developing the genome maps of meat animal species, there are still very few commercially available genetic tests for traits of economic importance in meat animal species (reviewed in Dekkers 2004). The majority of these tests are for mutations in major genes. For example, in Hampshire swine, a dominant arginine to glutamine mutation in the *PRKAG3* gene causes about a 70% increase in muscle glycogen and results in meat with low ultimate pH (acid meat), low water-holding capacity, and reduced yield after cooking (Milan et al. 2000). This mutation increased in frequency in the Hampshire breed because of beneficial effects of the gene on growth rate, which is easy to select for based on phenotype. This is an example of where selection for a gene with antagonistic, pleiotropic effects on two different traits can have unintended consequences. A direct test of the causative mutation is commercially available to select against those swine that carry the R200Q mutation.

The general lack of availability of commercial genetic tests for quantitative traits to date has stemmed from the difficulty in progressing from a QTL location to a direct or indirect test of the underlying causative mutation (reviewed in Dekkers 2004). Causality is difficult to prove so most commercial tests for QTL will use markers that are in linkage disequilibrium (non-randomly associated) with the causative mutation (QTN). Because many studies utilized between-breed differences in gene frequencies to identify QTL, additional research is typically needed to identify markers that are associated with the QTL on a within-breed basis so that these markers can be applied in selective breeding programs. In swine and cattle, commercial populations have been used to verify the significance of QTL identified using divergent intercrosses (e.g., Evans et al. 2003; Moser et al. 2004). These studies demonstrated that some of the QTL, including some for carcass and meat traits, do segregate in commercial populations so markers associated with these QTL could be used for selection. The predictive ability of these indirect tests is dependent on the strength of the association between the marker and underlying gene, which typically implies that the marker is very close to the true location of the causal gene. If single gene tests are used in selection, then the improvement in the selected trait will be proportional to the variation accounted for by that single gene, which may only be 5% to 10% for a quantitative trait. In cattle, a number of commercially available tests for carcass quality traits have been independently validated to verify company claims regarding the efficacy of the tests (Van Eenennaam et al. 2007).

Now that commercial assays of thousands of SNP markers are becoming available for meat animal species as an outcome of genome sequencing projects, the expectation is that many markers associated with QTL will be commercialized. How to simultaneously incorporate information from multiple markers into selection decisions to capture the majority of genetic variation in traits of interest is the next challenge. Genomic selection, or marker-assisted selection on a genome-wide scale (Meuwissen et al. 2001), is starting to be implemented and will enable the prediction of accurate breeding values for young animals, which will lead to increased genetic gain for traits that are difficult to select for, like carcass quality.

15.3 MAPPING MUSCLE-SPECIFIC GENES

Of genes specific to, or affecting skeletal muscle traits, one of the earlier investigations was the assignment of the halothane sensitivity locus (HAL) in swine to a genetic linkage group including several enzyme markers (Hanset et al. 1983; Gahne and Juneja 1985; Grashorn and Muller 1985). The locus associated with muscular hypertrophy in sheep was mapped when the callipyge gene was assigned to sheep chromosome 18 (Cockett et al. 1994). Mapping of skeletal muscle-specific genes followed in cattle. For example, skeletal muscle alpha-actin (*ACTA1*) was assigned to chromosome 28 by synteny mapping (Threadgill et al. 1994) and by linkage analysis (Riggs et al. 1997). Myogenic factor 5 (*MYF5*) was mapped by linkage to bovine chromosome 5 (Bishop et al. 1994), and the skeletal muscle-specific isoform for cytochrome c oxidase subunit VIII (*COX8B*) was mapped to chromosome 29 (formerly linkage group U7; Lomax et al. 1995). Most notably during the same year, the locus for double-muscling in cattle was mapped to chromosome 2 by linkage analysis (Charlier et al. 1995). Other muscle-related genes included *MRF6* (aka *MYF4* and herculin) mapped to bovine chromosome 5 (Riggs 1996), and myogenin (*MYOG*) mapped to cattle chromosome 16 (Beever et al. 1997a, b). Numerous related genes have been added to the genome maps of cattle and other species during the past decade. Now that whole genome sequences are available for many species, all known muscle-related genes are essentially "mapped" for those species. The current challenge is to identify how genetic variation in the gene coding and regulatory sequences affects phenotypes, and how networks of genes and gene products work together in an interrelated fashion.

15.3.1 Myogenic Transcription Factors

In a review of potential methods for optimizing production of lean meat, Wray-Cahen et al. (1998) discussed nutrients, hormones, and genes in meat production, and examined the possibility of genome alteration for improvement of muscle traits. Suggested candidates for control of muscle development in the prenatal environment included the family of myogenic regulatory factors, including *MYOD1*, *MYF5*, *MYOG*, and *MYF6*. While direct genomic modification of genes associated with myogenesis has not presently become a method of choice for altering muscle characteristics of food animal species, polymorphisms that result in altered skeletal muscle phenotypes have been identified for many myogenic transcription factors and could potentially be utilized in marker-assisted selection of breeding stock. Commercial tests for transcription factor DNA variants, such as the inhibitor of myogenic differentiation, myostatin (*MSTN*), are currently marketed to animal breeders.

15.3.1.1 Myogenin

In two lines of Yorkshire pigs, variability at the *MYOG* locus was estimated to account for 4% of the total phenotypic variance of birth weight, carcass weight, and growth rate (te Pas et al. 1999b). The authors hypothesized that regulation of *MYOG* expression was altered by SNPs. Delays in the onset of *MYOG* gene expression could increase the number of myoblasts, ultimately increasing the number of myofibers and overall birth weight. For this population of Yorkshire swine, selection for the favorable homozygous *MYOG* genotype increased growth rate, lean weight, and birth weight, thus reducing perinatal mortality without affecting backfat thickness. Delays during myogenesis resulting in increased myofibers have also been observed in other species, including quail (Coutinho et al. 1993). Additional evidence for the importance of timing of muscle regulatory factor expression was demonstrated in a study where satellite cells were collected from embryonic turkey pectoralis major muscle at different stages of development. Compared to an unselected control population, cells from turkeys selected for increased breast weight had an altered pattern of expression of both *MYOD1* and *MYOG* that likely contributed to the phenotypic differences (Liu et al. 2005).

15.3.1.2 MYF5

In contrast, a study of genetic variation in the porcine *MYF5* locus did not show an association with meat traits (te Pas et al. 1999a). Although three microsatellites and two RFLPs were identified in the promoter or near the coding sequence of porcine *MYF5*, no association was found with birth weight, growth rate, carcass weight, or backfat thickness.

15.3.2 Myostatin

15.3.2.1 Myostatin and Double-Muscling in Cattle

The double-muscled phenotype in cattle has been widely studied, and these animals are known to possess approximately twice as many myofibers as other cattle (Swatland and Kiefer 1974). Mice with a knockout mutation of the *GDF-8* gene were produced and the animals exhibited greatly increased skeletal muscle mass. The GDF-8 transcription factor was renamed myostatin because of its effect on myogenic differentiation (McPherron et al. 1997). The muscular phenotype of the myostatin knockout mice was not unlike that of double-muscled cattle, and the search began for mutation in the bovine myostatin gene. A deletion in the *MSTN* gene was described as the causative mutation for the muscular hyperplasia in Belgian Blue cattle (Grobet et al. 1997; Kambadur et al.

1997), and additional mutations in other European breeds were suggested as an explanation for the genetic heterogeneity observed in the double-muscling phenotype (Grobet et al. 1998). In Belgian Blue double-muscled cattle, an 11-bp deletion of the coding sequence results in a premature stop codon and loss of functional MSTN activity (Grobet et al. 1997). The mutation in Piedmontese cattle was identified as a G-A transition that converted a critical cysteine residue to a tyrosine, resulting in disrupted MSTN function (Kambadur et al. 1997). The *MSTN* gene was characterized in other vertebrates, including pig, sheep, turkey, and chicken, as well as 11 breeds of cattle, and mutations reported for Belgian Blue and Piedmontese cattle were verified, but no loss of function mutations were observed in the other species (McPherron and Lee 1997).

Other myostatin mutations have also been identified in cattle. Association of *MSTN* and *MYF5* polymorphisms with growth traits was observed for three Chinese cattle breeds: Nanyang, Qinchuan, and Jiaxian cattle (Zhang et al. 2007). Significant QTL affecting growth and carcass traits were identified in a Wagyu-Limousin F_2 population (Alexander et al. 2007). Included in this study was a QTL for longissimus muscle area on chromosome 2 that had been previously identified near the MSTN gene (Casas et al. 1998). The authors suggested that the Limousin MSTN F94L amino acid substitution (Grobet et al. 1998; Dunner et al. 2003) was a likely candidate for the QTL on the Wagyu background. The effect of the same amino acid substitution on carcass yield was examined in another study (Sellick et al. 2007). A major QTL for meat percentage was found on bovine chromosome 2 in a genome scan of a Limousin-Jersey backcross population, and four haplotypes of different *MSTN* SNPs segregated in the six sire families in the study. Estimation of genetic effects of the *MSTN* haplotypes and the haplotype combinations indicated that one haplotype found only within Limousin cattle had the largest effect on the observed traits. One SNP (g.433C>A) caused substitution of a leucine for a phenylalanine at amino acid position 94 (F94L). This substitution was believed to be a conservative change, leaving the *MSTN* gene functional. However, evidence from this study indicated that the F94L MSTN, or a closely linked mutation, is responsible for a least a partial loss of MSTN function in Limousin, resulting in a more muscular phenotype that is intermediate to the double-muscling phenotype (Sellick et al. 2007). A different study suggested that South Devon cattle in the UK have also been selected for an intermediate phenotype despite a relatively high incidence of the MSTN deletion observed in Belgian Blue cattle (Smith et al. 2000a).

15.3.2.2 Myostatin in Sheep

A search for carcass trait QTL associated with the *MSTN* gene region on sheep chromosome 2 led to the identification of a QTL affecting muscle and fat traits (Johnson et al. 2005a). This QTL was associated with increased muscling and decreased fat content in Texel sheep, and *MSTN* was proposed as the candidate gene for the traits. Meat quality, as assessed by ultimate pH, color, and Warner-Bratzler shear force measurements, was not affected by this QTL (Johnson et al. 2005b). A point mutation was identified in the 3′ UTR of the *MSTN* gene in Texel sheep that created a target site for two microRNAs that are abundant in skeletal muscle. The mutation resulted in translational inhibition of *MSTN* and was believed to contribute to the muscular hypertrophy observed in the Texel sheep (Clop et al. 2006). Six SNPs associated with *MSTN* were observed in a survey of 326 sheep, including several Australian breeds (Kijas et al. 2007). The previously described loss of function mutation had the greatest effect on carcass traits. While no effect was observed on physically measured quality traits, the mutation was not favored in a consumer sensory test. Zhou et al. (2008) also identified sequence variation within the coding region and found three *MSTN* alleles in crossbred Romney sheep.

15.3.2.3 Myostatin in Poultry

Polymorphic DNA markers for *MSTN* have been identified in poultry. Breast muscle weight was shown to be affected by variation in the chicken *MSTN* gene in F_2 populations of Broiler-Silky chickens (Zhiliang et al. 2004). In other studies, SNPs were identified in *MSTN* across nine breeds of indigenous Indian chickens and red jungle fowl (Kumar 2007) and in Wenling grass chicken (Zhu et al. 2007). Additional SNPs were also observed in the *MSTN* gene in three commercial broiler chicken lines, and five SNPs were associated with body weight and other traits (Ye et al. 2007). A preliminary study also identified polymorphisms in the *MSTN* gene of domestic geese (Yang et al. 2007).

15.3.2.4 Myostatin DNA Tests

DNA tests for myostatin functional status are commercially available for cattle. The North American Piedmontese Cattle Association requires documentation of *MSTN* gene status for animal registration [http://www.piedmontese.org/NAPAservices.html]. Despite the current understanding of how *MSTN* loss of function mutations result in increased muscularity and the trait's popularity in certain European markets, beef producers have not necessarily rushed to introduce this mutation into herds of cattle. Along with increased muscularity and yield of lean beef, fat is decreased — effectively reducing the meat quality grade. Double-muscled animals can exhibit increased birth weight along with elevated incidence of dystocia in some breeds. Casas et al. (1998) suggested that utilization of breeding schemes that maximize production of double-muscling heterozygotes and minimize homozygotes may be useful if production of leaner, heavily muscled carcasses are desired. Assuming the accessibility of high-throughput genotyping, Keele and Fahrenkrug (2001) developed a series of mating systems designed to utilize *MSTN*-defective genotypes. Depending on variables including combinations of biological and economic factors such as genotype and costs associated with genotyping, assisted calving, loss of quality grade due to leanness, and other factors, these schemes were optimized to maximize potential profit. Because meat from double-muscled animals tends to be more tender, the authors suggested that selection for nonfunctional *MSTN* alleles could be profitable to producers in a market that values a tenderness-based meat grading scheme, or in a situation where prices for marbling-based USDA Standard and Select grade beef approach the price for Choice grade.

15.3.3 Callipyge

A ram with low body fat and muscular hypertrophy of its hindquarters was identified in a production flock of Dorset sheep. Initial breeding studies were conducted with affected rams, and the phenotype of these animals was different from double-muscling observed in cattle. The mode of inheritance appeared to be dominant, and the myofiber hypertrophy restricted mainly to the pelvic limbs became evident only after birth. This mutation was dubbed "callipyge," from the Greek for beautiful buttocks (Cockett et al. 1994). The callipyge mutation (*CLPG*) was mapped to sheep chromosome 18. Additional studies revealed an unusual pattern of inheritance described as polar overdominance. The callipyge phenotype is only observed in heterozygous offspring who inherited the paternal *CLPG* allele. All other combinations of alleles at this locus appear normal (Cockett et al. 1996; Freking et al. 1998a). Eventually, the *CLPG* mutation was identified as an A-to-G transition within a 12-bp sequence motif found in an imprinted gene cluster (Freking et al. 2002; Smit et al. 2003). The imprinted gene cluster is characterized by both maternally and paternally imprinted genes, including *BEGAIN, DLK1, DAT, MEG3, RTL1, PEG11AS*, and *MEG8*, and the paternal allele up-regulates expression of *DLK1* and *RTL1* (reviewed in Cockett et al. 2005; Fleming-Waddell et al. 2007; Vuocolo et al. 2007). While the mechanism by which the mutant phenotype develops is not yet fully understood, it appears that up-regulation of DLK1 expression may alter signaling networks associated with skeletal muscle satellite cells, ultimately giving rise to muscular hypertrophy in affected animals (Vuocolo et al. 2007; White et al. 2008). Interestingly, polar overdominant inheritance of a *DLK1* variant was associated with lean muscle mass in swine (Kim et al. 2004). Paternal expression of *DLK1* and maternal expression of *MEG3* was also observed in crosses of Korean native pigs with Yorkshire pigs, although polar overdominant transmission was confirmed only for *DLK1* (Li et al. 2008).

Despite more than a decade of continued research effort from the initial linkage mapping of the *CLPG* locus to some understanding of a unique genetic mechanism, selection for the callipyge trait is not widely utilized by producers in the United States. Callipyge lambs appear normal at birth, and hypertrophy can be observed beginning around 3 weeks of age. Although overall animal size is not changed, *CLPG* animals have 30% to 40% more muscle mass, less fat, and more compact skeletal structure (Koohmaraie et al. 1995; Jackson et al., 1997a,b; Freking et al. 1998b). The mutation does not convey any apparent negative reproductive consequences (Freking and Leymaster 2006). Unfortunately, the increased muscularity is also associated with tougher meat and reduced consumer acceptance (Koohmaraie et al. 1995; Goodson et al. 2001; Bunch et al., 2004; Abdulkhaliq et al. 2007).

15.3.4 Calpain and Calpastatin Genes

Although many different factors affect the tenderness of meat, the calcium-dependent protease system is believed to play a significant role in postmortem tenderization and also affects myoblast proliferation and fusion (reviewed in Goll et al. 1998). The system consists of calcium-activated proteases, micromolar calpain or μ-calpain (*CAPN1*), and millimolar or m-calpain (*CAPN2*), along with calpastatin (*CAST*), which inhibits calpain activity. A skeletal muscle isoform (*CAPN3*) has also been identified, along with many other forms (reviewed in Goll et al. 2003)). In skeletal muscle, the activated proteases degrade structural substrates, and also undergo autolysis. *CAPN1* also autolyzes in postmortem muscle and is associated with postmortem tenderization (reviewed in Koohmaraie and Geesink 2006).

A number of commercial DNA tests for meat tenderness in cattle are based on polymorphisms in *CAPN1* and *CAST*. *CAPN1* was mapped to bovine chromosome 29 (Smith et al. 2000b) in a region associated with meat tenderness. Polymorphisms within exons 9 and 14 of *CAPN1* have been demonstrated to predict tenderness phenotypes in *Bos taurus* cattle (Page et al. 2002; 2004). An additional *CAPN1* tenderness marker was identified that segregates in both *Bos taurus* and *Bos indicus* cattle (White et al. 2005; Casas et al. 2006). A study of Korean Hanwoo cattle suggested that a polymorphism in the 3′UTR of *CAPN1* was associated with marbling score (Cheong et al. 2008). In chickens, three SNPs were identified in three exons of the *CAPN1* gene. Specific haplotype combinations of these SNPs were associated with live weight, carcass weight, breast muscle weight, and leg muscle weight. The effect of *CAPN1* haplotype was also influenced by the strain-specific genetic strain background of the birds (Zhang et al. 2008b).

A tenderness test based on polymorphisms in *CAST*, which inhibits calpain-induced protein degradation and tenderization during postmortem ageing, was patented by Barendse (2002). Subsequent studies have also associated *CAST* variations with tenderness. An SNP resulting in a G-to-C substitution in the *CAST* gene was identified in crossbred beef cattle in commercial feedlots in a study of multiple carcass traits (Schenkel et al. 2006). In this study, the *CAST* SNP affected shear force during postmortem ageing. Schenkel et al. (2006) determined that the CC genotype was associated with beef that was more tender than the GG genotype, and the CG allele resulted in an intermediate phenotype. CAST is also a candidate gene for tenderness in pork based on QTL analysis and fine mapping of pig chromosome 2 (Ciobanu et al. 2004; Meyers et al. 2007).

The function of *CAPN3* is not clear (Sorimachi et al. 1989). Originally believed to be a skeletal muscle-specific isoform of calpain, *CAPN3*, was mapped in the pig to SSC1q15-17 (Briley et al. 1996), and in cattle to BTA10 (Riggs 1996). The bovine and ovine skeletal muscle calpain genes (*CAPN3*) were cloned and characterized. The mapping to bovine chromosome 10 was confirmed and the sheep *CAPN3* was mapped to chromosome 7 (Nonneman and Koohmaraie 1999). A restriction-fragment polymorphism in the pig *CAPN3* gene was associated with tenderness (Larsen et al. 1998). More recently, an additional tenderness QTL for cattle has been mapped to bovine chromosome 10, and *CAPN3* is a putative gene candidate (Davis et al. 2008).

15.3.5 Porcine Stress Syndrome

The incidence of "watery pork" was known to be variable among different breeds of pigs long before its genetic cause was known (Bendall and Lawrie 1964). Stress was shown to contribute to unsatisfactory meat condition described as pale, soft, exudative (PSE) musculature (Briskey 1964; Butterworth et al. 1967; Gahne 1968). Harrison (1972) suggested that porcine stress syndrome, malignant hyperthermia (or halothane sensitivity), and PSE meat all reflected the same disorder. Sensitivity to halothane was shown to be inherited as a recessive trait (Minkema et al. 1977). Genes for several enzymes were shown to be linked to the *HAL* locus in swine, and one closely linked gene, glucosephosphate isomerase (*GPI*), was mapped by *in situ* hybridization to chromosome 6 (Davies et al. 1988).

The ryanodine receptor gene (*RYR1*) was proposed as a candidate for malignant hyperthermia in humans because *RYR1* was mapped near *GPI*, and both a linkage study and physiological data supported an association between the disorder and a potentially defective calcium release channel (MacLennan et al. 1990). A point mutation in the pig skeletal muscle ryanodine receptor (*RYR1*) was associated with porcine stress syndrome in five breeds of lean, heavily muscled swine (Fujii et al. 1991), as well as British Landrace pigs (Otsu et al. 1991). The *RYR1* gene was assigned to the *HAL* linkage group (Mariani et al. 1992) and was mapped to chromosome band SSC6q12 (Chowdhary et al. 1994).

Phenotype screening for halothane gene-negative pigs had been inefficient because the *HAL* gene frequency was low and most animals were heterozygotes (Van Arendonk and Kennedy 1990). Selection for animals heterozygous for the *HAL* gene continued because the locus was also associated with increased lean meat quantity (Rothschild et al. 2004). Currently, commercial tests are available for assessment of RYR1 gene status, and the National Swine Registry of the United States requires negative stress test status as a condition of animal registration (http://www.nationalswine.com/rulesandregulations/dnastresstestpolicy.html).

In poultry, a meat quality issue similar to PSE pork exists. While slaughter age, increased body weight, and feed conversion rates have improved in poultry, selection for these traits has affected muscle texture and flavor to reduce meat quality (Anthony 1998). Unlike genetic analysis useful for eradicating PSE in the pork industry, such DNA tests currently seem to lag behind for poultry species. Ottini et al. (1996) showed that chicken alpha- and beta-isoforms of the ryanodine receptor are orthologs to mammalian *RAR1* and *RAR3*. Wang et al. (1999) demonstrated differential ryanodine binding capacity between commercial turkeys and a genetically unimproved population. Transcript variants were characterized in the turkey skeletal muscle ryanodine receptor alpha-isoform (α-*RYR*) (Chiang et al. 2004). Later, additional alternative splicing variants were identified (Chiang et al. 2007). Ten alternative splicing transcript variants were found in α-*RYR*, and for β-*RYR*, Chiang and colleagues identified ten SNPs and eight haplotypes. These synonymous SNPs resulted in altered predicted secondary structures of the corresponding mRNAs. Expression of one variant was also associated with heat stress. These data suggested divergent mechanisms might affect calcium channel activity of either of these receptor isoforms, but show promise for potential molecular marker tests for stress syndrome in poultry.

15.3.6 Rendement Napole (*RN*$^-$)

Another type of poor meat quality in pork distinct from PSE was described as the Hampshire effect due to its incidence in the Hampshire breed (Monin et al. 1984). Postmortem meat samples from affected animals have a low ultimate pH ("acid pork") and an increased glycogen content. Cooked meats, particularly noted in ham, suffer water loss and a reduction in yield. It appears that this mutation was originally selected for when producers selected animals for increased lean meat quantity. However, the meat quality was reduced in heterozygous *RN*$^-$ carriers, and various enhancement procedures reduced the quality even further (Carr et al. 2006).

The napole technological yield method was used for measuring the post-cooking meat weight to fresh weight (Naveau 1986), and the "rendement napole locus" (*RN*$^-$) most likely represented a dominant gene with unfavorable effects on meat quality (Le Roy et al. 1989). These reports led to the initiation of linkage studies in families of pigs in which the *RN*$^-$ locus was segregating, and the trait was soon mapped to chromosome 15 near marker Sw936 (Milan et al. 1995; Mariani et al. 1996; Milan et al. 1996). Potential candidate genes were suggested based on the position of markers and comparative gene maps. However, identification of the gene required additional fine mapping within a 2.5-Mb region on chromosome 15. A substitution of glutamine for arginine at position 200 of PRKAG3, a muscle-specific isoform of the gamma-subunit of adenosine monophosphate activated protein kinase (*AMPK*), was identified as the causative mutation in *RN*⁻ pigs (Milan et al. 2000). Additional mutations in the *PRKAG3* gene that also affect glycogen quantity in pork were subsequently identified (Ciobanu et al. 2001). DNA tests for *RN* gene status are available through the National Swine Registry (USA).

15.3.7 Boar Taint

Off-flavor in pig meat from boars also affects pork quality. High levels of skatole have been shown to accumulate in the fat of boars, resulting in off-flavor in pork, or boar taint (Babol et al. 2004). Babole et al. (2004) observed significant differences between breeds, demonstrating the existence of genetic effects on skatole levels in uncastrated pigs. A candidate gene that may influence skatole levels is *CYP2E1* (cytochrome p450 2el), located on pig chromosome 14. Some association was found between an SNP in the *CYP2E1* promoter and skatole deposition in a commercial population (Skinner et al. 2005). In another report, five SNPs were identified in the coding region of *CYP2E1* (Lin et al. 2006). One SNP resulted in an amino acid substitution (A475T) that reduced *CYP2E1* expression. Lin and colleagues also identified a frameshift mutation in the *CYP2A6* gene (Lin et al. 2004a) and a substitution mutation in the *SULT1A1* coding region (Lin et al. 2004b), and suggested that both could be associated with increased skatole in fat tissue. A PCR test was developed for an SNP in the cytochrome b5A (*CYB5A*) gene correlated with reduced androstenone levels in pigs (Peacock et al. 2008). The test developers found that a G>T polymorphism had modest effects on androstenone in fat of Landrace, Yorkshire, and Large White × Duroc crossbred pigs, as well as on fat skatole in Duroc and Sire Line pig breeds. A study conducted in Swedish pig populations indicated that the effect of *CYB5A* genotype was only observed in pigs greater than 115 kg live weight (Zamaratskaia et al. 2008). Like many meat quality traits, boar taint is likely a multigenic trait influenced by breed-specific genetic background. These candidate genes may eventually be used as selection tools for reducing off-flavor in pork.

15.4 MARBLING AND OTHER MEAT QUALITY TRAITS

Many genome scan analyses have been conducted, typically relating genetic markers to a battery of carcass and meat quality traits. Several companies market DNA tests based on QTL marker combinations that have been strongly associated with meat quality. For many meat characteristics, however, the specific mechanisms and genes controlling these traits remain unclear at this time. Typically, multiple QTLs have been identified as significantly associated with meat quality in genetic analyses of carcass traits. Likely, many gene products interact, each contributing small effects to the overall genetic component of the ultimate tenderness, tastiness, and juiciness of meat.

15.4.1 QTL Databases

All the QTLs associated with meat traits cannot be adequately reviewed in this chapter because of the sheer volume and constant addition of new data. A variety of reports are briefly discussed, and the reader is referred to species-specific QTL databases for current catalogs of traits and associated QTL markers. QTL browsers for swine and cattle were developed to make data from the literature widely accessible (Hu et al. 2005; Polineni et al. 2006). A review noted the establishment of a chicken QTL database containing about 700 curated QTLs from previously published works (Abasht et al. 2006). A multi-species QTL database for livestock has also been reported (Hu et al. 2007; Hu and Reecy 2007).

A listing of QTL databases follows:

Chicken:	http://www.animalgenome.org/QTLdb/chicken.html
Cattle:	http://genomes.sapac.edu.au/bovineqtl/
	http://www.vetsci.usyd.edu.au/reprogen/QTL_Map/
	http://www.animalgenome.org/QTLdb/cattle.html
Sheep:	http://sphinx.vet.unimelb.edu.au/QTLdb/
Swine:	http://www.animalgenome.org/QTLdb/pig.html

15.4.2 QTL Studies

Hundreds of QTL studies for meat quality traits can be found in the scientific literature. For example, a genome scan of a Meishan × Yorkshire F_2 swine population confirmed a number of previous genome-wide significant QTLs (Paszek et al. 2001) for pork traits. QTL markers affecting carcass traits have been identified on nearly every pig autosome. A few QTLs were associated with known gene loci, such as the growth hormone locus on chromosome 12 that is associated with loin eye area and percentage lean meat, and a QTL on chromosome 6 near the heart fatty acid binding protein that is associated with intramuscular fat (Paszek et al. 2001). In another study, a QTL for linoleic acid percentage was identified on pig chromosome 4 (Clop et al. 2003). Polymorphisms were identified for *DECR1*, a mitochondrial 2,4-dienoyl CoA reductase gene located within the QTL region. These SNPs appeared to be in linkage disequilibrium with the trait but did not appear to be the causal mutation (Amills et al. 2005).

In a bovine QTL study of fatty acid composition, a candidate gene was identified in Angus cattle, as well as in a Jersey × Limousin backcross population (Morris et al. 2007; Zhang et al. 2008a) Polymorphisms were identified in the thioesterase domain of the bovine fatty acid synthase (*FASN*) gene. In both studies, variations in *FASN* affected the fatty acid composition of phospholipids. In another study typical of QTL analyses in cattle (a population produced by a Brahman × Hereford sire on *Bos taurus* cows), QTLs were identified for hot carcass weight, weaning weight, marbling, loin area, and Warner-Brazler shear force (tenderness) (Casas et al. 2003).

Demonstrating the concept that multiple genes likely interact to result in the final meat quality phenotypes, Wimmers et al. (2007) examined ten genes for association with carcass traits in 1700 performance-tested fattening pigs of commercial purebred and crossbred herds of Duroc, Pietrain, Pietrain × (Landrace × Large White), Duroc × (Landrace × Large White), as well as in an experimental F_2 population based on a reciprocal cross of Duroc and Pietrain pigs. The genes tested for association included ankyrin1, a non-coding STS, carbonic anydrase III, erythropoietin receptor, high-mobility group AT-hook 2, myopalladin, non-metastatic cells 1 protein, platelet-derived growth factor receptor, ELKS/RAB6-interacting/CAST family member 1, and titin (*ANK1, bR10D1, CA3, EPOR, HMGA2, MYPN, NME1, PDGFRA, ERC1, TTN*), for association. Nine of these genes showed association with meat-quality and carcass traits at a nominal *P*-value of 0.05. The remaining gene, *PDGFRA* was not significantly associated with any of the measured traits. The genes *HMGA2, CA3, EPOR, NME1,* and *TTN* affected meat color, pH, and conductivity of loin 24 hr postmortem. Ham weight and lean content were associated with *CA3* and *MYPN* at *P*-values of < 0.003. None of the genes showed significant associations for a particular trait across all populations of animals. These data show that effects of individual genes may or may not influence the measured phenotype, depending upon genetic background of the animal.

15.4.3 Thyroglobulin

One of the first bovine commercial meat-quality DNA tests to arise from QTL analyses was the TG5 gene test (Barendse et al. 2004). This test is based on an SNP in the 5′ upstream region of the thyroglobulin (*TG*) gene and was associated with marbling scores. In a different population of cattle representing diverse breeds, the authors failed to associate *TG* with marbling except in Wagyu-derived animals (Casas et al. 2007). In a validation of several commercial DNA tests for beef quality traits, the TG5 SNP did associate with meat quality grade (Van Eenennaam et al. 2007).

15.5 GLOBAL EXPRESSION

15.5.1 mRNA Profiling

Whole genome sequences are now available for several meat animal species. With these genetic roadmaps, research can begin to explore the mechanisms behind the interaction of networks of gene products and regulatory molecules that contribute to the overall phenotype of an animal. Whole genome oligo microarrays are one tool for rapid assessment of global gene expression profiling. A typical microarray consists of thousands of unique oligonucleotides located in individual spots on a glass slide. RNA from a test sample can be reverse-transcribed and labeled with a fluorescent dye (e.g., Cy3), and RNA from a control sample can be labeled with another dye of a different fluorescent wavelength (e.g., Cy5). Both labeled samples are co-hybridized to the array slide and fluorescence is detected after laser excitation of the dyes, producing a pseudo-colored image as in Figure 15.5. The quantity of mRNA for each "probe" sequence can be estimated by relative fluorescence intensity of the two dyes. For example, a spot that contained abundant mRNA signal from the test subject and little mRNA from the control might be visualized as a green spot on the slide, and the opposite case would be represented as a red spot. Genes that are approximately equal in expression for the two samples would be visualized as a yellow spot. This technique can be used to profile expression across developmental timepoints or for comparison of different treatments or phenotypes. More specialized arrays can be used to profile specific targets such as individual exons for examination of alternative splicing patterns. Microarrays can also be useful for candidate gene discovery and system-wide signal transduction pathway analysis, especially when utilized in conjunction with QTL analysis.

Mechanisms by which muscle-related phenotypes develop are gradually beoming more clear. Studies of global expression profiles in the context of whole-animal physiology should enable

FIGURE 15.5 (See color insert following page 148.) Microarray. This image represents a two-color bovine oligo microarray. RNA samples were obtained from skeletal muscle from two different steers. One sample was labeled with Cy3 and the second was labeled with Cy5 and co-hybridized to the array. Genes that were expressed at a greater level in the first animal are pseudo-colored green, and those expressed higher in the second animal are red. Genes expressed at equivalent levels are yellow. (*Source:* Image provided by R.N. Vaughn and P.K. Riggs.)

methods for the selection of breeding animals based on systems genetics. Recent reviews indicate examples of how integrated efforts might be utilized, particularly for elucidation of myogenesis and other developmental pathways important for meat quality (Lehnert et al. 2006; te Pas et al. 2007). Several microarray studies have focused on gene expression during myogenesis. Deato and Tjian (2007) found that during differentiation of myoblasts to myotubes, a switch occurs that alters the composition of binding proteins in the promoter transcriptional complex. This finding demonstrates the complexity of gene regulation important in skeletal muscle development. In addition, gene expression profiles in developing longissimus muscle tissue in both pig (te Pas et al. 2005) and bovine (Lehnert et al. 2007) fetuses have been described. Lehnert and colleagues identified differentially expressed genes related to different sire breeds. In another study, Yu and colleagues (2007) compared gene expression between longissimus and semimembranosus muscles in Korean cattle at slaughter to identify several genes that were differentially expressed between the two muscle types that may be important for muscle growth and fat metabolism.

15.5.2 Protein Profiling

As tools have become available to allow association of muscle phenotypes with genetic variation, rapidly improving proteomics tools are likely to play an important role in efforts to understand the complex networks of developmental signals that ultimately determine the meat phenotype of an animal. Proteomics techniques ranging from well-established two-dimensional gel electrophoresis (2DGE) methods to sophisticated mass spectroscopy efforts have research applications ranging from early growth and development to postmortem events (reviewed in Bendixen 2005). These techniques involve isolation of proteins from muscle tissue. The total cellular protein can be isolated from the entire cell or from subcellular compartments, and can be further fractionated to reduce complexity. Differential protein expression and/or protein modification can be identified by comparing protein 2DGE profiles from different animals, or the protein profiles can be analyzed directly by mass spectroscopy. As a step toward understanding proteins associated specifically with meat quality, a reference map of proteins of bovine semitendinosus muscle from Charolais cattle at slaughter was generated from 2DGE experiments (Bouley et al. 2004). In total, 129 variants of 75 different proteins were identified in samples obtained from muscle. Later improvements in characterization of alkaline proteins allowed identification of 32 additional gene products (Chaze et al. 2006). Global proteomics methods are gradually being applied to meat animal research and, along with mRNA profiling, will become important tools for dissecting molecular mechanisms related to muscle growth and development.

15.6 SUMMARY

It will be important for animal breeders to consider individual genes in the context of the entire genotype of an animal. As our understanding of how sequence variations within a given gene, or regulatory region of that gene, ultimately result in individual phenotypic differences for a particular trait, we can begin to explain mechanisms underlying those variations. For multigenic traits, the task is more complicated because each associated gene contributes only a portion of the phenotype, and these genotypic differences may result in different outcomes in different genetic backgrounds or breeds. Understanding the implications of a particular genotype will require an understanding of the whole animal system and genetic interactions across diverse gene signaling pathways. The new era of systems genetics, combining informatics with functional genomic and proteomic research, is just beginning. As we begin to understand the genes that regulate biological and physiological mechanisms underlying phenotypes of interest, we should be able to go back to the DNA to develop DNA tests for specific genes. These tests should be more specific than many current marker-based tests for beef cattle, and should result in better selection tools for producers.

ACKNOWLEDGMENTS

This chapter presented only a small survey of the large body of literature regarding gene and QTL mapping, and potential marker tests available for use in marker assisted selection; the authors regret omission of all relevant citations that were not included. Many topics given cursory coverage or omitted completely (such as growth factors) are discussed in greater detail in other chapters in this book.

The authors were supported, in part, by Texas AgriLife Research.

REFERENCES

Abasht, B., J.C. Dekkers, and S.J. Lamont. 2006. Review of quantitative trait loci identified in the chicken. *Poult. Sci.* 85:2079–2096.

Abdulkhaliq, A.M., H.H. Meyer, J.R. Busboom, and J.M. Thompson. 2007. Growth, carcass and cooked meat characteristics of lambs sired by Dorset rams heterozygous for the callipyge gene in Suffolk and Texel rams. *Small Ruminant Res.* 71:92–97.

Alexander, L.J., T.W. Geary, W.M. Snelling, and M.D. MacNeil. 2007. Quantitative trait loci with additive effects on growth and carcass traits in a Wagyu-Limousin F2 population. *Anim. Genet.* 38:413–416.

Amaral, M.E., K.E. Owens, J.S. Elliott, C. Fickey, A.A. Schaffer, R. Agarwala, and J.E. Womack. 2007. Construction of a river buffalo (*Bubalus bubalis*) whole-genome radiation hybrid panel and preliminary RH mapping of chromosomes 3 and 10. *Anim. Genet.* 38:311–314.

Amills, M., O. Vidal, L. Varona, A. Tomas, M. Gil, A. Sanchez, and J.L. Noguera. 2005. Polymorphism of the pig 2,4-dienoyl CoA reductase 1 gene (*decr1*) and its association with carcass and meat quality traits. *J. Anim. Sci.* 83:493–498.

Anthony, N.B. 1998. A review of genetic practices in poultry: efforts to improve meat quality. *J. Muscle Foods* 9:25–33.

Babol, J., G. Zamaratskaia, R.K. Juneja, and K. Lundstrom. 2004. The effect of age on distribution of skatole and indole levels in entire male pigs in four breeds: Yorkshire, Landrace, Hampshire and Duroc. *Meat Sci.* 67:351–358.

Barendse, W., R. Bunch, M. Thomas, S. Armitage, S. Baud, and N. Donaldson. 2004. The TG5 thyroglobulin gene test for a marbling quantitative trait loci evaluated in feedlot cattle. *Austr. J. Exp. Agric.* 44:669–674.

Beever, J.E., S.R. Fisher, G. Guerin, and H.A. Lewin. 1997a. Mapping of eight human chromosome 1 orthologs to cattle chromosomes 3 and 16. *Mamm. Genome* 8:533–536.

Beever, J.E., S.R. Fisher, and H.A. Lewin. 1997b. Polymorphism identification in the ACADM, AT3, IL10, MYOG and TSHB genes of cattle. *Anim. Genet.* 28:373–374.

Bendall, J.R. and R.A. Lawrie. 1964. Watery pork. *Anim. Breeding Abstr.* 32:1–8.

Bendixen, E. 2005. The use of proteomics in meat science. *Meat Sci.* 71:138–149.

Bishop, M.D., S.M. Kappes, J.W. Keele, R.T. Stone, S.L. Sunden, G.A. Hawkins, S.S. Toldo, R. Fries, M.D. Grosz, J. Yoo, et al. 1994. A genetic linkage map for cattle. *Genetics* 136:619–639.

Botstein, D., R.L. White, M. Skolnick, and R.W. Davis. 1980. Construction of a genetic linkage map in man using restriction fragment length polymorphisms. *Am. J. Hum. Genet.* 32:314–331.

Bouley, J., C. Chambon, and B. Picard. 2004. Mapping of bovine skeletal muscle proteins using two-dimensional gel electrophoresis and mass spectrometry. *Proteomics* 4:1811–1824.

Bradford, G.E. 1967. Genetic and economic aspects of selecting for lamb carcass quality. *J. Anim. Sci.* 26:10–15.

Briley, G.P., P.K. Riggs, J.E. Womack, D.L. Hancock, and C.A. Bidwell. 1996. Chromosomal localization of the porcine skeletal muscle calpain gene. *Mamm. Genome* 7:226–228.

Briskey, E.J. 1964. Etiological status and associated studies of pale, soft, exudative porcine musculature. *Adv. Food Res.* 13:89–178.

Bunch, T.D., R.C. Evans, S. Wang, C.P. Brennand, D.R. Whittier, and B.J. Taylor. 2004. Feed efficiency, growth rates, carcass evaluation, cholesterol level and sensory evaluation of lambs of various hair and wool sheep and their crosses. *Small Ruminant Res.* 52:239–245.

Butterworth, M.H., D. Steinhauf, and J.H. Weniger. 1967. Stress tolerance as a production character in the pig. 2. Methods of determining stress proneness and first results on relationship between stress proneness and production characters in the pig. *Zuchtungskunde* 39:283–294.

Carr, C.C., J.B. Morgan, E.P. Berg, S.D. Carter, and F.K. Ray. 2006. Growth performance, carcass composition, quality, and enhancement treatment of fresh pork identified through deoxyribonucleic acid marker-assisted selection for the Rendement Napole gene. *J. Anim. Sci.* 84:910–917.

Casas, E., J.W. Keele, S.D. Shackelford, M. Koohmaraie, T.S. Sonstegard, T.P. Smith, S.M. Kappes, and R.T. Stone. 1998. Association of the muscle hypertrophy locus with carcass traits in beef cattle. *J. Anim. Sci.* 76:468–473.

Casas, E., S.D. Shackelford, J.W. Keele, M. Koohmaraie, T.P. Smith, and R.T. Stone. 2003. Detection of quantitative trait loci for growth and carcass composition in cattle. *J. Anim. Sci.* 81:2976–2983.

Casas, E., S.N. White, S.D. Shackelford, T.L. Wheeler, M. Koohmaraie, G.L. Bennett, and T.P. Smith. 2007. Assessing the association of single nucleotide polymorphisms at the thyroglobulin gene with carcass traits in beef cattle. *J. Anim. Sci.* 85:2807–2814.

Casas, E., S.N. White, T.L. Wheeler, S.D. Shackelford, M. Koohmaraie, D.G. Riley, C.C. Chase, Jr., D.D. Johnson, and T.P. Smith. 2006. Effects of calpastatin and micro-calpain markers in beef cattle on tenderness traits. *J. Anim. Sci.* 84:520–525.

Castle, W.E. 1920. *Genetics and Eugenics. 2nd ed.* Harvard University Press, Cambridge.

Charlier, C., W. Coppieters, F. Farnir, L. Grobet, P.L. Leroy, C. Michaux, M. Mni, A. Schwers, P. Vanmanshoven, R. Hanset, et al. 1995. The *MH* gene causing double-muscling in cattle maps to bovine chromosome 2. *Mamm. Genome* 6:788–792.

Chaze, T., J. Bouley, C. Chambon, C. Barboiron, and B. Picard. 2006. Mapping of alkaline proteins in bovine skeletal muscle. *Proteomics* 6:2571–2575.

Cheong, H.S., D.H. Yoon, B.L. Park, L.H. Kim, J.S. Bae, S. Namgoong, H.W. Lee, C.S. Han, J.O. Kim, I.C. Cheong, and H.D. Shin. 2008. A single nucleotide polymorphism in CAPN1 associated with marbling score in Korean cattle. *BMC Genet.* 9:33.

Chiang, W., C.P. Allison, J.E. Linz, and G.M. Strasburg. 2004. Identification of two αryr alleles and characterization of αryr transcript variants in turkey skeletal muscle. *Gene* 330:177–184.

Chiang, W., H.J. Yoon, J.E. Linz, J.A. Airey, and G.M. Strasburg. 2007. Divergent mechanisms in generating molecular variations of αryr and βryr in turkey skeletal muscle. *J. Muscle Res. Cell Motil.* 28:343–354.

Chowdhary, B.P., T. Raudsepp, D. Honeycutt, E.K. Owens, F. Piumi, G. Guerin, T.C. Matise, S.R. Kata, J.E. Womack, and L.C. Skow. 2002. Construction of a 5000(rad) whole-genome radiation hybrid panel in the horse and generation of a comprehensive and comparative map for ECA11. *Mamm. Genome* 13: 89-94.

Chowdhary, B.P., P.D. Thomsen, I. Harbitz, M. Landset, and I. Gustavsson. 1994. Precise localization of the genes for glucose phosphate isomerase (gpi), calcium release channel (crc), hormone-sensitive lipase (lipe), and growth hormone (gh) in pigs, using nonradioactive *in situ* hybridization. *Cytogenet. Cell Genet.* 67:211–214.

Churchill, G.A. and R.W. Doerge. 1994. Empirical threshold values for quantitative trait mapping. *Genetics* 138:963–971.

Ciobanu, D., J. Bastiaansen, M. Malek, J. Helm, J. Woollard, G. Plastow, and M. Rothschild. 2001. Evidence for new alleles in the protein kinase adenosine monophosphate-activated gamma(3)-subunit gene associated with low glycogen content in pig skeletal muscle and improved meat quality. *Genetics* 159:1151–1162.

Ciobanu, D.C., J.W. Bastiaansen, S.M. Lonergan, H. Thomsen, J.C. Dekkers, G.S. Plastow, and M.F. Rothschild. 2004. New alleles in calpastatin gene are associated with meat quality traits in pigs. *J. Anim. Sci.* 82:2829–2839.

Clop, A., F. Marcq, H. Takeda, D. Pirottin, X. Tordoir, B. Bibe, J. Bouix, F. Caiment, J.M. Elsen, F. Eychenne, C. Larzul, E. Laville, F. Meish, D. Milenkovic, J. Tobin, C. Charlier, and M. Georges. 2006. A mutation creating a potential illegitimate microRNA target site in the myostatin gene affects muscularity in sheep. *Nat. Genet.* 38:813–818.

Clop, A., C. Ovilo, M. Perez-Enciso, A. Cercos, A. Tomas, A. Fernandez, A. Coll, J.M. Folch, C. Barragan, I. Diaz, M.A. Oliver, L. Varona, L. Silio, A. Sanchez, and J.L. Noguera. 2003. Detection of QTL affecting fatty acid composition in the pig. *Mamm. Genome* 14:650–656.

Cockett, N.E. 2003. Current status of the ovine genome map. *Cytogenet. Genome Res.* 102:76–78.

Cockett, N.E., S.P. Jackson, T.L. Shay, F. Farnir, S. Berghmans, G.D. Snowder, D.M. Nielsen, and M. Georges. 1996. Polar overdominance at the ovine callipyge locus. *Science* 273:236–238.

Cockett, N.E., S.P. Jackson, T.L. Shay, D. Nielsen, S.S. Moore, M.R. Steele, W. Barendse, R.D. Green, and M. Georges. 1994. Chromosomal localization of the callipyge gene in sheep (*Ovis aries*) using bovine DNA markers. *Proc. Natl. Acad. Sci. U.S.A.* 91:3019–3023.

Cockett, N.E., M.A. Smit, C.A. Bidwell, K. Segers, T.L. Hadfield, G.D. Snowder, M. Georges, and C. Charlier. 2005. The callipyge mutation and other genes that affect muscle hypertrophy in sheep. *Genet. Sel. Evol.* 37(Suppl. 1):S65–S81.

Coutinho, L L., J. Morris, H.L. Marks, R.J. Buhr, and R. Ivarie. 1993. Delayed somite formation in a quail line exhibiting myofiber hyperplasia is accompanied by delayed expression of myogenic regulatory factors and myosin heavy chain. *Development* 117:563–569.

Cover, S., O.D. Butler, and T.C. Cartwright. 1956. The relationship of fatness in yearling steers to juiciness and tenderness of broiled and braised steaks. *J. Anim. Sci.* 15:456–463.

Cover, S., T.C. Cartwright, and O.D. Butler. 1957. The relationship of ration and inheritance to eating quality of the meat from yearling steers. *J. Anim. Sci.* 16:946–956.

Cox, D.R., M. Burmeister, E.R. Price, S. Kim, and R.M. Myers. 1990. Radiation hybrid mapping: a somatic cell genetic method for constructing high-resolution maps of mammalian chromosomes. *Science* 250:245–250.

Cully, G. 1807. *Observations on Livestock: Containing Hints for Choosing and Improving the Best Breeds of the Most Useful Kinds of Domestic Animals,* 4th edition. G. Woodfall, Paternoster-Row, London.

Davies, W., I. Harbitz, R. Fries, G. Stranzinger, and J.G. Hauge. 1988. Porcine malignant hyperthermia carrier detection and chromosomal assignment using a linked probe. *Anim. Genet.* 19:203–212.

Davis, G.P., S.S. Moore, R.D. Drinkwater, W.R. Shorthose, I.D. Loxton, W. Barendse, and D.J. Hetzel. 2008. QTL for meat tenderness in the *M. Longissimus lumborum* of cattle. *Anim. Genet.* 39:40–45.

Deato, M.D.E. and R. Tjian. 2007. Switching of the core transcription machinery during myogenesis. *Genes Dev.* 21:2137–2149.

Dekkers, J.C. 2004. Commercial application of marker- and gene-assisted selection in livestock: strategies and lessons. *J. Anim. Sci.* 82(E-Suppl.):E313–E328.

Dekkers, J.C. and F. Hospital. 2002. The use of molecular genetics in the improvement of agricultural populations. *Nat. Rev. Genet.* 3:22–32.

Dunner, S., M.E. Miranda, Y. Amigues, J. Canon, M. Georges, R. Hanset, J. Williams, and F. Menissier. 2003. Haplotype diversity of the myostatin gene among beef cattle breeds. *Genet. Sel. Evol.* 35:103–118.

Evans, G.J., E. Giuffra, A. Sanchez, S. Kerje, G. Davalos, O. Vidal, S. Illan, J.L. Noguera, L. Varona, I. Velander, O.I. Southwood, D.J. de Koning, C.S. Haley, G.S. Plastow, and L. Andersson. 2003. Identification of quantitative trait loci for production traits in commercial pig populations. *Genetics* 164:621–627.

Fleming-Waddell, J.N., L.M. Wilson, G.R. Olbricht, T. Vuocolo, K. Byrne, B.A. Craig, R.L. Tellam, N.E. Cockett, and C.A. Bidwell. 2007. Analysis of gene expression during the onset of muscle hypertrophy in callipyge lambs. *Anim. Genet.* 38:28–36.

Freking, B.A., J.W. Keele, C.W. Beattie, S.M. Kappes, T.P. Smith, T.S. Sonstegard, M.K. Nielsen, and K.A. Leymaster. 1998a. Evaluation of the ovine callipyge locus. I. Relative chromosomal position and gene action. *J. Anim. Sci.* 76:2062–2071.

Freking, B.A., J.W. Keele, M.K. Nielsen, and K.A. Leymaster. 1998b. Evaluation of the ovine callipyge locus. II. Genotypic effects on growth, slaughter, and carcass traits. *J. Anim. Sci.* 76:2549–2559.

Freking, B.A. and K.A. Leymaster. 2006. Evaluation of the ovine callipyge locus. Iv. Genotypic effects on reproductive traits. *J. Anim. Sci.* 84:311–316.

Freking, B.A., S.K. Murphy, A.A. Wylie, S.J. Rhodes, J.W. Keele, K.A. Leymaster, R.L. Jirtle, and T.P. Smith. 2002. Identification of the single base change causing the callipyge muscle hypertrophy phenotype, the only known example of polar overdominance in mammals. *Genome Res.* 12:1496–1506.

Fries, R., R. Hediger, and G. Stranzinger. 1986. Tentative chromosomal localization of the bovine major histocompatibility complex by *in situ* hybridization. *Anim. Genet.* 17:287–294.

Fujii, J., K. Otsu, F. Zorzato, S. de Leon, V.K. Khanna, J.E. Weiler, P.J. O'Brien, and D.H. MacLennan. 1991. Identification of a mutation in porcine ryanodine receptor associated with malignant hyperthermia. *Science* 253:448–451.

Gahne, B. 1968. Pale exudative meat (pse) in pigs — a symptom of stress. *Svenska Svinavelsforeningens Tidskrift* 58:41–46.

Gahne, B. and R.K. Juneja. 1985. Prediction of the halothane (hal) genotypes of pigs by deducing Hal, Phi, Po2, Pgd haplotypes of parents and offspring: results from a large-scale practice in Swedish breeds. *Anim. Blood Groups Biochem. Genet.* 16:265–283.

Gaye, P., J.C. Mercier, G. Petrissant, J.L. Vilotte, and P. Popescu. 1986. Structure of milk protein complementary DNA: application to study of the genes and their chromosomal sites. *Reproduct., Nutr., Dev.* 26:573–574.

Goll, D.E., V.F. Thompson, H. Li, W. Wei, and J. Cong. 2003. The calpain system. *Physiol. Rev.* 83:731–801.

Goll, D.E., V.F. Thompson, R.G. Taylor, and A. Ouali. 1998. The calpain system and skeletal muscle growth. *Can. J. Anim. Sci.* 78:503–512.

Goodson, K.J., R.K. Miller, and J.W. Savell. 2001. Carcass traits, muscle characteristics, and palatability attributes of lambs expressing the callipyge phenotype. *Meat Sci.* 58:382–387.

Grashorn, M. and E. Muller. 1985. Relationships between blood groups, isozymes and halothane reaction in pigs from a selection experiment. *Anim. Blood Groups Biochem. Genet.* 16:329–335.

Grobet, L., L.J. Martin, D. Poncelet, D. Pirottin, B. Brouwers, J. Riquet, A. Schoeberlein, S. Dunner, F. Menissier, J. Massabanda, R. Fries, R. Hanset, and M. Georges. 1997. A deletion in the bovine myostatin gene causes the double-muscled phenotype in cattle.[see comment]. *Nat. Genet.* 17:71–74.

Grobet, L., D. Poncelet, L.J. Royo, B. Brouwers, D. Pirottin, C. Michaux, F. Menissier, M. Zanotti, S. Dunner, and M. Georges. 1998. Molecular definition of an allelic series of mutations disrupting the myostatin function and causing double-muscling in cattle. *Mamm. Genome* 9:210–213.

Groenen, M.A., H.H. Cheng, N. Bumstead, B.F. Benkel, W.E. Briles, T. Burke, D.W. Burt, L.B. Crittenden, J. Dodgson, J. Hillel, S. Lamont, A.P. de Leon, M. Soller, H. Takahashi, and A. Vignal. 2000. A consensus linkage map of the chicken genome. *Genome Res.* 10:137–147.

Haley, C.S. and S.A. Knott. 1992. A simple regression method for mapping quantitative trait loci in line crosses using flanking markers. *Heredity* 69:315–324.

Hallerman, E.M., A. Nave, Y. Kashi, Z. Holzer, M. Soller, and J.S. Beckmann. 1987. Restriction fragment length polymorphisms in dairy and beef cattle at the growth hormone and prolactin loci. *Anim. Genet.* 18:213–222.

Hanset, R., P. Leroy, C. Michaux, and K.N. Kintaba. 1983. The hal locus and its linkage with the PHI and 6-PGD loci in the Belgian Pietrain and Landrace breeds. In: *34th Annu. Meet. Eur. Assoc. Anim. Production*, Madrid, Spain, p. 130–131.

Harrison, G.G. 1972. Pale, soft, exudative pork, porcine stress syndrome and malignant hyperexia — an identity. *J. South African Vet. Assoc.* 43:57–63.

Hawken, R.J., J. Murtaugh, G.H. Flickinger, M. Yerle, A. Robic, D. Milan, J. Gellin, C.W. Beattie, L.B. Schook, and L.J. Alexander. 1999. A first-generation porcine whole-genome radiation hybrid map. *Mamm. Genome* 10:824–830.

Hu, Z.L., S. Dracheva, W. Jang, D. Maglott, J. Bastiaansen, M.F. Rothschild, and J.M. Reecy. 2005. A qtl resource and comparison tool for pigs: PigQTLDB. *Mamm. Genome* 16:792–800.

Hu, Z.L., E.R. Fritz, and J.M. Reecy. 2007. AnimalQTLdb: a livestock qtl database tool set for positional QTL information mining and beyond. *Nucleic Acids Res.* 35:D604–D609.

Hu, Z.L. and J.M. Reecy. 2007. Animal QTLdb: beyond a repository. A public platform for QTL comparisons and integration with diverse types of structural genomic information. *Mamm. Genome* 18:1–4.

Ihara, N., A. Takasuga, K. Mizoshita, H. Takeda, M. Sugimoto, Y. Mizoguchi, T. Hirano, T. Itoh, T. Watanabe, K.M. Reed, W.M. Snelling, S.M. Kappes, C.W. Beattie, G.L. Bennett, and Y. Sugimoto. 2004. A comprehensive genetic map of the cattle genome based on 3802 microsatellites. *Genome Res.* 14:1987–1998.

Jackson, S.P., M.F. Miller, and R.D. Green. 1997a. Phenotypic characterization of Rambouillet sheep expressing the callipyge gene. II. Carcass characteristics and retail yield. *J. Anim. Sci.* 75:125–132.

Jackson, S.P., M.F. Miller, and R.D. Green. 1997b. Phenotypic characterization of Rambouillet sheep expression the callipyge gene. III. Muscle weights and muscle weight distribution. *J. Anim. Sci.* 75:133–138.

Johnson, P.L., J.C. McEwan, K.G. Dodds, R.W. Purchas, and H.T. Blair. 2005a. A directed search in the region of GDF8 for quantitative trait loci affecting carcass traits in Texel sheep. *J. Anim. Sci.* 83:1988–2000.

Johnson, P.L., J.C. McEwan, K.G. Dodds, R.W. Purchas, and H.T. Blair. 2005b. Meat quality traits were unaffected by a quantitative trait locus affecting leg composition traits in Texel sheep. *J. Anim. Sci.* 83:2729–2735.

Kambadur, R., M. Sharma, T.P. Smith, and J.J. Bass. 1997. Mutations in myostatin (GDF8) in double-muscled Belgian Blue and Piedmontese cattle. *Genome Res.* 7:910–916.

Keele, J.W. and S.C. Fahrenkrug. 2001. Optimum mating systems for the myostatin locus in cattle. *J. Anim. Sci.* 79:2016–2022.

Kijas, J.W., R. McCulloch, J.E. Edwards, V.H. Oddy, S.H. Lee, and J. van der Werf. 2007. Evidence for multiple alleles effecting muscling and fatness at the ovine gdf8 locus. *BMC Genet.* 8:80.

Kim, K.S., J.J. Kim, J.C. Dekkers, and M.F. Rothschild. 2004. Polar overdominant inheritance of a DLK1 polymorphism is associated with growth and fatness in pigs. *Mamm. Genome* 15:552–559.

Koohmaraie, M. and G.H. Geesink. 2006. Contribution of postmortem muscle biochemistry to the delivery of consistent meat quality with particular focus on the calpain system. *Meat Sci.* 74:34–43.

Koohmaraie, M., S.D. Shackelford, T.L. Wheeler, S.M. Lonergan, and M.E. Doumit. 1995. A muscle hypertrophy condition in lamb (callipyge): characterization of effects on muscle growth and meat quality traits. *J. Anim. Sci.* 73:3596–3607.

Kumar, S.T. 2007. Genetic relationship among chicken populations of India based on SNP markers of myostatin gene (gdf 8). *Int. J. Poul. Sci.* 6:684–688.

Larsen, N.J., S. Kenealy, C.K. Tuggle, and M.F. Rothschild. 1998. A HincII polymorphism in the porcine calpain, large polypeptide L3 (capn3) gene. *J. Anim. Sci.* 76:918–919.

Le Roy, P., N.J., J.M. Elsen, and P. Sellier. 1989. Evidence for a new major gene influencing meat quality in pigs. *Genet. Res.* 55:33–40.

Lebreton, C.M. and P.M. Visscher. 1998. Empirical nonparametric bootstrap strategies in quantitative trait loci mapping: conditioning on the genetic model. *Genetics* 148:525–535.

Lehnert, S.A., A. Reverter, K.A. Byrne, Y. Wang, G.S. Nattrass, N.J. Hudson, and P.L. Greenwood. 2007. Gene expression studies of developing bovine longissimus muscle from two different beef cattle breeds. *BMC Dev. Biol.* 7:95.

Lehnert, S.A., Y.H. Wang, S.H. Tan, and A. Reverter. 2006. Gene expression-based approaches to beef quality research. *Austr. J. Exp. Agric.* 46:165–172.

Li, X.P., K.T. Do, J.J. Kim, J. Huang, S.H. Zhao, Y. Lee, M.F. Rothschild, C.K. Lee, and K.S. Kim. 2008. Molecular characteristics of the porcine DLK1 and MEG3 genes. *Anim. Genet.* 39:189–192.

Lin, Z., Y. Lou, and E.J. Squires. 2004a. Molecular cloning, expression and functional characterization of the cytochrome p450 2A6 gene in pig liver. *Anim. Genet.* 35:314–316.

Lin, Z., Y. Lou, and E.J. Squires. 2004b. Molecular cloning and functional analysis of porcine SULT1A1 gene and its variant: a single mutation SULT1A1 causes a significant decrease in sulfation activity. *Mamm. Genome* 15:218–226.

Lin, Z., Y. Lou, and E.J. Squires. 2006. Functional polymorphism in porcine CYP2E1 gene: its association with skatole levels. *J. Steroid Biochem. Mol. Biol.* 99:231–237.

Liu, C., D.C. McFarland, and S.G. Velleman. 2005. Effect of genetic selection on MyoD and myogenin expression in turkeys with different growth rates. *Poult. Sci.* 84:376–384.

Lomax, M.I., P.K. Riggs, and J.E. Womack. 1995. Structure and chromosomal location of the bovine gene for the heart muscle isoform of cytochrome C oxidase subunit VIII. *Mamm. Genome* 6:118–122.

MacLennan, D.H., C. Duff, F. Zorzato, J. Fujii, M. Phillips, R.G. Korneluk, W. Frodis, B.A. Britt, and R.G. Worton. 1990. Ryanodine receptor gene is a candidate for predisposition to malignant hyperthermia. *Nature* 343:559–561.

Mariani, P., M. Johansson, H. Ellegren, I. Harbitz, R.K. Juneja, and L. Andersson. 1992. Multiple restriction fragment length polymorphisms in the porcine calcium release channel gene (crc): assignment to the halothane (hal) linkage group. *Anim. Genet.* 23:257–262.

Mariani, P., K. Lundstrom, U. Gustafsson, A.C. Enfalt, R.K. Juneja, and L. Andersson. 1996. A major locus (RN) affecting muscle glycogen content is located on pig chromosome 15. *Mamm. Genome* 7:52–54.

McPherron, A.C., A.M. Lawler, and S.J. Lee. 1997. Regulation of skeletal muscle mass in mice by a new TGF-beta superfamily member. *Nature* 387:83–90.

McPherron, A.C. and S.J. Lee. 1997. Double muscling in cattle due to mutations in the myostatin gene. *Proc. Natl. Acad. Sci. U.S.A.* 94:12457–12461.

Meuwissen, T.H., B.J. Hayes, and M.E. Goddard. 2001. Prediction of total genetic value using genome-wide dense marker maps. *Genetics* 157:1819–1829.

Meuwissen, T.H.E. and M.E. Goddard. 2002. Mapping multiple qtl by combining linkage disequilibrium/linkage analysis in outbred populations. In: *7th World Congr. Genetics Applied to Livestock Production,* Montpellier, France, p. 669–672.

Meyers, S.N., S.L. Rodriguez-Zas, and J.E. Beever. 2007. Fine-mapping of a QTL influencing pork tenderness on porcine chromosome 2. *BMC Genet.* 8:69.

Milan, D., J.T. Jeon, C. Looft, V. Amarger, A. Robic, M. Thelander, C. Rogel-Gaillard, S. Paul, N. Iannuccelli, L. Rask, H. Ronne, K. Lundstrom, N. Reinsch, J. Gellin, E. Kalm, P.L. Roy, P. Chardon, and L. Andersson. 2000. A mutation in PRKAG3 associated with excess glycogen content in pig skeletal muscle. *Science* 288:1248–1251.

Milan, D., P. Le Roy, N. Woloszyn, J.C. Caritez, J.M. Elsen, P. Sellier, and J. Gellin. 1995. The RN locus for meat quality maps to pig chromosome 15. *Genet. Sel. Evol.* 27:195–199.

Milan, D., N. Woloszyn, M. Yerle, P. Le Roy, M. Bonnet, J. Riquet, Y. Lahbib-Mansais, J.C. Caritez, A. Robic, P. Sellier, J.M. Elsen, and J. Gellin. 1996. Accurate mapping of the "acid meat" RN gene on genetic and physical maps of pig chromosome 15. *Mamm. Genome* 7:47–51.

Minkema, D., G. Eikelenboom, and P. van Eldik. 1977. Inheritance of M.H.S.-susceptibility in pigs. In: *Proc. Third Inter. Conf. Production Disease in Farm Animals,* Wageningen, The Netherlands, p. 203–207.

Monin, G., J. Gruand, D. Laborde, and P. Sellier. 1984. The Hampshire effect on meat quality in pigs. *Journees de la Recherche Porcine en France* 1984:p. 59–64.

Morisson, M., A. Lemiere, S. Bosc, M. Galan, F. Plisson-Petit, P. Pinton, C. Delcros, K. Fève, F. Pitel, V. Fillon, M. Yerle, and A. Vignal. 2002. ChickRH6: a chicken whole-genome radiation hybrid panel. *Genet. Sel. Evol.* 34:521–533.

Morris, C.A., N.G. Cullen, B.C. Glass, D.L. Hyndman, T.R. Manley, S.M. Hickey, J.C. McEwan, W.S. Pitchford, C.D. Bottema, and M.A. Lee. 2007. Fatty acid synthase effects on bovine adipose fat and milk fat. *Mamm. Genome* 18:64–74.

Moser, D.W., R.M. Thallman, E.J. Pollak, M.E. Dikeman, C.A. Gill, S.R. Koontz, T.R. Holm, and E.W. Dressler. 2004. Meeting consumer demands through genetic selection: the NCBA carcass merit project. In: *Proc. 36th Annu. Meeting Beef Improvement Federation,* Sioux Falls, SD, p. 42–44.

Mullis, K., F. Faloona, S. Scharf, R. Saiki, G. Horn, and H. Erlich. 1986. Specific enzymatic amplification of DNA *in vitro*: the polymerase chain reaction. *Cold Spring Harb. Symp. Quant. Biol.* 51(Pt. 1):263–273.

Naveau, J. 1986. The genetic determination of meat quality in pigs. Heritability of the technological yield of meat prepared by the "Napole" method. *18emes Journees de la Recherche Porcine en France.*

Nonneman, D. and M.Koohmaraie. 1999. Molecular cloning and mapping of the bovine and ovine skeletal muscle-specific calpains. *Anim. Genet.* 30:456–458.

Oliver, W.M. and T.C. Cartwright. 1969. *Double Muscling in Cattle: A Review of Expression, Genetics, and Economic Implication.* Texas Agricultural Experiment Station, College Station.

Olsen, H.G., S. Lien, M. Gautier, H. Nilsen, A. Roseth, P.R. Berg, K.K. Sundsaasen, M. Svendsen, and T.H. Meuwissen. 2005. Mapping of a milk production quantitative trait locus to a 420-kb region on bovine chromosome 6. *Genetics* 169:275–283.

Otsu, K., V.K. Khanna, A.L. Archibald, and D.H. MacLennan. 1991. Cosegregation of porcine malignant hyperthermia and a probable causal mutation in the skeletal muscle ryanodine receptor gene in backcross families. *Genomics* 11:744–750.

Ottini, L., G. Marziali, A. Conti, A. Charlesworth, and V. Sorrentino. 1996. Alpha and beta isoforms of ryanodine receptor from chicken skeletal muscle are the homologues of mammalian RYR1 and RYR3. *Biochem. J.* 315(Pt. 1):207–216.

Page, B.T., E. Casas, M.P. Heaton, N.G. Cullen, D.L. Hyndman, C.A. Morris, A.M. Crawford, T.L. Wheeler, M. Koohmaraie, J.W. Keele, and T.P. Smith. 2002. Evaluation of single-nucleotide polymorphisms in CAPN1 for association with meat tenderness in cattle. *J. Anim. Sci.* 80:3077–3085.

Page, B.T., E. Casas, R.L. Quaas, R.M. Thallman, T.L. Wheeler, S.D. Shackelford, M. Koohmaraie, S.N. White, G.L. Bennett, J.W. Keele, M.E. Dikeman, and T.P. Smith. 2004. Association of markers in the bovine CAPN1 gene with meat tenderness in large crossbred populations that sample influential industry sires. *J. Anim. Sci.* 82:3474–3481.

Paszek, A.A., P.J. Wilkie, G.H. Flickinger, L.M. Miller, C.F. Louis, G.A. Rohrer, L.J. Alexander, C.W. Beattie, and L.B. Schook. 2001. Interval mapping of carcass and meat quality traits in a divergent swine cross. *Anim. Biotechnol.* 12:155–165.

Peacock, J., Y. Lou, K. Lundstrom, and E.J. Squires. 2008. The effect of a c.-8G>T polymorphism on the expression of cytochrome b5A and boar taint in pigs. *Anim. Genet.* 39:15–21.

Perucatti, A., S. Floriot, G.P. Di Meo, D. Soglia, R. Rullo, S. Maione, D. Incarnato, A. Eggen, P. Sacchi, R. Rasero, and L. Iannuzzi. 2006. Comparative fish mapping of mucin 1, transmembrane (MUC1) among cattle, river buffalo, sheep and goat chromosomes: comparison between bovine chromosome 3 and human chromosome 1. *Cytogenet. Genome. Res.* 112:103–105.

Polineni, P., P. Aragonda, S.R. Xavier, R. Furuta, and D.L. Adelson. 2006. The bovine qtl viewer: a web accessible database of bovine quantitative trait loci. *BMC Bioinformatics* 7:283.

Riggs, P.K. 1996. Development and Characterization of a Bovine Interspecific Hybrid Backcross Panel: Linkage Analysis, Gene Mapping, and Cytogenetics. Ph.D. dissertation. Texas A&M University, College Station.

Riggs, P.K. and K.A. Hahn. 1998. Fluorescence *in situ* hybridization. In: M. Hanausek and Z. Walaszek (Eds.), *Methods in Molecular Medicine No. 14: Tumor Marker Protocols.* Humana Press, Totowa, NJ, p. 245–254.

Riggs, P.K., K.E. Owens, C.E. Rexroad, 3rd, M.E. Amaral, and J.E. Womack. 1997. Development and initial characterization of a *Bos taurus* x *B. Gaurus* interspecific hybrid backcross panel. *J. Hered.* 88:373–379.

Rohrer, G.A. 2004. An overview of genomics research and its impact on livestock reproduction. *Reprod. Fertil. Dev.* 16:47–54.

Ron, M. and J.I. Weller. 2007. From Qtl to qtn identification in livestock — winning by points rather than knock-out: a review. *Anim. Genet.* 38:429–439.

Rothschild, M.F., B.J.P., and D.C. Ciobanu (Eds.). 2004. *Genome Analysis of QTL for Muscle Tissue Development and Meat Quality. Muscle Development of Livestock Animals.* CABI Publishing, Oxfordshire, p. 247–262.

Saiki, R.K., S. Scharf, F. Faloona, K.B. Mullis, G.T. Horn, H.A. Erlich, and N. Arnheim. 1985. Enzymatic amplification of beta-globin genomic sequences and restriction site analysis for diagnosis of sickle cell anemia. *Science* 230:1350–1354.

Schenkel, F.S., S.P. Miller, Z. Jiang, I.B. Mandell, X. Ye, H. Li, and J.W. Wilton. 2006. Association of a single nucleotide polymorphism in the calpastatin gene with carcass and meat quality traits of beef cattle. *J. Anim. Sci.* 84:291–299.

Seaton, G., C.S. Haley, S.A. Knott, M. Kearsey, and P.M. Visscher. 2002. QTL express: mapping quantitative trait loci in simple and complex pedigrees. *Bioinformatics* 30:339–340.

Sellick, G.S., W.S. Pitchford, C.A. Morris, N.G. Cullen, A.M. Crawford, H.W. Raadsma, and C.D. Bottema. 2007. Effect of myostatin F94l on carcass yield in cattle. *Anim. Genet.* 38:440–446.

Skinner, T.M., E. Doran, J.D. McGivan, C.S. Haley, and A.L. Archibald. 2005. Cloning and mapping of the porcine cytochrome-p450 2e1 gene and its association with skatole levels in the domestic pig. *Anim. Genet.* 36:417–422.

Smit, M., K. Segers, L.G. Carrascosa, T. Shay, F. Baraldi, G. Gyapay, G. Snowder, M. Georges, N. Cockett, and C. Charlier. 2003. Mosaicism of solid gold supports the causality of a noncoding A-to-G transition in the determinism of the callipyge phenotype. *Genetics* 163:453–456.

Smith, J.A., A.M. Lewis, P. Wiener, and J.L. Williams. 2000a. Genetic variation in the bovine myostatin gene in UK beef cattle: allele frequencies and haplotype analysis in the South Devon. *Anim. Genet.* 31:306–309.

Smith, T.P., E. Casas, C.E. Rexroad, 3rd, S.M. Kappes, and J.W. Keele. 2000b. Bovine CAPN1 maps to a region of BTA29 containing a quantitative trait locus for meat tenderness. *J. Anim. Sci.* 78:2589–2594.

Sorimachi, H., S. Imajoh-Ohmi, Y. Emori, H. Kawasaki, S. Ohno, Y. Minami, and K. Suzuki. 1989. Molecular cloning of a novel mammalian calcium-dependent protease distinct from both m- and mu-types. Specific expression of the mRNA in skeletal muscle. *J. Biol. Chem.* 264:20106–20111.

Sturtevant, A.H. 1913. A third group of linked genes in *Drosophila ampelophila*. *Science* 37:990–992.

Swatland, H.J. and N.M. Kiefer. 1974. Fetal development of the double muscled condition in cattle. *J. Anim. Sci.* 38:752–757.

Tautz, D. 1989. Hypervariability of simple sequences as a general source for polymorphic DNA markers. *Nucleic Acids Res.* 17:6463–6471.

te Pas, M.F.W., A.A.W. De Wit, J. Priem, M. Cagnazzo, R. Davoli, V. Russo, and M.H. Pool. 2005. Transcriptome expression profiles in prenatal pigs in relation to myogenesis. *J. Muscle Res. Cell Motil.* 26:157–165.

te Pas, M.F.W., F.L. Harders, A. Soumillion, L. Born, W. Buist, and T.H. Meuwissen. 1999a. Genetic variation at the porcine MYF-5 gene locus. Lack of association with meat production traits. *Mamm. Genome* 10:123–127.

te Pas, M.F.W., I. Hulsegge, A. Coster, M.H. Pool, H.H. Heuven, and L.L. Janss. 2007. Biochemical pathways analysis of microarray results: regulation of myogenesis in pigs. *BMC Dev. Biol.* 7:66.

te Pas, M.F.W., A. Soumillion, F.L. Harders, F.J. Verburg, T.J. van den Bosch, P. Galesloot, and T.H. Meuwissen. 1999b. Influences of myogenin genotypes on birth weight, growth rate, carcass weight, backfat thickness, and lean weight of pigs. *J. Anim. Sci.* 77:2352–2356.

Threadgill, D.S., D.W. Threadgill, Y.D. Moll, J.A. Weiss, N. Zhang, H.W. Davey, A.G. Wildeman, and J.E. Womack. 1994. Syntenic assignment of human chromosome 1 homologous loci in the bovine. *Genomics* 22:626–630.

Uleberg, E. and T.H. Meuwissen. 2007. Fine mapping of multiple QTL using combined linkage and linkage disequilibrium mapping — a comparison of single QTL and multi QTL methods. *Genet. Select. Evolut.* 39:285–299.

Van Arendonk, J.A. and B.W. Kennedy. 1990. Efficiency of selection strategies for halothane-negative gene. *J. Anim. Sci.* 68:1569–1576.

Van Eenennaam, A.L., J. Li, R.M. Thallman, R.L. Quaas, M.E. Dikeman, C.A. Gill, D.E. Franke, and M.G. Thomas. 2007. Validation of commercial DNA tests for quantitative beef quality traits. *J. Anim. Sci.* 85:891–900.

Vuocolo, T., K. Byrne, J. White, S. McWilliam, A. Reverter, N.E. Cockett, and R.L. Tellam. 2007. Identification of a gene network contributing to hypertrophy in callipyge skeletal muscle. *Physiol. Genomics* 28:253–272.

Wang, L.J., T.M. Byrem, J. Zarosley, A.M. Booren, and G.M. Strasburg. 1999. Skeletal muscle calcium channel ryanodine binding activity in genetically unimproved and commercial turkey populations. *Poult. Sci.* 78:792–797.

White, J.D., T. Vuocolo, M. McDonagh, M.D. Grounds, G.S. Harper, N.E. Cockett, and R. Tellam. 2008. Analysis of the callipyge phenotype through skeletal muscle development; association of DLK1 with muscle precursor cells. *Differentiation* 76:283–298.

White, S.N., E. Casas, T.L. Wheeler, S.D. Shackelford, M. Koohmaraie, D.G. Riley, C.C. Chase, Jr., D.D. Johnson, J.W. Keele, and T.P. Smith. 2005. A new single nucleotide polymorphism in CAPN1 extends the current tenderness marker test to include cattle of *Bos indicus*, *Bos taurus*, and crossbred descent. *J. Anim. Sci.* 83:2001–2008.

Wimmers, K., E. Murani, M.F. te Pas, K.C. Chang, R. Davoli, J.W. Merks, H. Henne, M. Muraniova, N. da Costa, B. Harlizius, K. Schellander, I. Boll, S. Braglia, A.A. de Wit, M. Cagnazzo, L. Fontanesi, D. Prins, and S. Ponsuksili. 2007. Associations of functional candidate genes derived from gene-expression profiles of prenatal porcine muscle tissue with meat quality and muscle deposition. *Anim. Genet.* 38:474–484.

Womack, J.E. 1988. Molecular cytogenetics of cattle: a genomic approach to disease resistance and productivity. *J. Dairy Sci.* 71:1116–1123.

Womack, J.E. 1996. Integrating the syntenic, physical, and linkage maps. In: R.H. Miller, V.G. Pursel, and H.D. Norman, Editors, *Beltsville Symposia in Agricultural Research XX: Biotechnology's Role in the Genetic Improvement of Farm Animals,* No. 20. p, 91–102. American Society of Animal Science, Savoy, IL.

Womack, J.E. 2005. Advances in livestock genomics: opening the barn door. *Genome Res.* 15: 1699–1705.

Womack, J.E., J.S. Johnson, E.K. Owens, C.E. Rexroad, 3rd, J. Schlapfer, and Y.P. Yang. 1997. A whole-genome radiation hybrid panel for bovine gene mapping. *Mamm. Genome* 8:854–856.

Womack, J.E. and Y.D. Moll. 1986. Gene map of the cow: conservation of linkage with mouse and man. *J. Hered.* 77:2–7.

Wray-Cahen, C.D., D.E. Kerr, C.M. Evock-Clover, and N.C. Steele. 1998. Redefining body composition: nutrients, hormones, and genes in meat production. *Annu. Rev. Nutr.* 18:63–92.

Wu, C.H., K. Nomura, T. Goldammer, T. Hadfield, J.E. Womack, and N.E. Cockett. 2007. An ovine whole-genome radiation hybrid panel used to construct an RH map of ovine chromosome 9. *Anim. Genet.* 38:534–536.

Yang, F.P., Y.Q. Chen, S.P. Li, Q.S. Li, and J.Y. Wang. 2007. Study on the single nucleotide polymorphism of myostatin gene's coding region in three domestic goose. *J Yangzhou University, Agricultural and Life Sciences Edition* 28:29–32.

Ye, X., S.R. Brown, K. Nones, L.L. Coutinho, J.C. Dekkers, and S.J. Lamont. 2007. Associations of myostatin gene polymorphisms with performance and mortality traits in broiler chickens. *Genet. Sel. Evol.* 39:73–89.

Youatt, W. 1834. *Cattle: Their Breeds, Management, and Diseases.* Baldwin and Craddock, Paternoster Row, London.

Yu, S.L., H. J. Chung, B.C. Sang, C.S. Park, J.H. Lee, D.H. Yoon, S.H. Lee, and K.D. Choi. 2007. Identification of differentially expressed genes in distinct skeletal muscles in cattle using CDNA microarray. *Anim. Biotechnol.* 18:275–285.

Zamaratskaia, G., Y. Lou, J. Peacock, L. Rydhmer, H. Andersson, R.K. Juneja, G. Chen, K. Lundstrom, and E.J. Squires. 2008. Effect of polymorphism in the porcine cytochrome b5 (CYB5A) gene on androstenone and skatole concentrations and sexual development in Swedish pig populations. *Animal* 2:190–196.

Zhang, R.F., H. Chen, C.Z. Lei, C.L. Zhang, Y.D. Lan, H.J. Zhang, B. Bao, H. Niu, and X.Z. Wang. 2007. Association between polymorphisms of MSTn and MYF5 genes and growth traits in three Chinese cattle breeds. *Asian-Austr J. Anim. Sci.* 20:1798–1804.

Zhang, S., T.J. Knight, J.M. Reecy, and D.C. Beitz. 2008a. DNA polymorphisms in bovine fatty acid synthase are associated with beef fatty acid composition. *Anim. Genet.* 39:62–70.

Zhang, Z.R., Y.P. Liu, X. Jiang, H.R. Du, and Q. Zhu. 2008b. Study on association of single nucleotide polymorphism of CAPN1 gene with muscle fibre and carcass traits in quality of chicken populations. *J. Anim. Breed Genet.* 125:1–7.

Zhiliang, G., Z. Dahai, L. Ning, L. Hui, D. Zuemei, and W. Changxin. 2004. The single nucleotide polymorphisms of the chicken myostatin gene are associated with skeletal muscle and adipose growth. *Science in China, Series C, Life Sciences* 47:25–30.

Zhou, H., J.G. Hickford, and Q. Fang. 2008. Variation in the coding region of the myostatin (GDF8) gene in sheep. *Mol. Cell Probes* 22:67–68.

Zhu, Z., D.J. Wu, and N.Y. Xu. 2007. SNPS myostatin gene and its genetic effects on carcass traits in chicken. *Hereditas (Beijing)* 29:593–598.

16 Animal Agriculture and Animal Welfare

An Essay on How They Benefit Each Other in the Twenty-First Century

Joe M. Regenstein

CONTENTS

16.1 INTRODUCTION

Believe nothing that you think.
—Contemporary bumper sticker

In recent years, animal welfare has become more important to consumers, both in the U.S. and in many other developed (and also in some developing) countries. I make this distinction because some commentators have postulated that only in societies with enough wealth and leisure do issues like animal welfare even arise. In addition, animal welfare and its relationship to animal agriculture have become part of the "environmental/sustainability" dialog that is characterizing the larger social/political fabric of twenty-first century living.

As a food scientist, I believe that it is important to look critically at the long-term issues that are being raised in these discussions, as the choices we make concerning animal agriculture and its role in society will strongly impact the nature and amount of food in the future. Those of us whose subspecialty is in the area of meat science are often on the front line with respect to many of these controversial issues. Teaching courses in Food Law, Managing Food Waste without Trashing the Environment, and Environmental Stewardship on the Cornell Campus, along with responsibilities in the introductory meat science course, has helped me to develop a more holistic view of the issues challenging animal agriculture that need to be addressed and has even given me some ideas of how I might consider addressing these issues. Teaching a unique course in kosher (Jewish) and halal (Muslim) dietary laws, which focus on animal issues, including slaughter, has provided me with the opportunity to focus on some different aspects of the debate, especially as I have been asked to participate in national (U.S.) animal welfare standards development efforts that are described below. In 2006, working with a number of colleagues at Cornell, we initiated an animal science course on Animal Welfare. And, the previous year (2005), I had the honor of being a judge at the National Intercollegiate Animal Welfare Judging Contest held at the University of Wisconsin, Madison. This is an excellent program designed to help students on campuses around the country focus critically on evaluating animal welfare concerns using a set of detailed paired comparisons of information obtained from two facilities using very different production systems producing the same agricultural animal.

This chapter focuses on a number of different issues that are part of debate and, I believe, can contribute to reasoned thinking about the problem. Looking at all of these together will also, I hope, help increase the sophistication of the discussion. This chapter's goal is to help the general public understand many of the issues involved in this debate. Hopefully, readers will glean some new perspectives that will, hopefully, help them think about the issues in a less simplistic mode. As this chapter is targeted at a general audience, I have chosen not to include references and footnotes, except for direct quotes.

How to begin? I would like to start by framing the chapter within an ethical framework. All of us like to think of ourselves as ethical persons; yet, even when each of us works hard to be ethical, we cannot always agree on what is right and what is wrong. Why? Because we must remember that our ideas about right and wrong are formed not solely from our scientific and technical training, but also just as importantly, from our total education, both formal and informal. Our ethical beliefs reflect who we are and reflect both our formal and informal education — that is, life experiences.

However, as a society, we need to address ethical issues using rational arguments and reasoning. Thus, the rationale must be stated. We must place these ethical concepts into the "Marketplace of Ideas" to ensure that they meet a higher standard than simply reflecting our own "thinking" as challenged in the starting bumper sticker quote above. A look at some materials from a formal ethics text might also be helpful.

"Philosophy, like morality itself, is the first and last an exercise in reason — the ideas that should come out on top are the ones that have the best reasons on their sides" (Rachels, *The Elements of Moral Philosophy*, 1999, p. xii). So the challenge for all of us is to truly use rational arguments when discussing controversial ethical principles.

But what are rational arguments? This is not always easy to determine because even this simple idea requires judgments on which we can disagree. However, one common approach to framing arguments definitely does not meet the rational standard. That is, if something specific is wrong with a particular system (i.e., such as the food system), how do we address it? I believe that it rationally means that we will need to work harder to correct the specific problem identified as a problem. That is, the first approach is correction. What it does not mean is that the particular system is fatally flawed and should not exist, and/or that another system is the obvious replacement. If all attempts at a solution fail, then we might have to consider one of possibly many drastic alternatives.

For example, if there are problems with our food system, specifically with animal agriculture, the people involved in animal agriculture need to address these issues and find rational, acceptable solutions. Pointing out these specific problems, even if there are a large number of such problems, does not provide a reasonable argument for accepting alternative agricultural systems, such as veganism, which is the absence of all products of animal agriculture. The alternative system(s), which might in fact be a better alternative for all or some people, must stand on its (their) own merit as the most rational choice. So, if veganism is the answer, it must show that it is the best choice in its own right and that the problems that cannot be solved with respect to plant-based agriculture are smaller and have less long-term impact than those that cannot be solved by animal agriculture.

So, in light of the above discussion, I would like to propose an ethical statement that I hope is general enough to provide an acceptable framework for this chapter: The responsibility to treat animals properly (which eventually needs to be defined carefully) is an ethical responsibility that falls to all humans, but most particularly to those who actually are directly involved with animals. My hope is that although we may disagree with the details of how to carry out this ethical charge, we all can accept this as an umbrella within which animal agriculture can be expected to operate.

As a member of the broadly defined animal agriculture community, I hold myself and those I work with accountable for the welfare of farm animals, and hope that others who are more removed from direct animal involvement will be supportive of a stronger effort toward rational, scientifically based animal welfare improvements through their actions. One of the best ways to support improved animal welfare of farm animals is by learning more about animal welfare and then purchasing products that have met appropriate standards. At present there are definitely practices used by commercial animal agriculture that do not meet a minimal scientifically defined animal welfare standard. So, the goal is to use education of both those involved and of the public, along with appropriate market-driven actions to bring about the necessary changes. (I also believe that plant agriculture will need to go through a similar process to determine the appropriate standards for plant production systems.)

The idea behind a scientifically based animal welfare standard is to start by focusing on those procedures that can be measured and critically evaluated so that their benefits and cost or other downside impacts can be determined. Other issues of concern are often treated as ethical mandates, sometimes because they cannot be evaluated scientifically, and may, in fact, represent a personal or group's belief system. Some of these concerns reflect an attitude toward a specific issue or to a broader framework of the nature of animal agriculture that may not meet a "reason-based" standard. They are, therefore, "anthropomorphic" (i.e., human) choices based on feelings and perceptions. There is nothing wrong with these concepts, and they may be powerful motivators for individual or group purchasing or political action. But that still does not mean that they meet the standard of being "the ideas that should come out on top [as they] are the ones that have the best reasons on their sides."

At this point I would like to address a series of relevant issues as rationally as possible, understanding that they often engender strong emotional responses, and I, like all of us, have a point of view that you may not agree with but I will try to present a balanced perspective.

By way of an apology: I realize that many of my examples in this chapter are coming from the poultry industry. There are two reasons for this situation. First, having spent five summers on a poultry farm and as a former member of a poultry department, I know more about chickens than other animals. And second, poultry production is often the most intensive form of animal agriculture, so they are often the first to deal with some of the more controversial issues associated with intensive indoor animal agriculture, i.e., what some people call "factory farming."

16.2 ANIMAL BEHAVIOR

The concerns for the welfare of animals are often summarized by "The Five Freedoms for Animals" originally put forth by the Bramble Commission (1965) in the UK:

1. Freedom from hunger, thirst, and malnutrition
2. Freedom from fear and distress
3. Freedom from physical and thermal discomfort
4. Freedom from pain, injury, and disease
5. Freedom to express normal patterns of behavior

Many of us probably wish we could be assured of all of the above for ourselves. Some research, however, suggests that some stress (in moderation) may actually be beneficial. In the future, we may actually breed for better adjustment to stress. The first four ideas certainly represent reasonable goals and ideals for animals under the stewardship of humans. However, it is probably impossible to meet all four of these goals all the time, and there does remain the challenge of defining when one has stepped over the line. Obviously, an animal in the wild is certainly not assured of any of the above!

But what constitutes natural or normal behavior? Some normal animal behaviors are based on the need for the animal to protect itself in the wild. So, these behaviors may become unnecessary and may not make sense in the environment of commercial agriculture, especially if the husbandry is able to meet the first four concerns.

The issue of animal behavior might be better thought about in terms of what does the animal want and need? Which behaviors does it wish to express within the circumstances it finds itself? And how do we determine these wants and needs? Animals can often tell us when something is wrong. Sometimes in animal agriculture it takes the form of productivity decreases, i.e., fewer eggs, less milk, less weight gain, etc. In other cases, it may lead to questionable types of behavior, i.e., those that are referred to as stereotypies — they represent active manifestations of abnormal behavior generally because an animal has some need that has not been fulfilled. For example, an animal that is constantly biting a metal pipe is not happy.

Modern research methods are beginning to develop even more sophisticated measures of animal behavioral needs. This research will need to be incorporated into our thinking about these issues.

In general, as suggested above, production variables decrease when an animal is stressed. So as a working hypothesis, I would suggest that an unhappy animal showing decreased productivity is counterproductive to the goal of modern commercial animal agriculture to optimize productivity. Note that the baseline for optimal productivity must be based on the specific animal's potential — if it is not meeting its potential, then something is not right. For example, if particular strains of egg-laying chickens normally produce an average of 300 eggs per year per bird housed under the best available commercial practice, then any lesser production suggests that the optimum has not been reached. (I am using best available commercial practice as a standard. This is more realistic than production in an isolated, ideal academic or research setting, where expensive and specialized approaches are possible but not realistic for commercial practice — many of which may in fact underestimate productivity in some animals that are now specifically bred to thrive in commercial situations.) Given the reality of striving for maximum productivity, farmers should have a strong incentive to minimize these stresses. Other factors — ranging from economics to the willingness to make an effort to improve their operations, to a lack of knowledge of how to optimize their systems — are preventing some farmers from obtaining maximum productivity. So improvements are needed.

I would also like to explore a component of the natural behavior argument with which I am still struggling. According to what I have learned about animal rights advocates, the neutering of animals is acceptable to many of them because it, in fact, makes the animal's life better (less fighting, less stress, less being "raped") and appears to reduce certain cancer risks. And, I believe this analysis of the animal welfare implications is correct. Yet, taking away the sex drive and act seems to violate the ultimate in natural behavior. I also wonder if our pet animals, which are naturally predators, rather than farm animals, which are naturally prey animals, are often kept indoors most of their lives, not permitted to hunt or function in packs also violated natural behavior. So I find that I must question the basic "rational" basis of this standard of the Bramble Commission. Are *natural behavior* and *best animal welfare* in conflict? However, I also want to be clear that animal behavior is a powerful tool to evaluate animal welfare, and more good research in this area is needed.

16.3 BREEDING

In discussing the potential of an animal, a key component underlying that potential is the "genetics" of the animal, i.e., what it brings to this world from its parents. The environment can build on this base but cannot undo inherited problems.

One of the legitimate criticisms leveled against some modern animal agriculture systems, which also applies to some companion animals, is that the animals have been over-bred. That is, some of the traits the animal displays are counter-productive to animal welfare and/or to productivity. In breeding for a limited number of characteristics using classical breeding (i.e., carefully choosing which animals breed with each other), we may end up with animals that are not ideal for modern animal agriculture. For example, we have bred broiler birds that are too heavy for their leg structure. We also have bred animals that no longer can mate!

Why does this happen? With classical breeding, even in the computer age, a few key traits that are considered important are monitored by the breeders and the selection of "successful" animals (i.e., those animals that will be permitted to reproduce) is based on having the desired version of these traits. But when one selects for any one trait, which is controlled by either one or a few genes, one is also always selecting a chromosome that holds the gene. A gene can be thought of as the basic unit of biological information for a particular reaction; i.e., one gene provides the information to make one protein, usually an enzyme. Many genes are found on one chromosome without any obvious relationship to each other, so the chromosome is much larger than the gene. Thus, the breeder's desirable gene may also bring along with it some other very undesirable traits (genes) that are on the same chromosome and that, therefore, were brought along. And when there is too much focus on breeding for one or a few traits over a number of generations, classical breeding leads to problems with the "brought along" traits. And if these "brought along" traits get bad enough, their negative impact on productivity may finally create incentives to fix them. But by then they have to be pretty bad! However, sometimes another fix, less animal welfare friendly, may be used as a temporary (although often this turns into a long-term) solution. Social and political pressures, such as those related to animal welfare, could eventually force these changes to be dealt with even if the agricultural or pet breeder does not want to address them.

To illustrate an example of where the social/political needs for improved animal welfare led to a change in the breeding paradigm shows that surprising results can be obtained more easily than breeders and farmers might want to believe. A recent set of experiments at Purdue University described below demonstrates what can be done if one is serious about improving a particular situation that is rooted in poor selection criteria in conjunction with classical breeding. In all cases, careful attention by the breeder is needed that one is not improving one trait while negatively impacting one or more other traits.

Two of the issues being debated with respect to the animal welfare of egg-laying chickens are the issues of whether debeaking should be permitted and the issue of whether it is acceptable to have extreme feather loss by the end of a single lay period. Hens have a peck order and use "pecking" to maintain this order. Sometimes this pecking will lead to bleeding. Once a chicken begins to bleed, the other birds will essentially peck it to death. Not a very pretty sight. So at a very young age, which is believed to be less stressful, the animal may have some part of its beak removed with a hot, cauterizing knife. If done properly, the acute stress is believed to be minimal and the communal benefits make the procedure acceptable to most animal welfare scientists. Please note that I believe as part of my original ethical framework that a most important point in evaluating *any* procedure done to animals is that when looking at the procedure, it must be done properly. If not, the issue is how to get it done properly. Doing it right means that the equipment needs to be functioning properly, that it must be used properly by the people who are trained to use it, and that a high level of success (i.e., minimal failures) must be obtained throughout the entire day through management attention.

The other issue of concern to some is feather loss. While living in cages, the hens lose and break feathers and thus look "awful" after a while. Now this is an interesting issue — the birds' productivity and chemical stress indicators do not seem to change, so it is not an animal welfare problem. This appears to be mainly a "human" perception problem, not a "reasoned" animal-focused problem. But the researchers chose to tackle this issue also because the feather loss does bother many people.

Researchers noted that the chickens that are used for breeding purposes are considered very valuable and thus are given slightly larger housing than is typical for caged chickens, and they are housed one bird per cage, which means that these birds have no opportunity to socialize and that there is no selection pressure put on the breeders for either feather retention or not pecking. So the researchers obtained some fairly high quality breeder stock and put them into typical multiple bird cages — and guess what — in five generations of classical selection for the two traits that we are discussing, they had a flock of hens that did not need to be debeaked and that had most of their feathers intact when they finished the lay period.

This is exactly the type of breeding changes that we need to institutionalize as part of improving and changing animal agriculture. If a trait is undesirable, one needs to actively breed it out of the animal. It also illustrates that sensible solutions can be found to eliminate some of the animal welfare problems we may be concerned about.

What are some other areas of breeding that need to be worked on? The items below are just a few items that are worth noting. Pigs are often cruel to each other, which makes group housing more difficult. So a gentler pig would be beneficial, especially if group housing is a desired goal. There is the question of whether that arrangement is the pig's preference or if they care about whether they are in group housing or single housing. This has not been determined thus far.

Dairy cows have probably been over-bred with respect to milk yield versus their body conformation. Such a high milk yield often means that the cow's udders are too big for her body and this causes problems in mobility — so we may have to back off from the highest possible milk yields to produce healthier cows. Lameness in milking cows is another problem that may be amenable to improvement through breeding, although management remains an important component of dealing with this problem.

The inability of turkeys to breed on their own may reflect an excess of breeding that ought to be reversed, although this may not in itself be an animal welfare issue. Leg weaknesses in chickens (broilers) and turkeys used for meat are certainly a reflection of a larger bird whose total biology has not kept up with the increased productivity traits and should be subjected to genetic selection for stronger legs.

To Believe With Certainty, We Must Begin With Doubting.

—King Stanislaus (I) of Poland

16.4 GROUP VERSUS INDIVIDUAL OPTIMIZATION

The above changes brought about through breeding example indicated the potential benefit of putting breeding birds together in cages. Let us now look at one of the other issues that the use of group cages for egg layers has generated. In this case, we need to consider the issue of group optimization measurements versus individual optimization measurements. The paradigm used by many farmers at this point is that the overall agricultural endeavor should be optimized, usually meaning that productivity is optimized based on an economic model. I believe that this paradigm has to change within the ethical framework I have set out for animal agriculture.

One of the key concepts emerging from various animal welfare efforts is that each individual animal needs to be permitted to optimize his/her productivity. To illustrate: one of the most common laying cages for chickens in the U.S. was originally designed for five laying hens. Using such a cage system, each hen can lay about 300 eggs per bird per year. (That may seem like a lot of eggs for one bird but it is a realistic figure in modern agriculture, and so we will use that number for the "model birds" we are discussing. This will be taken as the potential productivity of a laying hen when she is not stressed.) But given the economic investment in facilities (i.e., buildings, cages, ventilation systems, etc.), some farmers figured out that if they put a sixth bird into that cage, they could make more money. Now, instead of getting 1500 eggs per cage, they get (as an example) 1700 eggs per cage, but not the 1800 eggs that would be expected of six happy hens. So at least the lowest bird or two in the peck order is demonstrably stressed — and they are not producing at their optimum. So this is where a major paradigm shift is needed — by implementing animal-by-animal optimization, the acceptable situation changes. If farmers want to put six birds in a cage, it would require doing some serious breeding of the birds for the many traits that will make this possible — that is, to yield 1800 eggs per cage in this example. Otherwise, at this time, it is clear that for the above example, five birds is acceptable in these cages but putting in a sixth bird is simply unacceptable.

I therefore expect that an emphasis on social traits related to group behavior will become a more important selection criterion for a number of agricultural animals. To illustrate how I believe a scientific standard for animal welfare might operate in this situation is that the standard would support the use of cages for egg layers, but oppose the current practice of putting six birds in the standard U.S. cage that is designed for five birds! And if we use hens that have been bred in conjunction with the earlier successful research example, we may not need to de-beak the birds and even their feathers may look nicer in the future.

As a bit of an aside, I want to comment on the issue of cage versus cage-free egg layers. The latter involve more elaborate (read expensive) housing, which takes up more space, still does not guarantee access to the outdoors, and uses more building materials, thus raising challenges of keeping the eggs and the birds out of manure, that is, which if not successful can possibly increase the presence of pathogenic bacteria on the eggs, and make heating/cooling less efficient. With larger groups, the peck order can also become more detrimental to hens on the lower end of the scale. So, the eggs are more expensive and it is not clear that either the birds or humans actually benefit from the change.

16.5 INDOOR AGRICULTURE

The previous example leads naturally to a broader discussion of the issue of indoor agriculture. This may lead us to an important paradigm shift, this time a need for an attitude shift by consumers. This may be difficult to accept for the many people brought up on the idea of the ideal farm being "Old MacDonald's" multi-species outdoor farm. By this, I simply mean the farm that is thought about in the children's song. A farm with a few each of many different animals wandering around outdoors is often the image associated with the traditional "family" farm.

The move to indoor agriculture is a profound change that has occurred in the past century but it is a much more positive change than perceived by most consumers. Bringing animal agriculture indoors permits the farmer to control many more environmental factors for the benefit of both the animal and the owner. It permits the farmer to "see" his/her animals more regularly, to optimize feed (often at the individual or small group level rather than the herd level, which means fewer animals are excluded from obtaining proper feed), to obtain proper medical care for the animals, and to protect the animals from predators and disease transmission. With this increased control through indoor agriculture came not only higher productivity, but also a much greater responsibility for the complete stewardship of the animals. However, from the animal's point of view, it creates an environment that is less stressful for animals than outdoors — e.g., fewer predators and better temperature control. (Do we really want to be outdoors when it is –20°F or 105°F?) So, several "survival" behaviors may simply become completely unnecessary! The decrease in stress has led to productivity increases related to environmental improvements separate from any genetic improvements in the animals. So, from an animal's point of view, going indoors may be a good thing.

Let me elaborate a bit on "productivity" as an animal welfare criterion. Although productivity is not a direct measure of animal welfare, it does seem that in most cases it correlates with animal welfare. The less stressed an animal is, the better its productivity. Until it can be proven that some forms of productivity increase with stress, the canons of science dictate that we go with the simplest explanation, i.e., increased productivity equals decreased stress. (This is "officially" referred to as Okham's razor.)

In trying to determine the animal's stress level, there are chemical measures (e.g., blood chemistry) that can be looked at; however, these suffer from a few major concerns. The first is that the very act of making the measurement (e.g., drawing blood) may be a stressor. However, some of these markers do require finite response times before they are seen, although this can be both an advantage and a disadvantage. The second is that to really be meaningful, a time course of these chemical markers is needed, and this adds to both the stress on the animal and the cost of such measurements. And third, these are subject to other confounding aspects such as the animal's daily circadian (24-hr) clock and the impact of feeding and other short-term changes in their conditions. It should be noted that there is a hypothesis suggesting that a little stress might be beneficial. So, the question of whether no stress is actually the best goal remains.

I also want to elaborate a bit more on the issue of "anthropomorphizing." We must be wary of "anthropomorphizing" issues of animal welfare. What looks bad to people may not be stressful to animals. Just because we do not like the way something looks, does not mean that the animal is being stressed by that situation. So one has to think about the standards one wants to use to evaluate animal welfare. Our evaluation must be guided by the animals' behavior, as understood by properly trained experts, plus chemical and physical (physiological) measurements, such as the level of stress hormones in the blood — and not by our intuitive reactions. One example of where this conflict may arise is the issue of light levels in indoor, closed poultry housing. The major reason these houses are closed (i.e., light tight) is to be able to control lighting. Egg-laying birds will only lay eggs when the amount of daylight is either increasing or unchanged. They will stop laying eggs when the daily amount of light is decreasing. But what is the amount of light they need to continue to lay eggs? It turns out to be a very low light, what to humans would still be considered darkness. But this is sufficient light to set their daily clock and the low level also minimizes pecking. So, most of the time, the house is kept at these low light levels and full lighting is only used when the human caretakers are present so that they can see.

It is also interesting that some farmers have gone to 24-hr lighting — this is believed to be detrimental to the birds and so most current animal welfare standards call for at least 4 hr of darkness. However, I understand that some scientists are revisiting this issue and suggesting that continuous lighting is acceptable.

My take on "factory farming" is that there are many good points about such a system — but that too many abuses have been allowed into the system and these must be addressed and corrected. I do not consider "factory farming" to be a dirty word, but I also recognize that the industry itself has been a part of the problem. Like everything else, factory farming must be done properly and well.

It Is Better To Know Nothing Than To Know What Ain't So

—Josh Billings

16.6 PAIN AND SUFFERING

When humans evaluate animal welfare, one of the issues of legitimate concern is pain and suffering — but these are both "loaded" words. As someone not in this field of research, I have tried to understand the issues using simplified definitions of these two key words that permit some important distinctions to be made. *Pain* is an immediate and acute response of the brain to negative stimuli. *Suffering* is a more complex brain processing operation that takes the pain information along with other information where the brain uses prior knowledge to predict and be concerned with the future from a negative point of view, e.g., to realize that a bad situation (e.g., pain) may continue into the future.

The research data shows that higher animals feel pain. Mammals and birds certainly fall into the category of animals where pain is an issue. We are still learning where the line is between those animals that do and those animals that do not feel pain. For example, the research literature about the issue of whether fish feel pain is currently producing conflicting results, enough so that I do not believe a definitive answer is available at this time. However, prudence suggests that the possibility of their feeling pain be taken into account when designing new systems for handling fish, especially with respect to aquaculture. So, the animal welfare requirements for the care of mammals and birds need to take into account the role of both acute and chronic pain. However, whether pain is a concern with fish remains to be determined. Animals below fish in the animal classification scheme may not feel pain, and thus there is a rational basis for establishing a different set of animal welfare requirements for these animals.

What, then, do we know about animal suffering? The first question is whether animals can suffer and, if so, which animals can suffer? The jury is also still out on this issue, although some research seems to suggest that primates may, in fact, suffer. But where this line will be drawn is not clear, i.e., which animals do or might suffer needs to be studied. Why? The standards for animals that suffer must, I believe, be higher than the animal welfare for animals that feel pain but do not suffer.

But let us look at a situation that has been presented to the public to suggest animal suffering but which is in fact being misused by organizations that find it convenient to show such pictures to humans for fund-raising purposes, even if they know these are not accurate depictions. They are playing to human emotions and not to human reason.

Are pictures of baby seals crying a reflection of suffering? Very possibly not! Because constant tearing is the way that they keep their eyes moist and clean, i.e., it is a normal routine biological function, it is not crying! To humans who do not understand this simple difference between us and seals, it makes for an emotionally wrenching picture, but one can get that picture with a baby seal that is with its mother and is totally happy. It is not a criterion for evaluating the humaneness of the seal hunt in Eastern Canada. This hunt has taken quite a beating over the years and has for many become so emotional that a rational discussion is not possible. The fact is that there are many other issues involved in evaluating seal hunting — most of these issues have been distorted by some of the "environmental groups" to serve their purposes.

Some of the other issues include (1) the impact of this hunt on the overall seal population, which is prone to getting too big and then crashing; (2) the impact on fish populations, which have been heavily impacted by both over-fishing, which is probably the major problem initially, but which is aggravated by the large amounts of these same fish consumed by seals and other marine mammals, and which is probably a major cause of the failure of the fish stocks to rebound after fishing has been stopped; (3) the impact of the hunt on the environment, where the question of managing wildlife on the basis of a single species and/or as an ecological community continues to challenge wild-life managers; and (4) finally, the impact of the absence of the hunt on the human community's social structure, where the hunt is an important source of income for a population trying to exist in a harsh environment. All these issues need to be considered in evaluating the pluses and minuses of the seal hunt. Overall, a controlled hunt might, in fact, be justified and the use of the wooden "club," when done properly, is believed by many animal welfare experts to be a quick and humane form of stunning, as it is believed that a sharp blow to the head in the proper place with a properly designed instrument is a humane way of stunning. The special wooden club used in this hunt, mainly by people who have hunted using this technique over centuries, is probably more humane than the typical death that occurs after being shot with a gun, and probably competes well with some of the official humane slaughter methods! So, as we learn more about these complex issues, we will have to make ethical value judgments but, hopefully, these will be well informed and will balance a wide variety of conflicting input information. We may not all agree about the seal hunt but it is certainly not the brutal, barbaric practice as it is often pictured.

16.7 ANIMAL RIGHTS

In looking at the needs of animals thus far in this discussion, the emphasis has been on animal welfare. This orientation clearly affects my judgment of what is important. However, other people come at the same issues from different philosophical points of view. The more animal-centered view with respect to this topic is often referred to as the animal rights approach. I will try to raise a few points about animal rights that hopefully will help the reader think more about these issues. But recognize that I am much more pragmatic than philosophical, and I certainly cannot debate the validity of various philosophical constructs.

Tom Regan, a well-known North Carolina State University professor and animal rights philosopher, wrote that "Species are not individuals...the rights view does not recognize the moral rights of species." What does this mean? It means that all individual animals are created equally and are potentially subjected to the same "rights." First, this philosophical framework gives no rights to plants. So what might this mean? I take it to mean that a "Sequoia" tree has no rights, while a worm or a mosquito or a slime mold has significant rights. And a domesticated cow has the same individual rights as a Giant Panda. This is definitely a challenging concept to consider carefully. Actually, I suspect that Regan's definition of animal may be more limited than that of a traditional biologist. However, I have problems dividing the world into species beyond humans that have individuals with rights versus those that do not. And do we really mean the same "rights" for all such species, or "rights" appropriate to the biological nature of the animal in question? And can one have "rights" without "responsibilities" — a concept that I believe applies uniquely to humans. And what exactly does "having rights" mean? Do humans have any right to use animals for food, for research, for recreation and companionship, etc.? It seems that their answer would be "no" — and I simply cannot imagine a world without animals interacting with and being used by humans — and I certainly cannot picture a world where everyone is a vegan.

However, I should also indicate that Tom Regan, while philosophically committed to only doing for animals what is good for them, strongly supports practical animal welfare efforts as important steps in the right direction.

So, I return to my starting ethics of our responsibility to animals and note that even within the complex academic disciplines of ethics and philosophy, the issues are debated. For example, Jeremy Bentham, the philosophical father of Utilitarianism (i.e., the idea of providing the most good for the most people [and animals]), was a carnivore. And to quote him: "We are the better for it, and they are never the worse...The death they suffer in our hands commonly is, and always may be, a speedier and, by that means, a less painful one than that which would await them in the inevitable course of nature." But I would argue that we can make it even less painful by paying attention to animal welfare. Next time you watch a wildlife program with a lion "bringing down" a zebra or a zebu on TV, remember that is what life in the wild is really like — not so idyllic! This is what it means to be a prey animal. Interestingly, all our farm animals are prey animals, yet our major pets (i.e., cats and dogs) are predators as we discussed earlier.

I would like to now share with you some text that was sent to me by an advisor to the group People for the Ethical Treatment of Animals (PETA). He introduced me to the pragmatic animal rights approach using the "Equality of Consideration" concept: "...each animal's suffering should be equally considered in making ethical decisions. This does not entail the view that all animals suffer equally or have equal moral value. ... As you can imagine, animal rights groups have to regularly make decisions about what animals are most worth our time and so we are constantly making decisions that value one animal more than another. An animal rights ethic can be worked out in different ways, but I think it virtually always ends up that individuals reserve the highest degree of moral concern for the human animal."

As suggested by a student, another point to consider when thinking about the potential equality of animals is the fact of their biological hierarchy that is evident in their development, anatomy, and physiology. This suggests we would be going against nature to say they are all *created equal* because they are not created equally.

We Fear Things in Proportion to Our Ignorance of Them

—Livy

At this point, I believe it would be helpful, having had a discussion of animal welfare/animal agriculture, to address some broader issues that need to be considered in evaluating the role of animal agriculture going into a, hopefully, more sustainable future. I define sustainability as the ability to continue to produce human needs from the environment with the least impact on the environment so that future generations can have the same opportunities as we have. The first of these considerations will be to think about human population changes and the implications that this will have. This will then move into considerations of other human needs, like diet.

16.8 A POPULATION PERSPECTIVE

The population of the world is currently a little over 6 billion people, and it is anticipated to grow to 9 billion people during the next 50 years. The good news according to some population demographers is that it may possibly peak at just above 9 billion people rather than continue to grow — but that will take an improvement in a number of factors that are beyond the scope of this chapter. Others, who are less optimistic, predict that the future population could go as high as 11 billion people by the end of the century.

Regardless of the final figures, this population growth means that we will need to increase the efficiency of agriculture. We will have more people requiring more food but we will have less land available for agriculture. So, for example, I believe that indoor agriculture (factory farming?) can make a substantial contribution to the total food supply by increasing the yield per acre and by using marginal land rather than prime agricultural, residential, or park land. Indoor animal agriculture is also able to go multi-tiered; i.e., even the most efficient solar-based greenhouse system must be one "layer" tall — plant agricultural facilities can be put on top of a building (an important future resource in the urban area) but only one story at a time without needing very expensive and energy intensive grow lights. Obviously, in the future we will need to think about how to better integrate the two major agricultural systems, i.e., plants and animals. More water will need to be shared between plants and animals, and the delivery to both will need to be optimized to minimize waste. Dual use of water is actually being done now to a certain extent with aquaponics — combining hydroponics (raising plants in water) with aquaculture (raising fish).

16.9 SUSTAINABLE AGRICULTURE AND OLD MACDONALD'S FARM

In a sustainable world, what will the farms of the future look like? I am concerned that for many folks, particularly those in urban areas, the only form of agriculture they think about is the Old MacDonald's farm as their model and ideal, although this is often a highly inefficient form of agriculture. The multipurpose family farm was designed to feed the family — it struggled to feed a lot of people. It was a hard life, such that many of the young people left the farm to seek fame and fortune elsewhere. Yet it appears that this inefficiency is being replicated in some of the current "alternative" agricultural systems. It may be romantic to "return to manual farm labor" — but for most of society's history, the goal has been to minimize hard labor. In California, they have passed a law that prohibits the use of human labor for weeding because it is so hard on the human body and no longer considered necessary — but the law has a specific exemption for organic farmers! Farming is very hard and dangerous work, and the drive has been to make it easier and safer.

So I hope to suggest that a sustainable agricultural system will be one that uses land, people, and resources in the best possible way to create a good life for animals, farmers, farm workers, and consumers. However, it is difficult to picture a planet-wide sustainable system without animal agriculture that is consistent with the lifestyle of those of us in the U.S. who are mostly in the top 5% of the world's income scale. In this sense, I believe a "right-sized" and better-operated animal agricultural system will be an integral part of sustainable agriculture for many years to come.

One area where we are currently failing to optimize animal agriculture is with respect to the collection and use of manure. We are therefore creating some serious environmental problems rather than reaping the benefits that are there: Manure can be used for many different purposes — everything from fertilizer, including for organic crops, to energy production. But the systems for handling manure need to be developed, optimized, and used. Ironically, the larger concentrated animal feeding operations (CAFOs) are generally taking the lead in using new systems as they can generate the economy of scale needed to beneficially use these newer systems. Thus, I believe that in the future, the large factory farms will be more sustainable than many of the other models. We are already getting reports that the total energy input for organic foods often exceeds those of regular foods.

Another benefit we gain as we move animals indoors, particularly ruminants, is that we have the opportunity to capture the methane they generate and use the methane as a fuel or starting material for the synthesis of some of the chemicals we now make by using petrochemical feedstocks — so today's problems with animals as methane generators become future opportunities. In a totally closed barn, the gas venting system might incorporate systems to capture the methane and use it as a fuel. Given that methane is about 21 times more detrimental than carbon dioxide, burning methane is a very viable option; i.e., if one molecule of methane gives one molecule of carbon dioxide, the planet is 20 times better off and we have obtained some useful energy besides! In bakeries, we are already capturing and re-using the carbon dioxide released during the baking process. But if this is ever to be realistically applied remains to be determined.

16.10 RUMINANTS AS PART OF A SUSTAINABLE PLANET

What is the unique role of ruminants (cows, sheep, and goats — i.e., animals that chew their cud)? Ruminants are currently the only practical way to transform grassland into a useful resource for feeding people — and it is a very short food chain — grass to ruminant to people. Frances Moore Lappe actually pointed this out more than 35 years ago in her book *Diet for a Small Planet* by noting that cattle (ruminants) can convert inedible plants to protein and may also neutralize human toxins in these plants. And a lot of the Earth's surface is grassland — which can grow on a very "shallow" topsoil and which, if managed correctly, can limit erosion. Ruminants can also digest many other interesting by-products of modern plant-based agriculture (and even those from animal agriculture if the disease issues can be address, although some would argue that it is not ideal to feed waste from a particular species to animals of the same species). These might otherwise not be used as effectively. Ruminants can also, for example, be used to clear away underbrush in forests and other wilderness areas, possibly more efficiently overall than humans tramping the ground with gasoline-powered equipment. The key in many situations is to control grazing. To the extent that the charge of overgrazing, particularly on public lands in the U. S., is valid, this practice needs to be reigned in. But plowing up grasslands for crops may not be the most sensible thing to do.

Interestingly, in recent years some managers of modern pasture management systems are finding that growing the "pasture" and then harvesting it (with power equipment?) is actually more efficient than having the animals themselves harvest it — one does not have the problem of overgrazing but in the process more petroleum-based fuels are used. A benefit, however, is that the animals could be kept indoors and thus more of their methane and manure captured. Further analysis of this change is needed to determine if it really makes sense as a long-term sustainable practice. On the other hand, could we do more to harvest grass (e.g., grass clippings and yard trimmings) in urban and suburban areas and make these available as ruminant feed?

Knowledge Is Proud That She Knows So Much.
Wisdom Is Humble That She Knows So Little.

16.11 LAND USE ISSUES

The discussion of ruminants and their social role is part of the bigger question of land use around the world, i.e., what is the optimum use of any parcel of land?

Turning grassland into cropland can be very environmentally damaging. Turning other land use areas into pasture (like rain forests) I believe is wrong. We will need to determine how to address this on a global basis rather than allow local exploitive practices to harm the biosphere. How to do this, while still respecting many competing rights such as land ownership and job/wealth creation, remains a challenge.

By the way, the following numbers might be of interest as we think about land use specifically in the U.S.: Of the 2.3 billion acres of land in the U.S., 52% is used for agriculture, according to the

USDA's Economic Research Service. Two-thirds (~34% of the total land in the U.S.) of this agriculture land is used for grazing. Of the remaining U.S. agricultural land, 29% (~14% of the total) is forests, 26% (~12% of the total) is grassland pasture and rangeland, and 20% (~10% of the total) is crops. So we certainly have a lot of land that is appropriate for animal agriculture and less so for food crops and it might be properly used in that capacity.

I also want to look at the broader issue of the appropriateness of different feeds for animals, whether they are ruminants or not.

16.12 FEED AND SUSTAINABILITY

Some of the "foods" fed to agricultural and companion animals are obtained from the by-products of modern food production. These by-products are not necessarily "grain" that could be fed to people in lieu of feeding it to animals. They are by-products of our plant-based agriculture — like wheat hulls, peanut shells, and brewers' by-products and more recently distiller solids, a by-product of corn ethanol production. Like grass, the most valuable way to use these already created biological molecules is probably to feed them to animals, although more valuable uses may be discovered in the future. And looking at feeds obtained as by-products raises an interesting question: Should we be feeding these to pigs instead of dogs, to cattle instead of cats? If one looks at all the pet food in the modern supermarket, is that really the most efficient use of the resources they represent?

Some grain land that is used to produce food for animals might be better used for human food. What might some of the consequence of such a change be? Should we be going to more grass-fed beef, i.e., animals that are only raised on pasture crops? I would suggest possibly a partial yes answer. But grass-fed beef is not quite as tender or quite as "red" as grain-fed beef. Part of the explanation is that it (grass-fed beef) is leaner (healthier for us to consume?) but not as tender. So, are we ready to make that compromise to improve the world and our health — instead of giving up beef? On the other hand, the by-products we mentioned might still permit some animals to be "finished" — but possibly for a much shorter time on the feedlot or maybe all those by-products might be best fed to animals like pigs and poultry that cannot exist on a grass-only diet.

Another way to look at feeding animals agricultural by-products is that by using these by-products for feed, we get a better use of cropland, i.e., we get more benefit per acre by feeding to animals the part that humans do not use. In other countries and even sometimes in the U.S., after the "cash" crop is harvested, food animals are allowed to "clear" the fields. This seems quite sensible. I think we need to work harder to develop the beneficial synergies between plant and animal agriculture.

It has been pointed out that my "Old MacDonald" farm discussion might suggest that I think farms should be specialized and single-crop. Specialized — yes; single crop – no. Often, the most efficient process would be one that has the right mix of crops, probably animal and plants together, as many farmers still do — but still limited to a small enough number of "crops/animals" that the farmer can be an expert on the crops he chooses to produce and can have the expertise to do a really good job.

I have been involved in both research and practical work on the composting of food waste. This process is certainly more beneficial than landfilling as a waste-management alternative for organic/biological wastes. But again, is this the best long-term (i.e., sustainable) approach? What about feeding these organic wastes to pigs and poultry? These animals are capable of eating many of the organic wastes from human food production. That has been their traditional role in many different societies. Have you ever seen chickens go after potato peels — raw or cooked? They love them. That was one of my favorite tasks on my aunt and uncle's chicken farm — it certainly made their day, a very anthropomorphic response.

And in modern times, the necessary heat treatment systems are available for use with organic waste to prevent pathogen problems. (Can we use a source of waste heat nearby to minimize the cost, i.e., co-generation heat?) Feeding food waste to animals was the traditional way of handling organic waste, so why not return to this practice? This might be a case where nostalgia for the past coupled with modern knowledge can lead to a much better approach to handling these materials, specifically retaining and optimizing the use of complex organic molecules.

16.13 VEGETARIANS AND VEGANS

Although I have argued that ruminants (and other animals) have important roles to play, some people feel strongly that we should not be using animals (vegetarians who usually do consume milk, eggs, and honey) or even animal products (vegans who eat no animal products) at all. Some extend this into concern for the use of leather, fur, and even animal-derived medicines.

Some folks object to the killing (slaughter) of animals for food and/or other purposes. That is a personal judgment reflecting an individual's philosophy expressed through their food choices.

Interestingly, some recent research has appeared that suggests that developing croplands, with modern agricultural methods (i.e., heavy machinery) may actually lead to the killing of more animals than grassland or land used for animal agriculture. Running plows, disks, harrows, etc. through a field kills a lot of small animals and, sometimes, even birds. The point is that crop production comes with its own price in terms of killing animals — and it is often not as "nice" a death as slaughter.

In fairness, it should be noted that there seems to be a debate about whether these deaths should be measured as number of animals killed/acre (crops are higher) or whether we should consider the issue as the number of animals killed to provide food for one person (food animals are higher because so many broilers are slaughtered). Although this is a valid discussion point, it does not change the fundamental fact that all food for humans is likely to harm some other creatures, either directly or through loss of habitat.

The issue of whether, in fact, a vegan diet would be better for the planet is one that I will return to shortly. But first it may be useful to look at some overall issues of the human diet. As a food scientist, it would be great to do this in much more detail, but I am going to try to keep it to an overview that is relevant to this chapter.

16.14 HUMAN DIET ISSUES

Animal food products, eaten in moderation, provide many essential nutrients, some of which are more difficult to obtain from a vegan diet with absolutely no animal products (e.g., easily absorbable iron, vitamin B12). Moderation in consumption of food in general, and of animal foods specifically, should be an educational and societal goal. Many medical conditions have been documented as due to excess consumption of animal products. However, that does not necessarily extrapolate to lesser, modest amounts. Furthermore, total abstinence from animal products requires a great deal of attention to one's diet, a commitment that many people are not willing to make at this time.

For example, in the U.S. it is clear that we have an obesity epidemic, which is due to the fact that we eat too much food and do not get enough exercise, and this is probably exacerbated by the fact that in this country many of us probably eat more meat than we need to. The question is whether it is appropriate (ethical?) for us to insist that people in the developing world should not have access to animal agricultural products, even when their standard of living increases? Hopefully, a sustainable agricultural system will provide modest amounts of animal protein foods for everyone.

Having admitted that an excess of meat products may be unhealthy, it should also be noted that overeating sugar is unhealthy as well. Too many French fries in no-trans fatty acid soybean oil are also unhealthy. So we need to learn a lot more about human nutrition to create a healthy population, which should then play a much greater role in determining the mix of agricultural products that ought to be grown worldwide. But it is also important to recognize that we eat for more reasons than just health, so a range of options and daily choices, rather than a single perfect diet, will always be required.

It is unfortunate that too many individual examples of "bad practices" in animal agriculture do exist; these require correction in a timely fashion. But even many unacceptable practices in animal agriculture do not, per se, negate the value of animal agriculture. These concerns alone do not prove that a vegan diet is better for people or the planet. That remains to be shown in much more detail before I believe such dramatic changes can be recommended. Plant agriculture will also need to deal with its own shortcomings. So in the meantime, we as a society continue to have an ethical responsibility to use animals in a wise and caring way.

16.15 ANIMAL WELFARE AUDITING SYSTEM

In moving toward the end of this chapter, I would like to sum up and give another example or two specifically focused on animal welfare.

Those of us in animal agriculture must be able to justify all of our animal welfare/animal care practices on a sound scientific basis and, where necessary, do the research to develop better methods. One recent example, where a focused effort to improve animal welfare has led to the research necessary to justify a major change in practice, has been with the practice of molting of chickens at the end of their first laying period (about 1 year). After a year of laying eggs, egg production drops. If the chicken is given a rest before starting another egg-laying cycle, egg production will increase. During molting, the birds lose all their feathers and grow new ones. (Molting occurs naturally in all birds at various times in their life cycles.) Molting has generally been done in commercial egg-laying facilities using a practice of feed and water withdrawal for as long as 14 days. This length of starvation, especially water withdrawal, is simply unacceptable. Therefore, there has been pressure to ban the practice. But if this is the solution, each chicken gets slaughtered at a younger age and more chickens are needed to produce the egg supply. For example, in 3 years if a flock is slaughtered every year, three birds are slaughtered. If the life of each bird is extended to even 1.5 years, then over 3 years, only two birds are slaughtered. So there is a trade-off between the short-term stresses of molting versus the number of birds slaughtered. And this does not count the resources required in the first 20 weeks to raise each bird before she reaches sexual maturity. So again, the investment in resources is decreased by molting birds.

The industry and the research community, when seriously challenged by the supermarkets and chain restaurants (working through their trade associations [FMI, the Food Marketing Institute and the NCCR, National Council of Chain Restaurants], which have begun to develop an animal welfare purchasing specification program) developed a high-fiber diet that permits chickens to have some feed and unlimited water throughout the molting period. Thus, fewer egg-laying chickens need to be slaughtered each year and fewer replacement chickens need to be grown. The new solution, I believe, is a better long-term solution and is an example of pragmatically addressing the problem head-on.

The FMI/NCCR program, introduced in the previous paragraph, is also indicative of how the food industry, both representing the production side and the marketing side, can make a real impact for the benefit of all consumers. The retail world, acting nominally on behalf of the consumer, is asking animal agriculture to update and upgrade animal welfare standards to address many of the problems of modern animal agricultural practices from an animal welfare point of view that have been identified by various parties. MacDonalds started the process of meeting animal welfare standards as a purchasing requirement, and Burger King and Wendy's have followed with strong standards, focused on the aspects of the production processes that are controlled by their immediate, direct suppliers.

How much did pressure from animal activists groups drive this agenda? Probably some, but we are also seeing that the animal agriculture and food industries are stepping up and trying to do better in meeting the legitimate scientific issues related to animal welfare.

The FMI/NCCR program is currently establishing certification standards for individual production units through the use of independent third-party audit systems (professionally trained auditors) that can be used by supermarkets and chain restaurants as part of their purchasing specifications. That is, unless an audit is passed according to the buyer's requirements, the buyer (i.e., the supermarket or chain restaurant) will not buy the product. These standards are slowly being rolled out after a lot of work by both the buyers and the animal agricultural trade associations. Hopefully, these new standards will contribute to better animal welfare and, with the support of consumers, animal agriculture can begin to improve on many of the practices that have been criticized, thus leading to significant progress in defining an animal welfare friendly, high production system, while simultaneously moving to a more sustainable animal agriculture. Support for these programs from consumers will be critical in moving the process forward.

16.16 SUMMARY

In many countries, various constituent (consumer) groups are making their requests for changes in animal agriculture based on their values, impressions, and ethical concerns. We must be prepared to correct the practices for which there are legitimate complaints and to provide good, clearly presented scientific information to explain to these stakeholders why some of their requests may not be the best choices. In other cases, the market has and will probably need to offer consumers the ability to select products that represent different production/processing systems, clearly and honestly labeled. People, especially those who can afford it and feel strongly about one or more specific production practices, even if these practices do not meet a reasoned, rationale argument standard, need to be served. (The customer is always right!) This complex consumer marketing environment is the challenge that animal agriculture will continue to need to address.

Hopefully, this chapter has provided some new information to think about regarding these complex issues.

ACKNOWLEDGMENTS

The author would like to thank Mark Cook in the Animal Science Department at the University of Wisconsin – Madison for suggesting that I turn my animal welfare talk into a book chapter. The 2006 and 2007 classes at Cornell on Animal Science 310, "An Introduction to Animal Welfare," were most helpful in critically reviewing this chapter and making many suggestions that have in many cases been successfully incorporated. In the end, however, responsibility for this chapter must be mine.

SUGGESTED RESOURCES

American Meat Institute: www.meatami.com

Cybertower: www.cybertower.cornell.edu: 1. Animal Welfare, Joe M. Regenstein; 2. Animal Welfare, Temple Grandin; and 3. Introduction to Kosher and Halal, Joe M. Regenstein.

Food Marketing Institute: www.fmi.org

Institute of Food Technologists: www.ift.org, go to Publications, go to Comprehensive Reviews, go to Vol. 2, Issue 3, p. 111. (Introduction to Kosher and Halal Food Regulations.)

National Council of Chain Restaurants: www.nccr.net

Index

D

E

F

G

M

Q

R

S

Z